20 019 565 17

AF598429

Thin Films

Frontiers of Thin Film Technology

Volume 28

Recent volumes in this serial appear at the end of this volume

Thin Films

Frontiers of Thin Film Technology

Edited by

Maurice H. Francombe

Department of Physics
Georgia State University
Atlanta, Georgia

Associate Editors

Colin E.C. Wood
A.G. Unil Perera
H.C. Liu
Phillip Broussard
J. Douglas Adam
Deborah Taylor

VOLUME 28

ACADEMIC PRESS

A Harcourt Science and Technology Company

San Diego San Francisco New York Boston
London Sydney Tokyo

This book is printed on acid-free paper 

ACADEMIC PRESS
A Harcourt Science and Technology Company
515 B Street, Suite 1900, San Diego, CA 92101-4495, USA
http://www.academicpress.com

Academic Press
Harcourt Place, 32 Jamestown Road, London, NW1 7BY, UK
http://www.academicpress.com

International Standard Serial Number: 1079-4050
International Standard Book Number: 0-12-533028-6

Printed in the United States of America
00 01 02 03 04 COB 9 8 7 6 5 4 3 2 1

Contents

Antimony-Based Infrared Materials and Devices

C.E.A. Grigorescu and R.A. Stradling

HgCdTe Infrared Detectors

Arvind I. D'Souza, P.S. Wijewarnasuriya and John G. Poksheva

Synthesis and Characterization of Superconducting Thin Films

Chang-Beom Eom and James M. Murduck

Fabrication of Superconducting Devices and Circuits

James M. Murduck

Microwave Magnetic Film Devices

Douglas B. Chrisey, Paul C. Dorsey, J. Douglas Adam and Harry Buhay

Ferroelectric Thin Films: Preparation and Characterization

S.B. Krupanidhi

Integration Aspects of Advanced Ferroelectric Thin-Film Memories

Deborah J. Taylor

Contributors

Epitaxial Film Growth and Characterization: Ian T. Ferguson, Alan G. Thompson, EMCORE Corporation, Somerset, New Jersey, USA

Epitaxial Film Growth and Characterization: Scott A. Barnett, Materials Science Department, Northwestern University, Evanston, Illinois, USA

Epitaxial Film Growth and Characterization: Fred H. Long, Department of Chemistry, Rutgers, The State University of New Jersey, Piscataway, New Jersey, USA

Epitaxial Film Growth and Characterization: Zhe Chuan Feng, Institute of Materials Research and Engineering, National University of Singapore, Singapore

Field Effect Transistors: FETs AND HEMTs: Prashant Chavarkar, Umesh Mishra, Department of Electrical and Computer Engineering, University of California, Santa Barbara, California, USA

Antimony-Based Infrared Materials and Devices: C.E.A. Grigorescu, R.A. Stradling, Blackett Laboratory, Imperial College of Science, Technology and Medicine, London, United Kingdom

HgCdTe Infrared Detectors: Arvind I. D'Souza, Boeing Sensor and Electronic Products, Anaheim, California, USA

HgCdTe Infrared Detectors: P.S. Wijewarnasuriya, Rockwell Science Center, Thousand Oaks, California, USA

HgCdTe Infrared Detectors: John G. Poksheva, Analysis Associates, Whittier, California, USA

Synthesis and Characterization of Superconducting Thin Films: Chang-Beom Eom, Department of Mechanical Engineering and Materials Science, Duke University, Durham, North Carolina, USA

Synthesis and Characterization of Superconducting Thin Films: James M. Murduck, TRW, Space and Electronics Group, Redondo Beach, California, USA

Fabrication of Superconducting Devices and Circuits: James M. Murduck, TRW, Space and Electronics Group, Redondo Beach, California, USA

Microwave Magnetic Film Devices: Douglas B. Chrisey, Plasma Processing Section, Naval Research Laboratory, Washington, DC, USA

Microwave Magnetic Film Devices: Paul C. Dorsey, Komag, Inc., Milpitas, California, USA

Microwave Magnetic Film Devices: J. Douglas Adam, Northrop Grumman STC, Baltimore, Maryland, USA

Microwave Magnetic Film Devices: Harry Buhay, Northrop Grumman STC, Pittsburgh, Pennsylvania, USA

Ferroelectric Thin Films: Preparation and Characterization: S.B. Krupanidhi, Materials Research Center, Indian Institute of Science, Bangalore, India

Integration Aspects of Advanced Ferroelectric Thin-Film Memories: Deborah J. Taylor, Motorola, Austin, Texas, USA

Preface

Volume 28 of the book series *Thin Films*, titled *Frontiers of Thin Film Technology*, focusses primarily on recent developments in those technologies that are critical to the successful growth, fabrication, and characterization of newly emerging solid-state thin film device architectures. The device structures considered include not only the dominant and rapidly evolving semiconductor integrated circuit components, but also structures that depend for their function upon novel photonic properties, as well as superconducting, magnetic, and ferroelectric behavior.

The nine review articles included in this volume have been selected from a new five-volume work, *Handbook of Thin Film Devices*, now being prepared for publication by Academic Press. This handbook, from which the chapters are drawn, provides a comprehensive, multi-topical scientific and engineering source embracing key aspects of a field that is basic to all commercial, defense, and space high-technology systems. *Thin Films* Volume 28, *Frontiers of Thin Film Technology*, is a condensed sampler, authored and edited by well-known experts, and offered in a convenient format for use by professional scientists, engineers, and students involved with the materials, design, fabrication, diagnostics, and measurements aspects of these important new devices.

In Chapter 1, Ian T. Ferguson, Alan G. Thompson, Scott A. Barnett, Fred H. Long, and Zhe Chuan Feng address the strengths and weaknesses of the modern non-equilibrium epitaxial methods of MBE and MOCVD for semiconductor compound growth and techniques for characterization of quality and parameter control and feedback. The more advanced devices now emerging can use the different properties caused by varying the composition or elastic strain of the epitaxial layer to effect changes in bandgap, refractive index, or carrier concentration. In addition, growth of very thin layers and quantum confinement have facilitated precise modification of electronic properties of compound semiconductor structures. This chapter provides a broad overview of the growth and characterization approaches needed for epitaxial III-V structures used in fabrication of superior high electron mobility transistors (HEMTs), Hetero-bipolar transistors (HBTs), and optical devices.

Prashant Chavarkar and Umesh Mishra illuminate the important technologies and performance possibilities of FETs and HEMTs in Chapter 2. HEMTs, which use the two-dimensional electron gas (2DEG) as the current conducting channel, have proved to be excellent candidates for microwave and millimeter wave analog applications and high-speed digital applications. The authors stress that to optimize performance it is crucial to understand the principles of device operation, to consider the effect of scaling in designing a microwave or millimeter wave HEMT device, and to appreciate the advantages and limitations of the materials system involved.

Chapter 3, by C.E.A. Grigorescu and R.A. Stradling, reviews the status of antimony-based infrared materials and devices previously confined mainly to defense (imaging and tracking) scenarios in the mid-wavelength MWIR (3-5 micron) spectral range. Thin film research studies, for example, of strained superlattices and of metastable alloy compositions, have led to Sb-based detector structures demonstrating IR sensitivity extending into the long wavelength LWIR (8–12 micron) range. Both detectors and emitters are discussed, covering the basic device physics and mechanisms limiting the performance as well as materials properties.

HgCdTe (mainly as a photoconductor) has long been the incumbent detector technology for military and space applications, ranging in wavelength from 2 to beyond 16 microns. Chapter 4, by Arvind D'Souza, Priyalal Wijewarnasuriya and John Poksheva, gives a summary review of this technology, covering the aspects of material preparation, junction-device characteristics, photovoltaic architectures, and recent developments in focal plane arrays (FPA). Over the last decade, with significant developments in Europe and the US in low-temperature molecular beam epitaxy (MBE), HgCdTe has advanced substantially as a large area FPA technology, with device producibility and uniformity ensuring its dominant presence in the high-end IR market.

The growth and characterization of the thin films and multilayers needed for low- and high-T_c superconducting devices is described by Chang-Beom Eom and Jim Murduck in Chapter 5. Since this area serves as the foundation of the device process, the authors give a careful overview of the growth techniques, discussing strengths and weaknesses as well as the standard characterization tools needed to validate the film's properties before continuing to a device fabrication phase.

In Chapter 6, Jim Murduck presents the guidelines for device fabrication for the standard materials used in industry: niobium, niobium nitride, and $YBa_2Cu_3O_7$. The layout of the standard processing and analysis steps to insure high quality devices are laid out, as well as discussing the difference in the state of the art for the low and high Tc production lines.

Ferrite devices play a key role in most microwave and millimeter wave systems where they provide duplexing, isolation, switching, phase shifting and power limiting functions. While much effort has been directed towards the size reduction and integration of active semiconductor devices, relatively little work has been directed towards achieving comparable size and cost reductions for ferrite devices. Douglas Chrisey, Paul Dorsey, and J. Douglas Adam provide an overview of recent developments in microwave magnetic film devices in

Chapter 7. Approaches used to deposit ferrite films are reviewed and compared, and the state of the art in the formation of garnet, spinel, and hexaferrite films is described. This chapter concludes with a detailed description of exploratory work on the integration of ferrite film devices with semiconductors. Although several problems remain to be solved, thin film ferrite devices appear attractive for millimeter wave applications in communications and radar systems.

In Chapter 8, S.B. Krupanidhi provides a comprehensive review of preparation and characterization techniques used in developing ferroelectric thin films for device applications. Deposition methods, uniquely suited to the controlled fabrication of such films, employ either physical growth with low energy bombardment (e.g. magnetron sputtering from a single or multiple target source, multi-ion beam reactive sputtering, and pulsed laser ablation) or chemical routes that involve no such bombardment (e.g. sol-gel, chemical vapor deposition, and metal organic chemical vapor deposition). The structure processing relationship of some ferroelectric oxide films that are being developed for high-performance memories and microelectromechanical systems (MEMS) are described. Finally, the reader is provided with a useful summary of the key techniques employed in electrical characterization of ferroelectric films for device applications.

Practical realization of stable, high-performance ferroelectric random access memories (FeRAMs) also depends critically on successful control of integration and processing parameters. Chapter 9, by Deborah Taylor, addresses the important issues related to the design and fabrication of the memory cells that are implemented in high-density FeRAMs and ultra-dense DRAMs. Among the items discussed are approaches for forming and patterning the capacitor stack, the damaging effects that hydrogen-containing ambients have on ferroelectric capacitors, the impact of ferroelectric processing on the silicon devices, and equipment issues for the commercial manufacturing of ferroelectric film memories. Finally, the author presents a summary and an outlook on the future of these ferroelectric film memories, which have the potential to capture a larger share of the total memory market, estimated for 1999 to be worth over $60 billion.

Maurice H. Francombe

Thin Films

Frontiers of Thin Film Technology

Volume 28

Epitaxial Film Growth and Characterization

IAN T. FERGUSON AND ALAN G. THOMPSON

EMCORE Corporation, Somerset, New Jersey, USA

SCOTT A. BARNETT

Materials Science Department, Northwestern University, Evanston, Illinois, USA

FRED H. LONG

Department of Chemistry, Rutgers, The State University of New Jersey, Piscataway, New Jersey, USA

ZHE CHUAN FENG

Institute of Materials Research and Engineering, National University of Singapore, Singapore

1.1. Introduction

Over the last 3–5 yr the market for compound semiconductor based devices has continued to expand and mature, and much of the commercial promise of the late 1980s for these materials has been realized. Many devices have now reached the stage of significant manufacturing volumes, including light emitting diodes (LED), laser diodes (LD), solar cells and electronic devices, such as high electron mobility transistors (HEMT) and heterojunction bipolar transistors (HBT). All of these devices require the deposition of thin epitaxial layers, and these layers often have lower defect and impurity levels as compared to bulk materials. The deposition of these epitaxial layers has used various deposition techniques such as vapor phase epitaxy (VPE), liquid phase epitaxy (LPE), molecular beam

ISSN 1079-4050
Vol. 28
ISBN 0-12-533028-6/$35.00

epitaxy (MBE), and metalorganic chemical vapor deposition (MOCVD). Of these, MBE and MOCVD have become dominant because they are capable of reproducibly generating the advanced device structures that require very thin layers and monolayer abrupt transitions in composition. The last 3 yr in particular have seen the formalization of these growth techniques to higher capacity (multiwafer) tools as manufacturing volumes have increased and, in parallel, sophisticated *in situ* monitoring tools have been developed. MBE has tended to dominate the growth of electronic devices (HEMT, HBT, etc.) where volumes are relatively low and a premium is placed on interface control. MOCVD has tended to dominate the growth of optoelectronics devices (high brightness (HB) LED and solar cells) where cost is more important and high capacity tools are required. During the same time period there has been a similar formalization in characterization techniques with most users now buying commercial equipment rather than building their own. High throughput production has raised a new challenge for whole wafer and nondestructive material characterization that is quite different from traditional single point and destructive measurements. In a production environment the necessity of reliable and rapid turn-around whole nondestructive wafer mapping characterization techniques has been become apparent and is currently being developed.

The production of cutting edge compound semiconductor devices requires the growth of high quality epitaxial layers. The word "epitaxy" is derived from the ancient Greek words "epi," meaning on, and "taxis," meaning arrangement. Thus an epitaxial layer is one that takes the same structure as the substrate it is deposited on, that is, the same crystal symmetry and lattice constant. If the layer is the same material as the substrate it is said to be *homo*epitaxial (GaAs/GaAs); if the layer is a different material it is *hetero*epitaxial (AlGaInP/GaAs, InSb/GaAs). Other derivatives include *strained-layer* epitaxy (GaInAs/ GaAs, etc.), where elastically strained layers of different lattice constant also exist.

All devices require spatial control of some parameter in at least one dimension. A simple case is the change from *n*-type to *p*-type doping as a function of depth that forms a *p-n* junction. More advanced devices can use the different properties caused by varying the composition or elastic strain of a layer to cause changes in the bandgap, refractive index, or carrier concentration. In this manner, carriers and photons may be confined or guided [1]. The ability to vary these properties while still maintaining the in-plane lattice constant of the substrate is referred to as *bandgap engineering*. In addition, the growth of very thin layers defined at atomic layer resolution and quantum confinement allowed for the precise modification of the electronic properties of compound semiconductors. A revolution occurred in the area of III-V compound semiconductor device design with the ability to realize structures that exhibited bandgap engineering and quantum confinement. New devices such as HEMT, where electrons are confined in a region having few scattering centers, and semiconductor laser

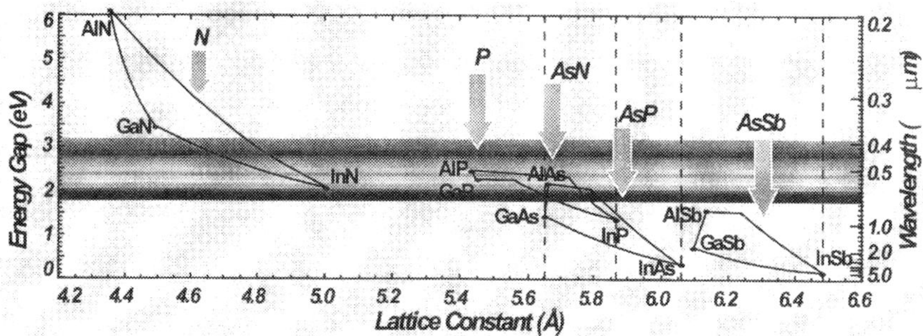

FIG. 1.1. A plot of alloy bandgap vs lattice constant illustrating the range of different ternary and quaternary alloy systems that can be lattice-matched to binary substrates [2]. (See color figure.)

diodes, where many different schemes are used to increase efficiency, shape the output beam, etc., were realized. All of the technologically important III-V compounds (Al, Ga, In) (N, P, As, Sb), Fig. 1.1, and most of their ternary and quaternary alloys have been grown by MBE and MOCVD [3]. Typically, there are up to four elements that need to be controlled for stoichiometry and effective bandgap engineering; see Table 1.1.

The purpose of this chapter is to provide the reader with a broad overview of the current status of epitaxial growth technology and the characterization of the deposited material. The focus is on III-V-based compound semiconductor materials and is not intended to be comprehensive because many of the details will be addressed elsewhere within this volume. In addition, there are several excellent books and review articles on many aspects of the subjects covered that will be referenced as necessary. This work will concentrate on current research, technology, and applications in an attempt to provide an overview of this subject area as it stands today.

TABLE 1.1
TYPICAL APPLICATIONS FOR III-V COMPOUND SEMICONDUCTOR MATERIALS

Material	Substrate	Applications
$Al_xGa_{1-x}As$	GaAs	Lasers and LED, HEMT, HBT, solar cells, photocathodes
$In_{0.53}Ga_{0.47}As$	InP	IR detectors for satellites, fiberoptic communications
$In_{1-x}Ga_xAs_{1-y}P_y$	InP	Lasers and LED for 1.3- and 1.55-μm fiberoptics
$In_{0.49}Ga_{0.51}P$	GaAs	LEDs, solar cells, HBT
$In_{0.49}(Ga_{1-x}Al)_{0.51}P$	GaAs	LED, lasers
$Al_xGa_{1-x-y}In_yN$	Sapphire	Blue/UV detectors, LED

1.2. Epitaxial Deposition Techniques

1.2.1. INTRODUCTION

The two principal techniques in widespread use today for the deposition of compound semiconductor materials are MBE and MOCVD. The latter technique is also referred to, and used interchangably with, MOVPE/OMVPE (metalorganic/organometallic vapor phase epitaxy). MOCVD is a broader term that is applicable to the deposition of crystal, polycrystalline and amorphous materials. Both MBE and MOCVD have produced a wide range of very high-purity semiconductor materials with excellent optical and electrical properties. Most research and development has centered on the growth of III-V semiconductor binary, ternary and quaternary alloys, with greatest emphasis on GaAs, (AlGa)As, and (GaIn)As, (Fig. 1.1). There has been a developing interest in, Al-free, P-containing alloys and narrow-bandgap Sb-containing alloys for optoelectronic applications. The last few years has also seen the emergence of III-nitrides for UV/blue emitters and high-power electronics. There has also been renewed interest in Si- and Si-Ge-based devices and these will be reviewed elsewhere in this volume. In this introduction an overview of various thin film deposition techniques will be completed before considering the MBE and MOCVD techniques in more detail.

In MBE, elements (Ga, In, etc.) evaporate from effusion cells (ovens) in the form of molecular beams onto a heated substrate. This takes place in ultrahigh vacuum (UHV) so that the beams are not scattered, and background contamination is reduced to an acceptable level. The biggest advantage of this technique is the ability to access the growing layers with a variety of diagnostic tools, such as reflection high energy electron diffraction (RHEED). Much has been learned about crystal growth processes and surface chemistry using these diagnostic tools, and they can be used to control the growth process to define layer thickness and composition. Under optimum conditions, MBE layers can be grown with excellent purity and with very abrupt interfaces. For example, the ability to accurately control the interfaces in devices such as HEMT structures has resulted in MBE taking the lead in this area.

In MOCVD, compounds of the desired materials (metalorganics, hydrides, etc.) are transported to a heated substrate, where a chemical reaction takes place at the surface. MOCVD growth is conducted at a pressure between 20 mtorr and atmospheric pressure, and the equipment is generally quite simple, especially for atmospheric growth. The chemistry is much more complex than MBE, although reactions can be accurately controlled by the correct selection of precursors, operating conditions, and reactor design. Moreover, sophisticated in situ monitoring tools are now being developed. MOCVD is a very versatile technique and has been used to deposit materials that are difficult to grow by MBE, such as phosphides and nitrides. However, MOCVD requires the storage and use of large quantities of hydrides for the group V source, sophisticated gas handling systems for gas delivery, and scrubbing systems for postgrowth effluents.

There are a number of hybrid techniques that combine the features of MBE and MOCVD. These all typically use the ultrahigh vacuum (UHV) environment of MBE but utilize materials other than the elements for sources. Most of these hybrid techniques were developed to overcome some limitation of MBE (frequent requirement to break vacuum to load sources, limited chemistry, etc.) and generally incorporated elements of MOCVD. One example is gas-source MBE (GSMBE), where solid evaporation sources are replaced by gas sources. In III-V semiconductors, this technique was developed to avoid difficulties in handling solid phosphorus by using PH_3 as the group V source. The gas is introduced into the chamber through a cracker that generates the molecular beam. This avoids having to open the chamber to replace the solid source and also has the advantage of being able to rapidly change the delivery rate by changing flow instead of the oven temperature. Si has also been grown using disilane (Si_2H_6) as the precursor rather than solid Si. Similarly, metalorganic sources commonly used in CVD processes can be used for the group-III source in MBE chambers. In metalorganic MBE (MOMBE), metalorganics are used with solid group V sources. This increases the chemical versatility but often results in high carbon contamination in the layers since, unlike MOVPE, there are no H radicals to displace the alkyl ligands that are left after the metalorganic compound has cracked. In chemical beam epitaxy (CBE), hydrides are used as group V sources in conjunction with metalorganic group-III sources to overcome some of the carbon problems of MOMBE. In principle, the growth chamber never needs to be opened to replace source materials in CBE. However, the use of metalorganic sources complicates the growth reactions. Hence, the processes are less well understood. Another derivative technique is plasma-assisted MBE. The plasma, typically from an electron-cyclotron-resonance source, is generally used to increase the reactivity of stable molecules such as N2 for wide bandgap III–V nitrides.

Other epitaxial techniques have also been developed for the epitaxial growth of III–V compound semiconductors, but all have limitations that have restricted their use to simpler devices and they do not have the extensive use of MBE and MOCVD for more advanced structures. These include liquid phase epitaxy (LPE) [4], vapor phase epitaxy (VPE) [5], and even magnetron sputtering [6]. A direct comparison of these different growth techniques is not simple because each technique has its own strengths and weaknesses. LPE, for example, has been widely used in research and has achieved many firsts, such as growing the first semiconductor laser diode. LPE is an equilibrium growth technique and the thermodynamics of the process are very well understood. It utilizes simple equipment and achieves high purity easily because of the stoichiometric control that results from depositing from a saturated (and dilute) melt. LPE is still widely used to produce LED and lasers, but is gradually being replaced by MBE and MOCVD for more sophisticated devices because it can be difficult to obtain the sharp interfaces and thinner layers required for quantum well structures. VPE usually involves a process in which one or more elements is transported by halides. For example, in hydride VPE, the group III material is transported as the chloride while hydride gases supply the group Vs. This technique is still widely used to prepare layers of GaAsP for red, yellow and green LEDs. Chloride VPE uses elemental group III and chloride group V (e.g., $Ga + AsCl_3$), and has produced very high-purity GaAs. Both techniques use hot wall reactors and have high growth rates that can be difficult to control, making the reproducibility of thin layers difficult to attain. Moreover, aluminum-containing compounds are problematic to grow, thus excluding technologically important materials such as AlGaAs and InGaAlP.

1.2.2. MOLECULAR BEAM EPITAXY

Molecular beam epitaxy (MBE) was first developed by Arthur [7] and Cho [8] for the controlled growth of III-V semiconductor epitaxial layers. In MBE, neutral thermal energy molecular or atomic beams (Ga, Al, As_4, etc.) provide the source for growth when they impinge on a hot crystalline substrate maintained in an ultrahigh vacuum (UHV) environment. The evaporants are called molecular beams when their mean free paths are much greater than the source to substrate distance (Knudsen regime), that is, when the pressure is $< 10^{-4}$ torr. The growing layer derives its crystalline orientation from the underlying substrate material.

The primary advantage of MBE is the capability for controlled growth of heterostructures with layer thicknesses down to a single molecular layer (ML). MBE was the main driver for the development devices that used bandgap engineering and quantum confinement. The well defined interfaces are a result of the low, well-controlled growth rates, $\sim 1\,ML/s$, combined with almost instantaneous interruption of growth using shutters over each molecular beam source. Typical growth temperatures provide sufficient surface diffusion to allow layer-by-layer growth, and provide extremely flat interfaces between layers, with minimal bulk interdiffusion [9]. In addition, MBE has the flexibility needed for various pulsed-growth procedures that can further improve interface flatness such as growth interruption (GI) and atomic layer epitaxy (ALE).

One important feature that has distinguished MBE from other growth methods has been the availability of UHV-compatible *in situ* surface diagnostics and characterization techniques. Modulated-beam mass spectrometry has been extensively used to provide understanding of the surface processes associated with MBE growth [10, 11]. RHEED provides information on crystal perfection and during growth for parameters such as growth rate and alloy composition [12].

The primary disadvantage of MBE is the need periodically to open the chamber to air to add source materials to the evaporation sources. Following the exposure of the MBE chamber to air, extensive high-temperature baking of the growth system over a week or so is required to obtain UHV operating conditions. There are also limitations on the materials that can be used in MBE, most notably phosphorus. Finally, the low growth rate means that it can take extended periods of time to grow thick structures such as vertical cavity surface emitting lasers (VCSEL). Most of these disadvantages, along with materials-specific requirements, have led to a number of technological modifications to the MBE growth process, see Section 1.2.1 [13]. However, some of these changes (e.g., MOMBE and CBE) involve such radical changes that they can almost be considered as separate growth techniques.

The purpose of this section is to provide the reader with an introduction to the technology of MBE. Compared to MOCVD growth (described in the next section), the optimal configuration and components of MBE growth (the growth chamber, sources, and *in situ* diagnostics) are well defined. Numerous books [14, 15, 16] and extensive review articles [12, 17, 18] have been published on MBE and the reader should refer to these and other references for more details as necessary.

1.2.2.1. Growth Chamber

Figure 1.2 shows a schematic drawing of typical MBE systems used for III–V growth [19]. The system consists of UHV growth and sample preparation chambers. A sample transfer and load-lock mechanism is used to introduce samples into the system and transfer them between chambers. Various mechanisms for inserting and removing substrates from the substrate holder are used in MBE systems. The sample preparation chamber often houses surface science capabilities for detailed characterization of the substrates and deposited films. MBE production systems are obtained by a relatively simple scaling to larger sizes without any major modification to the schematic shown in Fig. 1.2.

The chambers are typically pumped by a combination of an ion pump, cryopump, and a Ti sublimation pump with LN_2-trapped diffusion pumps, and/or turbomolecular pumps when high vapor pressures sources are being used. Liquid nitrogen-cooled shrouds are used for reducing water-vapor partial pressure. Base pressures of typically 1×10^{-10} torr are important for obtaining high purity semiconductors, because the arrival rate of impurity molecules is then 10^{-4} times lower than typical growth rates of $\sim$1 ML/s. Numerous reports show that semiconductor optical and electrical properties are strongly dependent upon the base vacuum. The source flange on MBE systems is mounted on the bottom, side, or at an angle, taking advantage of gravity to keep the source materials in the

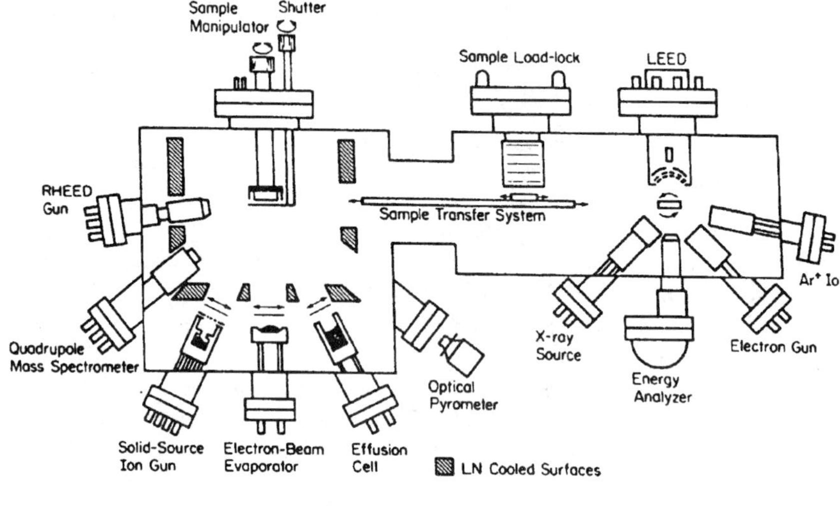

FIG. 1.2. Schematic of the MBE system typically used for III-V growth [19].

crucibles. Because of the long, narrow shape of Knudsen cell crucibles, they can be mounted in a nearly horizontal orientation without the evaporant coming out of the source. Water, alcohol, or liquid nitrogen-cooled shrouds are used around effusion cells to minimize heating and subsequent desorption of impurities from surfaces by radiation from hot parts.

The substrate-holder assembly includes a substrate heater and capability for translating and rotating the substrate. Substrate heating is usually accomplished using a resistance heater placed behind the substrate. The substrate is mounted on a molybdenum block, in some cases using indium metal to form a good thermal contact. The block assembly can be inserted onto and removed from the heater/manipulator assembly using the sample transfer apparatus. Substrate rotation is always used to minimize lateral variations in growth rate and composition.

1.2.2.2. Sources

By far the most common type of source used in MBE systems is a thermal evaporation source known as a Knudsen cell [14]. In its ideal form, a Knudsen cell is a heated cavity with an orifice small enough that it does not disturb the equilibrium vapor pressure inside. The effusion rate from the orifice then depends only on the vapor pressure of the evaporant, and not on the amount of evaporant.

Practical Knudsen cells have large openings to allow useful deposition rates, and the evaporation rate therefore varies with fill level. The cell contains a crucible, typically PBN with a thermocouple mounted on the back side to measure temperature, and is heated by resistive heater wires. The distribution of the evaporated flux depends on crucible shape and varies with the evaporant fill level [20]. For example, as the fill level drops, the evaporated beam is collimated by the sides of the crucible such that narrower distributions are obtained. This has led to the use of conical-shaped crucibles which eliminates beam collimation for typical wafer sizes to minimize the need for frequent calibration of evaporation rates [21].

The actual flux distributions obtained in MBE systems tend to be asymmetric because the sources are necessarily mounted off the substrate normal axis. For large source-to-substrate separations, the angular separation of the sources is reduced. Hence, the flux distributions become similar. This leads to more uniform composition profiles, but the deposition rate decreases rapidly as the spacing increases resulting in a trade-off between film uniformity and the deposition rate. Knudsen cells also have a relatively large thermal mass (i.e., the crucible and evaporant,) which makes rapid changes in cell temperature and evaporation rate problematic. Typically two cells are used with the same source material to produce sharp heterointerfaces between different alloys such as GaInAs and AlInAs. Furthermore, it is difficult to achieve gradual, programmed changes in composition. Although some groups have carried out detailed thermal analysis of the cells in order to predict cell temperature variations, and therefore flux variations, allowing the growth of gradually varying compositions [22].

In some cases, effusion cells are designed to provide a higher temperature at the open end of the crucible. These are usually termed two-zone, hot-lipped, or low-defect cells. The temperature gradient is maintained by a nonuniform heat source, or by special crucibles that absorb heat more effectively near the tip. The temperature gradient minimizes the accumulation of material near the tip, which is often a serious problem for Sb. The name low-defect cell refers to the possibility of reducing the oval defect problem, presumably by minimizing the amount of the evaporant (e.g., Ga) accumulated near the source tip.

Most group III sources (Ga, In, Al) evaporate as monomers, but both commonly used group V sources As and P (arsenic and phosphorus) evaporate predominantly as tetramers. There are advantages to MBE growth using dimer species as they react more efficiently on the surface and thus require lower overpressures and improve the properties of the film [23]. Figure 1.3 shows a schematic of a typical cracker cell [24]. It consists of two zones: a low-temperature sublimator crucible in which the tetramer is evaporated, and a high-temperature cracking furnace in which the dimer is produced prior to exiting the source. The cracking efficiency varies depending upon the temperature, the material used in the cracker region, and the geometry, which determines

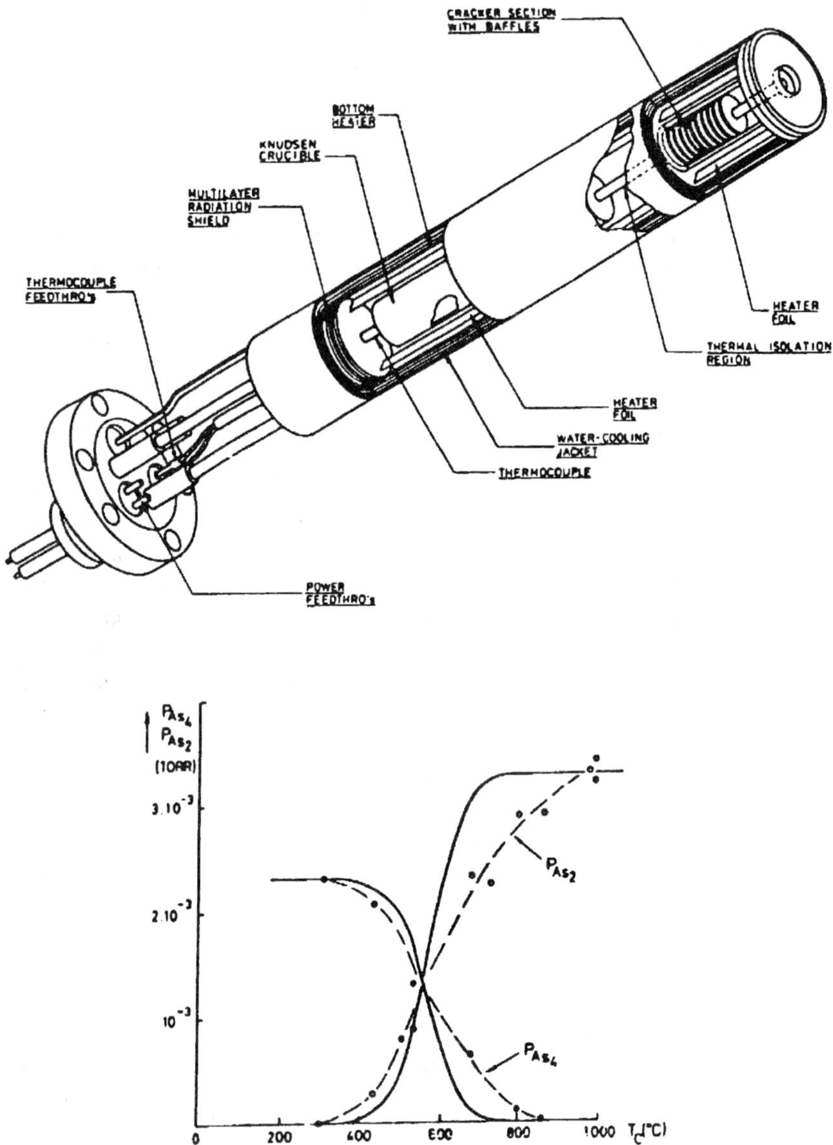

FIG. 1.3. (a) A schematic of a cracker cell consisting of two zones: a low-temperature sublimator crucible and a high-temperature cracking furnace. (b) Relative abundance of arsenic molecules evaporated through a Ta cracking tube [14].

the number of collisions molecules make with cracker surfaces. Systematic studies of various materials have shown that the arsenic cracking efficiency decreases in the order Pt, Pt-Rh, Re, Ta, Mo, W-Re, graphite [24]. Re was perhaps the most effective catalyst, providing 95% conversion efficiency at 700 °C, which was lower than the 800–1000 °C values used with conventional Ta or graphite. Pt was the most efficient at low temperatures; however, it reacts with As to form $PtAs_2$ and, hence, cannot be used.

A recent development for the evaporation of high-vapor-pressure species, such as As and P, is the valved cracker. These cells feature a valve between the sublimator and cracking sections that allows flux modulation without changing the temperature in the sublimator, which allows more controllable flux variations. This has the advantage of allowing abrupt or gradual changes in composition. The valved cracker sources have recently been shown to eliminate many of the practical difficulties with growing phosphides by solid-source MBE, and to allow the growth of P-containing heterostructures with excellent optical properties [25]. Heterostructures containing both As and P using two valved crackers have also been produced [26].

The advent of III-nitrides for blue lasers, LED, and high-temperature electronic devices has driven the search for nitrogen sources that are compatible with the MBE environment. Early attempts to adapt conventional MBE systems to nitride growth have suffered due to the limitations of available nitrogen sources. For example, using NH_3 with a conventional cracker source could not provide the necessary reactive nitrogen species. The approach now taken is the use of RF-excited plasmas using N_2 as the source of nitrogen. Optimized coupling of RF power to the plasma is important as the production of nitrogen molecular ions has been reported to cause damage during MBE-grown III-nitrides. High-quality MBE-grown GaN has now been produced using nitrogen plasma sources with growth rates of up to 0.8 μm/hr [27].

1.2.2.3. In Situ *Diagnostics*

The UHV environment of the MBE system has allowed a number of different techniques to be used for *in situ* diagnostics and monitoring of MBE growth. The quantities of interest include source and substrate temperatures, fluxes at the substrate, fluxes desorbing from the substrate, crystal structure and composition, and growth rate. The primary tools are mass spectrometry and RHEED. RHEED has been extensively used to provide information on crystal perfection, surface flatness, and surface reconstruction. Observations of oscillations in RHEED intensity during growth provide information on surface migration, growth rate, and alloy composition. Other techniques, such as ellipsometry [15] and reflectance difference spectroscopy [28, 29], have also been employed (Section 1.2.4). The sample preparation/introduction chamber can also contain UHV-compatible diagnostic systems for characterization of the substrate material and the grown layer. Auger electron spectroscopy (AES) and x-ray photoemission spectroscopy (XPS) are common examples.

Mass spectrometry is used to observe residual gases and to monitor molecular beam fluxes. In addition, the technique has been used to measure the fluxes desorbed from the substrate and film during growth. Applications include measurement of surface stoichiometry and characterization of the reaction mechanisms on substrate surfaces [11], observation of dopant-surface interactions [30], observation of surface segregation during InGaAs growth [31] and the correction of flux transients upon shutter operation [32]. Modulated-beam mass spectrometry (time of flight measurements), where the fluxes desorbed and/or reflected from the substrate surface are monitored, has been extensively used to improve the understanding of the surface processes associated with MBE growth.

RHEED is the most widely used diagnostic in MBE. Figure 1.2 shows the orientation of the RHEED system in the growth chamber. The electron beam is focused on the sample surface and the resulting diffracted beams are projected onto a phosphor-coated screen. As the RHEED system does not interfere with the molecular beams, it can be used to monitor growth as well as static surfaces. However this requires an on-axis wafer and may not be available for production systems. An imaging system is usually used to record the RHEED pattern.

A simple kinematic description of electron scattering can be used to explain the basic features of RHEED [33]. Figure 1.4 shows a schematic illustration of the scattering geometry and the corresponding Ewald sphere construction [34]. The Ewald sphere construction is a graphical representation of the Bragg diffraction condition. A diffracted intensity is expected whenever the Ewald sphere coincides with the reciprocal lattice rods. As the surface is not perfectly planar, but consists of a random distribution of terraces and steps, then the rods broaden and their intersection with the Ewald sphere consists of streaks (Fig. 1.5). When the surface is very rough, for example, 3D islands are present, electrons can penetrate through protrusions, resulting in a bulk-like spot pattern. If the islands form facets, arrowhead shaped spots and streaks perpendicular to facet surfaces can be observed in RHEED patterns. For example, the initial stage of growth of InSb on GaAs forms flat topped islands with {111} and {113} planes showing diffuse lines of intensity in these directions [37]. A review by Daweritz and Ploog [38] focuses on the structure and atomic scale morphology of GaAs surfaces using RHEED.

The spacing between RHEED streaks can be related to the spacing of surface features d by Bragg's law, $\lambda = 2d \sin \theta$. Additional streaks are commonly observed in RHEED patterns, indicating the presence of a surface periodicity that is larger than the bulk interplanar spacings. These surface reconstructions are important as they can reflect the surface composition and geometry such as step heights. The use of surface reconstructions has been used extensively in MBE to

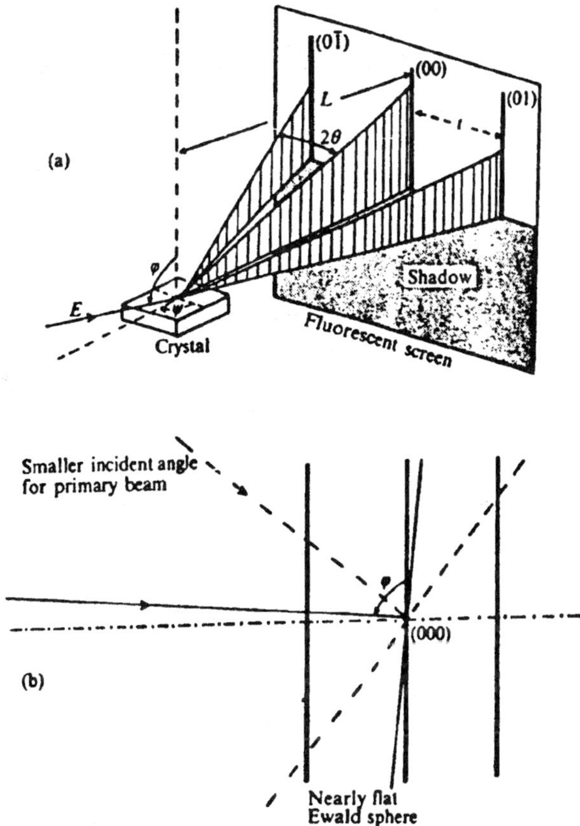

FIG. 1.4. (a) A schematic illustration of the RHEED scattering geometry and (b) the corresponding Ewald sphere construction [34].

map optimum growth conditions, see Fig. 1.5 [39]. RHEED can also resolve surface features with larger spacings on the surface, such as terraces separated by step edges. [40] have discussed vicinal surfaces with high densities of step edges. When the steps are well-ordered, splitting of diffraction streaks occurs, with the splitting distance giving the average terrace width.

One very important feature of RHEED is the ability to observe oscillations in the intensity corresponding to the growth of atomic or molecular layers. RHEED oscillations are a consequence of periodic smoothening and roughening of the growth surface due to the 2D nature of the nucleation and growth during MBE. Normally the specular beam in the RHEED pattern is monitored and oscillations are observed due to diffuse scattering as the surface periodically roughens [41]. A number of other features are observed in typical RHEED oscillations, as

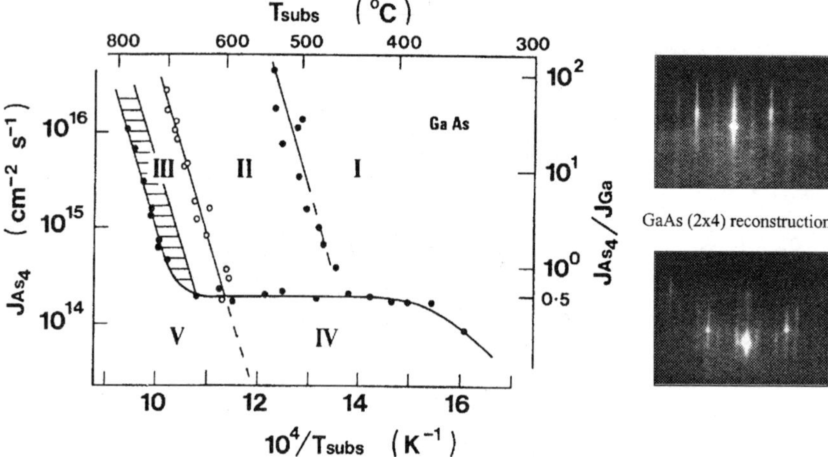

FIG. 1.5. The surface phase diagram of (001) GaAs for a fixed growth rate of 0.65 mm/hr [35]. The reconstructions are: I-mixed (2 × 4)c(4×4), II-(2 × 4), III-(1 × 1) bulk streaks, IV-(4 × 2) and V-c(× 2). The inset is a (2 × 4) RHEED pattern taken under group-V stabilized conditions. In the ⟨110⟩ directions, additional streaks are observed indicating an increased periodicity of the interplanar spacing in that direction [36].

illustrated for GaAs in Fig. 1.6. As the period corresponds to one monolayer, the oscillations can be used to measure growth rates and control the thicknesses of layers in the ML range. The intensity of oscillations dampens with time and, within the kinematic approximation, this indicates that the growth front progressively roughens to a point where no further temporal changes are detectable.

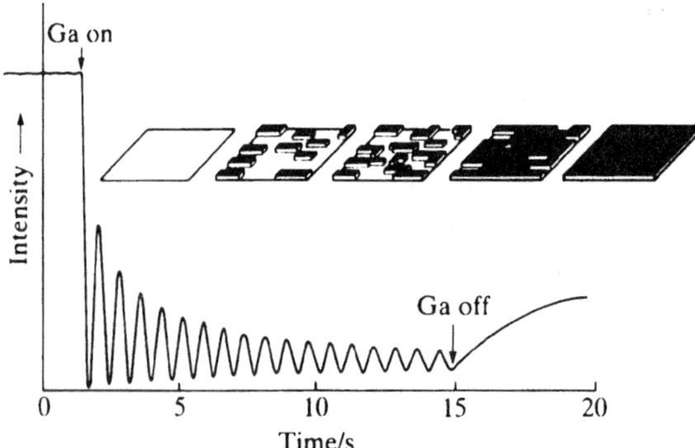

FIG. 1.6. Typical GaAs RHEED oscillation. Oscillations are a consequence of the periodic smoothening and roughening of surfaces due to 2D growth mode of MBE. Note the recovery in intensity of the specular spot when growth is terminated [42].

Another feature in the RHEED intensity is the recovery of intensity of the specular spot when growth is terminated. This indicates that surface diffusion smoothens the growth-roughened surface. Changes in the oscillation period on going from pure GaAs growth to AlGaAs growth have been used to measure the alloy composition [41]. The details of the RHEED oscillations depend critically on the diffraction conditions; careful choice of azimuth and in particular, angle of incidence of the electron gun can minimize these effects [43].

1.2.3.4. MBE Growth Mechanism

The extensive use of *in situ* diagnostics during the MBE growth of III-V compound semiconductors has meant that the MBE growth technique is among the best understood. Fortunately, much of what has been learned about growth kinetics has not been limited to MBE and has a much wider applicability to other growth techniques including MOCVD.

The growth of GaAs on GaAs(001) has been the most studied. The experimental techniques that have dominated the drive to develop and construct growth models are mass spectrometry and RHEED. The MBE growth of GaAs, and most III–V compounds, can be understood as follows, Ga evaporates in monomer form. Arsenic evaporates as the tetramer As_4 from elemental sources and as the dimer As_2 from a cracker cell. The GaAs is unstable above a noncongruent evaporation temperature ($\sim$600 °C) so excess arsenic is used to avoid non-stoichiometric growth. Above a critical T_s value of 300 °C, the excess group V species are desorbed to achieve correct stiochiometry. Arthur [7] first showed that the sticking coefficient of Ga is essentially unity, so the growth rate is primarily determined by the Ga arrival rate. Foxon and Joyce [10, 11] used modulated-beam mass spectrometry to develop the currently accepted model for As_2 and As_4 incorporation into GaAs. The model developed showed that the As_2 sticking coefficient is unity if a monolayer of gallium exists on the surface and growth occurs by a simple process of dissociative chemisorption of As_2 on a surface gallium atom. For As_4, the sticking coefficient somewhat surprisingly never exceeds 0.5. There is also evidence that growth occurs from gallium clusters on the surface [44, 45].

The most straightforward method to determine if an arsenic flux is appropriate for GaAs growth at different temperatures is to observe the surface reconstruction using RHEED. Various surface reconstructions are observed on GaAs(001) as a function of substrate temperature and the ratio of the arsenic and gallium (Fig. 1.5) [39]. The $(2 \times 4)/c(2 \times 8)$ reconstruction has been probed in the greatest detail, because optimized growth in most III-V material systems occurs with this reconstruction. Angle-resolved photoemission spectroscopy measurements showed that the 2-fold periodicity is due to an asymmetric dimerization of the arsenic bonds on the surface [46]. UHV scanning tunneling microscopy (STM) has shown the existence of the arsenic dimer and that the 4-fold periodicity was due to a unit cell consisting of three arsenic dimers and a missing dimer [47]. The anisotropy of the (2×4) reconstruction has important consequences for understanding the surface growth kinetics of MBE and MOCVD, see Section 1.2.5. However, it should be stressed that, because there is a differing surface stoichiometry, even within one reconstruction, there is no simple relationship to material quality [48].

The growth mechanisms of other compounds are similar to GaAs but with some differences. The growth of InAs has proved problematic because even at normal temperatures there is evidence that the growth mode favors the formation of microscopic droplets of free indium on the surface [49]. Phosphorous-based compounds, in particular InP, have not been extensively grown by MBE because the high vapor pressure makes it difficult to produce well-collimated beams [50]. One unique feature of InSb is that stoichiometric growth does not occur with a large excess of group V flux. Excess antimony can incorporate into the InSb layer because of the relatively low Sb vapor pressure (comparable to In) [51]. Ternary and quaternary alloys also grow with a similar growth mechanism but the effects of sublimation and group V desorption rates from the surface must be accounted for [52, 53]. For example, sublimation affects not only growth rate but the alloy composition because the higher-atomic-weight group-III elements sublime more rapidly.

1.2.3. Metalorganic Chemical Vapor Deposition

Manasevit pioneered the initial use of MOCVD in 1968 and has reported a first-hand account of the early days of MOCVD [54]. The development of MOCVD growth was primarily driven by limitations in other growth techniques, such as LPE and VPE. LPE was successfully used to fabricate many novel devices, but the problems of scaling up this technique to produce uniform films on large area substrates was never completely overcome. While chloride and hydride VPE has made many strides for Ga(As,P) device structures, there were fundamental problems in producing the AlGaAs layers needed for lasers and other heterojunction devices. It was not until 1977, when Dupuis and Dapkus [55] reported the achievement of room temperature lasers that the MOCVD technique was finally considered to be capable of filling its early promise.

Figure 1.9 shows a schematic of a MOCVD gas handling system and the two main reactor geometries. The gas handling system controls the incoming gases and directs them to the entrance of the reactor, using pressure regulators, mass flow controllers (MFCs), and valves. The metalorganic sources (TMG, TMI, etc.) are placed in temperature-controlled baths and also have their own MFC and pressure controllers. A carrier gas, usually hydrogen for semiconductor materials, is used to transport the reactants to the substrates and to carry away the byproducts of the reaction. All these flows are directed into either the reactor or a vent line, via a fast switching manifold. The hydrides (AsH_4 and PH_3) normally have a separate injection system to minimize prereactions in the gas lines before the growth chamber. Most MOCVD systems are known as cold wall reactors because the substrate and the susceptor are significantly hotter than any other part of the reactor. The substrates or wafers are typically placed on a susceptor, which is made of a material that is compatible with the reactants and does not contaminate the wafers. It is usually heated by RF induction, resistively or by infrared radiation from lamps. The system is designed so that deposition occurs only on a substrate placed on the susceptor.

The three major mechanisms that govern MOCVD are thermodynamics, kinetics, and hydrodynamics. As MOCVD is an exothermic process, the maximum possible growth rate will be limited by thermodynamic forces trying to restore equilibrium, and will decrease as the temperature of the reaction site (i.e., the heated substrate) increases. On the other hand, if kinetics dominate, the reaction rates will limit the growth rate and these increase as the temperature increases. Last, if the growth rate is limited by the mass transport of reactants to the substrate surface, then the process is relatively temperature-independent. These three regimes typically lead to the "inverted bathtub" curve of growth rate vs temperature shown in Fig. 1.7. This has been confirmed by experiment, with the mass transport limited region occurring between ~550 and 750 °C for GaAs

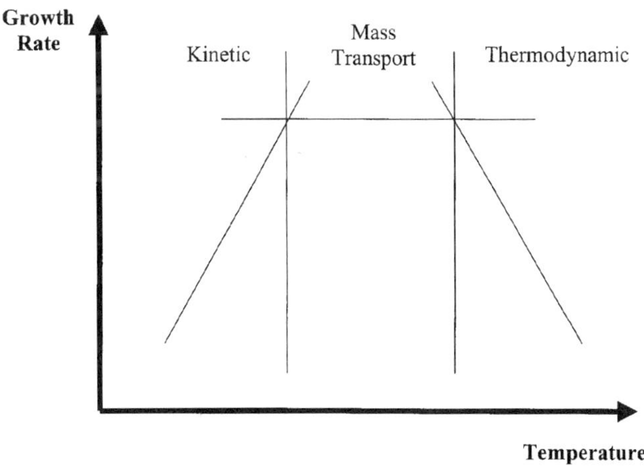

FIG. 1.7. The growth rate of III-V material grown by MOCVD vs the growth temperature. The MOCVD growth normally takes place in the mass transport region so that the process is controlled by the flow rates of the reactants and not by temperature.

[3]. Most MOCVD processes are operated in this regime so that the growth rate may be reproducibly controlled by the flow rates of the reactants, and not by temperature critical or area sensitive processes.

The concept of a boundary layer is useful in understanding the gas flow kinetics [56]. The velocity of a fluid at the substrate or a constraining wall must be zero, while in the bulk of the fluid it is some uniform value. The region in which the velocity is changing due to the presence of the wall or the substrate is called the boundary layer. As the gas velocity increases, this layer becomes thinner. Ideally the bulk flow should be smooth or laminar, but in practice, changes in the cross-sectional area, density gradients, turbulence, and especially temperature gradients can all cause recirculation cells to occur. Recirculation cells trap reactants and slowly release them to the main flow, making rapid changes in doping or composition difficult to attain.

The design of an MOVPE reactor is constrained by many factors: materials properties and wafer capacity are the most important. Numerous configurations of cold wall reactors have been used for the MOCVD of III–V semiconductor compounds. However, there are two main geometries used for deposition: the horizontal and vertical configurations, and most reactor chambers may have small variations to these two basic designs. Most of the modifications to reactors and the associated process have been developed empirically, but modeling efforts are beginning to shed some light on the important fundamentals.

Simple modeling and flow visualization was a critical element in early MOCVD reactor design [57]. Modeling is used to find flow patterns that show no recirculation and temperature isotherms that are very uniform with a sharp temperature gradient perpendicular to substrate. Good uniformity of the temperature isotherms is necessary to achieve good compositional uniformity, and a sharp temperature gradient (or thin thermal boundary layer) allows the reactant gases to come close to the disk before pyrolyzing. Fotiadis *et al.* [58] have published an excellent paper, which comprehensively discusses all the parameters that affect this geometry, particularly the reactor geometry, flow, operating pressure, temperature, wall materials, and rotation, etc. A good correlation between a computer model of the flow pattern in a rotating disk chamber operating under proper conditions, together with a flow visualization pattern obtained for the same conditions, is shown in Fig. 1.8 [59].

So far, only deposition processes that are driven by the temperature of the substrate, sometimes known as thermal CVD, have been considered. Some CVD processes are assisted by other excitation mechanisms, in order to lower the growth temperature (to avoid diffusion or other reactions in the substrate material), or to help break down a stable precursor. The most widely used are plasma enhanced (or excited), optically enhanced (using lasers or lamps to break chemical bonds) and electron beams. These techniques are more commonly used for the deposition of amorphous and polycrystalline films [60].

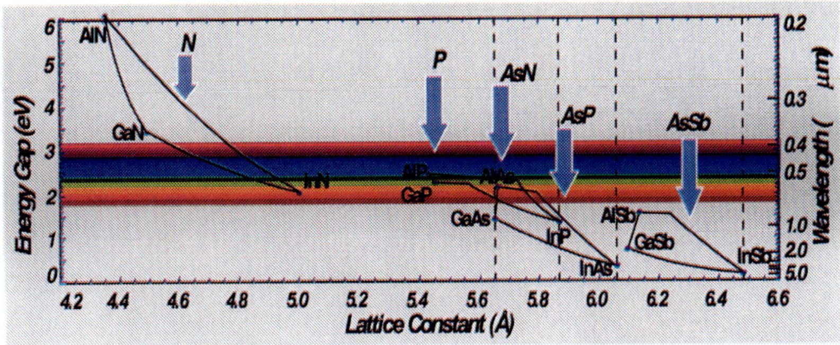

FIGURE 1.1. A plot of alloy bandgap vs lattice constant illustrating the range of different ternary and quaternary alloy systems that can be lattice-matched to binary substrates [2].

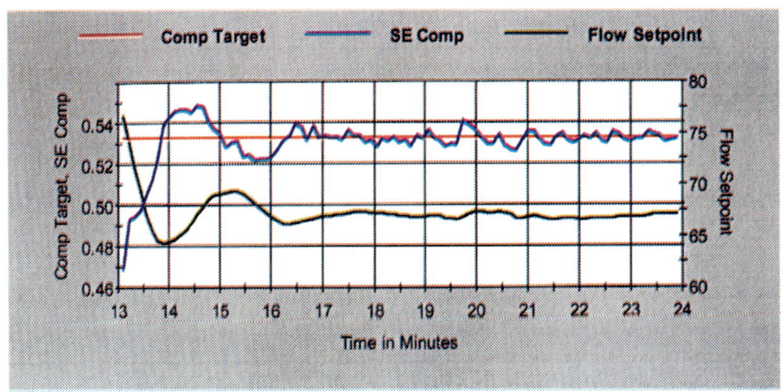

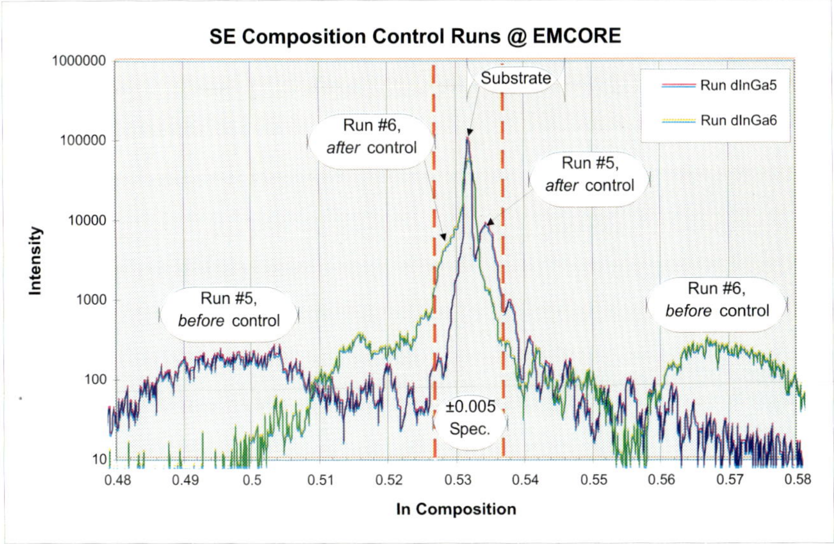

FIGURE 1.13. The *in situ* ellipsometric feedback control of the growth of GaInAs. The top graph shows a composition control experiment in which the SE composition signal was used to automatically adjust the TMIn flow to achieve the target lattice matched $In_xGa_{1-x}As$ composition of $x = 0.532$. X-ray diffraction scans were used to verify the accuracy of the SE composition control. Using SE feedback control of the growth, a bulk film with a composition within the lattice-matching specification was achieved for initially indium-rich and gallium-rich growth.

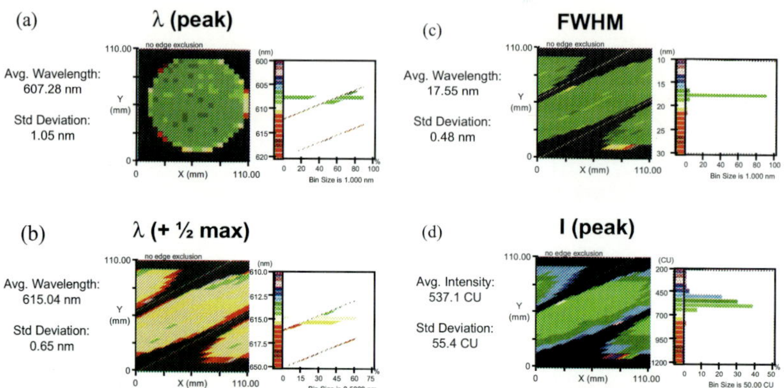

FIGURE 1.22. Room-temperature photoluminescence maps of a $Ga_{1-x}Al_x)_{0.5}In_{0.5}P$ film with $x \sim 0.23$ grown on 100-mm diameter GaAs wafer. The uniformity data for λ(peak), $\lambda(+1/2 \text{ max})$, FWHM and I(peak) is summarized in Table 1.3.

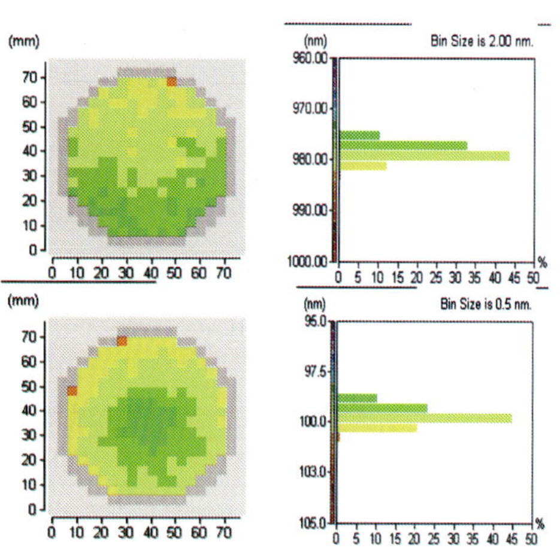

FIGURE 1.33. (a) RT photoluminescence peak wavelength map (wavelength: 977.9 ± 1.8 nm) and (b) a FWHM map (FWHM: 99.6 ± 0.44 nm) of the InGaAs channel of an optimized *p*HEMT structure.

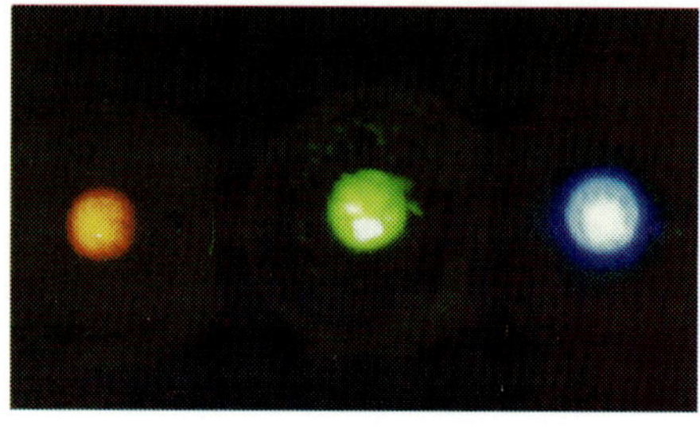

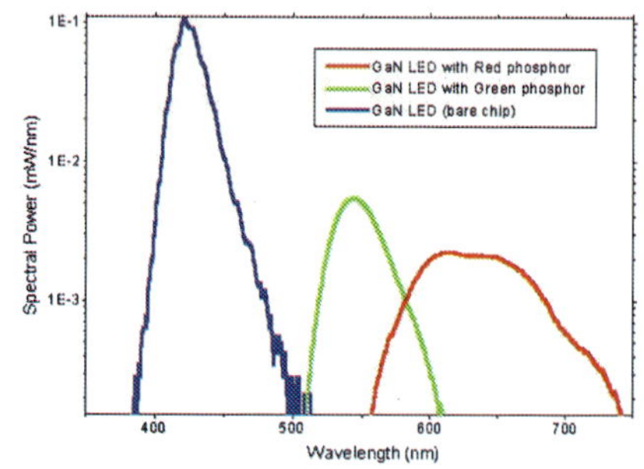

FIGURE 1.35. Red, green and blue (RBG) emission from down-converting phosphors and a III-nitride LED die. The peak emission at 425 nm is from the LED while those at 550 nm and 625 nm are from the phosphors $SrGa_2S_4$: 4%Eu and $Zn_{0.25}Cd_{0.75}S$: AgCl, respectively. (See color figure.)

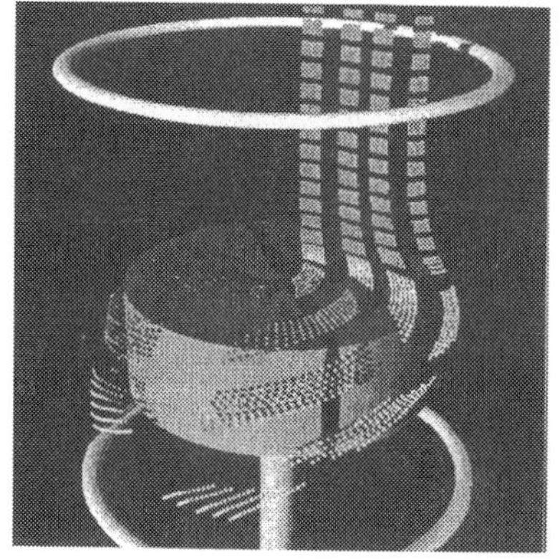

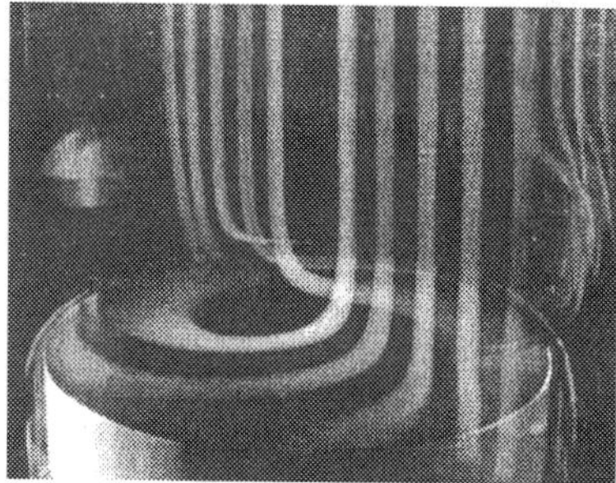

FIG. 1.8. A comparison of computer-generated flow patterns and smoke trails showing the reactant flow patterns in a rotating disk reactor [59]. The flow patterns shows no recirculation and significant flow at the center of the disk.

In the next section the horizontal and vertical reactor configurations will be reviewed because growth systems based on both these geometries are widely used within the MOCVD growth community. Important features will be detailed, along with the practical scaling up to production type systems. A number of books describing MOCVD materials growth and technology are available [3, 61]. Numerous review papers give an account of current progress in the field of MOCVD [62, 63, 64].

1.2.3.1. Horizontal Reactors

The horizontal configuration is probably the most widely used design for MOCVD growth, particularly for small, research-scale reactors, see Fig. 1.9. The substrate wafer sits on a susceptor in a horizontal tube and reactants flow over it, parallel to the wafer surface. In practice, the gas inlet region is tapered slowly from the small inlet tube up to the final cross-section shape, to allow a laminar flow to develop and to avoid the creation of recirculating cells of gas [65]. The susceptor is usually recessed into the floor to divert all the gas across the wafer and to avoid abrupt edges that could cause turbulence. The susceptor is wedge-shaped or tilted to compensate for depletion of the reactants as they flow over the wafer. This increases the gas velocity, thinning the boundary layer, and thereby increasing the concentration of the remaining reactants. The geometry, the type of material being grown, and the carrier flow are all interactive. The precise geometry is usually selected from experience, published designs [66], and modeling [58]. The temperature range is dictated by the materials being grown and the carrier flow is then adjusted for optimum uniformity or other desired properties.

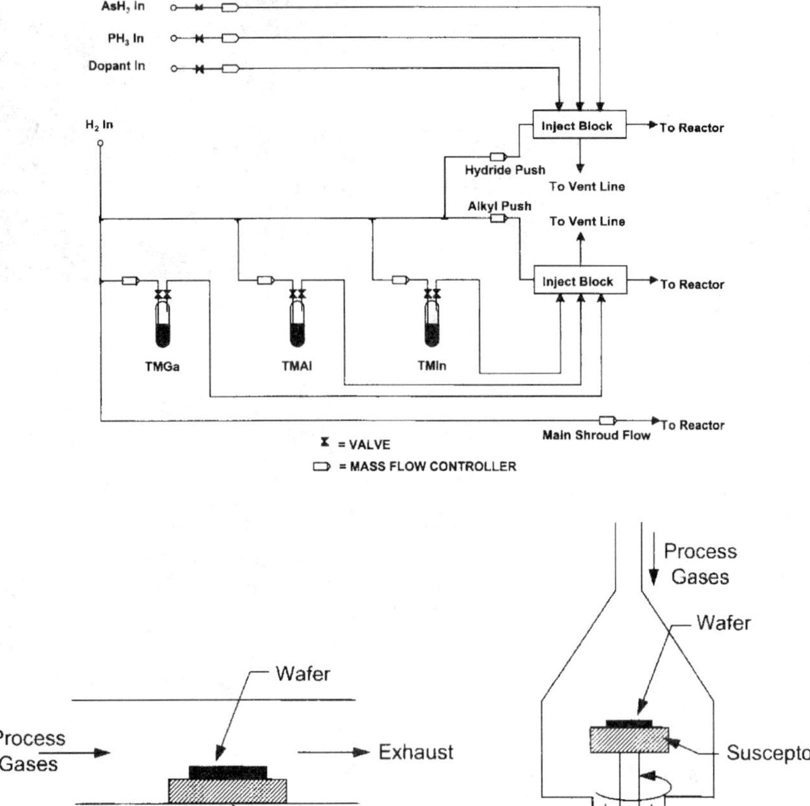

FIG. 1.9. Schematic of the MOCVD gas delivery system and the (a) horizontal and (b) vertical configurations of MOCVD deposition systems.

Disadvantages of this design include: the depletion effect, a transverse nonuniformity due to the side walls, deposits on the ceiling, which affect uniformity and surface quality, and a tendency to exhibit recirculation cells. However, most of these disadvantages have been mitigated, to some extent, by engineering or process optimization. The boundary layer uniformity can be enhanced by increasing the gas velocity. This is typically achieved by lowering the pressure in the reactor from atmospheric to about 100 mbar. The uniformity can be further improved by rotating the wafer (by gas foil or direct mechanical means). In this way, the linear depletion that occurs longitudinally can be further averaged out. The transverse, parabolic, nonuniformity due to the sidewalls can only be lessened by making the tube and susceptor widths considerably larger than the wafer. Uniformities of $\pm1\%$ can be achieved in a fully optimized design, but in general, only for single wafer systems. For critical materials, such as the Al containing III–Vs, a glove box filled with inert gas can be used at the end of the reactor to avoid exposing it to oxygen during tube removal for cleaning, and while loading the wafer. Recirculation cells that are caused by convection can be minimized using an inverted geometry, [67] but in general modern horizontal reactor designs that are operated in the correct flow and pressure region are not too prone to this type of recirculation [65].

The commercial reduction of the horizontal reactor is known as a planetary reactor [68], see Fig. 1.10. At first glance, this reactor looks like a vertical reactor (see next section), but the distinguishing feature is that the gases are injected from

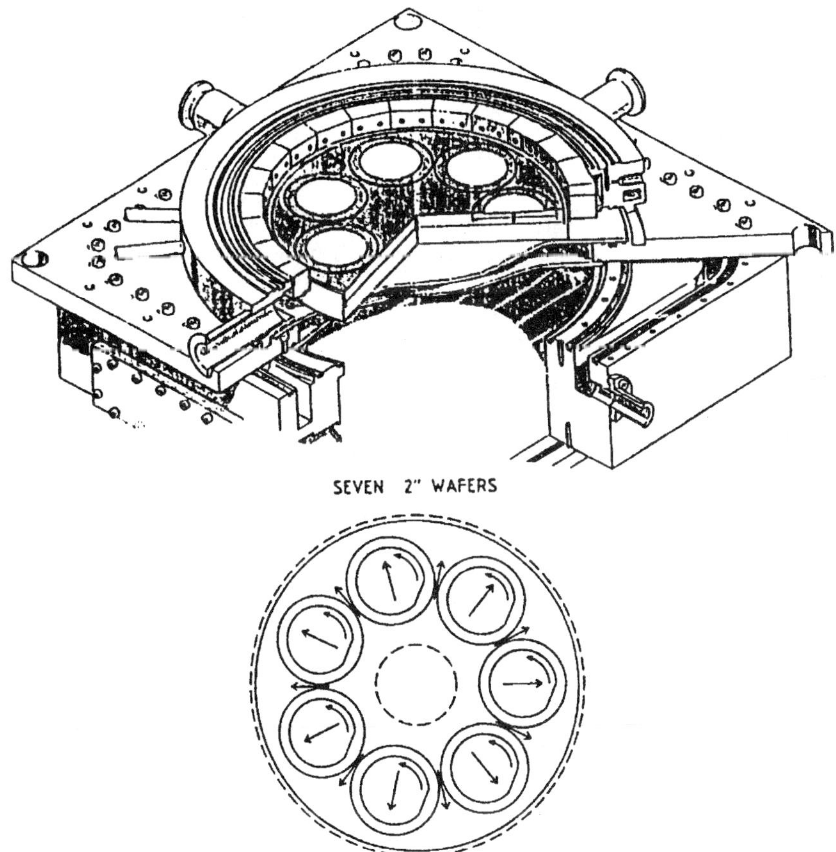

FIG. 1.10. A schematic of a planetary reactor [68]. The planetary system is characterized by radial flow of the reactants from the center of the reactor and rotation of both the growth platter and individual wafers.

the center and flow parallel to the wafer surface from one side to the other. The growth chamber is metal, with some quartz containment surfaces. In this reactor, the large susceptor is rotated, plus each individual wafer is mounted on a small, rotating susceptor that is recessed into the large one. Rotation is achieved by gas that is introduced behind each wafer carrier and is constrained by spiral grooves; this causes the carrier to float and rotate. This rotation is frictionless and does not cause particle generation if clearances are properly maintained. By the correct choice of geometry and carrier gas flow, the depletion across the susceptor can be made approximately linear, so that with the wafer rotation, good thickness uniformities of ± 1–2% can be achieved. However, gas foil rotation is complex and the large planet/susceptor therefore takes a long time to equilibrate at different temperatures. Despite this, when correctly optimized, many state-of-the-art structures have been grown simultaneously on multiple wafers at a time in this type of system.

1.2.3.2. Vertical Reactors

The first experiments to deposit III-V compounds, using metalorganic sources, were performed using a vertical reactor. In this geometry, the wafer is mounted in a vertical tube, typically made of quartz, and the reactants are directed perpendicularly to the wafer surface. The reactants are therefore incident over the entire surface and thereby avoid the longitudinal depletion effect of the horizontal geometry, Fig. 1.9(b). The carrier gas and reactants enter at the top, impinge on the wafer, and exit at the bottom. The entrance region may either be gently tapered out to the full diameter to avoid recirculating flows, as in the horizontal geometry, or the reactants can be spread out uniformly over the tube diameter by a diffusing element such as a porous plug. The susceptor can be RF or resistively heated, and is rotated to even out geometrical and heating inhomogeneities.

In this reactor, modeling studies have shown that there are three cases in which the boundary layer over the wafer will be uniform [59]. These are: stagnation point flow; impinging jet flow; and rotating disk flow. The latter is the case in which rotation is sufficiently rapid to affect the flow pattern (>500 rpm), unlike that used to improve uniformity. Most other vertical reactors attempt to emulate the stagnation point model. As in the horizontal geometry, in order to overcome recirculating flows and buoyancy effects from the hot susceptor, the carrier gas flow must be carefully balanced for each operating condition to achieve true stagnation point flow. The impinging jet condition is not used much in practice because it is difficult to avoid recirculation cells.

Advantages of the vertical geometry are that the system is easy to construct, it may be operated at atmospheric pressure, and reasonable uniformities ($\pm 2\%$) over a small wafer are easily attainable. However, for stagnation point flow, careful flow balancing is necessary, particularly at atmospheric pressure operation, to avoid recirculation cells. It is difficult to scale up beyond a single 75-mm wafer for this type of reactor. There are many home-built vertical reactors in use throughout the world that give good results. Reducing the pressure of operation helps avoid recirculation cell problems and increases gas velocity, thereby improving layer uniformity. The effects of buoyancy induced recirculation can be mitigated by using an inverted geometry [69, 70]. This configuration is attractive for single wafer systems, but mounting the wafer is not straightforward.

The most successful commercial manifestation of the vertical reactor geometry does not use stagnation point flow but the rotating disk flow, where the susceptor is rotated at high-rotation speeds (see Fig. 1.11) [71]. The effect of the increased rotation is to create a pumping action that pulls the gases down and across the wafers on the disk. Under a quite wide range of conditions, the result is that the boundary layer is uniform, and a forced convection flow is established that overcomes the tendency to form recirculation cells [59, 72]. Also, no wall deposits occur above the plane of the susceptor (disk), meaning that the necessity for frequent reactor cleaning is minimized or negated. Also, the design is highly scalable to very large dimensions, offering the possibility of processing large batches of wafers simultaneously with inherent uniformity. One of the major parameters of current interest involves increasing the size of the disk in order to increase the reactor capacity. The RDR design is fundamentally scalable due to two factors. First, the same flow patterns and temperature isotherms can be

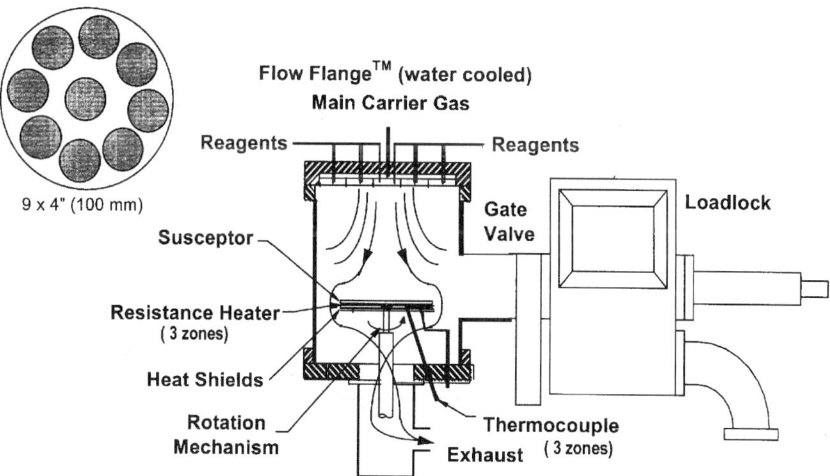

FIG. 1.11. A schematic of a rotating disk reactor. The reactant and carrier gases flow vertically downwards towards the high-speed rotating disk through a diffusing screen. The growth reactor uses a three-zone stationary graphite heater to ensure uniformity. The loadlock system allows fully automated transfer of platters without breaking the vacuum of the reaction chamber.

maintained for the different sizes by keeping the dimensionless constants the same [59]. Second, as a vertical reactor, the depletion effects can be minimized by feeding reactants along the radius of the disk. The ability to transfer processes directly from a small research scale system to a manufacturing operation is also unique to this geometry. Advantages include, the inherent uniformity, steep temperature gradients above the wafers (minimizing prereactions), and a high utilization of reactants. The ability to use metal construction for safety, reproducibility, and UHV load lock compatibility for automated wafer loading and unloading is important. Disadvantages include the need to operate at low pressures for reasonable carrier gas flow rates and the necessity to protect the high-speed rotation mechanism from particulates.

1.2.3.3. Other System Considerations

The reactor chamber is the primary component of the deposition system. Unlike MBE, however, the optimum MOCVD performance depends critically on other system elements. Most important is the gas handling system that controls the incoming gases and directs them to the entrance of the reactor, using pressure regulators, mass flow controllers (MFC), and valves. The metalorganic sources are placed in temperature controlled baths and also have their own MFC and pressure controllers. All these flows are directed into either the reactor or a vent line, via a fast switching manifold. The switching manifold uses low-volume valves to switch established gas flows without flow or pressure transients. This, together with a well-designed reactor, is essential in achieving the atomically abrupt transitions in composition and doping needed for advanced devices. The effluent leaving the reactor, which consists of hot gases, vapors, and particles, is also a concern. The exhaust system must trap or condition the gases and associated particulates before reaching the atmosphere in such a way that the lines do not become blocked and the vacuum pump and/or other components are protected [56].

An important issue that permeates all aspects of MOCVD system design is safety. As most processes use highly toxic gases (such as AsH_2, and PH_3), pyrophoric materials (the metalorganics), and hydrogen (H_2), the design must concentrate on keeping these materials away from the operators and exposure to air. Typically, the entire gas handling system, the reactor, and the exhaust system are contained in a single cabinet that is exhausted continuously to the outside. In addition, an all-metal chamber is much less likely to suffer a catastrophic breakage than a quartz reactor.

1.2.3.4. MOCVD Growth Mechanisms

The reader is referred to several review articles that address the extensive research into, and understanding of, the basic MOCVD growth process [63, 73, 74, 75]. A detailed description of the chemistry of MOCVD of compound semiconductors is beyond the scope of this review [3, 58, 62]. For example, epitaxial films of silicon have been grown in cold wall reactors from the materials SiH_xCl_y, where $x + y = 4$. As the proportion of Cl increases, the reaction can be pushed toward etching, rather than deposition. Understanding the chemistry and thermodynamics involved enables the operator to select the best starting concentrations and growth temperature to achieve the desired results. Sherman [76] discusses this point at some length in his book, and this is a good illustration of the complexity involved even for the growth of a single element material that has been widely studied. In this section, some comments will be made to highlight the status of the MOCVD precursors because these are critical to the growth process.

Group III Sources: The dominant Group III sources continue to be the trimethyl compounds (e.g., TMGa), with triethylgallium used for some processes. The purity of these sources is now at a satisfactory level for most device applications, although variations in quality have been observed. The alternatives are either more expensive or have less favorable properties, such as low vapor pressure or poor stability [7]. Efforts to reduce the oxygen content of aluminum compounds continue as these materials are still used for the growth of most optoelectronic devices. The cost of the major metalorganics, particularly TMGa, has been decreasing as the volume of use has increased, reflecting higher production volumes required by the suppliers.

Group V Sources: Despite their toxicity, the hydrides continue to be the most widely used Group V precursors. Their widespread use is due to a combination of relatively low cost and high purity. The convenience of a gaseous source is also a reason. The organic materials TBA and TBP are used less frequently; their lower toxicity is outweighed by their higher cost for many users. In the face of extensive research, ammonia still remains the only viable source to date for practical III-N growth [78].

Other Materials: The classical dopants, in either gaseous or liquid form, meet most current requirements for III–V MOCVD growth. These include SiH_4, H_2Se and DETe for *n*-type, and DMZn, DEZn, and Cp_2Mg for *p*-type. The latter, which is widely used for InGaAlP and the III-N materials, has shown variable quality, particularly when employed for GaN growth. The use of carrier gases other than hydrogen is increasing due to improved purification technology and some advantages in epitaxial layer quality [79]. Nitrogen is used during the growth of InGaN for example, as it gives higher In incorporation and improved morphology [80, 81].

1.2.4. *In situ* Monitoring and Process Control

In situ process monitors are becoming more sophisticated and are now exhibiting the potential of real-time control of the growth process [82]. The traditional approach to process control was to keep all critical growth parameters that could affect the layer properties as constant as possible. This requires frequent calibration runs and continuous tweaking of the growth process (growth rate, composition, doping, morphology) based on *ex situ* characterization. However, for this approach to be successful, good control of the thermal environment of the reactor and of all source temperatures, flows, pressures, etc., is necessary. Any changes in the thermal environment are particularly problematic because reactor coatings change emissivities and thermal conduction. This can affect growth rate and temperature dependent properties, such as composition or doping. Using the traditional approach, reproducibilities of ~1–2%, run to run, are achievable but only if 10–20% of the runs are dedicated to tests and calibration, rather than the end product.

Many different techniques, RHEED, optical, x-ray, etc., have been used in a diagnostic mode to investigate growth mechanisms by looking at surface structure during growth. Using *in situ* methods, it is now possible to directly control the parameter(s) of interest, such as growth rates and alloy composition. For example, mass spectrometer desorption measurements have been used for monitoring the composition of alloys and heterostructures during MBE growth [26]. Using a real-time feedback, this technique has been used to control the composition of (Al,Ga)As-, (Ga,In)As- and Ga(As,Sb)-based material systems and to control growth of complicated graded structures [83]. Most *in situ* probes used to monitor epitaxial growth utilize different optical techniques. Aspnes [84] has provided an excellent review of this topic. The optical techniques of reflectivity and ellipsometry have shown the most promise for the practical *in situ* control of layer properties. Reflection difference spectroscopy (RDS) has also been investigated and has provided useful information about surface kinetics but RDS has not emerged as a viable *in situ* control technique due to implementation difficulties.

Measuring the near-normal reflectance requires relatively simple equipment and has been reported by many groups. Breiland and Killeen [85] have reviewed the early work and showed how growth rates and compositional information can be extracted using the virtual interface method pioneered by Aspnes [84]. Growth rate accuracies reach 1% within a few nm of growth, and better than 0.1% for layer thicknesses of 20–40 nm. The composition of AlGaAs layers grown by MOCVD has been measured after deriving the optical constants at the growth temperature from layers of GaAs and AlAs. Figure 1.12 shows a reflectance plot of the early part of a run set up to obtain growth rates for different sources and alloy compositions. The growth of VCSEL with a reproducibility of ±0.3% over 10 runs, and a uniformity of ±0.2% over a 75-mm wafer has also been reported

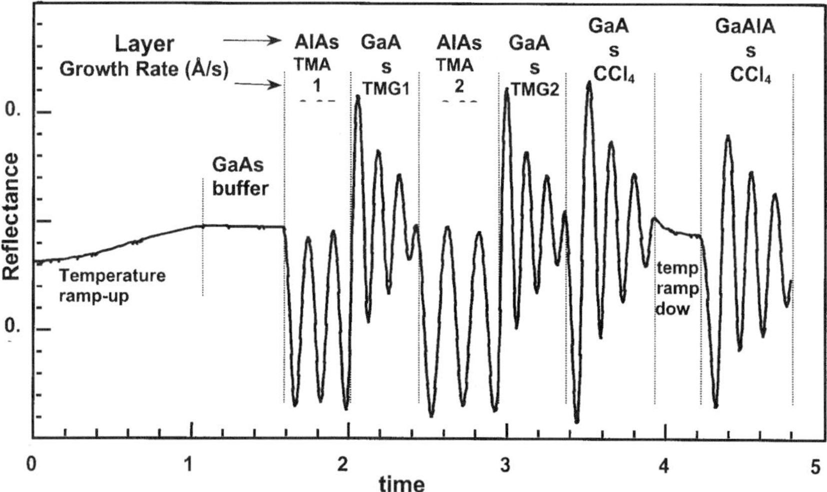

FIG. 1.12. Reflectance plot as a function of time showing the growth of different layers for calibration of growth rates and composition. Note the different growth rates for different MO sources and for heavily C-doped GaAs grown at slightly different temperatures (courtesy of Sandia National Laboratories).

using these techniques. Similar data have been obtained for the growth of GaAs/AlAs Bragg reflector stacks by MOCVD [86]. This group was also able to measure the Al content of AlGaAs films and obtained a reproducibility of <1% for a Bragg reflector optimized for 980 nm.

The ellipsometric technique has been applied to MBE growth by several groups [87, 88]. The equipment is more complex than for reflectance, and the windows must generally be strain-free. However, growth parameters can be derived more rapidly than for reflectance so thinner layers can be used for compositional control. This is particularly important for materials systems such as GaInAs due to critical lattice match conditions. Like reflectance, ellipsometry does not directly measure material properties, but a complex dielectric function that is then related to the material properties using *ex situ* measurements of test layers (produced at the growth temperature). The first report of accurate control of alloy composition by ellipsometry was for the growth of AlGaAs layers [87, 89]. A comparison of the study of the accuracy of reflectance and ellipsometry for thickness measurements has been completed [90]. It was found that for thinner layers ellipsometry is more accurate, but for films thicker than a few tens of nm, the accuracies are similar. The ellipsometric technique also allows the measurement of the substrate temperature to better than 1 °C, and to accurately observe the complete desorption of the oxide layer prior to growth.

More recently, ellipsometry has been used for the compositional control of InP-based GaInAs/AlInAs HEMT and HBT structures in a multiwafer MOCVD rotating disk reactor [82]. Most of the data reviewed here were obtained from single wafer systems and the sample rotation was low compared to that of a rotating disk reactor. Sophisticated structures, such as AlInAs/GaInAs on InP, require not only the development of accurate dielectric function libraries, but also rapid data processing and fast algorithms to allow critical lattice match conditions to be controlled. The control algorithms must also reflect the specifics of the growth reactions and be able to differentiate differences in the dielectric function due to changes in material composition and temperature. In addition, the measurements and the data stream must quickly follow changes in the input conditions such as any change in source arrival rate or sample temperature. The real-time monitoring and control of composition (hence lattice match conditions) has been achieved during growth of AlInAs and GaInAs. Figure 1.13 shows a composition control experiment in which the SE composition signal was used to automatically adjust the TMIn flow to achieve the target lattice matched $In_xGa_{1-x}As$ composition of $x = 0.532$. Figure 1.13 also shows the x-ray diffraction data used to verify the accuracy of the SE composition control. In both of these growths, the $In_xGa_{1-x}As$ composition intentionally commenced outside lattice match conditions (5% indium rich and 5% gallium rich). Using ellipsometric feedback control of the composition, a bulk film with a composition within the lattice matching specification was achieved. This *in situ* monitor is now being used to actively control the indium composition and layer thickness for InGaAs and InAlAs layers grown on InP.

1.2.5. Manufacturing Issues

The market for compound semiconductor devices has finally grown to the point where there are now several high volume commercial applications [91]. Much of the early research and development on III–V devices that established MBE and MOCVD as the dominant growth techniques were completed on relatively simple, single-wafer systems. Uniformity and reproducibility were not the dominant issues because the initial focus was on development of growth processes and proof of concept of device operation. This situation has evolved with the need for high-volume device production. The emphasis is now on the growth of high-quality films over large high-volume platters (up to 400-mm diameter), containing multiple wafers (see Fig. 1.14). This new focus requires growth systems in which the deposition is laterally uniform, abruptly switchable, and robust enough to withstand variations in process parameters. Moreover, when purchasing growth equipment, the focus is now on specifying the quality of the material produced rather than the details of the equipment. Some of the issues

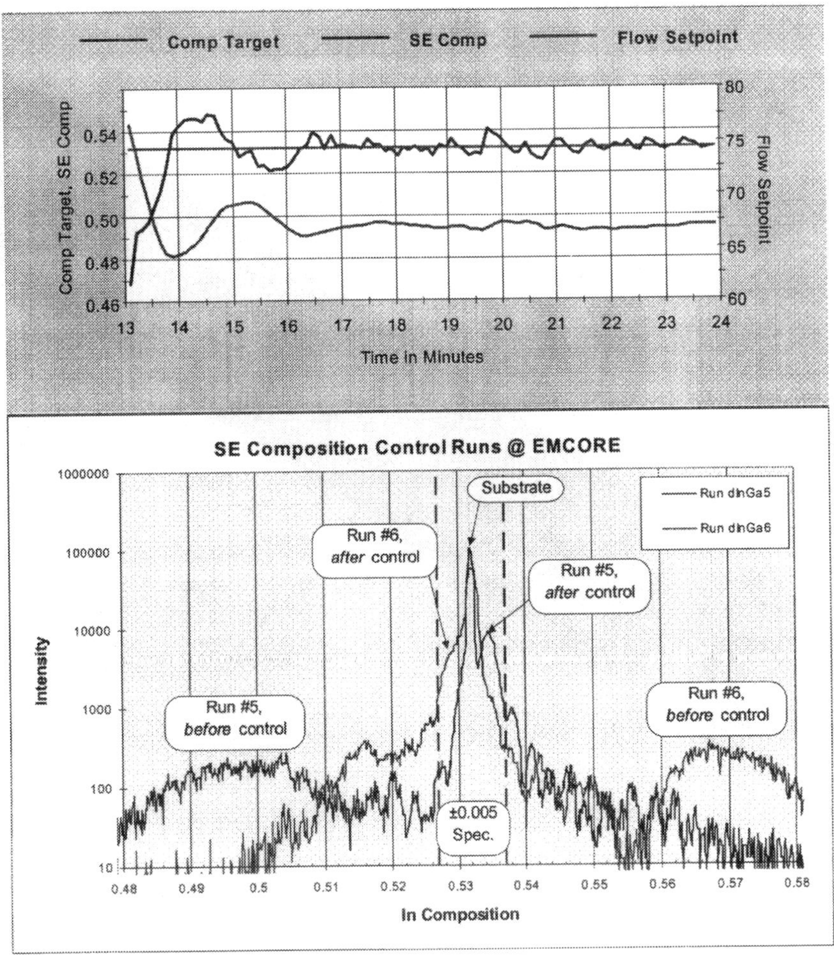

FIG. 1.13. The *in situ* ellipsometric feedback control of the growth of GaInAs. The top graph shows a composition control experiment in which the SE composition signal was used to automatically adjust the TMIn flow to achieve the target lattice matched $In_xGa_{1-x}As$ composition of $x = 0.532$. X-ray diffraction scans were used to verify the accuracy of the SE composition control. Using SE feedback control of the growth, a bulk film with a composition within the lattice-matching specification was achieved for initially indium-rich and gallium-rich growth. (See color figure.)

associated with the establishment of high-volume production of III–V compound semiconductor devices are reviewed in what follows.

The performance of any device while in production must be adequate to meet the intended end use application. Therefore the base material, prior to device fabrication, must meet a particular set of specifications. However, this is a

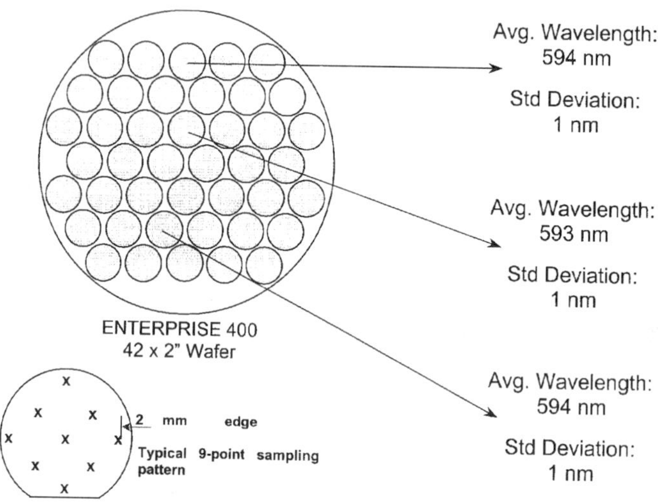

FIG. 1.14. Typical results obtained for InGaAlP light emitting diodes (LEDs) grown by MOCVD on a 400-mm platter that contains 42 50-mm wafers. The wavelength of the electroluminescence emission (at 20 mA) is shown for three randomly selected wafers for multiple devices as shown in the insert.

necessary but not sufficient condition for successful manufacturing, because, in practice, yield losses due to process variations and tolerances are also important. These occur because the final cost per wafer and device must be low enough to be competitive. If the yield loss is small, it will have little effect, but if the yield loss is large it will reduce the number of devices produced, thus increasing costs. A few working devices may show feasibility but this does not establish a robust production process or address the need for on-going material qualification. Typically, the specifications for each layer thickness, composition, and doping are critical; other factors such as lattice matching, interface abruptness, and surface morphology are also important.

To assure a reasonable yield, the uniformity and reproducibility of the epitaxial layers must be held within a certain statistical tolerance. This applies to wafer uniformity (the most widely quoted figure), from wafer to wafer within a batch for a multiwafer reactor, and run-to-run over a long period of time. In manufacturing, these three measurements are equally important, as the yield will suffer if the growth and device fabrication processes must be continually adjusted. During production, test and calibration runs must be performed to determine and adjust growth rates, compositions, and doping levels to continuously qualify the material. Very often these runs are on simple structures to facilitate measurements and do not produce a saleable product. This reduces the effective equipment utilization time, reducing throughput and increasing costs. The more reproducible the growth system, the fewer of these calibration and test runs will need to be completed.

The materials used in the growth process, substrate wafers, sources, gases, etc., are all rather expensive, which contributes substantially to the finished wafer cost. Therefore, any improvement that reduces materials consumption will help reduce overall costs. In MBE, using cracker sources to produce dimers improves the reaction efficiency of the group V by a factor of two. In MOCVD, decreasing the V/III ratio to the lowest feasible level will minimize hydride consumption. Moreover, having high alkyl (metalorganic) efficiency and an efficient wafer-packing density will ensure the maximum quantity of alkyls deposited on the wafers [92].

Cost of ownership (COO) models have shown that the largest system that is reasonably fully utilized usually results in the lowest epitaxial wafer cost [56]. For higher volume lines, this translates into a large batch size. The throughput can be maximized by using the highest possible growth rate and by shortening the nonproductive part of the growth process—the overhead time (which includes loading and unloading steps, heat up and cool down, etc.). The throughput is decreased any time the system is not available for growth. In addition to the nonproductive calibration and test runs, time is needed for preventive maintenance operations and unscheduled stoppages. The latter are normally due to system or facility breakdown. Decreasing any of these will improve the throughput, which is directly related to product cost. The growth system therefore needs to be as robust as possible and require minimum downtime for maintenance operations such as cleaning and source replacement. Once again, a high reproducibility minimizes the number of calibration runs and increases throughput and a good COO model is a critical tool for accurately assessing the throughput for a given tool.

With wafer throughput the key to achieving an efficient manufacturing operation, the compound semiconductor industry is transitioning towards the automatic movement of wafers similar to that done in the silicon industry. This has required the development of a multi-chamber cluster tool technology that is capable of receiving unprocessed wafers from one cassette and returning processed wafers to the same or another cassette, or chamber attached to a central wafer handling tool. The silicon industry has standards for wafer manipulation systems and growth chamber geometries. Many of these standards, such as those used in MESC-compatible cluster tools, are now being adopted by the compound semiconductor industry because they allow for easier integration of equipment from different manufacturers. Figure 1.15 shows the first-ever cluster tool constructed for the compound semiconductor industry. This cluster tool allows multiple chambers to operate simultaneously and complete different functions, which improves throughput.

FIG. 1.15. The first, MESC compatible, MOCVD cluster tool that conforms to the standards used within the silicon industry and that has been designed for the growth of compound semiconductors. A robotic handler can transfer platters between the loading cassette and the different growth chambers.

The application of *in situ* monitoring (Section 1.2.4) to a manufacturing system requires that optical access must be provided to the wafers in the growth environment. Optical ports or windows are needed that do not become coated during the growth process and are positioned at the optimum angles. The technique must be insensitive to wafer motion because in batch reactors the wafers are either individually or collectively rotated to achieve improved uniformity. The ultimate embodiment of an *in situ* control system will be the robust *intelligent* growth of the material or the device without the intervention of the grower.

1.2.6. MATERIAL ISSUES

The formation and structure of interfaces remains an important topic for III-V heterostructures. For semiconductor devices in particular, lattice-matched materials are usually used because the low-defect-density layers with planar surfaces obtained will provide high-quality heterojunction interfaces. Details of the growth mechanism are of considerable interest because they determine the interface morphology, which, in turn, determines the characteristics of devices, especially in cases where the device active layers are only a few monolayers thick.

A layer-by-layer (Frank-van der Merwe) growth mechanism is usually observed during growth of lattice-matched layers. Layer-by-layer growth is characterized by two limiting cases, see Fig. 1.16 [93]. One limit is where surface migration lengths are much smaller than substrate surface features, that is, terraces separated by step edges and growth occurs by the nucleation of 2D islands. The other limit is where migration lengths are much greater than surface terraces, where growth occurs by attachment of adatoms to the step edges. The step edges then propagate at a velocity dependent on the step density and growth rate; this limit is called step-flow growth. The transition between these two growth modes has easily been observed in RHEED oscillation data as seen in Fig. 1.16. A considerable anisotropy has been observed on GaAs(001) surfaces by misorientating towards either the [110] or $[\bar{1}10]$ direction, leading to nonequivalent $[\bar{1}10]$ or [110]-oriented terraces and step edges [94]. Scanning tunneling microscopy has shown that one type of step edge (As-terminated) is typically rougher than the other (Ga-terminated) [95] and the kinetics of adatom attachment at step edges are different for different step structures [96], see Fig. 1.17. Very often the substrate may be purposely cut at an angle to a major crystal plane to enhance step flow growth and thereby control the growth kinetics (Fig. 1.17 [97]).

(Al,Ga)As/GaAs tunneling superlattices, the first III–V device structure produced, was grown by MBE [98]. This material system was chosen because, in addition to having suitable band structures, GaAs and AlAs have the same structure and nearly identical lattice parameters within 1%. However, it was found

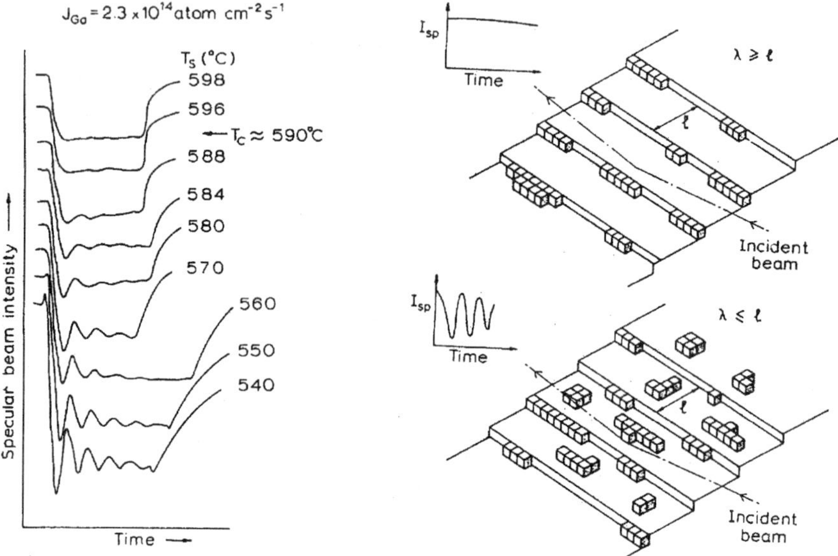

FIG. 1.16. A schematic showing step-flow growth when $\lambda > l$ and 2D nucleation when $\lambda < l$. The transition from step-flow growth to 2D island growth is indicated by the observation of RHEED oscillations [93].

that for even layer-by-layer growth the resulting heterojunction interfaces are not perfectly planar. This was especially important in semiconductor devices containing quantum wells and superlattices, because even a monolayer change in well thickness can significantly change electron energy levels.

The desire to obtain improved growth and interface morphology has led to a number of pulsed-growth techniques. These provide a means for obtaining additional control over the growth process, particularly adatom migration, and can improve interface morphology. Growth Interruption (GI) is a technique that takes advantage of the apparent morphological improvement of surfaces during growth pauses that was evidenced by the recovery of the RHEED specular beam during MBE growth [37]. Narrower luminescence linewidths are normally observed in the photoluminescence (PL) spectra of AlGaAs/GaAs quantum well structures when GI is employed [99]. However, it remains a contentious issue as to whether or not GI always results in the formation of improved interfaces [100, 101]. GI is also a necessary technique when transitioning between different layers in a heterojunction. Considerable effort must be invested to find the best switching mechanism to optimize the interface depending on the material systems involved.

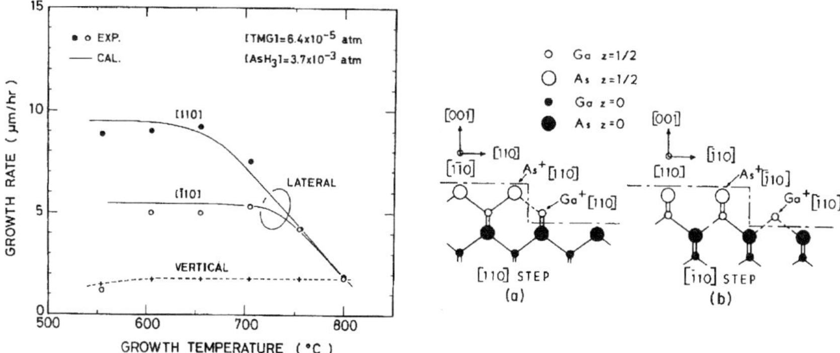

FIG. 1.17. The MOCVD growth of GaAs using TMG and AsH_3 on different vicinal planes exhibits a large difference in growth rate [97]. This indicates how the difference in the step-edge characteristics has a major effect on the growth kinetics.

Another flux modification technique that has been widely used is atomic layer epitaxy (ALE). ALE growth relies strongly on coverage-dependence incorporation probabilities to achieve self-limiting deposition of single monolayers, even though the impinging flux may exceed one monolayer [102]. A major advantage of this approach is extreme growth uniformity and control over exact thicknesses. Typically, the first atomic layer of material chemisorbs on the surface while any excess, weakly bonded, physisorbed states thermally desorb before the next deposition cycle starts. The ALE process does not generally work well unless a mechanism for self-limiting growth occurs [103]. The low overall growth rate tends to make the ALE technique unsuitable for the growth of thick layers and it is primarily used for thin buffer layers.

It is also possible to grow thin, defect-free, lattice-mismatched layers. The resulting strained layers exhibit substantial changes in their band structure and properties compared with the bulk [104]. As layer thicknesses are limited, it is possible to grow several thin strained layers into a superlattice structure. Examples include $In_xGa_{1-x}As/GaAs$ [105] and $In_xAl_{1-x}As/InP$ [106]. The maximum strain and layer thickness leading to high-quality layers are usually limited by two factors: introduction of misfit dislocations; and roughening of the growth surface. The relaxation of coherency strains by misfit dislocations has been studied in great detail, both theoretically and experimentally. A critical thickness is generally observed beyond which dislocations are introduced into the structure. Matthews and Blakeslee [107] and People and Bean [108] compared the energy of a strained film with a film relaxed by misfit dislocations, showing that a critical thickness h_c existed beyond which the relaxed structure was energetically favorable. In Fig. 1.18 the Matthews-Blakeslee and People-Bean

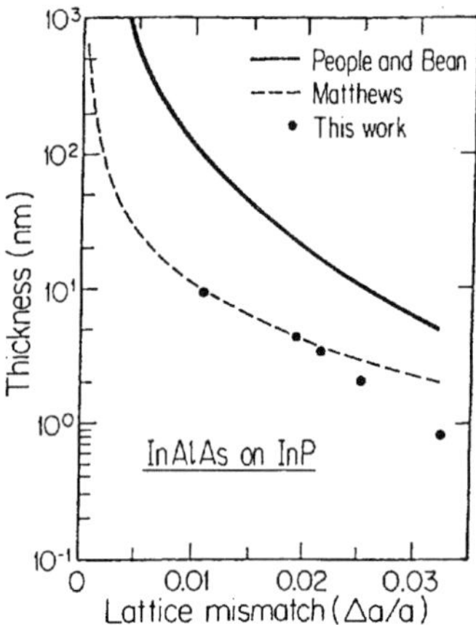

FIG. 1.18. Matthews-Blakeslee and People-Bean predictions of the critical thickness above which misfit dislocations begin to relax strain vs lattice mismatch. The predictions are compared with data for AlInAs alloys on InP. [106].

predictions of the critical thickness are compared with data for AlInAs alloys on InP [106].

There is a catastrophic transition, known as the Stranski-Krastanov growth mode, from the layer-by-layer growth mechanism to three-dimensional (3D) island growth at some critical thickness defined by the strain in the epitaxial film [109, 110]. This growth mode is technologically important for heteroepitaxial systems characterized by a large mismatch in crystal structure and/or lattice constant. Examples include GaAs on Si [111], InSb on GaAs [112], and GaN on sapphire [113]. In each of these cases, high densities of defects such as threading dislocations, low-angle grain boundaries, stacking faults, twins, and antiphase domains characterize the films. Defect densities can be reduced somewhat through optimization of deposition conditions. Strategies for growing high-quality films usually involve growing thick layers because high defect density near the substrate decreases rapidly with increasing thickness [14]. Other techniques include depositing defect *barriers* such as superlattices [115], low-temperature ALE-grown buffer layers [116] or the lateral epitaxial overgrowth (LEO) [117].

A number of semiconductor alloy systems have also been found to exhibit atomic ordering [118] or phase separation [119]. Ternary III–V semiconductor alloys such as GaAsSb [120], GaInP [63], and InAsSb [121] have shown ordering behavior. Ordering generally results in a modification of the band structure of the material, most noticeably a significant bandgap narrowing. It depends strongly upon the composition of the alloy and the specifics of the growth process. Phase separation in semiconductor alloys has also been widely documented and related to atomic size differences between the constituents [122]. A large number of alloys such as (Ga,In)(As,P) and In(As,Sb) exhibit substantial miscibility gaps over a wide range of compositions for typical growth temperatures [123]. Growth within miscibility gaps generally results in spinodal decomposition and occurs despite the fact that MBE and MOCVD are considered to be nonequilibrium growth processes [124]. This leads to degradation of semiconductor electrical and optical properties [125].

1.3. Materials Characterization

1.3.1. INTRODUCTION

Following the development of the growth techniques described in the last section, the next most important element of materials growth is the characterization techniques used in materials optimization. The characterization techniques used to study the different III–V material properties were developed parallel to the deposition techniques. At first, most of these techniques were specific to a particular laboratory and the measurements were mostly completed on especially constructed equipment. In addition, most of the techniques involved the destruction of the material investigated. As growth systems standardized, a formalization of different characterization techniques also occurred. This resulted in equipment manufacturers producing characterization equipment that became accepted as a standard by the grower, and ultimately the industry as a whole. In the last few years another change in the approach to compound semiconductor material characterization has also taken place: namely, the move from destructive to nondestructive characterization techniques and full wafer mapping (up to 15-cm diameter) capabilities. This transition has been driven by the need to move beyond the simple deposition of epitaxial layers to complicated structures over large areas that can be fabricated into operational devices. The need for mapping techniques has been driven by the requirement to monitor deposition over whole wafers and correlate the properties of material parameters, such as uniformity with yield. The industry continues to search for a correlation between some key device metric and some material property probed by a relatively simple nondestructive characterization technique.

Following the deposition of a III–V epitaxial structure, samples are characterized by multiple characterization techniques. These techniques typically include: photoluminescence (PL), high-resolution crystal x-ray diffraction (HXRD), Hall mobility, C-V profiling, sheet resistivity, Nomarski microscopy, scanning electron microscopy (SEM), transmission electron microscopy (TEM), optical transmission and reflectance, Raman scattering, and Fourier transform infrared (FTIR) spectroscopy. Of these techniques, PL and HXRD have emerged as the more important for compound semiconductors. The PL technique provides information on both alloy composition and, indirectly, crystal properties. For example, room temperature (RT) luminescence intensity of the bandedge emission normally gives a strong indication of the overall electroluminescence properties of the device. PL can be completed at low temperatures (< 77 K) to investigate dopant incorporation but RT measurements simplify the equipment needed. High-resolution x-ray diffraction is used to provide information on alloy composition, layer thickness, and lattice match conditions for heteroepitaxial growth [126]. Correlations between the structural properties and optical/electrical properties for process optimization of the deposition technique, and ultimately, the operational characteristics of the device are made.

As technological advances are made new characterization tools are being developed for techniques that would not have been previously viable. For example, newly designed Raman instruments using a charge coupled device (CCD) detector array, a single grating spectrometer (with a notch filter to depress the laser signals), and microscope focusing have been developed, all of which have greatly improved light throughput ability and spectral sensitivity [127]. Raman scattering has been recognized as a rapidly developing nondestructive characterization tool for semiconductors [128]. Raman spectroscopy can provide information on crystalline quality, impurity and defects, stress and strain, and other properties. Compared to PL measurements it is relatively simple to quantify doping levels with Raman scattering using plasmon-phonon coupling [129].

High throughput production also demands the large area deposition of epitaxial compound materials, which raises a new challenge for whole wafer nondestructive material characterization. Many of the characterization techniques listed in the preceding allow for whole wafer and nondestructive material characterization. These techniques, tightly coupled with the epitaxial processes, are necessary to realize the high quality and high uniformity growth of state-of-art materials in a III–V production environment. Nondestructive characterization techniques minimize the number of "good" wafers that have to destroyed that could be later processed into devices, thus increasing yields. Reliable whole wafer nondestructive characterization techniques are required to maintain a wafer-to-wafer repeatability to maximize the yield and minimization costs. These requirements are quite different from the single point and destructive measurements that dominated the development of III–V compound semiconductors. Nondestructive whole wafer mapping is becoming commonplace for wafers up to 10–15-cm diameters.

Numerous mapping techniques are currently being developed for compound semiconductors. Most, such as PL and XRD, tend to be an extension of pre-existing single point measurements, where rapid data acquisition and analysis techniques have allowed for the development of multipoint measurements. Others, such as measurement of sheet resistance, were developed primarily as a mapping technique. Sheet resistivity/conductivity mapping using a contactless eddy current technique has become a routine characterization and qualification process for epitaxial films and device structures. Other nondestructive mapping techniques primarily use optical methods (PL, UV/visible, and FTIR spectroscopy, etc.). Typical unformities for epitaxial film thickness, sheet resistivity, major PL band peak wavelength, and width are 1–2%. The uniformity is normally calculated as the standard deviation divided by the mean, but no standard currently exists. For techniques without automatic mapping capabilities, multiple point measurements were employed to obtain information over the entirety of the wafer.

In this section the specific details of the characterization technique will not be considered but rather the utility of different characterization techniques for various material systems. Examples are shown to demonstrate various characterization techniques and their application to III–V materials in the areas of substrate choice, epitaxial growth, and device optimization. Details about semiconductor characterization techniques can be found in Bullis *et al.* [130] and Cahn and Liftshin [131].

1.3.2. SiC AS A SUBSTRATE MATERIAL

SiC semiconductors are very attractive for high-power, high-temperature and high-radiation tolerance applications, due to their extraordinary material properties that include large bandgap, high electron saturation velocity, high breakdown field, and high thermal conductivity. High-quality wafers of both 4H- and 6H-SiC have been grown [132]. Various applications using these polytypes of SiC have emerged for devices produced by both ion-implantation and epitaxial growth [133]. Wafers of SiC are also a promising substrate for nitride semiconductor growth due to their compatible lattice structure and similar thermal expansion coefficients. However, the crystal quality is still rather poor (see Section 1.3.4). Both XRD and Raman spectroscopy have been used to measure the crystalline and doping properties of SiC prior to epitaxial growth. The spatial uniformity of SiC wafers is especially important if these materials are to be used in high-power electronics or as a substrate for nitride semiconductor growth [129].

Figure 1.19 shows the high resolution x-ray rocking diffraction (HRXRD) curves of a symmetric (0 0 12) and asymmetric (102) reflection of SiC. The full width at half maximium (FWHM) of the symmetric reflection is approximately 21 arcseconds and this is similar to that of GaAs and sapphire substrate materials. However, the FWHM for the (102) asymmetric reflection can not be determined because the rocking curve is the convolution of a number of relatively sharp

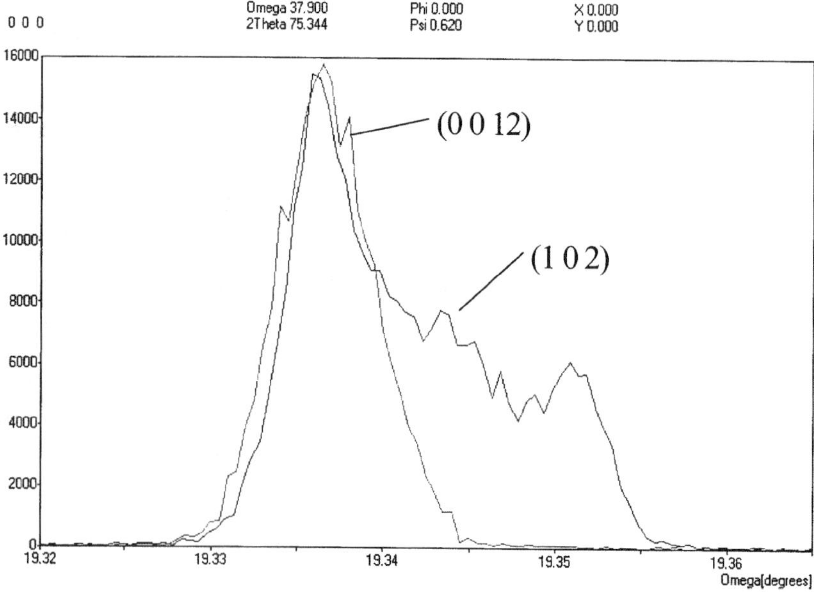

FIG. 1.19. SiC symmetrical (0 0 12) and asymmetrical (102) XRD rocking curves. The symmetrical reflection suggests that the SiC material quality is good and similar to other III–V materials. However, asymmetrical reflection shows that several crystal domains with different crystal orientations exist.

peaks, forming a larger broad peak. This indicates that the x-ray beam illuminates several crystal domains, with different crystal orientations. The beam size for this measurement was 0.5 mm^2. This means that crystal domains that have the same crystal orientation appear to be very small. The existence of these low-angle grain boundaries between these domains can have a strong effect on the properties of the epitaxial layers [134]. A further discussion of the application HRXRD asymmetric reflections applied to GaN can be found in Section 1.3.4.

Heavily doped SiC wafers are often not spatially uniform and higher carrier concentrations can be observed as a dark spot normally in the center of the wafer. Characterization techniques are required to quantify these nonuniformities because they can produce nonuniform heating and, subsequently, poor quality epilayers. Common measurement techniques for electrical characterization, such as Hall or 4-point resistivity measurements, require direct contact with the sample that prevents the latter growth of epilayers. Raman has been shown to be a useful probe of SiC [135, 136] and plasmon-phonon coupling can be used to estimate the spatial dependence of the carrier concentration across the SiC wafer [129]. This technique is nondestructive and the spatial resolution of the Raman scattering is determined by the spot size of the laser.

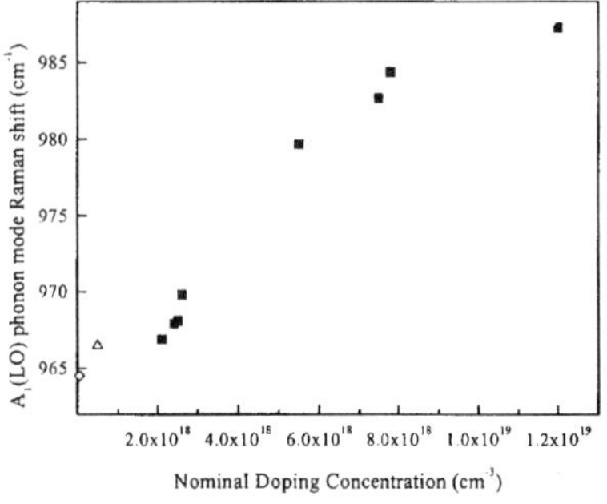

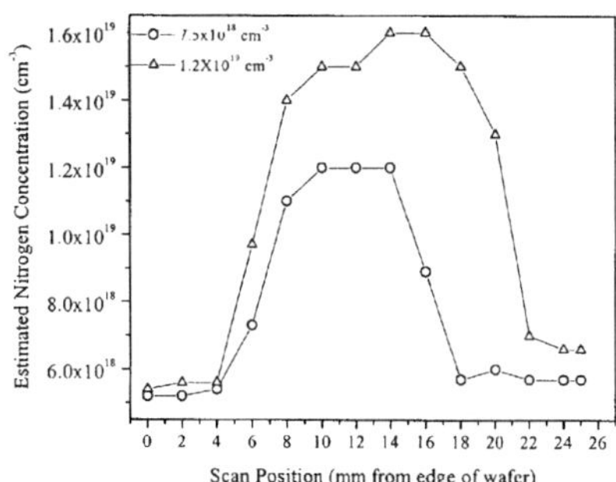

FIG. 1.20. (a) A_1(LO) phonon position vs nitrogen concentration in SiC. (b) The nonuniform carrier concentration profile across two heavily doped 4H-SiC wafers determined by Raman spectroscopy. The nominal nitrogen concentration is shown in the legend.

Raman scattering measurements have been completed at room temperature for semi-insulating 4H-SiC, and heavily doped 4H-SiC and 6H-SiC wafers. The wafers of SiC were *n*-type, nitrogen doped, with concentrations ranging from 2.1×10^{18} cm^{-3} to 1.2×10^{19} cm^{-3}. Nitrogen is well established as a donor in SiC [133]. A semi-insulating wafer of 4H-SiC was also examined as a reference. The A_1 longitudinal optical (LO) phonon lineshape was investigated as a function of nitrogen doping concentration. The change in nitrogen concentration from 2.1×10^{18} cm^{-3} to 1.2×10^{19} cm^{-3} in 4H-SiC produces a dramatic change in the position and shape of the A_1(LO) phonon, the phonon increases in frequency, and asymmetrically broadens. This large change in the frequency of the A_1(LO) phonon makes its position a sensitive probe of doping in this concentration ranges, (see Fig. 1.20).

To address the need for noncontact, *in situ* diagnostics for SiC wafers the spatial dependence of Raman scattering using the A1(LO) phonon was investigated. Figure 1.20 shows the nitrogen concentration across the wafer estimated from the A1(LO) phonon frequency. The spatial dependence is not the same for the two SiC wafers shown, and the concentration profile is often not symmetric with respect to the center of the wafer. The edges of both wafers had doping levels much lower than the nominal value specified for the wafer. Using Raman measurements a variation of approximately 26% was found in the nitrogen doping concentration over the center of the wafer.

1.3.3. III–V COMPOUND SEMICONDUCTORS

1.3.3.1. InGaAlP for Optoelectronic Applications

High-performance visible light emitting diodes (LED) based on InGaAlP/GaAs are now in production [137, 138]. InGaAlP is also emerging as a material for vertical cavity surface emitting laser (VCSEL) development and applications [139]. The quaternary $In_{0.5}Ga_{1-x}Al_x)_{0.5}P$ alloy compound is lattice-matched to a GaAs substrate over the entire composition range, and has a direct bandgap in the red-green wavelength region, up to 2.30 eV (539 nm) [140]. InGaAlP offers high radiative efficiencies in wavelength regions that cannot be realized with conventional ternary materials, such as GaAlAs and GaAsP. The growth of candela class high-brightness InGaAlP/GaAs LED has been accomplished by MOCVD in the orange-yellow [141], green [142], and yellow-green [143] wavelength regions. To achieve high-performance InGaAlP LEDs, it is important to grow these epitaxial materials with high crystalline quality under lattice-match conditions with precise control of composition, thickness, and electrical/optical properties. In order to produce InGaAlP LEDs of low unit cost, it is necessary to grow these materials over large areas and multiple wafers. High throughput systems that can hold 42 wafers of 2 inch diameter (Fig. 1.14) or 12 wafers of 100-mm diameter on one growth run require the rapid characterization of multiple wafers. The development of nondestructive mapping characterization techniques for the initial optimization of the InGaAlP, and the ongoing qualification of both the material and device structures is now a major challenge. Nondestructive characterization techniques are preferred as they minimize the number of wafers that have to be destroyed and which could be later processed into devices, thus increasing yields.

The optimization of the quaternary compound $In_{0.5}(Ga_{1-x}Al_x)_{0.5}P$ semiconductor materials requires the use of multiple characterization techniques. When these epitaxial materials were first grown, a large range of both destructive and nondestructive techniques were utilized, as described in Section 1.3.1, and some of this data is summarized in Table 1.2. The most important techniques for the initial characterization of $In_{0.5}(Ga_{1-x}Al_x)_{0.5}P$ are high-resolution x-ray diffraction

TABLE 1.2
SUMMARIZED RESULTS OF WHOLE WAFER CHARACTERIZATION ON MOCVD InGaAlP/GaAs

Property	Sample No.	x(Al)	Average Value	Standard Deviation	Uniformity
Thickness	#1		0.773 μm	0.011 μm	1.5%
Sheet resistivity	#2	0.24	101.4 Ω/□	1.05 Ω/□	1.03%
	#3	0.60	217.9 Ω/□	4.3 Ω/□	1.97%
PL λ(peak)	#4	0.23	607.3 nm	1.05 nm	0.17%
$\lambda(+1/2\ \max)$			615.0 nm	0.65 nm	0.11%
FWHM			17.55 nm	0.48 nm	2.7%
I(peak)			537 CU	55 CU	10%
Raman $\omega_{LO,GaP}$-	#5	0.20	391.21 cm^{-1}	0.19 cm^{-1}	
$\omega_{LO,AlP}$-			456.00 cm^{-1}	0.34 cm^{-1}	
x variation		$\Delta x/\Delta\omega_{GaP\text{-}LO} \sim 0.03/cm^{-1}$		$\Delta x \sim 0.006$	
		$\Delta x/\Delta\omega_{AlP\text{-}LO} \sim 0.04/cm^{-1}$		$\Delta x \sim 0.014$	

that measures lattice match conditions, and PL that shows the correct emission wavelength and optimal intensity. These are the primary techniques that are used as the feedback mechanism for the optimization of growth parameters. Once completed, the characterization techniques are then minimized to nondestructive, wafer mapping, characterization techniques. Two of the more important techniques in this regard are PL (as it has a strong correlation to the overall electroluminescence properties of the LED), and sheet resistivity, which is related to doping uniformity and, hence, yield.

Figure 1.21 shows a typical single-point room temperature (RT) PL spectrum for a MOCVD grown layer of $In_{0.5}(Ga_{1-x}Al_x)_{0.5}P/GaAs$ with $x \sim 0.23$ taken during a wafer map. The InGaAlP film was n-type Si-doped at 1E18 cm^{-3} and is grown lattice-matched to a 100-mm GaAs (100) substrate oriented 10° off towards (110). Spectral information is obtained at each point of the map using rapid data analysis and includes peak intensity I(peak), peak wavelength λ(peak), full width at half maximum (FWHM), upper wavelength at half maximum intensity $\lambda(+1/2\ \mathrm{max})$ and lower wavelength at half maximum intensity $\lambda(-1/2\ \mathrm{max})$ (see Fig. 1.21). However, single-point measurements have minimal value for these materials. The PL maps are preferred because they can be used to directly predict the uniformity of the composition and crystalline quality of the $In_{0.5}(Ga_{1-x}Al_x)_{0.5}P$ over the GaAs substrate. The occurrence of misfit dislocations due to lattice mismatch reduces PL intensity and broadens the FWHM. Figure 1.22 shows room-temperature (RT) PL maps of the same InGaAlP wafer in Fig. 1.21 and four parameters: λ(peak), $\lambda(+1/2\ \mathrm{max})$, FWHM, and this data is then summarized in Table 1.3. This InGaAlP film possesses a FWHM uniformity of 2.7%, indicating a very good crystalline uniformity and lattice matching for the 100-mm wafer. Typical measurement results show that the $\lambda(+1/2\ \mathrm{max})$ has a standard deviation value smaller than λ(peak) because its position is easier to determine. In addition, it has been suggested that the $\lambda(+1/2\ \mathrm{max})$ value is closer to the bandgap energy than the λ(peak) from the RT PL band of compound semiconductors.

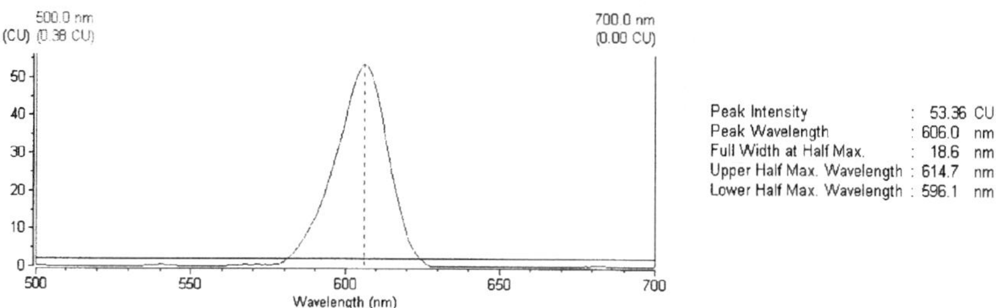

FIG. 1.21. A single-point, room-temperature, photoluminescence spectrum of the MOCVD-grown $(Ga_{1-x}Al_x)_{0.5}In_{0.5}P/GaAs$ with $x \sim 0.23$.

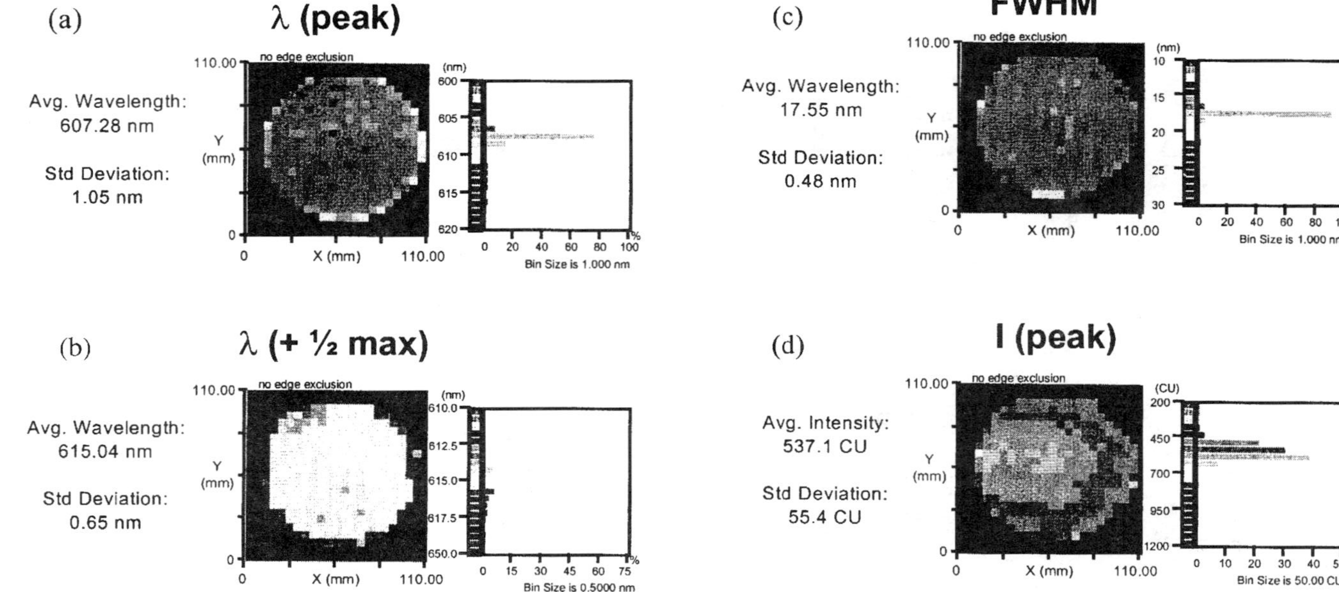

FIG. 1.22. Room-temperature photoluminescence maps of a $Ga_{1-x}Al_x)_{0.5}In_{0.5}P$ film with $x \sim 0.23$ grown on 100-mm diameter GaAs wafer. The uniformity data for λ(peak), $\lambda(+1/2$ max), FWHM and I(peak) is summarized in Table 1.3. (See color figure.)

TABLE 1.3
RT PL MAPPING DATA FOR AlIn GaP

	Average	Uniformity (%)
$\bar{\lambda}$ (peak) (nm)	607.3	0.17
$\overline{FWHM}$ (nm)	17.55	2.7
$\bar{I}$ (peak) (counts)	537.0	10.0
$\bar{\lambda}(+1/2\,\text{max})$ (nm)	615.0	0.11

Figure 1.23 exhibits the sheet resistivity maps of two 100-mm n-type, Si-doped $In_{0.5}(Ga_{1-x}Al_x)_{0.5}P$ epi-wafers, grown on GaAs, with different Al compositions. The sample with $x = 0.24$ has an average sheet resistivity of $101.4\,\Omega/\square$ and a standard deviation σ of $1.05\,\Omega/\square$ corresponding to a uniformity of 1.03%, see Fig. 1.23(a). The $In_{0.5}(Ga_{1-x}Al_x)_{0.5}P$ film of $x = 0.60$ in Fig. 1.23(b) has an average sheet resistivity of $217.9\,\Omega/\square$, a standard deviation σ of $4.3\,\Omega/\square$ and a wafer uniformity of 1.97%. The high Al compositions are required for shorter wavelength devices and are also a more stringent test of system integrity and source purity because the Al getters impurities such as oxygen. For this reason, it is usually more difficult to achieve high quality growth for the high Al composition, indirect gap, InGaAlP than for low x(Al) InGaAlP.

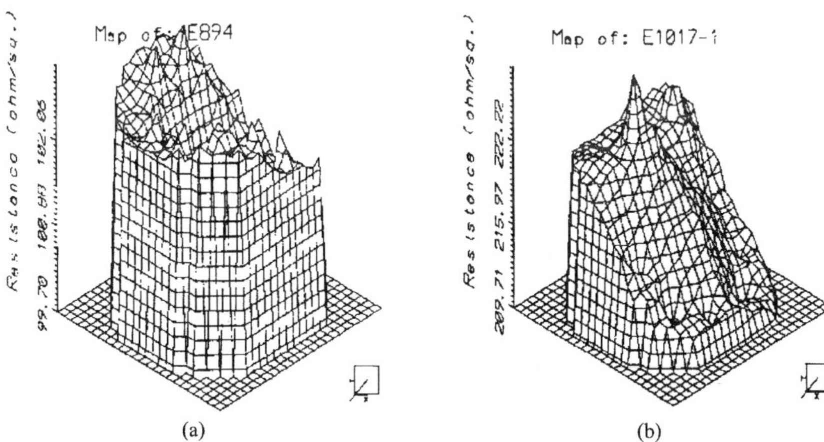

(a) (b)

FIG. 1.23. Sheet resistivity maps of two n-type Si-doped $Ga_{1-x}Al_x)_{0.5}In_{0.5}P$ epilayers on 100-mm GaAs with different Al compositions (a) with $x = 24\%$ (an average sheet resistivity of $101.4\,\Omega/\square$ and an uniformity of 1.03%) and (b) with $x = 60\%$ (an average sheet resistivity of $217.9\,\Omega/\square$ and an wafer uniformity of 1.97%).

1.3.3.2. InSb for MR Sensors

The InSb has the highest electron mobility and narrowest bandgap among III-V compound semiconductors and is attractive for applications in the high-speed electronic and optoelectronic devices in the infrared [144]. Various growth techniques have been applied to grow InSb on GaAs, including MBE [112], LPE [145], magnetron sputter epitaxy [146], and MOCVD [147, 148]. Interests in large area and multiwafer growth of InSb films on GaAs substrates are beginning to mature for their use in infrared applications and MR sensors [149].

Materials characterization for InSb films, grown on GaAs substrates, is critical. The large lattice mismatch of $\sim$14.5% between InSb and GaAs results in a high density of dislocations near the interface of InSb/GaAs, which can propagate throughout the entire InSb film [150]. A great deal of effort has been made to improve the InSb film crystalline quality by optimizing the various growth conditions, such as III-V ratio, pressure, growth temperature, growth rate, and film thickness. Traditionally, InSb films are routinely characterized by Hall measurements, X-ray diffraction, scanning electron microscopy (SEM), etc. The PL is problematic for InSb because the sample normally has to be cooled to at least 77 K, and wafer mapping requires a large bore cryostat. More characterization technologies are needed for the routine characterization of InSb films, in particular those nondestructive methods suitable for large diameter wafers. Multipoint Raman scattering microscopy was used to study the effects of III-V source ratios on the film crystalline quality and to optimize the growth parameters over a 100-mm wafer [136].

The InSb materials have been developed for magnetoresistors (MR) sensors for use in automotive applications and this has required high-volume multiple wafer growth runs. Early in the materials optimization cycle, it was found necessary to closely relate destructive and nondestructive characterization techniques to obtain a complete understanding of the properties of the epitaxial material. Typical values of mobility and carrier concentration for undoped InSb (1.2 μm thick) on 100-mm GaAs substrates were $>48{,}000\ \text{cm}^2/\text{v-s}$ and 1.8–$2.2 \times 10^{16}/\text{cm}^3$, respectively. X-ray data show a FWHM of 260–320 arcseconds. Again, there was a need to continuously qualify the material without reducing yield by destroying wafers that could be processed into devices. X-ray FWHM data and sheet resistance measurements have been developed as the primary nondestructive characterization technique. The mobility and thickness are sensitive to x-ray data. It is then possible to qualify the material by performing periodic destructive tests to verify that mobility and thickness data are consistent, using sheet resistance and x-ray measurements on all other wafers. Figure 1.24(a) shows a 55-point map of the sheet resistivity of a typical InSb epilayer grown on a 100 mm diameter GaAs substrate with uniformity $<1\%$ over the entire wafer. The sheet resistance and mobility data have been obtained from over 2900 wafers

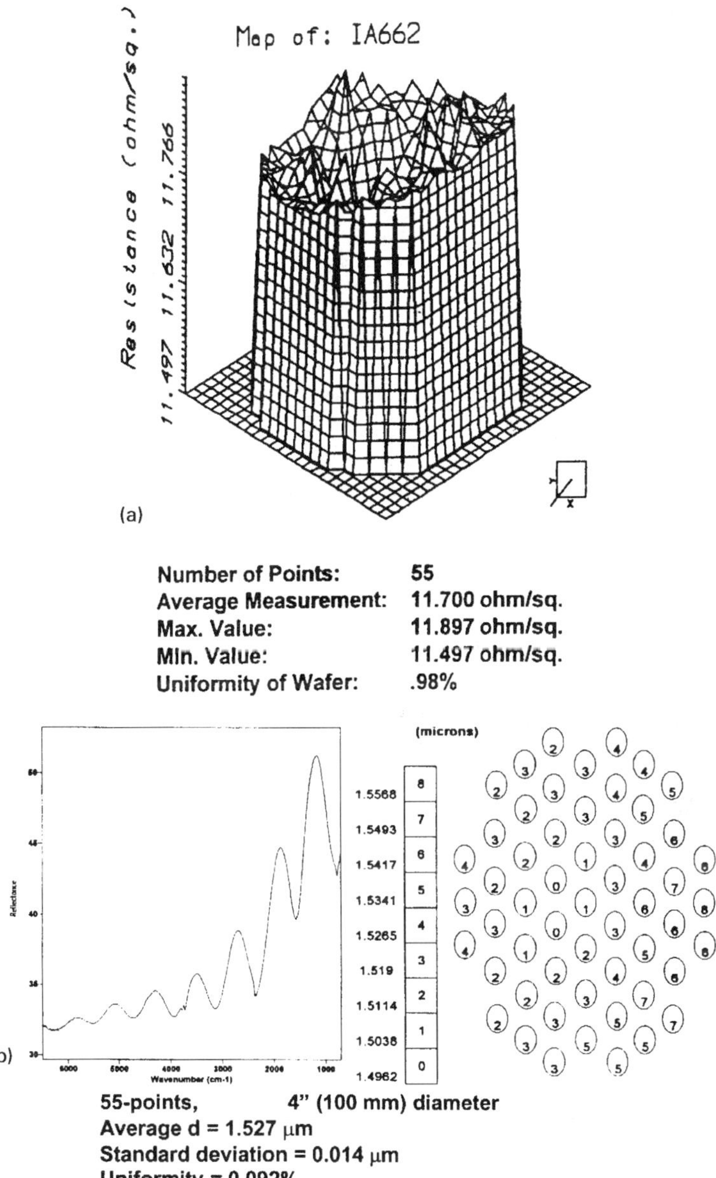

FIG. 1.24. (a) Sheet resistivity map of a MOCVD-grown 100-mm InSb films on GaAs. (b) FTIR spectrum and thickness map of a (100-mm) diameter InSb/GaAs.

TABLE 1.4
CHARACTERIZATION DATA FOR 100-MM InSb/GaAs WITH NONDESTRUCTIVE DATA FOR >2900 WAFERS AND DESTRUCTIVE DATA FOR >500 WAFERS

Test	Standard Deviation	Uniformity
Sheet resistance	0.93%	1.94%
Thickness	0.38%	1.63%
Mobility	1872 cm^2/V s	39,817 cm^2/V s
X-ray FWHM	19.4 arcsecond	151.2 arcsecond

while the thickness (SEM) and x-ray data are from about 500 wafers. Table 1.4 shows the summarized characterization data.

The determination of InSb epitaxial film thickness and distribution over the entire layer is critical in epilayer growth, especially in the case of large diameter wafer production. SEM is usually employed to measure the InSb film thickness, but this technique is destructive and is not considered particularly convenient in a production environment [151, 152]. Reflectance interference fringes are not observed in the ultraviolet to visible wavelength range from the InSb because of its narrow bandgap. However, reflectance interference fringes from InSb/GaAs can be observed in the infrared (IR) wavelength range of 800–8000 cm^{-1}, or 1.2–12 μm. Fourier transform infrared (FTIR) spectroscopy has been developed to measure the thickness of InSb, and the thickness distribution over the surface of a 100-mm wafer can be mapped, see Fig. 1.24(b) [149]. The FTIR spectrum on the left-hand side of Fig. 1.24(b) shows interference fringes that can be used to calculate the film thickness. A 55-point thickness map distribution is shown on the right-hand portion of Fig. 24(b). For this MOCVD-grown structure the average InSb film thickness was 1.532 μm with a uniformity of 0.92%, indicating excellent uniformity over the 100-cm substrate.

1.3.4. WIDE BANDGAP MATERIALS: III-NITRIDES

Progress in the area of GaN-based materials and devices is evidence that these aterials have significant and important applications in the areas of optoelectronics and high-temperature/high-power electronics [78, 153]. The large changes in physical properties, such as bandgap, crystal structure, phonon energy, and electronegativity difference between GaN and GaAs, demonstrate that nitride semiconductors are fundamentally distinct from traditional III-V semiconductors. Large native GaN substrates (>25 mm) are not available at this time and are unlikely to be commercially feasible within the foreseeable future. Consequently, two other substrate materials have been investigated, sapphire (Al_2O_3) and SiC, of which sapphire is the more widely used. Disadvantages of sapphire are the large lattice mismatch and large difference in thermal expansion coefficient, compared to GaN. However, sapphire substrates are available that have very good crystal quality at a relatively low cost. SiC, on the other hand, has physical properties that more closely match those of GaN, but the crystal quality remains rather poor. Due to the high lattice and thermal expansion mismatch between GaN and sapphire, a large number (10^{8-10} cm^{-2}) of threading dislocation defects are generated at the GaN sapphire interface that propagate through the film. It has been reported that these defects have a strong negative effect on the mobility and degree of compensation in the material. In addition, these defects have been found to completely quench the observation of electroluminscence from other III-V compound semiconductors. However, even at these dislocation densities, high brightness InGaN/GaN multiple quantum well (MQW) light emitting diodes (LED) have been successfully grown. While the exact mechanism of the intense emission from the InGaN is still a source of discussion, there is evidence that it is due to radiative recombination from the indium-rich InGaN clusters. Another III-nitride, $Al_xGa_{1-x}N$ (365 nm ($x = 0$) to 200 nm ($x = 1$)), is a promising material for UV photodetectors due to relatively high mobility, sharp cut-off wavelength, and high quantum efficiency.

Despite the impressive results obtained to date for III-nitrides [117, 154, 155], further optimization of device performance requires that the fundamental mechanisms upon which these devices operate must be better understood. Optimization of the GaN growth to reduce defect density is of paramount importance in achieving high-quality GaN based devices. In this section, characterization techniques that provide a basic understanding of the III-N material properties are highlighted.

The technique that is very often used to study the precise structural properties of III–V compound semiconductors is transmission electron microscopy (TEM). Coupled with transmission electron diffraction (TED), TEM provides both quantitative and qualitative information about the microstructure of thin epitaxial layers. However, TEM/TED requires extensive sample preparation that does not allow for the rapid feedback of information required for materials optimization, and TEM/TED is destructive. The x-ray diffraction data can provide similar information as for TEM/TED and in a timely manner. However, the complexity of the structural properties of the III-N means that a more sophisticated data collection and analysis is required beyond the simple on-axis (004) reflections that are normally investigated for III–V semiconductors.

It has been shown that the crystal structure of GaN is a hexagonal columnar structure of crystallites with a very small angular distribution in orientation [156]. There are two components to this distribution, due to the tilting and rotation of the columns relative to each other, and the *c*-axis (Fig. 1.25). The tilt of the

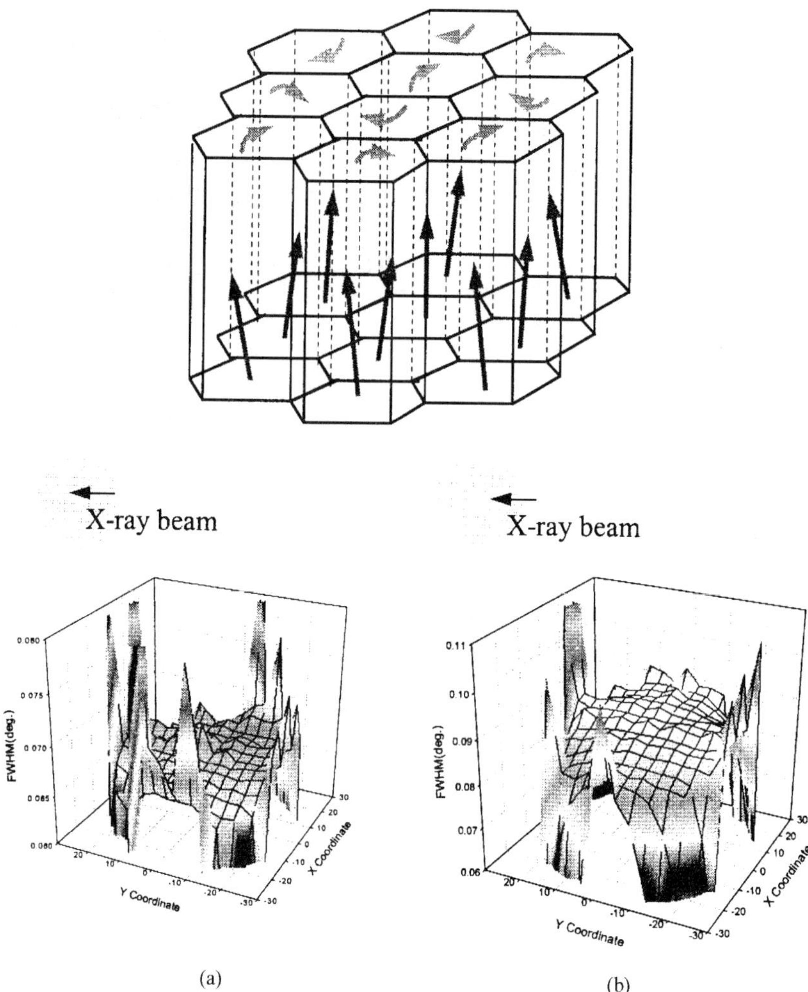

FIG. 1.25. A schematic of the columnar structure of GaN. HRXRD maps of (a) the symmetric (004) reflection and (b) the asymmetric (105) reflection of a GaN epilayer on sapphire wafer show the distribution of the 'tilt' and 'twist' of the columnar structure of GaN, respectively.

columns can be associated with screw dislocations. Screw dislocations with a [*00l*] dislocation line direction distort all (*hkl*) planes with *l* nonzero [157], for example a (002) lattice plane, which gives a symmetrical Bragg reflection. If the crystal is tilted, an angular movement of the vector perpendicular to the (*00l*) lattice planes (which coincides with the *c*-axis of the hexagonal crystal) causes a broadening of the symmetrical reflection in diffraction space, see Fig. 1.25(a). Edge dislocations have a state of plane strain such that, in a (*00l*) plane, they will distort only the (*hkl*) planes with either *h* or *k* nonzero. This means that symmetric (*00l*) rocking curves will be insensitive to the pure edge dislocation [157]. The presence of edge dislocations can be interpreted by a twist of the hexagonal columns. To see these dislocations, an asymmetrical Bragg reflection must be investigated, an example of which is a (102) lattice plane, see Fig. 1.25(b). The Bragg reflections are not only broadened by tilt and twist of the columns, but are also due to finite length scales in a direction perpendicular to the sample surface. The effects of microscopic tilt can be separated from those of finite length scales using a reciprocal space map [126]. Table 1.5 summarizes a complete data set obtained for a GaN epilayer on a sapphire substrate. It interesting to note that the finite length scale estimated from these measurements of 0.6 μm is similar to that measured for the minority carrier diffusion lengths.

When the GaN epilayer was profiled across the substrate, a continuous increase of the FWHM of the (105) reflection towards the wafer flat was observed—but not for the (002) reflection, see Fig. 1.26. A PL map of the GaN bandedge emission was also completed for the same region (Fig. 1.26). When comparing these data, a pronounced correlation between the FWHM and the PL intensity was found to exist. The narrowest FWHM occurs at the side of the wafer that is opposite to the flat side, and this corresponds to the region that gives the highest PL intensity. This shows that the reduction of the structural defects present in GaN films grown on sapphire is of critical importance in realizing high-performance electrical and optical devices.

Optimization of the GaN growth to reduce the dislocation density is consequently that of paramount importance in achieving high-quality GaN based devices. However, the x-ray data in Fig. 1.26 has shown that not every defect has a strong dependence on every material property. The quality of the epitaxial GaN layer and hence the dislocation density depends strongly on the buffer-layer

TABLE 1.5
RESULTS ON TILT, TWIST, AND FINITE LENGTH SCALES OF GaN AND Al_2O_3

Bragg Reflection	Tilt	Finite Length Scales	Twist
GaN (104) in plane	0.02°	6300 Å	—
GaN (114) in plane	0.013°	7025 Å	—
GaN (101) out of plane	—	—	0.12°
GaN (105) out of plane			0.09°
Sapphire (2 1 10)	0.006°	infinite	—
Sapphire (2 1 10)	—	—	0.006°

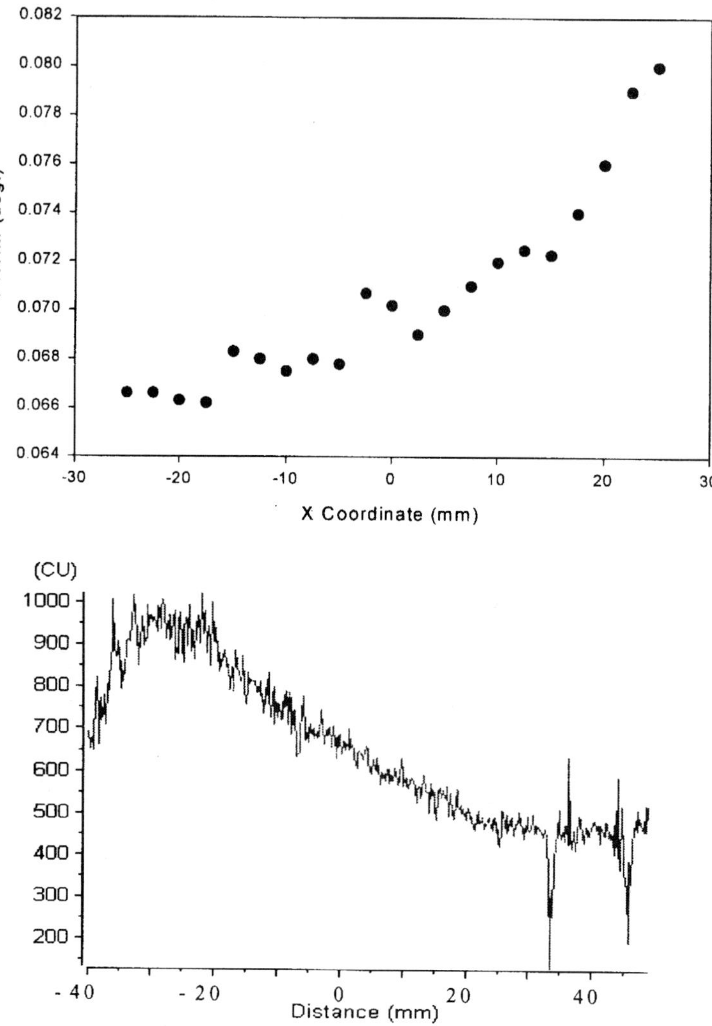

FIG. 1.26. A continuous increase of the FWHM of the (105) reflection towards the flat side of the wafer can be observed in Fig. 1.25. A pronounced correlation between the (105) reflection FWHM (i.e., the "twist" component) and the PL intensity exists that is not seen for the (004) reflection (i.e., the "til" component).

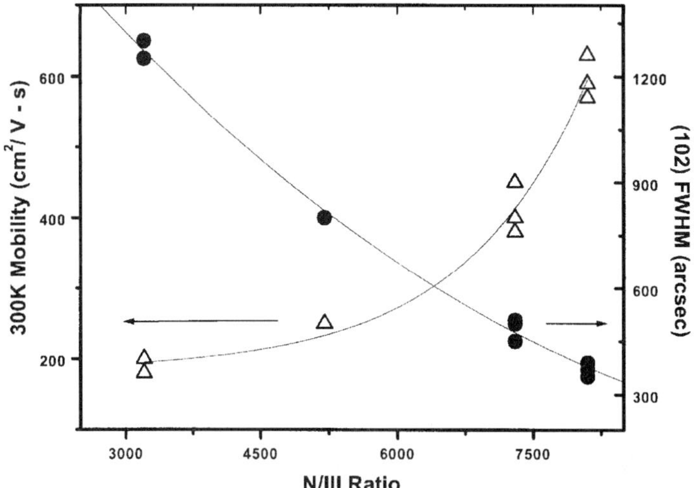

FIG. 1.27. 300 K mobility and the HRXRD (102) FWHM peak as a function of N/III ratio during the buffer-layer growth. All of the films were ~2.5 μm thick and had background carrier concentrations around 3-9E16 cm^{-3}.

growth conditions. A low-temperature GaN buffer layer of 20 nm, at approximately 500 °C, is generally grown prior to the thick high-temperature epitaxial GaN layer at 1050 °C [113, 158]. One of the strongest effects on buffer-layer quality is the N/III ratio during the buffer deposition. Figure 1.27 shows a plot of mobility and FWHM of the asymmetric (102) reflection as a function of N/III ratio during the low-temperature buffer-layer growth. All the (high-temperature) epilayers had a N/III ratio of ~2300 and the total layer thickness on which the measurements were completed was ~2.5 μm. An increased N/III ratio dramatically improves the mobility and the crystal quality but this correlation was not observed with (002) type reflections. This improvement in buffer-layer quality is likely due to an improvement in nitrogen incorporation for higher N/III ratios [159]. For the highest mobility film (~650 cm/V-s), the dislocation density has been estimated to be mid-$10^8/cm^2$ [160]. The reduction of dislocations has also been found to be a critical factor in achieving high hole concentrations for Mg-doped films.

To clarify the role that dislocations play in the performance of GaN-based devices, several MSM-type UV photodetectors were recently fabricated on AlGaN layers grown over both low- and high-N/III ratio buffer layers. The higher-defect GaN material exhibited a systematic breakdown with electrode spacing, and the breakdown field was estimated to be $\sim 10^5$ V/cm for various electrode spacings. These devices exhibited large gain but had a very slow

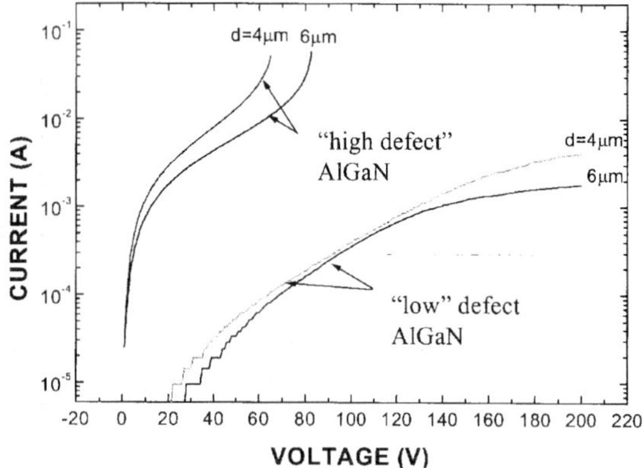

FIG. 1.28. I-V characteristics of MSM photodetectors based on high (typically 10^{9-10} cm^{-2}) and low (typically 10^{7-8} cm^{-2}) dislocation AlGaN films. For low dislocation films, the premature breakdown is suppressed.

temporal response of 90–120 s. The breakdown had a mixed character with components due to avalanche (the source of the gain and microplasmas) and Zener (the observed negative temperature dependence) mechanisms. For MSM devices based on lower defect material, no breakdown was observed, indicating that dislocations play a strong role in the breakdown and gain of these devices (Fig. 1.28). These devices had a relatively fast time response of 1–5 ms. The dislocation density must be as low as possible for high-quality GaN-based avalanche photodiodes to be realized [161].

The primary driver for III-nitrides has been the production of high brightness blue LED [162]. The active region of these devices is normally based on a InGaN/GaN multiple quantum well (MQW) structure. There is a need to understand the physical properties of the luminescence transition mechanism in the MQW in order to further improve the emission intensity of the LED [163, 164]. A much broader PL is typically observed from $In_xGa_{1-x}N$, compared to GaN that cannot be explained solely by alloy-broadening and has also been attributed to indium concentration fluctuations [165, 166]. Nanoscale fluctuations of the indium concentration (clustering) in the $In_{0.22}Ga_{0.89}N$ layers can be controlled by carefully varying the growth conditions. Figure 1.29(a) shows the (002) reflection obtained for three different MQW consisting of 10 periods of 3 nm $In_{0.22}Ga_{0.78}N$ and 11 nm GaN. These MQW were grown under different growth conditions to vary the degree of clustering. The well and barrier thicknesses were kept constant in addition to the average indium composition of the complete structure. The zero-order diffraction peak corresponds to an average indium composition of 5% for the entire MQW or 11% indium in the InGaN layers. Despite the differences in growth conditions no significant difference was observed in their (002) XRD spectra, see Fig. 1.29(a). However,

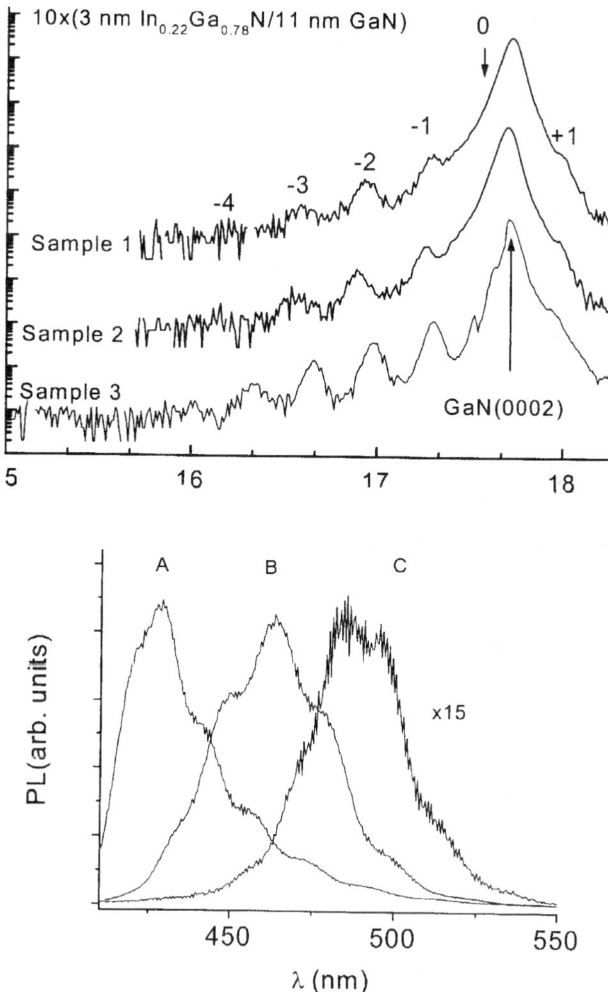

FIG. 1.29. (a) HRXRD spectra and (b) time-averaged PL spectra for three different MQWs, with the same nominal indium concentration grown under different conditions. The emission maximum is observed to vary from 430 to 480 nm. (The high-energy side of the PL from MQW A is partially cut off by the filter used to attenuate the laser light.)

both the PL and time-averaged PL for these three different MQW, A, B, and C showed a difference in the peak wavelength varying from 430 to 480 nm, respectively, Figure 1.29(b). Theoretical analysis of the role of the compositional fluctuations suggests that the magnitude of the nanoscale fluctuations increases from MQW A to C.

A dramatic wavelength dependence is observed in the emission kinetics for MQW C at room temperature [167], see Fig. 1.30. For wavelengths longer than 480 nm (2.58 eV), nanosecond decays are observed. As the emission wavelength is moved to shorter wavelengths, the lifetime is observed to have shortened and the kinetics are clearly not due to single exponential, (see Fig. 1.30). These changes in lifetime can be attributed to a tail in the density of states, due to disorder. According to Fermi's golden rule, the transition rate will be proportional to the density of states; therefore, if the frequency dependence of the electronic matrix element is neglected, an exponential tail in the density of states will be reflected in the energy dependence of the emission lifetime. The PL decays from all three MQW can be fitted to a stretched exponential function,

$$I(t) = I_0 \exp\left(-\left(\frac{t}{\tau}\right)^{\beta}\right)$$

where β is between 0 and 1, τ is the lifetime and $I(t)$ is the PL intensity as a function of time. Stretched exponentials have been used to describe the dynamics of disordered systems for over 100 yr. Although InGaN is not a heavily disordered

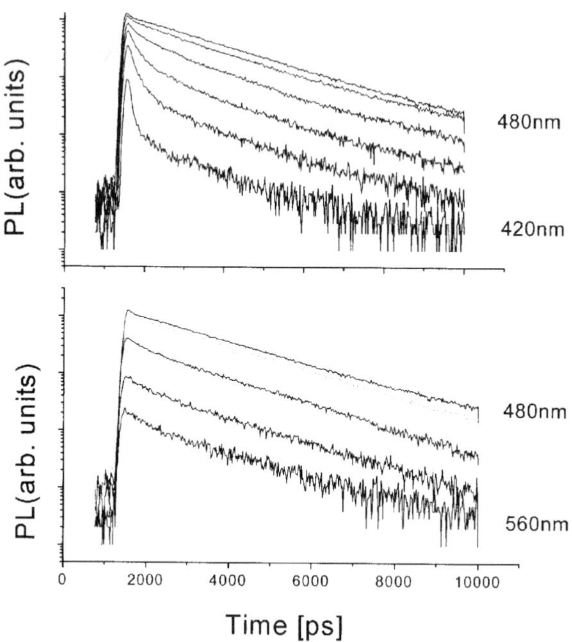

FIG. 1.30. Wavelength dependence of the emission kinetics for MQW C from 420 to 560 nm in 20-nm steps. In the top figure the emission lifetime is clearly observed to decrease at shorter wavelengths but little change is observed for wavelengths longer than 480 nm, see bottom figure.

semiconductor, it is likely that exciton localization and migration will take place. The β can be independently determined from τ by plotting the double logarithm of the signal vs the logarithm of the time. A plot of τ and β vs wavelength is shown in Fig. 1.31 for MQW C. For the three MQWs, β and τ were found to vary with emission energy. The β for the emission peak of the three MQW was observed to increase from 0.75 to 0.85 with apparently increasing indium phase segregation at longer wavelength emission. A higher degree of indium phase segregation is consistent with more isolated quantum dots inside the 2D quantum well.

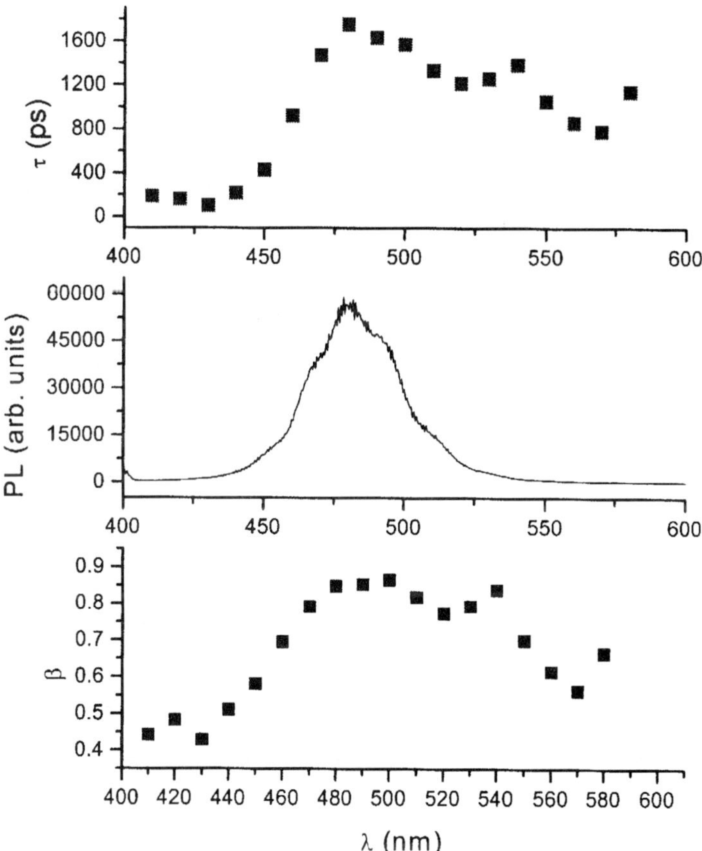

FIG. 1.31. β and τ vs emission wavelength for MQW C. Both β and τ are a maximum at the peak of the PL emission. The β parameter is obtained from a fit to a stretched exponential function and can be related to the clustering in the GaInAs MQWs.

1.3.5. GaInAs pHEMT MIMC Optimization by Device Fabrication

There is a large market for monolithic microwave-integrated circuits (MMIC) devices for microwave applications, such as cellular phones and digital broadcast service, to replace discrete and hybrid circuits that require expensive testing and tuning procedures. Typically, the starting material used has been epitaxial wafers grown by MBE [168]. More recently, MOCVD growth and process technology has been used to develop truly manufacturable HEMT materials with respect to performance, yield, and cost reduction at high-volume production. Initial results have shown that power MMIC devices produced from the MOCVD-grown materials have performance and yields equal to or exceeding those obtained on the same fabrication line with optimized MBE-grown wafers [75, 169]. For example, MOCVD-grown low-noise *p*HEMT wafers that were inserted into an MMIC production line achieved low-noise amplifier performance and yield comparable to similar MBE-grown wafers. The 38-GHz LNA MMIC with 20 dB gain, 4 dB noise figure, and RF and DC yields as high as 72% were fabricated [170].

The 35-GHz, 0.5-W power amplifier MMIC that can be applied to Ka-band applications were also fabricated. The structure of a typical double planar, power *p*HEMT structure is shown in Fig. 1.32(a). The layers were grown by MOCVD in a multiwafer reactor. The active layer is a pseudomorphically strained quantum well layer of GaInAs. The development and optimization of this type of device requires extensive characterization of the material deposited and intermediate structures. *In situ* control methodologies have been developed to try to reduce the optimization cycle and ongoing material qualification [82]. In general, nondestructive characterization mapping techniques are again used, since the wafer will later be fabricated into the devices and cross correlation with the basic material characterization will be completed.

In general, prior to the growth of *p*HEMT structures, it is necessary to recalibrate the growth of AlGaAs and GaInAs as part of a standard qualification procedure. A calibration growth is first completed to ensure that the buffer layer structure is highly resistive. A growth run is performed with only a 2000-Å-thick GaAs buffer layer and a 10-period superlattice, as shown in Fig. 1.32(a). Typically, less than 1 μA of current flowed when approximately 11 V were applied between two alloyed contacts on the layer. A typical sheet resistance uniformity of the AlGaAs layer with a 24% composition was 1.7%.

A series of $In_xGa_{1-x}As$ epitaxial films and $In_xGa_{1-x}As$-strained QW structures were grown on GaAs wafers (with $x \sim 12\%$). The InGaAs QW serves as the active channel for conduction in the *p*HEMT structure and the wavelength uniformity of the InGaAs quantum well provides an indication of the quality of this well. The distribution of PL emission from the InGaAs QW PL was <2 nm in optimized material. A typical variation in sheet resistance uniformity

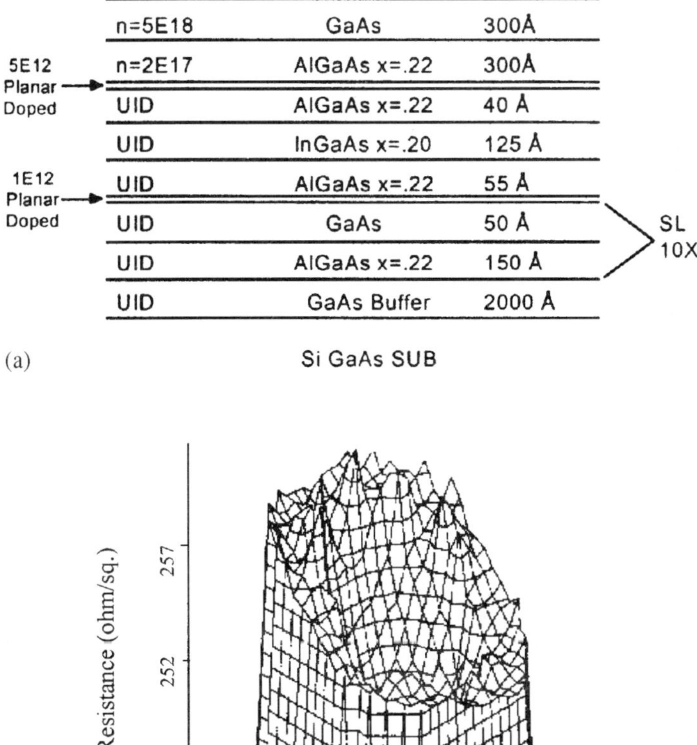

FIG. 1.32. (a) A schematic of a power *p*-HEMT structure diagram. (b) The sheet resistivity map of a *p* HEMT structure without the *p*-doped cap had an average sheet resistance of $253.0\,\Omega/\square$ and a uniformity of 1.6%.

was 1–2% across the wafer. Having obtained the optimal values for the InGaAs QW layer thickness and spacer-layer thickness, a test structure identical to that of Fig. 1.32(b) was grown except that the final cap layer was undoped. This allows for a more straightforward interpretation of the Hall measurements and the sheet carrier concentration in the InGaAs channel. Hall measurements performed at room temperature resulted in a sheet carrier concentration of $3.0 \times 10^{12}/\text{cm}^2$ with a mobility of $>5500\ \text{cm}^2/\text{Vs}$.

After optimization of the growth parameters and using feedback information from preliminary fabrication runs, high-quality *p*HEMT materials were grown. The silicon doping in the GaAs cap layer was adjusted to yield a sheet resistance for the structure between 250 and 280 Ω/□. A series of double-planar, power *p*HEMT were then grown and processed. The *p*HEMT exhibited very high uniformity. Figure 1.32(b) shows the sheet resistivity map of a 75-mm diameter *p*HEMT wafer, possessing an average sheet resistance of 253.0 Ω/□, a standard deviation (SD) of 4.09 Ω/□, and a uniformity of 1.6%. Figure 1.33(a) shows the room-temperature map of the main PL peak wavelength with an average value of 977.9 nm, with SD of 1.8 nm corresponding to a uniformity of 0.2%. A map of the FWHM (in nm) of the GaInAs in Fig. 1.33(b) shows an average of 99.6 nm at RT, SD of 0.44 nm, and a uniformity of 0.44%. These data provide good evidence

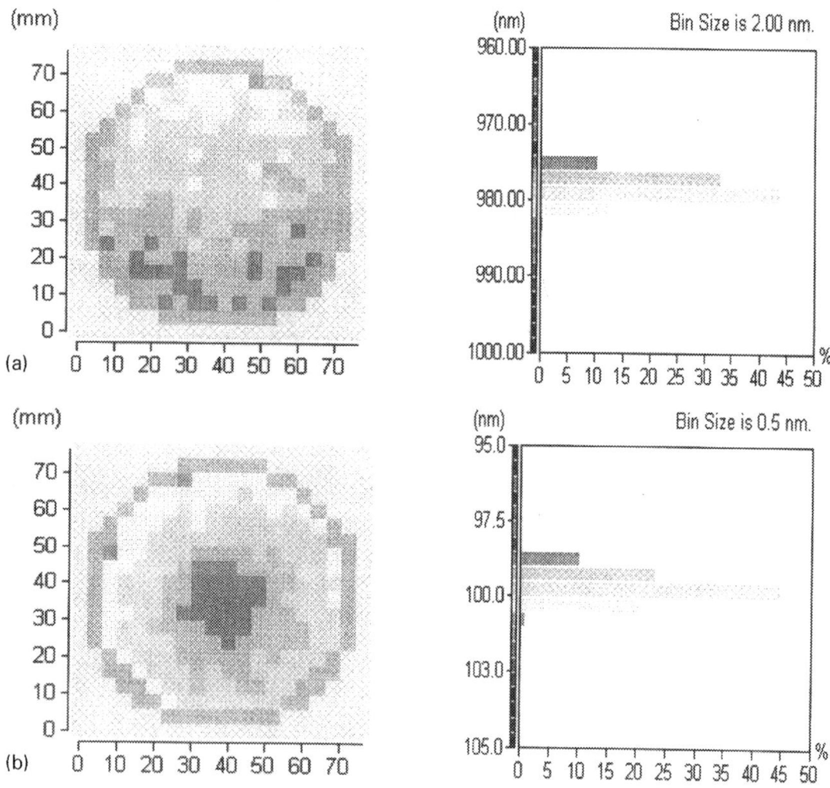

FIG. 1.33. (a) RT photoluminescence peak wavelength map (wavelength: 977.9 ± 1.8 nm) and (b) a FWHM map (FWHM: 99.6 ± 0.44 nm) of the InGaAs channel of an optimized *p*HEMT structure. (See color figure.)

for the high quality and high uniformity of optimized *p*HEMT structural materials prepared by MOCVD.

High-power amplifier MMIC require large wafer real estate, and therefore, deposition uniformity. One of the challenges of the design of a high-power amplifier is to minimize the MMIC dimension without compromising performance. Figure 1.34(a) shows the layout of the final chip with six (6) cells or modules with 5 dB gain. Each cell consists of two *p*HEMT with 600-μm gate-widths. The MMIC has a total of 4800 μm of power *p*HEMT gate-width at the output stage and uses Lange couplers at both the input and output to reduce the return loss and minimize combiner losses. The optimized *p*HEMT wafers with high material quality and excellent uniformity were fabricated into MMIC

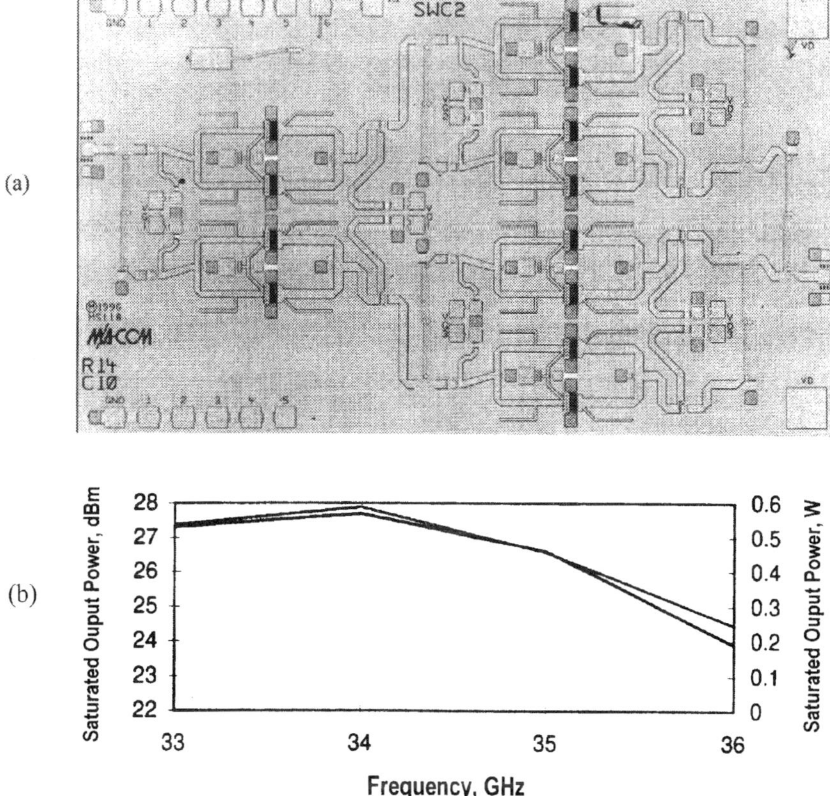

FIG. 1.34. (a) A photograph of a processed MMIC chip with 4.8 mm gate-width. (b) The saturated output vs power frequency for two of these MMIC chips.

structures. Completed wafers yielded 5.55 × 3.2-mm MMIC chips with saturated output power of over 0.50 W at 34 GHz and 0.46 W at 35 GHz, see Fig. 1.34(b). With a further design iteration it is believed that a 1 W output power could be obtained from the device. Very often, the ultimate optimization of a particular device structure requires detailed characterization of the fabricated device rather than that of the material or device structure before fabrication.

1.4. Future Directions

The next few years will see the continued maturation of the rapidly developing compound semiconductor industry. Much of the research effort will focus on the development of devices that will meet specific applications in various markets, (see Table 1.6). Many of these devices, such as LED, will become commodity items, and this will drive the next generation of the growth tools used to produce them. These tools, primarily MBE and MOCVD derivatives, will likely resemble tools used within the silicon industry and conform to those basic standards. Sample handling will be limited to transporting fluoride packs and inserting these into growth systems with single wafer handling capabilities. These tools will also require sophisticated *in situ* control systems to allow them to be operated by someone with minimal technical experience.

The cutting edge of compound semiconductors will likely see the further integration of III-V compounds with other material systems. One focus has been the deposition of III–V materials and device structures on silicon to better integrate into existing silicon-based electronics. Advances have allowed for the successful monolithic integration of GaInAs photodiodes with silicon readout circuits [171]. It may also be possible that germanium will emerge as a *universal* substrate for III-V materials based on its extensive use in III–V solar cell

TABLE 1.6
COMPOUND SEMICONDUCTOR MARKETS (from various sources)

Product	Application	1997 Est. Market Size ($ in millions)	1997–2000 CGAR (%)
Lasers	Data transmission and storage, consumer electronics	1,990	15
Sensors	Automotive, consumer electronics	1,500	30
Solar cells	Satellite communications	300	50
LED	Pagers, displays	1,950	16
MOCVD systems	CS materials/devices	100	35
GaAs IC	High-speed communications, DBS satellite dishes	1,485	15

applications. Another developing area of interest is the integration of frequency agile dielectric oxides such as $Ba_{1-x}Sr_xTiO_3$ (BST) and $PbZr_{1-x}Ti_xO_3$ (PZT) with III–V materials for use in adaptive electronics [172]. These hybrid devices are projected to be used as tunable notch filters and transmission/receive modules for wideband phased arrays for communication, radar, and navigation systems.

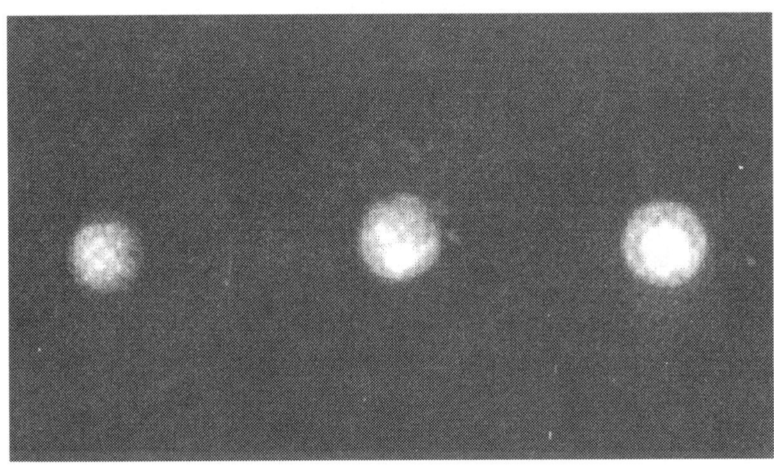

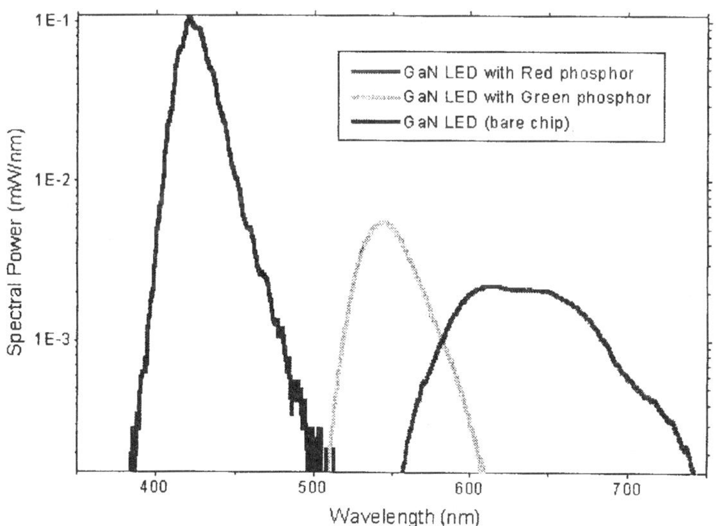

FIG. 1.35. Red, green and blue (RBG) emission from down-converting phosphors and a III-nitride LED die. The peak emission at 425 nm is from the LED while those at 550 nm and 625 nm are from the phosphors $SrGa_2S_4$: 4%Eu and $Zn_{0.25}Cd_{0.75}S$: AgCl, respectively. (See color figure.)

Wide bandgap materials, especially the III-nitrides and SiC, will continue to be a focus of widespread attention as markets continue to develop for these materials. The III-nitrides will find extensive use in optoelectronic applications and both GaN and SiC will be used for high-temperature and high-power applications. Perhaps one of the largest markets for the III-nitrides will be solid state lighting or display applications. This may represent a market much larger than that of laser diodes if applied to domestic and commercial lighting. The ability to produce visible radiation in a controlled manner is the basis for most indoor and outdoor lighting, as well as the production of all displays.

High-brightness III-nitride UV LED have been used to excite phosphors [173]. Upon absorption of the UV light, the phosphor converts the energy to visible radiation, including white light, depending on the type of phosphor used. Wide ranges of high efficiency phosphors are commercially available, having been used to fabricate television screens and fluorescent lighting for many decades. Figure 1.35 shows red and green phosphor emission and blue LED emission, thereby providing the three (RGB) primary colors. The LED came from the same wafer and *hence the same device structure*. The UV-pumped phosphor technology will create new energy-efficient-compact lighting and color display systems.

Acknowledgments

One of the authors (I.F.) would like to thank his coworkers at EMCORE, especially M. Schurman, C. Tran and M. Pelczynski, for many fruitful discussions and their continued collaboration in all the areas referenced in this chapter. As well, thanks go to T. Ryan and E. B. Fantner of Philips Analytical and J. Burton, S. Lukacs, and M. Pophristic of Rutgers University for their contributions to the work. In addition, thanks are due to Jo Ann McDonald for her careful proofreading of the manuscript. The work has been funded, in part, by BMDO, DARPA, ONR, AF, and NASA.

References

1. O'Reilly, E. P. (1989). *Semicond. Sci. Tech.* **4**: 121.
2. Cho, A. Y. (1985). *Molecular Beam Epitaxy and Heterostructures*, Chang, L. L. and Ploog, K., eds., Dordrecht: Martinus Nijhoff.
3. Stringfelow, G. B. (1989). *Organometallic Vapor Phase Epitaxy: Theory and Practice*, San Diego: Academic Press.
4. Dawson, L. R. (1972). *Prog. Solid State Phys.* **7**: 117.
5. Jones, M. E. and Shaw, D. W. (1976). *Treatise on Solid State Chemistry*, vol. 5, N. B. Hannay, ed., New York: Plenum.
6. Rousina, R. and Webb, J. B. (1991). *Semicond. Sci. Tech.* **6**: C42.

7. Arthur, J. R. (1969). *Structure and Chemistry of Solid Surfaces*, New York: Wiley.
8. Cho, A. Y. (1970). *J. Appl. Phys.* **41**: 2780.
9. Flynn, C. P. (1988). *J. Phys.* **F18**: L195.
10. Foxon, C. T. and Joyce, B. A. (1975). *Surf. Sci.* **50**: 434.
11. Foxon, C. T. and Joyce, B. A. (1977). *Surf. Sci.* **64**: 293.
12. Joyce, B. A., Vvedensky, D. S., and Foxon, C. T. (1994). *Handbook on Semiconductors*, vol. 3, 275 (1994).
13. Foxon, C. T. (1991). *Appl. Surf. Sci.* **50**: 28.
14. Parker, E. H. (1985). The Technology and Physics of Molecular Beam Epitaxy, New York: Plenum Press.
15. Herman, M. A. and Sitter, H. (1989). *Molecular Beam Epitaxy: Fundamentals and Current Status*, New York: Springer-Verlag.
16. Tsao, J. Y. (1991). *Material Fundamentals of Molecular Beam Epitaxy*, San Diego: Academic Press.
17. Foxon, C. T. and Joyce, B. A. (1981). *Current Topics in Materials Science* **7**: 1.
18. Kasper, E. and Bean, J. C. (1988). *Silicon Molecular Beam Epitaxy*, vols. I and II, Boca Raton: CRC Press.
19. Cho, A. Y. and Cheng, K. Y. (1981). *Appl. Phys. Lett.* **38**: 360.
20. Luscher, P. E. and Collins, D. M. (1979). *Prog. Cryst. Growth Charact.* **2**: 15.
21. Yamashita, T., Tomita, T., and Sakuri, T. (1987). *Jpn. J. Appl. Phys.* **26**: 1192.
22. Vlcek, J. C., and Fonstad C. G. (1991). *J. Crystal Growth* **111**: 56.
23. Kunzel, H., and Ploog, K. (1980). *Appl. Phys. Lett.* **36**: 416.
24. Lee, R. L., Schaffer, W. J., Chai, Y. G., Liu, D. and Harris, J. S. (1986). *J. Vac. Sci. Technol.* B **4**: 568.
25. Mobray, D. J., Kowalski, O. P., Skolnick, M. S., DeLong, M. C., Hopkinson, M., David, J. P., and Cullus, A. G. (1994). *J. Appl. Phys.* **75**: 2029.
26. Johnson, F. G., Wicks, G. W., Viturro, R. E. and LaForce, R. (1993). *J. Vac. Sci. Technol.* B **11**: 823.
27. Sánchez-García, M. A., Sánchez, F. J., Naranjo, F. B., Calle, F., Calleja, and E., Muñoz, E. (1992). *MRS. Internet J. Nitride Semicond. Res.* **3**: 32.
28. Harbison, J. P., Aspnes, D. E., Studna, A. A., and Florez, L. T. (1988a). *J. Vac. Sci. Technol.* B **6**: 740.
29. Harbison, J. P., Aspnes, D. E., Studna, A. A., Florez, L. T., and Kelly, M. K. (1988b). *Appl. Phys. Lett.* **52**: 2046.
30. Barnett, S. A., Winters, H. F., and Greene, J. E. (1985). *Surf. Sci.* **151**: 67.
31. Kao, Y. C., Celii, F. G. and Lin, H. Y. (1993). *J. Vac. Sci. Technol.* B **11**: 1023.
32. Celii, F. G., Kao, Y. C., Beam III, E. A., Duncan, W. M., and Moise, T. S. (1993). *J. Vac. Sci. Technol.* B **11**: 1018.
33. Larsen, P. K. and Dobson, P. J. (1987). *Proceedings of a NATO Advanced Research Workshop on Reflection High Energy Electron Diffraction and Reflection Electron Imaging of Surfaces*, New York: Plenum Press.
34. Prutton, M. (1975). *Surface Physics*, Oxford: Clarendon, p. 44.
35. see ref. 34.
36. see ref. 39.
37. Zhang, J., Dawson, P., Neave, J. H., Hugill, K. J., Gilbraith, I., Fawcett, P. N., and Joyce, B. A. (1990). *J. Appl. Phys.* **68**: 5595.
38. Daweritz, L. and Ploog, K. (1994). *Semicond. Sci. Tech.* **9**: 123.
39. Newstead, S. M., Kubiak, R. A., and Parker, E. H. (1987). *J. Cry. Growth* **81**: 49.
40. Hottier, F., Theeten, J. B., Masson, A., and Domage, J. L. (1977). *Surf. Sci.* **65**: 563.
41. Joyce, B. A., Zhang, J., Neave, J. H., and Dobson, P. (1988). *J. Appl. Phys.* A **45**: 255.

42. Joyce *et al.* (1984).
43. Zhang, J., Neave, J. H., Dobson, P. J., and Joyce, B. A. (1987) *Appl. Phys.* A **42**: 317.
44. Gibson, E. M., Foxon, C. T., Zhang, J., and Joyce, B. A. (1990). *Appl. Phys. Lett.* **57**: 1203.
45. Joyce *et al.* (1998).
46. Larsen, P. K., van der Veen, J. F., Mazur, A., Pollmann, M., Neave, J. H., and Joyce, B. A. (1982). *Phys. Rev.* BV **26**: 3222.
47. Pashley, M. D. (1990). *J. Cry. Growth* **99**: 473.
48. Oliveira, A. G., Parker, S. D., Droopad, R., and Joyce, B. A. (1990). *Surf. Sci.* **227**: 150.
49. Wang, P. D., Holmes, S. N., Le, Tan, Stradling, R. A., Ferguson, I. T., and Oliveira, A. G. (1992). *Semicond. Sci. Tech.* **7**: 767.
50. Baillargeon, J. N., Cho, A. Y., Fischer, R. J., Pearah, P. J., and Cheng, K. Y. (1994). *J. Vac. Sci. Technol.* B **12**: 1106.
51. Tsao, J. Y. (1991). *J. Cry. Growth* **110**: 595.
52. Kean, A. H., Stanley, C. R., Holland, M. C., Martain, J. L., and Chapman, J. N. (1991). *J. Cry. Growth* **111**: 189.
53. Joyce *et al.* (1981).
54. Manasevit, H. M. (1981). *J. Crystal Growth* **55**: 1.
55. Dupuis, R. D. and Dapkus, P. D. (1977). *Appl. Phys. Lett.* **31**: 466.
56. Thompson, A., Stall, R. A., Kroll, W., Armour, E., Beckham, C., Zawadzki, P., Aina, L., and Siepel, K. (1997) *J. Cryst. Growth* **170**: 92.
57. Wang, C. A. *et al.* (1986). *J. Crystal Growth* **77**: 136.
58. Fotiadis, D. I., Kieda, S., and Jensen, K. F. (1990). *J. Crystal Growth* **102**: 441.
59. Breiland, W. G. and Evans, G. H. (1991). *J. Electrochem. Soc.* **138**: 1806.
60. Eden, G. (1992). *Photochemical Vapor Deposition*, New York: John Wiley & Sons.
61. Razeghi, M. R. (1989). *MOCVD Challenge: A Survey of GaIn-InP for Photonic and Electronic Applications*, vol. 1, Bristol and Philadelphia: Adam Hilger.
62. Kuech, T. F. (1992). *Proc. IEEE* **80**: 1609.
63. Stringfellow, G. B. (1994). *Handbook of Crystal Growth*, vol. 3B, Chap. 12, Amsterdam: Elsevier Science.
64. Moon, R. (1997). *J. Cryst. Growth* **170**: 1.
65. Einset, E. O., Jensen, K. V., and Kleijn, C. R. (1993). *J. Crystal Growth* **132**: 483.
66. Takenaka, C. *et al.* (1988). *J. Crystal Growth* **91**: 173.
67. Kikkawa, T., Tanaka, H., and Komeno, J. (1991). *J. Crystal Growth* **107**: 370.
68. Frijlink, P. M. (1988). *J. Crystal Growth* **93**: 207.
69. Parson, J. D. (1992). *J. Crystal Growth* **116**: 387.
70. Gadgil, P. N. (1993). *J. Electronic Mat.* **22**: 171.
71. Tompa, G. S. *et al.* (1988). *J. Crystal Growth* **93**: 220.
72. Biber, C. R., Wang, C. A., and Motakef, S. (1994). *J. Crystal Growth* **123**: 545–554.
73. Omstead, T. R. and Jensen, K. F. (1990). *Chemistry of Materials* **2**: 39.
74. Kisker, D., Stephenson, G., Tersoff, J., Fuoss, P. H., and Brennan, S. (1996). *J. Crystal Growth* **163**: 54.
75. Thompson, A. G. (1997). *Materials Letts.* **30**: 255.
76. Sherman, A. (1987). *Chemical Vapor Deposition for Microelectronics*, New York: Noyes Publications.
77. O'Brien, P., Malik, M. A., Chunggaze, M., Trindade, T., Walsh, J. R., and Jones, A. C. (1997). *J. Crystal Growth* **170**: 23.
78. Dupuis, R. D. (1997). *J. Cryst. Growth* **178**: 56.
79. Hardtdegen *et al.* (1996). *The Electrochemical Society, Proceedings* **96-2**: 49.
80. Yuan, C., Salagaj, T., Kroll, W., Stall, R. A., Schurman, M., Hwang, C-Y., Li, Y., Mayo, W. E., Lu, Y., Krishnankutty, S., and Kolbas, R. M. (1996). *J. Electronic Materials* **25**: 749.

81. Piner, E. L., Behbehani, M. K., El-Masry, N. A., McIntish, F. G., Roberts, J. C., and Bedair, S. M. (1997). *Appl. Phys. Lett.* **70**: 461.
82. Johs, B., Hale, J. X., Herzinger, C., Doctor, D., Elliot, K., Olson, G., Chow, D., Roth, J., Ferguson, I. T., Pelczynski, M., Kuo, C. H., and Johnson, S. (1998). *MRS Proceedings* **502**: 3.
83. Evans, K. R., Kaspi, R., Jones, C. R., Sherriff, R. E., Jogai, V., and Reynolds, D. C. (1993). *J. Crystal Growth* **127**: 523.
84. Aspnes, D. E. (1994). *Surface Science*, **307–309**: 1017.
85. Breiland, W. G. and Killeen, K. P. (1995). *J. Appl. Phys.* **78**: 6726.
86. Raffle, Y., Kuszelewicz, R., Azoulay, R., Le Roux, G., Michel, J. C., Dugrand, L., and Toussaere, E. (1993). *Appl. Phys. Lett.* **63**: 3479.
87. Aspnes, D. E., Quinn, W. E., and Gregory, S. (1990) *Appl. Phys. Lett.* **57**: 2707.
88. Maracas, G. N., Edward, J. L., Shiralagi, K., Choi, K. Y., Droopad, R., Johs, B., and Woollam, J. A. (1992). *J. Vac. Sci. Tech.* A **10**: 1832.
89. Aspnes, D. E. (1995). *IEEE J. Select. Topics in Quant. Elect.* **1**: 1054.
90. Gilmore, W. and Aspnes, D. E. (1995). *Appl. Phys. Lett.* **66**: 1617.
91. Moon, (1997).
92. Thompson, A. G. (1994). *Processing of Advanced Mat.* **4**: 181.
93. Neave, J. H., Joyce, B. A., Dobson, P. J., and Zhang, J. (1985). *Appl. Phys. Lett.* **47**: 100.
94. Otha, K., Kojma, T., and Nakagawa, T. (1989). *J. Cry. Growth* **95**: 71.
95. Pashley, M. D., Haberern, K. W., and Gaines, J. M. (1991). *Appl. Phys. Lett.* **58**: 406.
96. Shitara, T., Zhang, J., Neave, J. H., and Joyce, B. A. (1992). *J. Appl. Phys.* **71**: 4299.
97. Asai, H. (1987) *J. Cryst. Grow.* **80**: 425.
98. Esaki, L. and Tsu, R. (1970). *IBM J. Res. Develop.* **14**: 61.
99. Tanaka, M., Sakaki, H., and Yoshino, J. (1986). *Jap. J. Appl. Phys.* **25**: L155.
100. Warick, C. A., Jan, W. Y., Ourmazd, A., and Harris, T. D. (1990). *Appl. Phys. Lett.* **56**: 2666.
101. Gammon, D., Shanabrook, B. V., and Katzer, D. S. (1990). *Appl. Phys. Lett.* **57**: 2710.
102. Suntola, T. and Antson, M. J. Finish Pat. 52359 (1974) and U.S. Pat. 4058430 (1977).
103. Herman, M. A., Vulli, M., and Pessa, M. (1985). *J. Crys. Growth* **73**: 403.
104. Osbourne, G. C. (1982). *J. Appl. Phys.* **53**: 1586.
105. Chen, T. R., Zhao, B., Zhuang, Y. H., Yariv, A., Ungar, J. E., and Oh, S. (1992). *Appl. Phys. Lett.* **60**: 1782.
106. Lievin, J-L. and Fonstad, C. G. (1987). *Appl. Phys. Lett.* **51**: 1173.
107. Matthews, J. W. and Blakesee, A. E. (1975). *J. Cry. Growth* **27**: 118.
108. People, R. and Bean, J. C. (1985). *Appl. Phys. Lett.* **47**: 322.
109. Price, G. L. (1991). *Phys. Rev. Lett.* **66**: 469.
110. Snyder, C. W. and Orr, B. G. (1992). *Europhys. Lett.* **19**: 33.
111. Houdre, R. and Morkoc, H. (1990). *Crit. Rev. Solid State and Mater. Sciences* **16**: 91.
112. Michel, E., Singh, G., Slivken, S., Besikci, C., Bove, P., Ferguson, I., and Razeghi, M. (1994). *Appl. Phys. Lett.* **65**: 3338.
113. Nakamura, S. (1991). *Jpn. J. Appl. Phys.* **30**: L1705.
114. Sheldon, P., Jones, K. M., Al-Jassim, M. M., and Jacobi, B. G. (1988). *J. Appl. Phys.* **63**: 5609.
115. Yamaguchi, H. and Horikoshi, Y. (1992). *Appl. Phys. Lett.* **60**: 2341.
116. Thompson, P. E., Twigg, M. E., Godbey, D. J., Hobart, K. D., and Simons, D. S. (1993). *J. Vac. Sci. Technol.* B **11**: 1077.
117. Nakamura, S., Senoh, M., Hagahama, S., Iwasa, N., Yamada, Y., Matsushita, T., Sugimoto, Y., and Kiyoku, H. (1996). *Appl. Phys. Lett.* **69**: 3034.
118. Norman, A. G., Seong, T-Y., Ferguson, I. T., Booker, G. R., and Joyce, B. A. (1993). *Semicond. Sci. Tech.* **8**: S9.
119. Ferguson, I. T., Norman, A. G., Seong, T.-Y., Thomas, R. H., Phillips, C. C., Zhang, X., Booker, G. R., Stradling, R. A., and Joyce, B. A. (1992). *Inst. Phys. Conf. Ser.* **120**: 395.

120. Ihm, Y-E., Otsuka, N., Klem, J., and Morkoc, H. (1987). *Appl. Phys. Lett.* **51**: 2013.
121. Seong, T.-Y., Norman, A. G., Ferguson, I. T., and Booker, G. R. (1994). *Appl. Phys. Lett.* **64**: 3593.
122. Henoc, P., Izrael, A., Quillec, M., and Launois H. 1982). *Appl. Phys. Lett.* **40**: 963.
123. Ishida *et al.* (1989).
124. Glas, F., Treacy, M. M., Quillec, M., and Launois, H. (1982). *J. de Phys.* **43**: C5.
125. Mackenzie, R. A., Liddle, J. A., and Grosvenor, C. R. (1991). *J. Appl. Phys.* **69**: 250.
126. Fewster, P. F. (1998). *Critical Reviews in Solid State and Material Sciences.*
127. Pollak, F. H. (1990). In *Analytical Raman Spectroscopy,* J. G. Grasselli and B. J. Bulkin, eds., New York: Wiley, p. 137.
128. Bassignana, I., Miner, C., and Puetz, N. (1989). *J. Appl. Phys.* **65**: 4299.
129. Burton, J. C., Sun, L., Pophristic, M., Long, F. H., Feng, Z. C., and Ferguson, I. (1998). *J. Appl. Phys.* **84**: 6268.
130. Bullis, W. M., Seiler, D. G., and Diebold, A. C. (1996). *Semiconductor Characterization: Present Status and Future Needs*, New York: AIP Press.
131. Cahn, R. W. and Bever, M. B. (1993). *Concise Encyclopedia of Materials Characterization*, Oxford-New York: Pergamon Press.
132. Hoffman, D., Eckstein, R., Kolbl, M., Makarov, Y., Muller, S. G., and Schmitt, E. (1997). *J. Crystal Growth* **174**: 669–674.
133. Choyke, W. J. and Pensl, G. (1997). *MRS Bulletin* **22**: 25–29.
134. Rupp, R., Widenhofer, A., Friedrichs, P., Peters, D., Schorner, R., and Stephani, D. (1998). *Mat. Sci. Forum* **264–268**: 89.
135. Feldman, D. W., Parker, J. H., Choyke, W. J., and Patrick, L. (1966). *Physical Review* **173**: 787.
136. Feng, Z. C., Ferguson, I. T., Armour, E., and Stall, R. A. (1996). *SPIE Proc.* **3279**: 161.
137. Bour, D. P. (1993). *Quantum Well Lasers*, P. S. Zoty, Jr. ed., San Diego: Academic Press 415.
138. Feng, Z. C., Pelczynski, M., Beckham, C., Cooke, P., Ferguson, I., and Stall, R. A. (1998). *MRS Proceedings* **484**: 13.
139. Hou, H. Q., Chui, H. C., Choquette, K. D., Hammons, B. E., Breiland, W. G., and Geib, K. M. (1996). *IEEE Photonics Tech. Lett.* **10**: 1285.
140. Kuo, C., Fletcher, R. M., Osentowski, T. D., Lardizabal, M. C., Graford, M. G., and Robbins, V. M. (1990). *Appl. Phys. Lett.* **57**: 2937.
141. Sugawara, H., Ishikawa, M., and Hatakoshi, G. (1992a). *Appl. Phys. Lett.* **58**: 1010.
142. Sugawara, H., Itaya, K., Nozaki, H., and Hatakoshi, G. (1992b). *Appl. Phys. Lett.* **61**: 1775.
143. Chang, S. J., Chang, C. S., Su, Y. K., Chang, P. T., Wu, Y. R., Huang, K. H., and Chen, T. P. (1997). *IEEE Photonics Tech. Lett.* **9**: 1199.
144. Michel, E., Singh, G., Slivken, S., Besikci, C., Bove, P., Ferguson, I., and Razeghi, R. (1996). *Appl. Phys. Lett.* **69**: 215.
145. Holmes, D. E. and Kamath, G. S. (1980). *J. Electron. Mater.* **9**: 95.
146. Webb, J. and Rousina, R. (1992). In *Semiconductor Interfaces and Microstructures* Z. C. Feng, ed., Singapore: World Scientific, 199.
147. Biefeld, R. M. and Hebner, G. A. (1990). *Appl. Phys. Lett.* **57**: 1563.
148. Choi, Y. H., Sudharsanan, R., Besikci, C., Bigan, E. and Razeghi, M. (1993). *MRS. Res. Soc. Symp. Proc.* **281**: 375.
149. Feng, Z. C., Beckham, C., Schumaker, P., Ferguson, I., Stall, R. A., Schumaker, N., Povloski, M., and Whitley, A. (1997). *Mat. Res. Soc. Symp. Proc.* **450**: 61.
150. Feng, Z. C., Perkowitz, S., Rao, T. S., and Webb, J. B. (1990). *J. Appl. Phys.* **68**: 5363.
151. McKee, M. A., Yoo, B-S., and Stall, R. A. (1992). *J. Crystal Growth* **124**: 286.
152. Yoo, B-S., McKee, M. A., Kim, S-G., and Lee, E. H. (1993). *Solid State Commun.* **88**: 447.
153. Neumayer, D. A. and Ekerdt, J. G. (1996). *Chem. Mat.* **8**: 9.

154. Itaya, K., Onomura, M., Nishio, J., Sugiura, L., Saito, S., Suzuki, M., Rennie, J., Nunoue, S., Yamamoto, M., Fujimoto, H., Kokomoto, Y., Ohba, Y., Hatakoshi, G., and Ishikawa, M. (1996). *Jpn. J. Appl. Phys.* **35**: 11315.
155. Akasaki, I., Sota, S., Sakai, H., Tanaka, T., Koike, M., and Amano, H. (1996). *Electron. Lett.* **32**: 1105.
156. Ponce, F. A. (1997). *MRS Bulletin* **22**: 51.
157. Heying, B. *et al.* (1996). *Appl. Phys. Lett.* **68** (5): 643.
158. Akasaki, I., Amano, H., Koide, Y., Hiramatsu, K., and Sawaki, N. (1989). *J. Cryst. Growth* **98**: 209.
159. Gaskill, D. K., Bottka, N., and Lin, M. C. (1986). *Appl. Phys. Lett.* **48**: 1449.
160. Keller, B. P., Keller, S., Kapolnek, D., Jiang, W-N., Wu, Y-F., Masui, H., Wu, X., Heying, B., Speck, J. S., Mishra, U. K., and Denbaars, S. P. (1995). *J. Electron. Mat.* **24**: 1707.
161. Tran, C. A., Karlicek Jr., R. F., Schurman, M., Osinsky, A., Merai, V., Li, Y., Eliashevich, I., Brown, M., Nering, J., Ferguson, I. T., and Stall, R. (1998). *J. Crystal Growth* **195**: 397.
162. Tran. (1998).
163. Sun, C. K., Vallee, F., Keller, S., Bowers, J. E., and DenBaars, S. P. (1997). *Appl. Phys. Lett.* **71**: 425.
164. Chichibu, S., Azuhata, T., Sota, T., and Nakamura, S. (1997). *Appl. Phys. Lett.* **70**: 2822.
165. Nakamura, S. (1997). *M.R.S. Internet Journal of Nitride Semiconductor Research* **2**: 5.
166. Minsky, M. S., Fleischer, S. B., Abare, A. C., Bowers, J. E., Hu, E. L., Keller, S., and DenBaars, S. P. (1998). *Appl. Phys. Lett.* **72**: 1066.
167. Pophristic, M., Long, F. H., Tran, C., Feng, Z. C., Karlicek, R. F., and Ferguson, I. T. (1998). *Appl. Phys. Lett.* **73**: 815.
168. Rogers, T. J., Nichols, K. B., Kopp, W. F., Smith, F. W., and Actis, R. (1996). *J. Vac. Sci. Technol.* B **14**: 2236.
169. Ferguson, I., Beckman, C., Feng, Z. C., Hou, H., Stall, R., Aina, L., and Seipel, K. (1998). *J. of Crystal Growth* **195**: 648.
170. Beckham, C., Stall, R. A., Thompson, A. G., Aina, L., and Seipel, K. (1996). *Proc. GaAs MANTECH Conf.*, 174.
171. Joshi, A. M., Brown, R., Fitzgerald, E. A., Wang, X., Ting, S., and Bulsara, M. (1997). *SPIE Proc.* **2999**: 211.
172. Li, T. K., Zhu, Y. F., Desu, S. B., Peng, C. H., and Masaya, N. (1996). *Appl. Phys. Lett.* **68**: 29.
173. Sato, S., Takahashi, N., and Sato, S. (1996). *Jpn. J. Appl. Phys.* **35**: L838.

Field Effect Transistors: FETs AND HEMTs

PRASHANT CHAVARKAR AND UMESH MISHRA

Department of Electrical and Computer Engineering, University of California, Santa Barbara, California, USA

Abstract

This chapter discusses the principles of operation and applications of field effect transistors. Field effect transistors, specifically high electron mobility transistors (HEMTs) or modulation-doped field-effect transistors are being extensively used in low noise and power amplifiers at microwave and millimeter-wave frequencies. The chapter begins with a discussion of current voltage and charge control mechanisms in HEMTs and this is followed by discussion of small-signal and large-signal equivalent circuit models of FET (which are used in circuit design). As operation at high frequencies is enabled by reduction in gate length, issues relating to device scaling are discussed. Issues related to operation of FETs as low-noise and power amplifier devices are discussed. The emergence of crystal growth techniques including molecular beam epitaxy (MBE) and metalorganic chemical vapor deposition (MOCVD) has enabled the fabrication of HEMTs using a variety of material systems. The selection of a material system depends on application and operating frequency. The AlGaAs/InGaAs pseudomorphic HEMT and the AlInAs/GaInAs HEMT are the two most widely used device structures. The performance of these devices as low-noise and power amplification devices is summarized. These devices have recently demonstrated their ability for insertion in high-volume commercial applications such as wireless and optical communication systems. This chapter therefore discusses the various design issues involving power amplifiers for wireless handsets and the suitability of GaAs *p*HEMTs for this application. The AlInAs/GaInAs HEMT with its high current density and superior high frequency performance has enabled the fabrication of digital circuits operating at 80 Gb/s. The various issues related to application of AlInAs/GaInAs HEMT for digital circuits are also discussed.

Vol. 28
ISBN 0-12-533028-6/$35.00
ISSN 1079-4050

2.1. Introduction

The concept of modulation doping was first introduced in 1978 (Dingle *et al.* 1978). In this technique electrons from remote donors in a higher bandgap material transfer to an adjacent lower gap material. The electrostatics of the heterojunction results in the formation of a triangular well at the interface, which confines the electrons in a two-dimensional (2D) electron gas (2DEG). The separation of the 2DEG from the ionized donors significantly reduces ionized impurity scattering resulting in high electron mobility and saturation velocity.

Modulation-doped field effect transistors (MODFETs) or high electron mobility transistors (HEMTs), which use the 2DEG as the current conducting channel have proved to be excellent candidates for microwave and millimeter-wave analog applications and high-speed digital applications. This progress has been enabled by advances in crystal growth techniques such as molecular beam epitaxy (MBE) and metalorganic chemical vapor deposition (MOCVD) as well as advances in device processing techniques, most notably electron beam lithography, which has enabled the fabrication of HEMTs with gate lengths down to 0.05 μm.

However, using a high electron mobility channel alone does not guarantee superior high-frequency performance. It is crucial to understand the principles of device operation and to take into consideration the effect of scaling to design a microwave or millimeter-wave HEMT device. The advantages and limitations of the material system used to implement the device also need to be considered. This chapter therefore begins with a discussion on the device operation of a HEMT in Section 2.2. This is followed by a discussion of scaling issues in HEMT in Section 2.3. These are of prime importance, as the reduction of gate length is required to increase the operating frequency of the device.

Sections 2.4 and 2.5 discuss the design issues for a low-noise HEMT and a power HEMT because the two major applications of a HEMT at microwave and millimeter-wave frequencies are low-noise amplification and power amplification. The first HEMT was demonstrated in the AlGaAs/GaAs material system in 1981. It demonstrated significant performance improvements over the GaAs MESFET at microwave frequencies. However, the high-frequency performance was not sufficient for operation at millimeter–wave frequencies. In the past ten years, the AlGaAs/InGaAs pseudomorphic HEMT on GaAs substrate (referred to as GaAs *p*HEMT) and the AlInAs/GaInAs HEMT on InP substrate (referred to as InP HEMT) have emerged as premier devices for microwave and millimeter–wave circuit applications. This highlights the importance of choosing the appropriate material system for device implementation. This will be discussed in Section 2.6.

Sections 2.7 and 2.8 will discuss the major advances in the development of the GaAs *p*HEMTs and InP HEMTs. Traditionally these devices have been used in low-volume, high-performance and high-cost military and space–based electronic systems. Recently the phenomenal growth of commercial wireless and optical fiber-based communication systems has opened up new areas of applications for these devices. This also means that new issues like manufacturing and operation at low bias voltage have to be addressed. Keeping this in consideration this chapter will focus on the application of GaAs *p*HEMTs for RF and wireless applications and the application of InP HEMTs in ultrahigh speed digital circuits for electronic front-ends in optical communication systems.

2.2. HEMT Device Operation and Design

2.2.1. LINEAR CHARGE CONTROL MODEL

The current control mechanism in the HEMT is control of the 2DEG density at the heterojunction interface by the gate voltage. Figure 2.1 shows the band diagram along the direction perpendicular to the heterojunction interface using the AlGaAs/GaAs interface as an example.

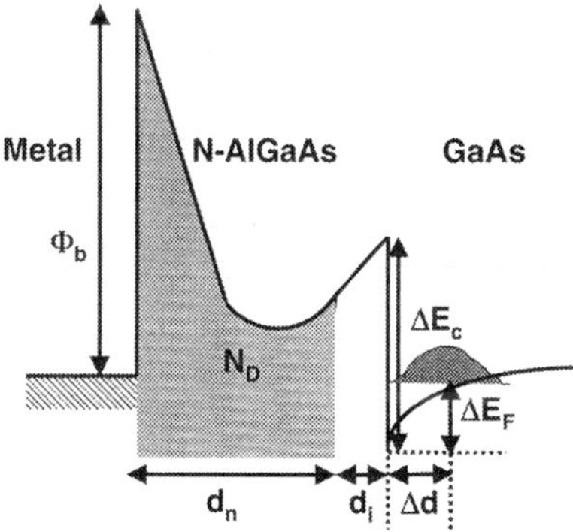

FIG. 2.1. Schematic of conduction band diagram at the AlGaAs/GaAs interface.

The first HEMT charge control model was proposed by Delagebeaudeuf and Linh in 1982. The potential well at the AlGaAs/GaAs interface is approximated by a triangular well. The energy levels in this triangular well and the maximum 2DEG density n_{sm} can be calculated by solving the Schrödinger equation in the triangular well and Poisson equation in AlGaAs donor layers (Drummond *et al.* 1986). For $0 < n_s < n_{\text{sm}}$, the sheet charge density n_s as a function of gate voltage V_g can be expressed as

$$qn_s = C_s(V_g - V_{\text{th}}) \tag{2.1}$$

where C_s is the 2DEG capacitance per unit area and is given by the following expression:

$$C_s = \frac{\varepsilon}{d_n + d_i + \Delta d} \tag{2.2}$$

here Δd is the distance of the centroid of the 2DEG distribution from the AlGaAs/GaAs interface and is typically or the order of 80 Å for $n_s \sim 10^{12}/\text{cm}^2$. Here V_{th} is the threshold voltage or pinch-off voltage and is given by

$$V_{\text{th}} = \phi_b - \frac{qN_D}{2\varepsilon} d_n^2 - \Delta E_c + \Delta E_F \tag{2.3}$$

where Φ_b, N_D and d_n are the Schottky barrier height on the donor layer, doping density, and doped layer thickness as illustrated in Fig. 2.1. Here ΔE_F is the Fermi potential of the 2DEG with respect to the bottom of the conduction band. It can be expressed as a function of 2DEG density as follows

$$\Delta E_F = \Delta E_{\text{FO}}(T) + an_s \tag{2.4}$$

where $\Delta E_{\text{FO}}(T) = 0$ at 300 K, $a = 0.125 \times 10^{16}$ V/m^2.

This simplified version of charge control is accurate only at low temperature. At room temperature, errors are introduced in the model due to parasitic charge modulation in the higher bandgap donor material. This results in premature saturation of the sheet charge and degradation of device performance.

A more accurate model for charge control, which self consistently solves Poiosson's and Schrödinger's equations in a self-consistent manner was proposed by Vinter (Vinter 1984). Apart from the 2DEG charge density n_s, the gate voltage also modulates the bound carrier density n_{bound} in the donor layer and the free electrons n_{free} in the donor layers.

2.2.2. Modulation Efficiency

As described in the previous section, the gate voltage also modulates the immobile and low mobility carriers in the higher bandgap donor layers in addition to the 2DEG charge in the channel. This reduces the efficiency of the gate voltage to modulate the drain current as the carriers in the higher bandgap donor layers do not contribute the drain current. The modulation efficiency of the FET is proportional to ratio between the change in drain current and the change in total charge required to cause this change (Foisy *et al.* 1988). This ratio is defined as follows,

$$\eta \alpha \frac{\delta \mathrm{I}_{\mathrm{ds}}}{\delta \mathrm{Q}_{\mathrm{tot}}} = \frac{\delta(q v_{\mathrm{sat}} n_s)}{\delta q(n_s + n_{\mathrm{bound}} + n_{\mathrm{free}})} \tag{2.5}$$

Dividing the numerator and denominator by δV_g we get the following expression

$$\frac{\delta \mathrm{I}_{\mathrm{ds}}}{\delta \mathrm{Q}_{\mathrm{tot}}} = v_{\mathrm{sat}} \frac{\delta(n_s)/\delta V_g}{\delta(n_s + n_{\mathrm{bound}} + n_{\mathrm{free}})/\delta V_g} \tag{2.6}$$

The modulation efficiency is defined as the ratio of the rate of the rate of change of the useful charge i.e. the 2DEG over that of the total charge,

$$\eta = \frac{\delta(n_s)/\delta V_g}{\delta(n_s + n_{\mathrm{bound}} + n_{\mathrm{free}})/\delta V_g} = \frac{\delta(n_s)/\delta V_g}{C_{\mathrm{TOT}}} = \frac{C_s}{C_{\mathrm{TOT}}} \tag{2.7}$$

The relation between the modulation efficiency and high frequency performance of the FET is evident in the following expressions

$$\begin{aligned} g_m &= \frac{\delta I_D}{\delta V_g} = \frac{\delta(q v_{\mathrm{sat}} n_s)}{\delta V_g} = q v_{\mathrm{sat}} \left(\frac{\delta n}{s \delta V_g} \right); \\ C_{\mathrm{gs}} &= C_{\mathrm{TOT}} L_g; \\ f_T &= \frac{g_m}{2\pi C_{\mathrm{gs}}} = \frac{q v_{\mathrm{sat}}(\delta n_s/\delta V_g)}{2\pi L_g C_{\mathrm{TOT}}} = \frac{v_{\mathrm{sat}}}{2\pi L_g} \eta \end{aligned} \tag{2.8}$$

Hence to improve the high frequency performance it is essential to improve the modulation efficiency.

Equation (2.8) must be used with caution in case of short gate length HEMTs. The saturation velocity v_{sat} due to high field and velocity overshoot effects. Using v_{sat} in this case may lead to values of modulation efficiency that are greater than 100%.

2.2.3. IMPACT OF THE $n_s\mu$ PRODUCT ON THE f_T OF A FET

Consider the operation of a FET in saturation mode. The channel of a FET can be divided into two parts, the source end of the channel which operates in the gradual channel mode and the drain end that operates in the saturated-velocity mode. In this mode the velocity of the electrons is proportional to the electric field in the channel. The voltage across the channel increases linearly until the channel is pinched-off near the drain side of the FET. The potential at the pinch-off point is the knee voltage or saturation voltage $V_{\rm DSS}$. The rest of the drain voltage exists between the pinch-off point and the drain contact in the drain depletion region of the FET. The value of $V_{\rm DSS}$ can be approximated by the maximum gate-bias swing $V_{\rm GSW}$, which can be applied without introducting carriers in the higher bandgap donor layers (parallel conduction). In other words, this is the gate voltage at which the maximum 2DEG ($n_{\rm sm}$) is achieved in the channel. Using Eq. (2.1) $n_{\rm sm}$ is expressed as

$$n_{\rm sm} = C_s \cdot V_{\rm GSW} \quad \text{where} \quad V_{\rm GSW} = V_{\rm gm} + V_T \quad \text{and} \quad C_s = \varepsilon/t \tag{2.9}$$

Here t is the thickness of the high bandgap barrier layer. Therefore the knee voltage can be expressed as $V_{\rm DSS} = n_{\rm sm}t/\varepsilon$, and the electric field E at the pinchoff point is given by $E = V_{\rm DSS}/L_g$. Now the effective electron velocity $v_{\rm eff}$ can be written as

$$v_{\rm eff} = \mu E = \mu n_{\rm sm} t/\varepsilon L_g \tag{2.10}$$

Finally, the current gain cutoff frequency f_T is given by

$$f_T = \frac{v_{\rm sat}}{2\pi L_g} = \frac{\mu n_{\rm sat} t}{2\pi\varepsilon L_g^2} = \frac{\mu n_{\rm sm}}{2\pi\varepsilon A L_g} \tag{2.11}$$

where A is the aspect ratio (L_g/t) for the device. It is clear from the preceding equation that for a given gate length and aspect ratio the f_T can be increased by maximizing $n_{\rm sm}\mu$ product. Note that this analysis assumes that the FET operates in a pure gradual channel mode. In a practical FET, the contribution of the $n_{\rm sm}\mu$ product in improvement of f_T depends on the extent of gradual channel conditions in the channel of the FET. Another advantage of a high $n_{\rm sm}\mu$ is the reduced parasitic access resistance of the device.

2.2.4. CURRENT-VOLTAGE (I–V) MODELS FOR HEMTs

By assuming linear charge control, gradual channel approximation and a 2-piece linear velocity-field model, the expression for the saturated drain current $I_{\rm DSS}$ in a HEMT is given by (Delagebeaudeuf *et al.* 1982),

$$I_{\rm DSS} = C_s v_{\rm sat}\left(\sqrt{(E_c L_g)^2 + (V_g - V_c(0) - V_{\rm th})^2} - E_c L_g\right) \tag{2.12}$$

Here E_c is defined as the critical electric field at which the electrons reach their saturation velocity $v_{\rm sat}$ and $V_c(0)$ is the channel potential at the source end of the gate. For a long gate length HEMT, Eq. (2.12) is valid until the onset of donor charge modulation, that is $0 < n_s < n_{\rm sm}$. The intrinsic transconductance of the device obtained by differentiating this expression with respect to the gate voltage and is expressed as follows:

$$g_{\text{mo}} = \frac{\delta I_{\text{ds}}}{\delta V_g} = C_s v_{\text{sat}} \frac{V_g - V_c(0) - V_{\text{th}}}{\sqrt{(V_g - V_c(0) - V_{\text{th}})^2 + (E_c L_g)^2}} \tag{2.13}$$

For a short gate length HEMTs, the electric field in the channel is much greater in magnitude than the critical electric field E_c. Assuming that the entire channel of the FET operates in saturated velocity mode, we can make the following assumption, that is, $V_g - V_c(0) - V_{\text{th}} \gg E_c L_g$. Then using Eq. (2.1), Eqs. (2.12) and (2.13) are reduced to the following:

$$I_{\text{DSS}} = q n_s v_{\text{sat}} \tag{2.14}$$

$$g_m = C_s v_{\text{sat}} \tag{2.15}$$

More insight can be obtained in terms of device parameters if the equation for charge control (Eq. (2.1)) is substituted in the expressions for I_{ds} and V_g as follows (Nguyen *et al.* 1992):

$$I_{\text{ds}} = q v_{\text{sat}} n_s \left[\sqrt{1 + \left(\frac{n_c}{n_s}\right)^2} - \frac{n_c}{n_s} \right] \tag{2.16}$$

$$g_{\text{mo}} = C_s v_{\text{sat}} \frac{1}{\sqrt{1 + (n_c/n_s)^2}} \tag{2.17}$$

where $n_c = E_c C_s L_g / q$ and $0 < n_s < n_{\text{sm}}$. Dividing both sides of Eq. (2.17) by $C_s V_{\text{sat}}$ the following expression for modulation efficiency is obtained:

$$\eta = \frac{1}{\sqrt{1 + (n_c/n_s)^2}} \tag{2.18}$$

Hence it is necessary to maximize the 2DEG density n_s to maximize the current drive, transconductance, and modulation efficiency of the HEMT. Although this is in contrast with the saturated-velocity model, it agrees with the experimental results. The foregoing results can also be used to select the appropriate material system and layer structure for the fabrication of high-performance microwave and millimeter-wave HEMTs.

Although the analytical model of device operation as was described here provides great insight into the principles of device operation and performance optimization, it fails to predict some of the nonlinear phenomena such as reduction of g_m at high current levels (g_m compression) and soft pinch-off characteristics.

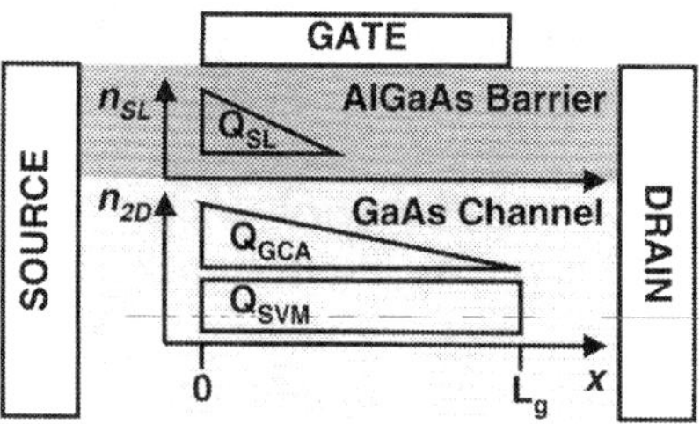

FIG. 2.2. Schematic diagram showing the location and distribution of Q_{SVM}, Q_{GCA} and Q_{SL} in a HEMT (Foisy *et al.* 1988). (© 1988 IEEE).

A model has been developed to explain these phenomena. The total charge in the HEMT is divided into three components. The first, Q_{SVM}, is the charge required to support a I_{ds} under the saturated velocity model (SVM). This charge is uniformly distributed under the gate.

In reality this is not the case as the electron velocity under the gate varies. To maintain the current continuity under the gradual channel approximation (GCA), extra charge under the channel has to be introduced. This is defined as Q_{GCA} and is maximum at the source end of the gate and minimum at the drain end.

The excess charge in the wide bandgap electron supply layer is denoted by Q_{SL}. Figure 2.2 shows the location and distribution of these charges in the HEMT. Only Q_{SVM} supports current density and thus contributes to the transconductance of the HEMT. The other two components contribute only to the total capacitance of the device. Hence the modulation efficiency (ME) of the HEMT in terms of these charges is expressed as

$$\eta = \frac{\delta Q_{\mathrm{SVM}}}{\delta(A_{\mathrm{SVM}} + Q_{\mathrm{GCA}} + Q_{\mathrm{SL}})} \tag{2.19}$$

and the transconductance can be expressed as $g_m = C_s v_{\mathrm{sat}} \eta$.

Figure 2.3 shows the variation of ME as a function of drain current density for an AlGaAs/GaAs HEMT and an AlGaAs/InGaAs *p*HEMT. At low current density, ME is low as most of the charge in the 2DEG channel has to satisfy the gradual channel approximation. This low value of ME results in low transconductance and soft pinch-off characteristics at low drain current densities. In the high current regime, modulation of Q_{SL} reduces the ME, resulting in gain compression. In the intermediate current regime the ME is maximum. However, if there exists a bias condition where both Q_{GCA} and Q_{SL} are modulated (as in the low band offset AlGaAs/GaAs system), it severely affects the ME.

For optimal high-power and high-frequency performance, it is necessary to maximize the range of current densities in which ME is high. The drop-off in ME due to parasitic charge modulation in the donor layers can be pushed to higher current density by increasing the maximum 2DEG density n_{sm}. The 2DEG

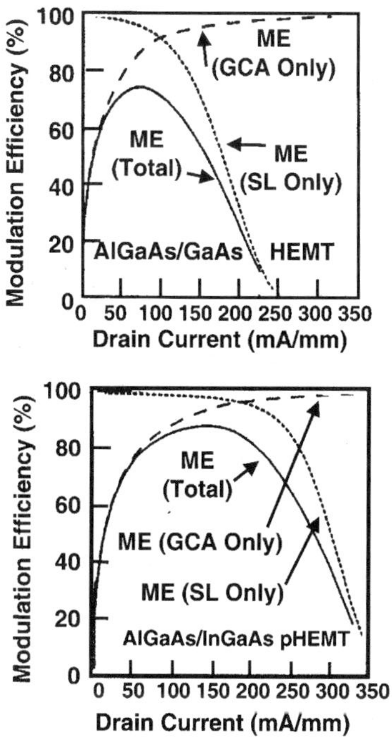

FIG. 2.3. Modulation efficiency as a function of current density for GaAs HEMT and GaAs pHEMT. (©1988 IEEE).

density can be maximized by using planar doping in the donor layer and by increasing the conduction band discontinuity at the barrier/2DEG interface. The drop-off in ME due to operation in gradual channel mode can be pushed to lower current densities by reducing the saturation voltage V_{Dsat}. This is achieved by increasing the mobility of the electrons in the 2DEG channel and by reducing the gate length. As seen from Fig. 2.3, higher modulation efficiency is achieved over a larger range of current density for the AlGaAs/InGaAs *p*HEMT, which has higher sheet charge density, mobility, and band discontinuity at the interface than the AlGaAs/GaAs HEMT.

2.2.5. SMALL SIGNAL EQUIVALENT CIRCUIT MODEL OF HEMT

The small signal equivalent circuit model of the HEMT is essential for designing HEMT-based amplifiers. The model also can provide insights into the role of various parameters in the high-frequency performance of the device. Figure 2.4

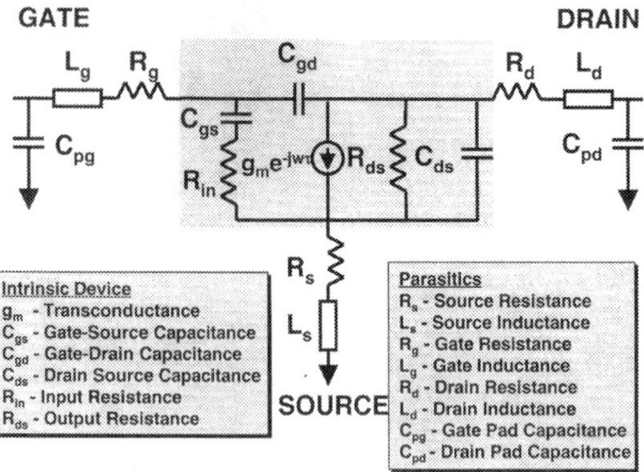

FIG. 2.4. Small signal equivalent circuit of a HEMT.

shows the small signal equivalent circuit for a HEMT. The Grey box highlights the intrinsic device.

The circuit elements in the preceding model are determined using microwave S-parameter measurements (Dambrine *et al.* 1988) (Berroth *et al.* 1990). The intrinsic circuit elements are a function of the DC bias, whereas the extrinsic circuit elements or parasitics are independent of it. The two measures of the high-frequency performance of a FET can now be defined in terms of the small signal model of the device as follows. The current gain cutoff frequency f_T can be defined as

$$f_T = \frac{g_m}{2\pi(C_{\text{gs}} + C_{\text{gd}})} \tag{2.20}$$

Hence to increase the current gain cutoff frequency it is essential to increase the g_m and reduce C_{gs} and C_{gd}. Referring to Eq. (2.8), it is clear that this can be achieved by increasing electron velocity in the channel and reducing gate length. The current gain cutoff frequency is mainly a physical measure of device performance. A more practical measure of high-frequency device performance is $f_{\max}$, the power gain cutoff frequency. This is the frequency at which the power gain of the FET is unity. It is defined as follows (Das 1985):

$$f_{\max} = \frac{f_T}{\sqrt{4g_{\text{ds}}\left(R_{\text{in}} + \dfrac{R_s + R_g}{1 + g_m R_s}\right) + \dfrac{4}{5}\dfrac{C_{\text{gd}}}{C_{\text{gs}}}\left(1 + \dfrac{2.5C_{\text{gd}}}{C_{\text{gs}}}\right)(1 + g_m R_s)^2}} \tag{2.21}$$

A simple form of Eq. (2.21) is:

$$f_{\max} = f_T \sqrt{\frac{R_{\text{ds}}}{4P_{\text{in}}}} = \frac{f_T}{\sqrt{4g_{\text{ds}}R_{\text{in}}}} \tag{2.22}$$

To improve the $f_{\max}$ of the device it is necessary to minimize the quantities in the denominator of Eq. (2.21). The crucial parameters here are the output conductance of the device g_{ds}, and the source and gate parasitic resistances R_s and R_g and the gate-drain feedback capacitance C_{gd} which need to be minimized. Reduction of g_{ds} can be achieved by appropriate vertical scaling (to be discussed in the next section). Reduction of R_s and R_g depends mainly on process technology. Reduction of C_{gd} can be achieved by proper design of the gate-drain region of the FET. The crucial parameter in the design of the gate drain depletion region is the gate-drain separation L_{gd} (Lester *et al.* 1988). Increasing L_{gd} reduces C_{gd} but also increases the effective gate length of the device, reducing the short channel effects. The optimum value of L_{gd} is 2.3 times that of the gate length L_g. Thus it is clear that $f_{\max}$ is a better measure of the high-frequency performance of a FET as it is determined not only by the material system used but also by the process technology and device design parameters.

2.2.6. LARGE SIGNAL MODELING

The small signal equivalent circuit of the HEMT is useful both for analysis of device operation and in circuit applications, such as low-noise amplifier, where the signal levels are small compared to the DC bias voltages. However, in many important applications like power amplifiers and oscillators, the HEMT operates under large signal conditions, that is the signal levels are comparable to the DC bias voltages. Under large signal conditions nonlinearity is introduced in the amplification process. Therefore, it is necessary to use the large signal nonlinear model of the HEMT when designing circuits for these applications. Nonlinear large signal models are used to predict gain compression and harmonic and intermodulation distortion in power amplifiers. Figure 2.5 shows the nonlinear equivalent circuit model of the HEMT.

The various parameters in the model are as follows: I_{gd} is the gate to drain avalanche current, which flows when the device approaches breakdown under large signal operation; I_{gs} represents the gate to source current when the gate is forward biased; and the drain to source current can be expressed as a function of input voltage $V_{\text{in}}(t)$ and output voltage $V_{\text{out}}(t)$ as follows (Curtice *et al.* 1985):

$$I_{\text{ds}} = (A_0 + A_1 V_1 + A_2 V_1^2 + A_3 V_1^3) \tanh(\gamma V_{\text{out}}(t)) \tag{2.23}$$

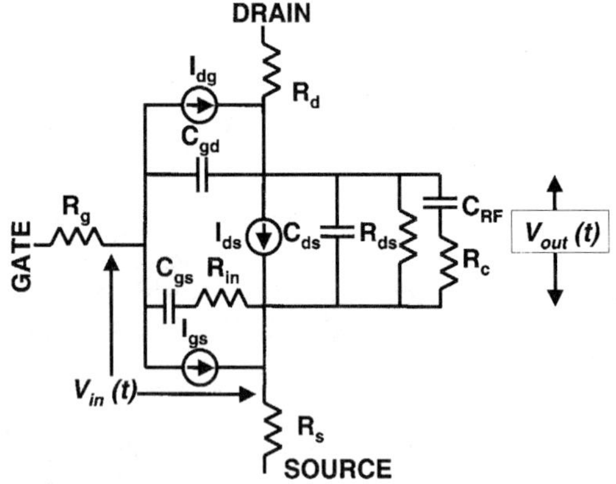

FIG. 2.5. Large signal nonlinear equivalent circuit of a HEMT.

the coefficients A_0 to A_3 are determined by fitting the model to measured data. The variable V_1 is given by the expression

$$V_1 = V_{\rm in}(t-\tau)[1+\beta(V_{\rm out}^o - V_{\rm out}(t)] \tag{2.24}$$

where

$$\tau = A_5 \cdot V_{\rm out}(t) \tag{2.25}$$

where β is the coefficient of pinch-off voltage change, $V_{\rm out}^o$ is the output voltage at which the coefficients A_0 to A_3 are determined and τ is the internal time delay of the FET. This expression accounts for the increase in the pinch-off voltage of the FET with drain bias.

The gate-drain avalanche current $I_{\rm dg}$ is given by the following expression:

$$I_{\rm dg} = \begin{Bmatrix} \dfrac{V_{\rm dg}(t)-V_B}{R_1}, & V_{\rm dg} > V_B \\ 0, & V_{\rm dg} > V_B \end{Bmatrix} \quad \text{where } V_B = V_{B0} + R_2 I_{\rm ds} \tag{2.26}$$

where R_1 is the breakdown resistance and R_2 is the resistance relating breakdown voltage to channel current.

The gate-source forward bias current $I_{\rm gs}$ is given by

$$I_{\rm gs} = \begin{Bmatrix} \dfrac{V_{\rm in}(t)-V_{\rm bi}}{R_F}, & V_{\rm in}(t) > V_{\rm bi} \\ 0, & V_{\rm in}(t) > V_{\rm bi} \end{Bmatrix} \tag{2.27}$$

where R_F is the effective value of forward bias resistance and $V_{\rm bi}$ is the builit-in voltage at the gate contact. The values of R_s, R_g, and R_d are determined by DC measurements. The large signal gate source capacitance C_{gs} and the large signal gate-drain capacitance C_{gd} are modeled as nonlinear junction capacitances using the following equations (Staudinger 1995):

$$C_{gs}(V_{gs}) = \frac{C_{gso}}{\left[1 - \frac{V_{gs}}{V_B}\right]^{MCGS}} \tag{2.28}$$

$$C_{gd}(V_{gd}) = \frac{C_{gdo}}{\left[1 - \frac{V_{gd}}{V_B}\right]^{MCGD}} \tag{2.29}$$

In these equations, V_{gs} and V_{gd} are the intrinsic gate-source and gate-drain voltages, respectively. The coefficients C_{gso}, *MCGS*, C_{gdo} and *MCGD* are extracted from *S*-parameter measurements at various bias conditions. The source-drain capacitance C_{ds} is assumed to be linear and exhibits no voltage dependence. The series RC network (C_{RF} and R_c) in the output side of the device is used to model the frequency dispersion of the output conductance. (See Section 2.2.7 for a discussion on frequency dispersion.)

The large signal transconductance and output conductance of the device is obtained by differentiating the expression for I_{ds} with respect to V_{gs} and V_{ds}. To model intermodulation distortion correctly it is necessary to model the derivatives of the expression for I_{ds} (Maas *et al.* 1990). An empirical model nonlinear model of the derivative was proposed by Angelov *et al.* (Angelov *et al.* 1992). It uses hyperbolic tangent functions as derivatives to model the variation of the transconductance with gate bias fairly accurately. In this model the drain current I_{ds} is a function of drain bias V_{ds} and is given by

$$I_{ds} = I_{pk}(1 + \tanh(\psi))(1 + \lambda V_{ds}) \tanh(\alpha V_{ds}) \tag{2.30}$$

where I_{pk} is the drain current at which peak transconductance is achieved, λ is channel length modulation parameter, α is the saturation voltage parameter, ψ is a power series function with V_{gs} as the variable and is centered at V_{pk}, the gate bias at which maximum transconductance is achieved. Here ψ can be expressed as follows:

$$\psi = P_1(V_{gs} - V_{pk}) + P_2(V_{gs} - V_{pk})^2 + P_3(V_{gs} - V_{pk})^3 + \cdots \tag{2.31}$$

Here P_1, P_2, P_3 are empirically determined coefficients and the number of terms in the power series is determined by the desired accuracy. Similarly the gate source capacitance C_{gs} and the gate drain capacitance C_{gd} can be modeled in terms of power series, details of which are presented in (Angelov *et al.* 1992). A simple first-order expression using the analysis in (Angelov *et al.* 1992) results in the following expressions:

$$C_{gs} = C_{gso}[1 + \tanh(P_{1gsg}V_{gs})][1 + \tanh(P_{1gsd}V_{ds})] \tag{2.32}$$

$$C_{gd} = C_{gdo}[1 + \tanh(P_{1dgd}V_{gs})][1 - \tanh(P_{1gdd}V_{ds})] \tag{2.33}$$

with C_{gso}, C_{gdo}, P_{1gsg}, P_{1gsd}, P_{1dgg} and P_{1gdd} as parameters. With this model good accuracy was demonstrated in modeling large signal characteristics of GaAs *p*HEMTs and InP HEMTs. As they account for nonlinear effects in device operation, large-signal models for HEMTs are based on numerical optimization using S-parameter measurements at various biases. Hence a tradeoff is involved between model accuracy and computational complexity.

2.2.7. Frequency Dispersion

Another important aspect of large signal modeling of a FET is the frequency dispersion of the transconductance and output resistance of the FET. Frequency dispersion characteristics have to be accounted in large-signal and wideband circuit applications of FET. Frequency dispersion is defined as variation of device parameters as a function of signal frequency. Electron trapping at the channel substrate interface and the surface of the device has been indicated as a mechanism for output resistance dispersion in GaAs MESFETs (Canfield *et al.* 1987). Transconductance dispersion is caused by the charging and discharging of surface states in the ungated region of the FET by the RF signal on the gate (Blight *et al.* 1986).

2.3. Scaling Issues in Ultrahigh-Speed HEMTs

The frequency at which a HEMT operates is limited by the electron transit time from the source ot the drain. Therefore to increase the frequency of operation it is necessary to reduce the gate length. However, as the gate length approaches 0.1 μm it is necessary to reduce the other parasitic delays in the device and take into account short channel effects to maintain the high-frequency performance of the HEMT.

2.3.1. Delay Time Analysis

The reduction of parasitic delays in a FET is essential to improve the high frequency performance as these delays can be as high as 45% of the intrinsic delay (Nguyen *et al.* 1989). Considering the small-signal model of a FET the total delay t_T in a FET can be expressed as follows (Nguyen *et al.* 1990):

$$t_T = t_{\text{pad}} + t_{\text{fringe}} + t_{\text{channel}} + t_{\text{transit}} + t_{\text{drain}} = 1/(2\pi f_T) \tag{2.34}$$

Here t_{pad} is the charging time for the parastic pad capacitance and is given by

$$t_{\text{pad}} = C_{\text{pad}}/g_m \cdot W \tag{2.35}$$

where C_{pad} is the pad capacitance and is typically 10 fF per 50 μm × 50 μm bonding pad, g_m is the extrinsic transconductance per unit gate width and W is the width of the device. To minimize t_{pad} it is necessary to have a high-gate-width, high-transconductance HEMT.

The gate fringe capacitance charging time (t_{fringe}) is given by

$$t_{\text{fringe}} = C_{\text{fringe}}/g_{\text{mo}} \tag{2.36}$$

where g_{mo} is the intrinsic transconductance of the HEMT and is related to the extrinsic transconductance (g_{m}) and source resistance R_s by the following expression:

$$g_m = g_{\text{mo}}/(1 + g_{\text{mo}} \cdot R_s) \tag{2.37}$$

The gate fringe capacitance C_{fringe} is typically 0.18 pF/mm, hence for a HEMT with an intrinsic transconductance of 1000–1500 mS/mm, t_{fringe} is approximately 0.1–0.2 ps.

Channel charging delay t_{channel} is associated with RC delays and is proportional to channel resistance. The channel charging delay is minimum at high current densities. The channel charging delay can be considered as a measure of the effectiveness of a FET to operate in the saturated velocity mode.

The transit delay of the FET, t_{transit}, can be expressed as the time required to traverse under the gate and is given by

$$t_{\text{transit}} = L_g/v_{\text{sat}} \tag{2.38}$$

The drain delay (t_{drain}) is the time required by the electron to traverse the depletion region between the gate and the drain and is a function of bias conditions (Moll *et al.* 1988). The drain delay increases with drain bias as the length of the depletion region beyond the gate increases. Drain delay is an important parameter for millimeter-wave power HEMTs. To increase the breakdown voltage of the device, gate-to-drain spacing has to be increased. When the device is biased at a high drain voltage to maximize the power output, it creates a drain depletion region which is of the order of gate length of the device. Thus the drain delay becomes a major component of the total delay in the device, and can limit the maximum f_T and f_{max}.

2.3.2. VERTICAL SCALING

Aspect ratio (the ratio between the gate length L_g and the gate-to-channel separation d_{Barrier}) needs to be maintained when gate length is reduced. Aspect ratio is a critical factor affecting the operation of the field effect transistor and should be maintained above five. As the gate length is reduced the distance between the gate and 2DEG has to be reduced so that the aspect ratio of the device is maintained. This distance between the gate and 2DEG (d_{Barrier}) is given by the quantity $d_n + d_i$ as seen in Fig. 2.1.

However, maintaining the aspect ratio alone does not guarantee improvement in device performance. This is clear if the variation of threshold voltage with the reduction in d_{Barrier} is examined. It is clear that d_i cannot be reduced, as it will result in degradation of mobility in the 2DEG channel due to scattering from the donors in the barrier layers. Therefore, to maintain aspect ratio the thickness of the doped barrier layer d_n has to be reduced. By examining Eq. (2.3) for threshold voltage, it is clear that this makes the threshold voltage more positive.

At first glance, this does not seem to affect device performance. The effect of the more positive threshold voltage is clear if the access regions of the device are considered. A more positive threshold voltage results in reduction of sheet charge in the access region of the device. This increases the source and drain resistance of the device, which reduces the extrinsic transconductance (see Eq. (2.37)) and also increases the channel charging time (due to increased RC delays). Thus the increased parasitic resistances nullify the improvements in speed in the intrinsic device.

The threshold voltage of the device then has to be kept constant with the reduction in d_n. From Eq. (2.3) it can be seen that the doping density in the high bandgap donor has to be increased. Since the threshold voltage varies as a square of the doped barrier thickness, a reduction in its thickness by a factor of 2 requires that the doping density be increased by a factor of 4.

High doping densities can be difficult to achieve in wide bandgap materials such as AlGaAs due to the presence of DX centers. Increased doping also results in higher gate leakage current, higher output conductance, and a lower breakdown voltage. Utilizing planar or delta doping wherein all the dopants are located in a single plane can alleviate these problems. This leaves most of the higher bandgap layer undoped and enables reduction of its thickness.

The threshold voltage of a planar-doped HEMT is given as follows (Chao *et al.* 1989):

$$V_T = \phi_B - \frac{qN_{2D}d_n}{\varepsilon} - \Delta E_c + \Delta E_F \tag{2.39}$$

where N_{2D} is the per unit area concentration of donors in the doping plane and d_n is the distance of the doping plane from the gate. In this case, the 2D doping density has to increase linearly with the reduction in barrier thickness. The transfer efficiency of electrons from the donors to the 2DEG channel also is increased, as all the dopant atoms are close to the 2DEG channel. Hence higher 2DEG sheet densities can be achieved in the channel and thus planar doping enables efficient vertical scaling of devices with reduction in gate length (Nguyen *et al.* 1989). From a materials point of view, efficient vertical scaling of a HEMT requires a high bandgap donor/barrier semiconductor that can be doped efficiently.

The voltage gain of the device (g_m/g_{ds}) can be considered as a measure of short channel effects in the device. The reduction of gate length and the gate to channel separation results in an increase in the transconductance of the device. However, to reduce the output conductance of the device it is also necessary to reduce the channel thickness, which then increases the carrier confinement in the channel. Enoki *et al.* have investigated the effect of the donor/barrier and channel layer thickness on the voltage gain of the device (Enoki *et al.* 1994). The gate to channel separation (d_{Barrier}) and the channel thickness (d_{channel}) were varied for a 0.08-μm gate length AlInAs/GaInAs HEMT. For a d_{Barrier} of 170 Å and a d_{channel} of 300 Å, the g_m was 790 mS/mm and g_{ds} was 99 mS/mm, resulting in a voltage gain of 8. When d_{Barrier} was reduced to 100 Å and d_{channel} was reduced to 150 Å, the g_m increased to 1100 mS/mm and g_{ds} reduced to 69 mS/mm; this doubled the voltage gain to 16. This illustrates the necessity to reduce the channel thickness to improve charge control in ultra-short gate length devices.

Subthreshold slope is an important parameter to evaluate short channel effects for digital devices. A high value of subthreshold slope is necessary to minimize the off-state power dissipation and to increase the device speed. Two-dimensional simulations performed by Enoki *et al.* indicate that reduction in channel thickness is more effective than the reduction in barrier thickness, for maintaining the subthreshold slope with reduction in gate length (Enoki *et al.* 1994).

The high-frequency performance of a device is a function of the electrical gate length $L_{g,\text{eff}}$ of the device. The electrical gate length of the device is larger than the metallurgical gate length L_g due to lateral depletion effects near the gate. The relation between $L_{g,\text{eff}}$ and L_g is given by (Chao *et al.* 1989),

$$L_{\text{g,eff}} = L_g + \beta(d_{\text{Barrier}} + \Delta d) \tag{2.40}$$

where d_{Barrier} is the total thickness of the barrier layers, Δd is the distance of the centroid of the 2DEG from the channel barrier interface and is of the order of 80 Å. The value of parameter β is 2.

Consider a long gate length HEMT ($L_g = 1$ μm) with a barrier thickness of 300 Å. Using Eq. (2.40), the value of 1.076 μm is obtained for $L_{g,\text{eff}}$. Thus the effective gate length is only 7.6% higher than the metallurgical gate length. Now consider an ultrashort gate length HEMT ($L_g = 0.05$ μm) with an optimally scaled barrier thickness of 100 Å. Using the same analysis, a value of 0.086 μm is obtained for $L_{g,\mathrm{eff}}$. In this case the effective gate length is 43% higher than the metallurgical gate length. Hence to improve the high-frequency performance of an ultrashort gate length HEMT, effective gate length reduction along with vertical scaling is required.

This requires optimization of the gate definition and gate recess etch processes as the lateral depletion is a strong function of gate recess width. The following is an excellent illustration of this process optimization. In 1994 Enoki *et al.* reported a 0.05 μm AlInAs/GaInAs HEMT with a f_T of 300 GHz (Enoki *et al.* 1994). The gate length definition and recess etching was done using conventional techniques. Recently the same group reported a 0.07 μm AlInAs/GaInAs HEMT with f_T of 300 GHz (Suemitsu *et al.* 1998). In this case an InP recess etch stop layer was incorporated in the layer structure. This reduced the extent of lateral etching during the gate definition process. Assuming that all other parameters remain constant, it can be assumed that the electrical gate lengths of the two devices are equal. Thus it can be concluded that a reduction of 0.02 μm in $L_{g,\mathrm{eff}}$ is achieved by optimizing the gate definition and gate recess process.

2.3.3. Horizontal Scaling

Reduced gate length is required for the best high-frequency performance. However, it should be kept in mind that the gate series resistance increases with the reduction in gate length. This problem can be solved with a T-shaped gate. This configuration lowers the gate series resistance while maintaining a small footprint. Another advantage of the T-shaped gate is reduced susceptibility to electromigration under large signal RF drive as the large gate cross section reduces current density. For a 0.1-μm gate length using a T-gate instead of a straight gate, reduces the gate resistance from 2000 Ω/mm to 200 Ω/mm.

The simplified expression for f_T as expressed in Eq. (2.20) does not include the effect of parasitics on the delay time in a FET. A more rigorous expression for f_T, which includes the effects of parasitics on f_T was derived by Tasker and Hughes and is given here (Tasker *et al.* 1989):

$$f_T = \frac{g_m/2\pi}{[C_{\mathrm{gs}} + C_{\mathrm{gd}}][1 + (R_s + R_d)/R_{\mathrm{ds}}] + C_{\mathrm{gd}} g_m (R_s + R_d)} \tag{2.41}$$

It is clear from Eq. (2.41) that it is necessary to reduce source and drain resistance R_s and R_d, respectively, to increase the f_T of a FET. Mishra *et al.* demonstrated a record f_T of 250 GHz for a 0.15-μm device with a self-aligned gate, which reduces the gate-source and gate-drain spacing and which then results in the reduction of R_s and R_d (Mishra *et al.* 1989). Equation (2.41) can be rearranged as follows (Tasker *et al.* 1989):

$$\frac{1}{2\pi f_T} = \frac{(C_{gs} + C_{gd})}{g_m} + \frac{(C_{gs} + C_{gd})(R_s + R_d)}{g_m R_{ds}} + C_{gd}(R_s + R_d) \tag{2.42}$$

where the first term on the right-hand side is the intrinsic delay of the device (τ_{int}) and the rest of the terms contribute to parasitic delay (τ_p). From this equation the ratio of parasitic delay to the total delay ($\tau_t = \tau_p + \tau_{int}$) is given as

$$\frac{\tau_p}{\tau_t} = g_m(R_s + R_d)\left[\frac{G_{ds}}{g_m} + \frac{1}{[1 + C_{gs}/C_{gd}]}\right] \tag{2.43}$$

Hence to improve the f_T of the device, the parasitic source and drain resistance have to be reduced as the gate length of the device is reduced. This minimizes the contribution of the parasitic delay to the total delay of the device.

2.4. Low-Noise HEMT Design

2.4.1. DEVICE NOISE FIGURE

The HEMT with its capability of providing high gain at microwave and millimeter wave frequencies is an ideal candidate for low-noise amplifiers. The noise figure F is a measure of low-noise performance of a device or a two-port network. The noise figure F is defined as the ratio between signal-to-noise at the input and the output terminals:

$$F = \frac{S_i/N_i}{S_o/N_o} \tag{2.44}$$

where S_i, N_i, S_o and N_o are the signal and noise powers at the input and output, respectively. If G_a and N_a are defined as gain of the device and total noise contributed by the device, then the noise figure can be expressed as

$$F = \frac{S_i/N_i}{S_o/N_o} = \frac{S_i/N_i}{G_a S_i/(N_a + G_a \cdot N_i)} = 1 + \frac{N_a}{G_a N_i} \tag{2.45}$$

If the input noise is assumed as thermal noise then the preceding expression changes to

$$F = 1 + \frac{N_a}{G_a k T_i B} \tag{2.46}$$

where k is the Boltzman constant, T_i is the ambient temperature, and B is the bandwidth of operation. This expression suggests that increasing the gain and reducing the noise contribution from the HEMT itself can minimize the noise figure of a HEMT because reduction in ambient temperature T_i reduces the noise figure of the device.

2.4.2. Fukui Noise Model

A simple expression relating the minimum noise figure of a FET, $F_{\min}$ with the device parameters, was proposed by Fukui (Fukui 1979),

$$F_{\min} = 1 + K_f \frac{f}{f_T} \sqrt{g_m(R_s + R_g)} \tag{2.47}$$

where f is the frequency of operation, f_T is the current gain cut-off frequency of the device, g_m is the transconductance and R_s and R_g are parasitic source and gate resistances. K_f is a frequency independent fitting factor. The preceding equation can also be written as

$$F_{\min} = 1 + 2\pi C_{\mathrm{gs}} K_f f \sqrt{\frac{(R_s + R_g)}{g_m}} \tag{2.48}$$

To reduce $F_{\min}$ it is necessary to minimize the gate source C_{gs}, the parasitic source and gate resistances R_s and R_g, and maximize the transconductance g_m. Reducing the gate length of the device improves the noise performance by simultaneously increasing the g_m and reducing C_{gs}. In other words, increasing the f_T of a device lowers its minimum achievable noise figure. Reduction in device parasitics is also crucial for low-noise performance. Using a recessed-gate geometry can reduce the parasitic source resistance. The gate resistance can be reduced by using a T-shaped gate, which lowers the gate resistance while maintaining a small footprint.

2.4.4. Noise-Temperature Model

Pospieszalski has proposed a model in which the noise sources in the intrinsic FET are represented in terms of thermal noise sources at the gate and drain terminals of the FET (Pospieszalski 1989). The noise contribution of these sources is represented in terms of noise temperatures T_a and T_d. Figure 2.6 shows the noise-temperature model of the intrinsic FET.

The thermal noise source at the gate is assumed to be at ambient temperature (T_a) and the thermal noise source at the drain is assumed to be at a fictitious drain

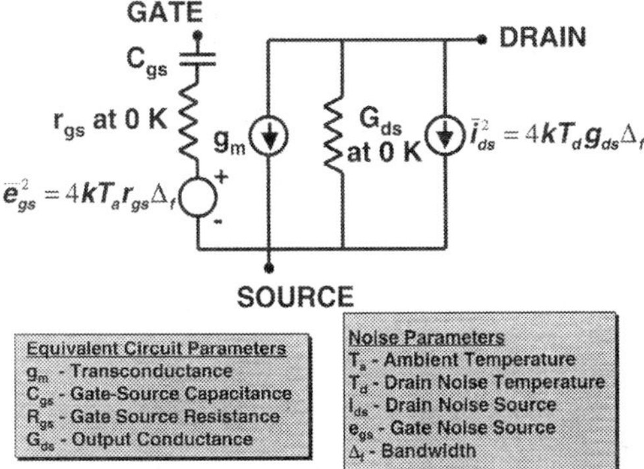

FIG. 2.6. Noise-temperature equivalent circuit model of an intrinsic FET.

temperature (T_d). Using these parameters, the four noise parameters of a FET are given by the following quantities:

$$X_{\text{opt}} = \frac{1}{2\pi C_{\text{gs}}} \tag{2.49}$$

$$R_{\text{opt}} = \sqrt{\left(\frac{f_T}{f}\right)^2 \frac{r_{\text{gs}} T_a}{g_{\text{ds}} T_d} + r_{\text{gs}}^2} \tag{2.50}$$

$$T_{\min} = 2\frac{f}{f_T}\sqrt{g_{\text{ds}} r_{\text{gs}} T_a T_d + \left(\frac{f}{f_T}\right)^2 (r_{\text{gs}} g_{\text{ds}} T_d)^2 + 2\left(\frac{f}{f_T}\right)^2 r_{\text{gs}} g_{\text{ds}} T_d} \tag{2.51}$$

$$g_n = \left(\frac{f}{f_T}\right)^2 \frac{g_{\text{ds}} T_d}{T_o} \tag{2.52}$$

where R_{opt} and X_{opt} are real and imaginary parts of the optimum source impedance Z_{opt}^N, which minimizes the noise figure $F_{\min}$, $T_{\min}$ is the minimum noise temperature of the FET, T_o is the standard noise temperature (290 K), g_n is the noise conductance. In other words when a source with an impedance of Z_{opt} is connected to the input of the FET, the noise at the output of the FET is minimized and a noise figure (temperature) of the FET is $F_{\min}$ ($T_{\min}$). The noise conductance g_n can be interpreted as measure of the sensitivity of the noise figure (temperature) to impedance mismatch at the source.

The noise parameters of the FET are determined as follows: first, the equivalent circuit parameters are determined from S-parameter measurements; then at a given frequency, the drain noise temperature is determined from noise measurements. The drain noise temperature is a fitted parameter and typically ranges from 300–700 K. Now the noise parameters of the FET can be determined at any frequency. For a frequency f that is much less than f_T, the expression for $T_{\min}$ can be simplified to:

$$T_{\min} = 2\frac{f}{f_T}\sqrt{g_{\text{ds}} r_{\text{gs}} T_a T_d} \tag{2.53}$$

The minimum noise temperature $T_{\min}$ can be related to the minimum noise figure $F_{\min}$ by

$$F_{\min} = 1 + \frac{T_{\min}}{T_o} = 1 + \frac{\sqrt{T_a T_d}}{T_o}\left(\frac{f}{f_T}\right)\sqrt{4 g_{\text{ds}} r_{\text{gs}}} \tag{2.54}$$

The preceding equations suggest that to minimize the noise figure, the f_T of the FET should be maximized and the input resistance should be minimized. This agrees well with the basic noise model presented in Section 2.4.2. Unlike the basic noise model, an additional parameter that needs to be minimized is the output conductance g_{ds} of the FET. In the foregoing analysis only the noise contribution of the instrinsic part of the FET was considered. In a real device, parasitic source, drain, and gate resistances (as shown in Fig. 2.4) contribute to increased thermal noise in the device.

The relation between the output conductance g_{ds} and the minimum noise figure $F_{\min}$ is clarified in a noise model by Hughes (Hughes 1992). Using Eq. (2.22) for $f_{\max}$, the expression for the minimum noise temperature $T_{\min}$ (Eq. (2.53)) can be written as

$$T_{\min} = (T_a T_d)^{1/2}\frac{f}{f_{\max}} \tag{2.55}$$

As the main objective of a low-noise FET is to amplify a signal, it is imperative to calculate the available gain of the FET when the FET is biased for minimizing the noise figure. We know that the input impedance of a FET (as seen in Fig. 2.6) is

$$Z_{\text{in}} = r_{\text{gs}} + \frac{1}{j\omega C_{\text{gs}}} \tag{2.56}$$

From elementary circuit theory, the source impedance Z^G_{opt}, which maximizes the gain of the FET is the complementary impedance of Z_{in} and is given by

$$Z^G_{\text{opt}} = r_{\text{gs}} + \frac{j}{\omega C_{\text{gs}}} \tag{2.57}$$

(It is important to note that the value of source impedance Z_{opt}^G, which maximizes the gain of the FET is different from Z_{opt}^N, which minimizes the noise at the output.) The maximum available gain G_a of the FET is given by

$$G_{a,\max} = \left(\frac{f_{\max}}{f}\right)^2 \tag{2.58}$$

At any other source impedance Z_g the available gain G_a from the FET is given by

$$\frac{1}{G_a} = \frac{1}{G_{a,\max}} + \frac{g_g}{R_g}|Z_g - Z_{opt}^G|^2 \qquad \text{where } g_g = \left(\frac{f}{f_T}\right)^2 g_{ds} \tag{2.59}$$

A simplified expression for the available gain under low-noise bias conditions, that is, when the source impedance is Z_{opt}^N is given by (Hughes 1992)

$$G_{A,\mathrm{opt}} = \left(\frac{4T_d}{T_g}\right)^{1/2} \frac{f_{\max}}{f} \tag{2.60}$$

Therefore, to minimize noise temperature T_{min} and maximize the available gain under low-noise bias conditions, it is necessary to maximize the f_{max} of the FET. A physical basis for the noise model can be understood if the product of the available gain under low-noise bias conditions $G_{A,opt}$ and the minimum noise temperature T_{min} is considered. Multiplying Eqs. (2.60) and (2.55) results in the following:

$$G_{A,\mathrm{opt}} T_{\min} = 2T_d \tag{2.61}$$

Equation (2.61) can be explained as follows: for a FET biased at Z_{opt}^N, the effective input noise power in a 1-Hz bandwidth is kT_{min}. The noise power at the output is the product of gain and the input noise power, which is $kT_{min}G_{A,opt}$. This is equal to $2kT_d$, which is twice the noise power from the output resistor. Therefore, under optimum low-noise bias conditions, at the output of the FET the contribution from input and output noise sources are equal. To minimize the noise at the output, it is also necessary to minimize the drain noise temperature T_d.

2.5. Power HEMT Design

2.5.1. POWER HEMT DEVICE DESIGN

The power performance of a FET at millimeter-wave frequencies is characterized in terms of the maximum output power or power density (P_{out}), associated gain (G_a), and power added efficiency (PAE).

In class A operation the theoretical maximum output power is given by the equation:

$$P_{\text{out}} = \tfrac{1}{8}(I_{\max} - I_{\min})(BV_{\text{gd}} - V_{\text{po}} - V_{\text{knee}}) \tag{2.62}$$

where $I_{\max}$ is the maximum channel current and $I_{\min}$ is the minimum channel leakage current due to gate-drain and/or source drain breakdown. BV_{gd} is the two-terminal gate to drain breakdown voltage, V_{po} is the pinch-off voltage and V_{knee} is channel knee voltage. Figure 2.7 shows a schematic representation of these quantities. The quantity $BV_{\text{gd}} - V_{po}$ can be expressed as 3-terminal off-state breakdown voltage BV_{ds}.

Therefore, to increase the power output of a FET it is clear that the maximum current density and breakdown voltage have to be increased and the channel knee voltage has to be reduced.

The PAE of a FET is a measure of how efficient it is in converting power from DC to microwave or millimeter-wave frequencies. PAE is defined as

$$\text{PAE} = \frac{P_o - P_i}{P_{\text{dc}}} = \frac{P_o(1 - 1/G_a)}{P_{\text{dc}}} \tag{2.63}$$

The ratio P_o/P_{dc} is defined as drain efficiency (DE) of the device and is expressed in terms of the drain bias voltage (V_{DD}) and the knee voltage (V_{knee}) as (Kushner 1989),

$$DE = \alpha\left(\frac{V_{\text{DD}} - V_{\text{knee}}}{V_{\text{DD}}}\right) \tag{2.64}$$

The parameter α is $\frac{1}{2}$ for Class A operation and $\pi/4$ for Class B operation. Therefore, in the limiting case of zero knee voltage and infinite gain, the maximum PAE that can be achieved for Class A operation is 50% in case of a device with infinite gain. For Class B the maximum possible PAE is 78% ($\pi/4$).

Another way to improve DE, PAE, and output power is to operate at high drain voltages. However, this requires devices with high breakdown voltages and also reduces the reliability of the device. For use in portable battery-operated wireless systems, a high PAE has to be achieved at a low drain bias voltage. In this case

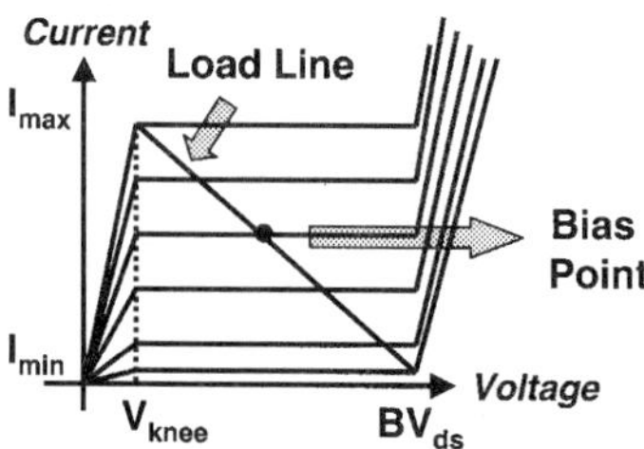

FIG. 2.7. Schematic representation and various parameters determining the power output of a FET.

the reduction of the knee voltage is crucial in achieving high PAE. Higher efficiency is achieved in Class B operation where the device is biased near pinch-off; therefore, high gain is required near pinch-off.

Long-term reliable CW operation of a power amplifier requires minimization of the channel temperature. The channel temperature of the device is directly proportional to the thermal power dissipated in the device (P_{diss}) and is given by

$$P_{diss} = P_{out}\left(1 - \frac{1}{G}\right)\left(\frac{1}{PAE} - 1\right) \tag{2.65}$$

Therefore, to reduce the dissipated thermal power at a given output power it is desirable to have a high-gain and high-power added efficiency.

The relation between device parameters and maximum available gain G_a at a particular frequency is given by the following equation:

$$\left(\frac{f_T}{f}\right)^2\left(\frac{R_{opt}}{R_{in}}\right) < G_a < \left(\frac{f_{max}}{f}\right)^2 = \left(\frac{f_T}{f}\right)^2\left(\frac{R_{ds}}{4R_{in}}\right)^2 \tag{2.66}$$

where R_{opt} is load resistance for maximum output power, R_{ds} and R_{in} are the output and input resistances of the FET, and f_T and f_{max} are the current gain and power gain cut-off frequencies of the device. Therefore, to increase the gain, it is necessary to increase the f_T by reducing the gate length. At the same time, short channel effects have to be minimized to maintain the output resistance of the device.

Another important parameter of a power HEMT is the large signal cut-off frequency f_c. This is the frequency when the large signal gain G_a is unity. Setting $G_a = 1$ in Eq. (2.66) we get the following relationship:

$$f_T\sqrt{R_{opt}/4R_{in}} < f_c < f_{max} = f_T\sqrt{R_{ds}/4R_{in}} \tag{2.67}$$

Hence HEMT devices capable of large signal operation at millimeter-wave frequencies need to have a high-current gain cutoff frequency f_T, high output resistance R_{ds}, and low input resistance R_{in}.

2.5.2. DEVICE LAYOUT CONSIDERATIONS FOR POWER HEMTS

The total power output and not the power density is the quantity of interest for millimeter-wave power applications. It is relatively easy to obtain high power densities from smaller devices; however, increasing the total output requires various other factors to be taken into consideration.

As the total power output of a small gate width FET is not enough for many microwave and millimeter-wave applications, larger gate width HEMTs combining many gate fingers are required. In many cases the total gate width may be in the region of 10 mm. Under these conditions, the device topology is critical as it determines both the RF and thermal characteristics of the device. This in turn also determines the device yield and long-term reliability. Typically a device layout with interdigitating gate fingers is used as it is most space efficient. Because at millimeter-wave frequencies the signal propagation delays in a large device approach the operating frequency, an appropriate distribution of the total gate width into gate fingers needs to be taken into consideration. The spacing between adjacent fingers is dictated by thermal constraints.

The source inductance of the FET limits the gain of the device and can be reduced by placing small vias under each source pad. In this case the fabrication technology of the source vias plays an important role in determining the device layout. The two technologies available for via fabrication are wet etching and reactive ion etching (RIE). Figure 2.8 compares the device layout of a power HEMT with wet-etched and RIE-etched vias (Smith *et al.* 1990).

It is obvious that RIE-etched vias result in a compact and uniform layout. In this case, a via can be placed under each source, thereby further reducing the source inductance and increasing the gain and efficiency of the device. Hur *et al.* have demonstrated significant reduction in source inductance by using individually grounded source finger vias (Hur *et al.* 1995). The reduction of source inductance results in smaller reverse transmission/feedback, which simplifies the matching network design. As all the source fingers are now connected through the

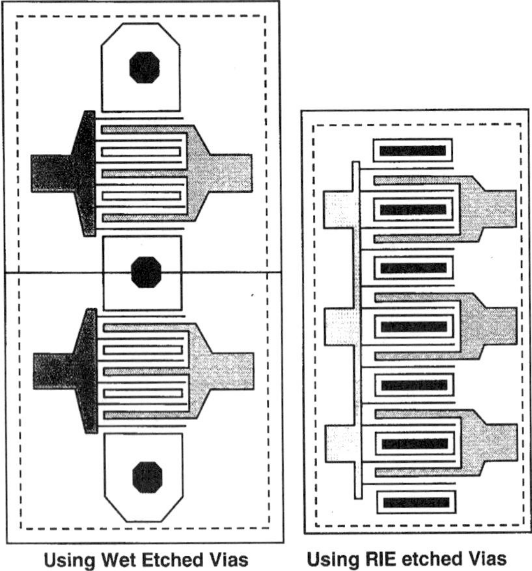

FIG. 2.8. Device layout of a power HEMT with wet-etched and RIE-etched vias. [47]

backside, no airbridge connections between source pads are required. This results in reduction of parasitic gate-source capacitance C_{gs} and gate-drain capacitance C_{gd}. Reduction in chip size also increases the device yield per wafer. An additional advantage of the RIE-etched via is the ability to metallize the via walls and reduce the thermal resistance of the device by providing an additional path for heat flow from the channel region.

Another effect that can degrade the performance of large transistors is the nonuniform distribution of RF signals. This problem can be alleviated by increasing the signal feedpoints per transistor cell as seen in Fig. 2.8 (Smith *et al.* 1994).

Substrate thickness is also an important issue in power HEMT design. Reduced substrate thickness helps in reducing the thermal resistance of the device and also enables efficient use of ground vias (which reduce source inductance, as was previously discussed). Teeter *et al.* have demonstrated an improvement in power performance of GaAs *p*HEMTs operating at 8 GHz by reducing the substrate thickness and optimizing the source via layout (Teeter *et al.* 1995).

Power requirements of millimeter-wave applications necessitate the use of large gate width devices. However, the increase in device width results in increased circuit losses due to degradation of amplifier bandwidth and low impedance of large devices. Thus it is essential to use proper device layout and fabrication techniques to prevent the degradation of output power density with gate width. Thus, the required output power can be achieved with minimal width devices.

2.6. Material Systems for HEMT Devices

The previous sections of this chapter discussed the various device parameters crucial to high-frequency performance of HEMTs. In this section the relationship between material and device parameters will be discussed. This will enable the selection of the appropriate material system for a particular device application. Table 2.1 illustrates the relationship between the device parameters and material parameters for the various constituent layers of the HEMT, namely the high bandgap donor and buffer layers, and the 2DEG channel. Figure 2.9 shows the schametic diagram of a HEMT, illustrating the material requirements from each component layer.

The first HEMT was implemented in the lattice-matched AlGaAs/GaAs system by Mimura *et al.* in 1981. The AlGaAs/GaAs HEMTs demonstrated significant improvement in low noise and power performance over GaAs MESFETs due to superior electronic transport properties of the 2DEG at the

TABLE 2.1
RELATIONSHIP BETWEEN DEVICE AND MATERIAL PARAMETERS

Device type	Device parameters	Material parameters	
		2DEG channel layer	Barrier/buffer layer
Short gate	High electron velocity	High electron velocity	
Length devices		High electron mobility	
	High aspect ratio		High doping efficiency
Power devices	High current density	High 2DEG density	
	Low gate leakage		High Schottky barrier
	High breakdown voltage	High breakdown field	High breakdown field
	Low output conductance		High quality buffer
	Good charge control	High modulation efficiency	
	Low frequency dispersion		Low trap density
Low-noise devices	Low Rs	High 2DEG density	
	High electron velocity	High electron velocity	
		High electron mobility	
Digital devices	Low gate leakage current		High Schottky barrier
	High current drive	High 2DEG density	

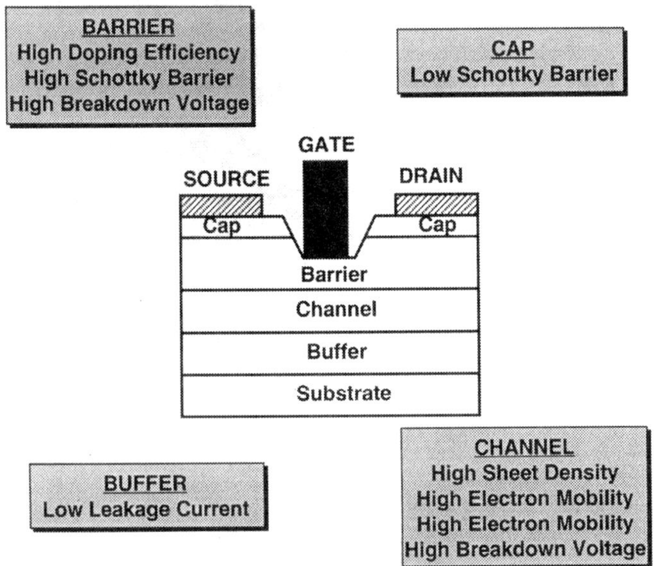

FIG. 2.9. Material requirements for HEMT devices.

AlGaAs/GaAs interface and better scaling properties. However, the limited band discontinuity at the AlGaAs/GaAs interface limits the 2DEG sheet charge density. Other undesirable effects, such as formation of a parasitic MESFET in the donor layer and real space transfer of electrons from the channel to donor, are prevalent however.

One way to increase band discontinuity is to increase the Al composition in AlGaAs. However, the presence of deep level centers (DX centers) associated with Si donors in AlGaAs prevents the use of high Al composition AlGaAs donor layers to increase the band discontinuity and also limits doping efficiency.

Problems relating to low band discontinuity can be solved by reducing the bandgap of the channel and using a material that has higher electron mobility and electron saturation velocity. The first step in this direction was taken by the implementation of an AlGaAs/InGaAs pseudomorphic HEMT (GaAs *p*HEMT) (Ketterson *et al.* 1986). In an AlGaAs/InGaAs *p*HEMT the electron channel consists of a thin layer of narrow bandgap InGaAs that is lattice mismatched to GaAs by 1–2%. The thickness of the InGaAs channel is thin enough (~ 200 Å) so that the mismatch strain is accommodated coherently in the quantum well, resulting in a dislocation free "pseudomorphic" material. However the Indium content in the InGaAs channel can be increased only up to 25%. Beyond this limit the introduction of dislocations due to high lattice mismatch degrades the electronic properties of the channel. The maximum Al composition that can be used in the barrier is 25% and the maximum In composition that can be used in the channel is 25%.

Using the $Al_{0.48}In_{0.52}As/Ga_{0.47}In_{0.53}As$ material system lattice matched to InP can simultaneously solve the limitations of the high bandgap barrier material and the lower bandgap channel material. The AlInAs/GaInAs HEMTs (InP HEMT) has demonstrated excellent, low-noise and power performance that extends well into the millimeter-wave range; they currently hold all the high-frequency performance records for FETs. The GaInAs channel has high electron mobility ($>10,000$ cm^2/Vs at room temperature), high electron saturation velocity (2.6×10^7 cm/s) and higher intervalley (Γ-L) energy separation. The higher conduction band offset at the AllInAs/GaInAs interface ($\Delta E_c = 0.5$ eV) and the higher doping efficiency of AlInAs (compared to AlGaAs) results in a sheet charge density that is twice that of the AlGaAs/InGaAs material system. Higher doping efficiency of AlInAs also enables efficient vertical scaling of short gate length HEMTs. The combination of high sheet charge and electron mobility in the channel results in low source resistance, which is necessary to achieve high transconductance. However, the low bandgap of the InGaAs channel results in low breakdown voltage due to high impact ionization rates. Various approaches have been investigated in recent years to solve this problem; they will be discussed later in this chapter.

TABLE 2.2
MATERIAL PARAMETERS OF AlGaAs/GaAs, AlGaAs/InGaAs AND AlInAs/GaInAs MATERIAL SYSTEMS

Material parameter	AlGaAs/GaAs	AlGaAs/InGaAs	AlInAs/GaInAs
ΔE_c	0.22 eV	0.42 eV	0.51 eV
Maximum donor doping	$5 \times 10^{18}/\mathrm{cm}^3$	$5 \times 10^{18}/\mathrm{cm}^3$	$1 \times 10^{19}/\mathrm{cm}^3$
Sheet charge density	$1 \times 10^{12}/\mathrm{cm}^2$	$1.5 \times 10^{12}/\mathrm{cm}^2$	$3 \times 10^{12}/\mathrm{cm}^2$
Mobility	8000 $\mathrm{cm}^2/\mathrm{Vs}$	6000 $\mathrm{cm}^2/\mathrm{Vs}$	12000 $\mathrm{cm}^2/\mathrm{Vs}$
Peak electron velocity	2×10^7 cm/s		2.7×10^7 cm/s
Γ-L valley separation	0.33 eV		0.5 eV
Schottky barrier	1.0 eV	1.0 eV	0.45 eV

Table 2.2 summarizes the material properties of the three main material systems used for the fabrication of HEMTs.

The emergence of growth techniques like metal organic chemical vapor deposition (MOCVD) and gas source molecular beam epitaxy (GSMBE) and continuing improvement in the existing growth tecniques like molecular beam epitaxy (MBE) enabled a new class of phosphorus-based material systems for fabrication of HEMTs. On the GaAs substrate, the GaInP/InGaAs has emerged as an alternative to the AlGaAs/InGaAs material system. GaInP has a higher bandgap than AlGaAs and hence enables high 2DEG densities due to the increased conduction band discontinuity (ΔE_c) at the GaInP/InGaAs interface. As GaInP has no aluminum it is less susceptible to environmental oxidation. The availability of high selectivity etchants for GaAs and GaInP simplifies device processing. However, the high conduction band discontinuity is achieved only for disordered GaInP, which has a bandgap of 1.9 eV. Using graded GaInP barrier layers and an $In_{0.22}Ga_{0.78}As$ channel, 2DEG density as high as $5 \times 10^{12}/\mathrm{cm}^2$ and a mobility of 6000 $\mathrm{cm}^2/\mathrm{Vs}$ were demonstrated (Pereiaslavets *et al.* 1996).

On InP substrates the InP/InGaAs material system can be used in place of the AlInAs/GaInAs material system. The presence of deep levels and traps in AlInAs degrades the low-frequency noise performance of AlInAs/GaInAs HEMT. Replacing the AlInAs barrier by InP or pseudomorphic InGaP can solve this problem. One disadvantage of using the InP-based barrier is the reduced band discontinuity (0.25 eV compared to 0.5 eV for AlInAs/GaInAs) at the InP/InGaAs interface. This reduces 2DEG density at the interface and modulation efficiency. Increasing the Indium content up to 75% in the InGaAs channel can increase the band discontinuity at the InP/InGaAs interface. The poor Schottky characteristics on InP necessitate the use of higher bandgap InGaP barrier layers or depleted *p*-type InP layers. A sheet density of $3.5 \times 10^{12}/\mathrm{cm}^2$

TABLE 2.3
FREQUENCY BANDS AND MILITARY AND COMMERCIAL APPLICATIONS

Frequency	Military/space	Commercial	Device technology
850 MHz–1.9 GHz		Wireless	Low noise—GaAs *p*HEMT Power—GaAs *p*HEMT
12 GHz (Ku-Band)	Phased array radar	Direct broadcast satellite	Low noise—GaAs pHEMT Power—GaAs *p*HEMT
20 GHz (K-Band)	Satellite downlinks		
27–35 GHz (Ka-Band)	Missile seekers	LMDS—Local multipoint distribution system	Low noise—GaAs *p*HEMT Power—GaAs *p*HEMT
44 GHz (Q-Band)	SATCOM ground terminals	MVDS—Multipoint video distribution system	Low noise—InP HEMT Power—GaAs *p*HEMT
60 GHz (V-Band)	Satellite crosslinks	Wireless LAN	Low noise—InP HEMT Power—GaAs *p*HEMT/InP HEMT
77 GHz		Collision avoidance radar	Low noise—InP HEMT Power—GaAs *p*HEMT/InP HEMT
94 GHz (W-Band)	FMCW radar		Low noise—InP HEMT Power—InP HEMT
100–140 GHz	Radio astronomy		Low noise—InP HEMT
Digital 10 Gb/s		Fiberoptic communication	GaAs *p*HEMT
Digital 40 Gb/s		Fiberoptic communication	InP HEMT

and mobility of 11,400 cm^2/Vs was demonstrated in an $InP/In_{0.75}Ga_{0.25}As/InP$ double heterostructure (Mesquida-Kusters *et al.* 1997).

Despite the large number of material systems available for fabrication of HEMTs, the GaAs *p*HEMT implemented in the $Al_xGa_{1-x}As/In_yGa_{1-y}As$ ($x \sim 0.25$; $y \sim 0.22$) material system and the InP HEMT implemented in the $Al_{0.48}In_{0.52}As/Ga_{0.47}In_{0.53}As$ material system have emerged as industry vehicles for implementation of millimeter-wave analog and ultrahigh-speed digital circuits. Hence the next two sections of this chapter will discuss the various performance aspects of the GaAs *p*HEMTs and InP HEMTs. Table 2.3 lists the various frequency bands and military and space applications in each band with the appropriate technology for each application.

2.7. AlGaAs/InGaAs/GaAs Pseudomorphic HEMT (GaAs *p*HEMT)

The first AlGaAs/InGaAs pseudomorphic HEMT was demonstrated in 1985 (Ketterson *et al.* 1985). Significant performance improvement over AlGaAs/GaAs HEMT was observed. 1-μm gate length devices had peak transconductance of 270 mS/mm and maximum drain current density of 290 mA/mm (Ketterson *et al.* 1986). The current gain cutoff frequency (f_T) was 24.5 GHz and the power gain cutoff frequency (f_{max}) was 40 GHz. An f_T of 120 GHz was reported for 0.2 μm gate length devices with $In_{0.25}Ga_{0.75}As$ channel (Nguyen *et al.* 1988). 0.1-μm gate length devices with a f_{max} of 270 GHz were demonstrated in 1989 (Chao *et al.* 1989).

2.7.1. Millimeter-Wave Power GaAs *p*HEMT

In the past few years, the GaAs *p*HEMT has emerged as a device of choice for implementing microwave and millimeter-wave power amplifiers. To achieve a high output power density, device structures with higher current density and consequently higher sheet charge are required. As the sheet charge density in a single heterojunction AlGaAs/InGaAs *p*HEMT is limited to $2.3 \times 10^{12}/cm^2$, a double heterojunction (DH) device structure has to be used to increase the sheet charge. In DH GaAs *p*HEMT carriers are introduced in the InGaAs channel by doping the AlGaAs barriers on both sides of the InGaAs channel. The AlGaAs barriers are doped with silicon using atomic planar doping to increase electron transfer efficiency. A typical charge density of $3.5 \times 10^{12}/cm^2$ and a mobility of 5000 cm^2/Vs is obtained for a double-heterojunction GaAs *p*HEMT structure. The high sheet charge thus obtained enables higher current drive and power handling capability. Figure 2.10 shows the layer structure of a typical millimeter-

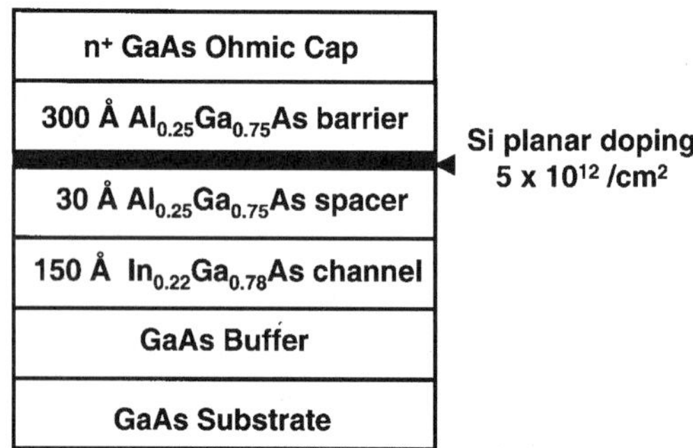

FIG. 2.10. Layer structure of a GaAs power *p*HEMT.

wave power GaAs *p*HEMT. In some cases a doped InGaAs channel is also used to increase sheet charge density (Smith *et al.* 1989) (Streit *et al.* 1991).

Breakdown voltage is an important parameter for power devices. A device with high breakdown voltage can be biased at high drain voltages, which increases the drain efficiency, voltage gain, and power added efficiency (PAE). Typical breakdown voltages of GaAs *p*HEMTs range from 8–15 V. The breakdown mechanism of a GaAs *p*HEMT can be either at the surface in the gate-drain of the device or in the channel (due to impact ionization).

There are several approaches used to increase the breakdown voltage of a GaAs *p*HEMT. The planar doping of AlGaAs barriers (as already described) helps in maintaining a high breakdown voltage, as most of the AlGaAs barrier is undoped. Another approach to increase the breakdown voltage uses a low-temperature grown (LTG) GaAs buffer below the channel. Using this approach, a 45% increase in channel breakdown voltage with a 12% increase in output power was demonstrated (Actis *et al.* 1995). Breakdown is surface-limited for the LTG buffer device whereas for the non-LTG-buffer device breakdown is channel limited.

Using a double recessed gate structure to tailor the electric field in the gate drain depletion region can also increase breakdown voltage. The increase in breakdown voltage is due mainly to reduction in the electric field at the gate edge by surface states in the exposed recess region. Huang *et al.* have investigated the effect of recess length or gate-drain separation L_{gd} on output power, PAE and breakdown voltage of a 0.2-μm gate length GaAs *p*HEMT (Huang *et al.* 1993). The gate-drain feedback capacitance C_{gd} and the output conductance g_{ds} was reduced as L_{gd} was increased from 0.38 μm to 0.98 μm. Although beneficial for high-frequency power performance of the device, it resulted in a decrease in f_T of the device. The decrease in f_T is due to the increase in the gate-drain depletion region, which increases the drain delay in the FET. The breakdown voltage of the devices increased from 12 to 17 V, the gain was constant at 5 dB, whereas the output power decreased from 26 to 20 dBm and PAE decreased from 29% to 12% as L_{gd} was increased from 0.38 μm to 0.978 μm. Thus a trade-off in device performance is involved in this approach.

The output power obtained from a HEMT also depends on the biasing conditions. To achieve high efficiency devices (as in Class B operation), the device is biased near pinch-off and, therefore, high gain is required near pinch-off. The mode of operation is ideally suited for *p*HEMTs, which typically have high transconductance near pinch-off due to their superior charge control properties. The effect for gate bias on the power performance of HEMT has been investigated (Danzilio *et al.* 1992). Higher gain is achieved under Class A conditions. Biasing the device at higher drain voltages can increase output power. Table 2.4 presents a summary of power performance of GaAs *p*HEMTs at various microwave and millimeter-wave frequencies.

TABLE 2.4
SUMMARY OF POWER PERFORMANCE OF GaAs *p*HEMTs

Frequency	Gate length	Gate width	Power density	Output power*	Gain**	PAE*	Device, drain bias (Reference)
12 GHz (Ku-Band)	0.45 μm	1.05 mm	0.77 W/mm	0.81 W	10.0 dB	60%	Double HJ*** $V_{ds} = 7$ V (Matsunaga *et al.* 1995)
	0.25 μm	1.6 mm	1.37 W/mm	2.2 W	14.0 dB	39%	Double HJ*** (Helms *et al.* 1991)
20 GHz (K-Band)	0.25 μm	600 μm	0.51 W/mm	306 mW	7.4 dB	45%	Prematched, $V_{ds} = 7$ V (Yarborough *et al.* 1994)
	0.15 μm	400 μm	1.04 W/mm	416 mW	10.5 dB	63%	LTG Buffer, $V_{ds} = 5.9$ V (Actis *et al.* 1995)
	0.15 μm	600 μm	0.12 W/mm	72 mW	8.6 dB	68%	$V_{ds} = 2$ V (Kao *et al.* 1996)
	0.15 μm	600 μm	0.84 W/mm	501 mW	11 dB	60%	$V_{ds} = 8$ V (Kao *et al.* 1996)
35 GHz (Ku-Band)	0.25 μm	500 μm	0.62 W/mm	310 mW	6.8 dB	40%	Double HJ, $V_{ds} = 5$ V Dow *et al.* 1992)
	0.15 μm	150 μm	0.63 W/mm *0.91 W/mm*	95 mW *137 mW*	*9.0 dB* 7.6 dB	51% 40%	Double HJ (Kao *et al.* 1989)
44 GHz (Q-Band)	0.25 μm		0.79 W/mm		5.1 dB	41%	Doped channel**** (Ferguson *et al.* 1989)
	0.15 μm	400 μm	0.5 W/mm	200 mW	9.0 dB	41%	Double HJ, $V_{ds} = 5$ V (Kasody *et al.* 1995)

	0.2 μm	600 μm	0.53 W/mm	318 mW	5.0 dB	30%	Double HJ, $V_{ds} = 5$ V (Boulais *et al.* 1994)
	0.15 μm	1.8 mm	*0.44 W/mm*	*800 mW*	5.8 dB	25%	Double HJ, $V_{ds} = 5$ V (Smith *et al.* 1994)
55 GHz (V-Band)	0.25 μm	400 μm	*0.46 W/mm*	*184 mW*	4.6 dB	25%	Doped channel, $V_{ds} = 5.5$ V (Tan *et al.* 1991)
	0.2 μm	50 μm	*0.85 W/mm*	*42 mW*	3.3 dB		Doped channel, $V_{ds} = 4.3$ (Saunier *et al.* 1998)
60 GHz (V-Band)	0.15 μm	150 μm	0.83 W/mm	125 mW	4.5 dB	32%	Double HJ
			0.55 W/mm	82 mW	4.7 dB	38%	(Kao *et al.* 1989)
	0.15 μm	400 μm	*0.55 W/mm*	*225 mW*	4.5 dB	25%	Double HJ, $V_{ds} = 5$ V
			0.44 W/mm	174 mW	4.4 dB	*28%*	(Lai *et al.* 1993)
94 GHz (W-Band)	0.25 μm	75 μm	*0.43 W/mm*	32 mW	3.0 dB	15%	Doped channel, $V_{ds} = 4.3$ V
				25 mW	4.0 dB	14%	(Smith *et al.* 1989)
	0.15 μm	150 μm	*0.38 W/mm*	*57 mW*	2.0 dB	16%	Double HJ
			0.30 W/mm	45 mW	*3.0 dB*	16%	(Kao et al. 1989)
	0.1 μm	40 μm	*0.31 W/mm*	*13 mW*	6.0 dB	13%	Doped channel $V_{ds} = 3.4$ V (Streit *et al.* 1991)
	0.1 μm	160 μm	*0.39 W/mm*	*63 mW*	4.0 dB	13%	Doped channel $V_{ds} = 4.3$ V (Streit *et al.* 1991)

**Output Power* in italics indicates device biased for maximum power output.
***PAE/Gain* in italics indicates device biased for maximum PAE/gain.
***Double HJ—Double heterojunction GaAs *p*HEMT.
****Doped channel—doped channel GaAs *p*HEMT.

From Table 2.4 it can be seen that at a given frequency the device power output and gain increases with decrease in gate length due to better high-frequency operation. Reduction in power gain is also observed for wider devices. This is due to the increase in source inductance, which increases with gate width and frequency and is due to the increase in gate resistance as a square of gate width. Low inductance via hole source grounding and proper gate layout is required to reduce these parasitics.

Reliability is important for space applications, typically a mean-time-to-failure (MTTF) of 10^7 h (1142 yr) is required for space applications. The GaAs *p*HEMTs have demonstrated MTTF of 1×10^7 h at a channel temperature of 125 °C. The main failure mechanism is the atmospheric oxidation of the exposed AlGaAs barrier layers and interdiffusion of the gate metallization with the AlGaAs barrier layers (gate sinking) (Chen *et al.* 1994). Using dielectric passivation layers to reduce the oxidation of AlGaAs can solve these problems. Using a refractory metal for gate contacts will minimize their interaction with the AlGaAs barrier layers. MTTF of 1.5×10^7 h at a channel temperature of 150 °C was achieved using molybdenum-based gate contact (Hori *et al.* 1995).

Traveling wave tubes (TWTs) have been traditionally used as multiwatt power sources for microwave applications up to K-band (20 GHz). Using GaAs *p*HEMTs in place of TWTs for these applications has many advantages, including lower cost, smaller size, smaller weight, and higher reliability. However, the typical power density of a GaAs *p*HEMT at 20 GHz is of the order of 1 W/mm. Hence the output power from a large number of devices has to be combined. To minimize combining losses, it is desirable to maximize power output of a single device. The output power density of a transistor is mainly decided by the intrinsic device parameters. When large devices (gate width of the order of mm) are used to increase the total power output, other factors such as device layout, signal distribution, and substrate thickness are of critical importance. The design of input signal distribution and output power combining networks is also crucial.

Several multiwatt GaAs *p*HEMT power modules recently have been demonstrated. A power module with 9.72-mm wide GaAs *p*HEMTs that delivered an output power of 4.7 W in the 18–21.2-GHz band with a PAE of 38% was demonstrated (Kraemer *et al.* 1994). Table 2.5 summarizes the recent results of multiwatt GaAs *p*HEMT power modulues.

2.7.2. LOW-NOISE GaAs *p*HEMTs

Figure 2.11 shows the structure of a generic low-noise GaAs *p*HEMT. As the drain current requirement for a low-noise bias is low, single-side doped heterojunctions are sufficient for low-noise devices. The emphasis here is on achieving higher mobility to reduce the parasitic source resistance.

TABLE 2.5
SUMMARY OF POWER PERFORMANCE OF MULTIWATT *p*HEMT POWER MODULES

Width	Frequency	Power	Gain	PAE	Reference
16.8 mm	2.45 GHz	10.0 W	13.5 dB	63.0%	$V_{ds} = 7$ V (Aucoin *et al.* 1993)
24 mm	2.45 GHz	11.7 W	14.0 dB	58.2%	$V_{ds} = 8$ V (Bouthillette *et al.* 1994)
32.4 mm	8.5–10.5 GHz	12.0 W	7.2 dB	40.0%	$V_{ds} = 7$ V (Kraemer *et al.* 1994)
16.8 mm	12 GHz	12.0 W	10.1 dB	48.0%	(Matsunaga *et al.* 1995)
25.2 mm	12 GHz	15.8 W	9.6 dB	36.0%	$V_{ds} = 9$ V (Matsunaga *et al.* 1996)
8 mm	12 GHz	6 W	10.8 dB	53.0%	$V_{ds} = 9$ V (Fu *et al.* 1994)
9.72 mm	18–21.2 GHz	4.7 W	7.5 dB	41.4%	$V_{ds} = 5.5$ V (Kraemer *et al.* 1994)

V_{ds} = drain bias for power measurements.

As already discussed, the AlInAs/GaInAs material system is the ideal choice for fabrication of low-noise microwave and millimeter-wave devices. However GaAs *p*HEMTs also find significant use in a millimeter-wave low-noise applications due to wafer size-, cost-, and process maturity-related advantages.

Henderson *et al.* first reported on the low noise performance of GaAs *p*HEMTs in 1986. Devices with 0.25-μm gate length had a noise figure of 2.4 dB and an associated gain of 4.4 dB at 62 GHz (Henderson *et al.* 1986). A 0.15-μm gate length $Al_{0.25}Ga_{0.75}As/In_{0.28}Ga_{0.72}As$ *p*HEMT with a noise figure of 1.5 dB and an associated gain of 6.1 dB at 61.5 GHz was demonstrated in 1991. (Tan *et al.* 1991). The reduction in noise figure is a direct result of reducing the gate length, which increases the f_T of the device.

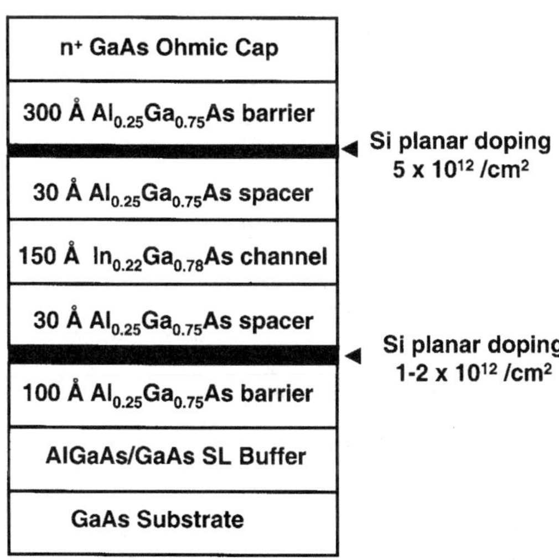

FIG. 2.11. Layer structure of a low-noise GaAs pHEMT.

Low-noise operation of a GaAs *p*HEMT was also demonstrated at 94 GHz (Chao *et al.* 1989). For a 0.1-μm gate length device a noise figure of 3.0 dB and associated gain of 5.1 dB was achieved. A noise figure of 2.1 dB and an associated gain of 6.3 dB was reported for a 0.1-μm gate length GaAs *p*HEMT at 94 GHz (Tan *et al.* 1990). The improvement in noise figure is attributed to the use of a T-shaped gate with end-to-end resistance of 160 Ω/mm by Tan *et al.*, compared to a trapezoidal gate with end-to-end resistance of 1700 Ω/mm used by Chao *et al.*. This further emphasizes the need to reduce parasitic resistances in low-noise devices.

One of the main system applications of low-noise GaAs *p*HEMTs is satellite direct broadcasting receiver systems (DBS) that are in increasing demand worldwide. Low-noise amplifiers operating at 12 GHz are a critical component in these systems. The low-noise performance of GaAs *p*HEMTs is more than adequate for these applications. A 0.25-μm gate length GaAs *p*HEMT with a noise figure of 0.6 dB and an associated gain of 11.3 dB at 12 GHz was reported by Tokue *et al.* (Tokue *et al.* 1991). Performance coupled with low cost packaging is one of the crucial factors in the high volume DBS market. Hwang *et al.* have demonstrated a 0.2-μm gate length GaAs *p*HEMTs in plastic packaging with a 1.0 dB noise figure and 9.9 dB associated gain at 12 GHz (Hwang *et al.* 1996). Recently a plastic packaged GaAs *p*HEMT device with a gate length of 0.17 μm demonstrated a noise figure of 0.35 dB, and 12.5 dB associated gain at 12 GHz (Hirokawa *et al.* 1996). Parasitic gate and source resistances are one of the most important factors determining the noise performance of a HEMT. Significant reduction in noise figure was achieved when a wide-head (1.35 μm) T-gate with a footprint of 0.13 μm was used. The noise figure at 12 GHz was 0.31 dB with an associated gain of 10.2 dB. The noise figure at 18 GHz was 0.45 dB (Lee *et al.* 1995). Table 2.6 summarizes the low-noise

TABLE 2.6
SUMMARY OF LOW-NOISE PERFORMANCE OF GaAs *p*HEMTs

Gate length	Frequency	F_{min} (dB)	Ga (dB)	Reference	Comments
0.25 μm	12 GHz	0.6 dB	11.3 dB	(Tokue *et al.* 1991)	
0.17 μm	12 GHz	0.35	12.5 dB	(Hirokawa *et al.* 1996)	Packaged device
0.25 μm	18 GHz	0.9 dB	10.4 dB	(Henderson *et al.* 1986)	
0.25 μm	62 GHz	2.4 dB	4.4 dB	(Henderson *et al.* 1986)	Improvement by reduction in L_g
0.15 μm	60 GHz	1.5 dB	6.1 dB	(Tan *et al.* 1991)	
0.10 μm	94 GHz	3.0 dB	5.1 dB	(Chao *et al.* 1989)	Improvement by reduction in R_g
0.10 μm	94 GHz	2.1 dB	6.3 dB	(Tan *et al.* 1990)	

performance of GaAs *p*HEMTs at various microwave and millimeter-wave frequencies.

2.7.3. GaAs *p*HEMTs FOR WIRELESS APPLICATIONS

The explosive growth of the wireless communication industry has opened up a new area of application for GaAs *p*HEMTs. Unlike the millimeter-wave military and space applications, the frequencies of operation of these applications are much lower. The frequencies used in typical cellular phones range from 850 MHz for the American Mobile Phone System (AMPS) to 1.9 GHz for the Japanese Personal Handy Phone System (PHS) and the Digital European Cordless Telephone (DECT). The following device parameters are of interest when considering device technologies for wireless applications (Halchin *et al.* 1997).

2.7.3.1. Operating Voltages

The battery dominates the size and weight of portable wireless phones in most cases. Typical cellular phones are powered by a single lithium ion battery that has a voltage of 3.6 V. Hence high output power and power-added efficiency have to be achieved at a drain bias less than 4 V. However, cost reduction can be achieved by using two 1.2-V Ni-H batteries. Hence power performance needs to be achieved at a drain bias of 2 V.

Most III–V semiconductor-based FETs have negative threshold voltages. This requires the addition of two additional circuits in the wireless handset. The first is a switch to cutoff the current to the transmitter power amplifier when the handset is in idle or receive mode. The other is a negative voltage generator to bias a power amplifier FET. Negative voltage production is one of the largest problems in the realization of very compact and low-cost handsets. Using enhancement mode devices eliminates the negative supply voltage generator and power-cutoff switch. The high gate turn-on voltage of an enhancement mode GaAs *p*HEMT, as compared to the enhancement mode GaAs MESFET, enables higher input voltage swing.

2.7.3.2. Power Added Efficiency and Power Density

The power amplifier in the RF transmitter block of a cellular phone consumes the most battery power. A device with higher power added efficiency (PAE) delivers the same amount of output power with less battery power consumption. Due to the finite on-resistance of the device, it is difficult to achieve high (PAE) at low bias voltages.

Higher power density in a device minimizes the device width required to achieve the desired output power. This is desired because wider gate width reduces yield and makes output matching difficult. The advent of digital cellular phones further increased the output power density requirements on power amplifiers. For a typical power density, the gate widths required for analog phones are of the order of 5 mm, whereas digital phones require devices with gate widths of the order of 17.5 mm.

The advent of newer direct-to-satelite systems such as Motorola IRIDIUM^{TM} imposes higher output power requirements on the power amplifier. A new specification to be satisfied is the noise figure at the receiving end of the phone as low-level satellite signals have to be acquired.

2.7.3.3. Linearity Requirements

Advanced digital wireless communication systems like Code Division Multiple Access (CDMA) use $\pi/4$-shifted quadrature phase shift keying (QPSK) modulation. In the QPSK modulation scheme, the signal information is contained in the amplitude and phase of the waveform. Hence high linearity is required in power amplification to minimize signal distortion. In a power amplifier, maximum power and power-added efficiency are achieved under bias conditions that result in nonlinear amplification. Therefore, it is critical to achieve high power output and PAE under linear amplification conditions. This problem is solved by circuit techniques that reduce signal distortion under nonlinear conditions by appropriate termination of higher harmonic frequencies in the signal.

Nonlinear amplification of a signal results in generation of spurious signals at higher frequencies that can occupy other channels (Inosako *et al.* 1995). Hence it is necessary to minimize power leakage into adjacent channels as this results in crosstalk. The spectral "spillover" of power into adjacent channels is denoted as Adjacent Channel Power Ratio (ACPR). The ACPR can be minimized having an amplifier with a linear amplification characteristic. As the transconductance of a FET varies with the gate bias point, selection of bias point also affects the ACPR. To minimize standoff current and maximize PAE, the FET is typically biased close to the pinch-off voltage. Hence, to minimize ACPR and maximize PAE and gain, the transconductance of the FET should rise rapidly just after pinch-off. This implies that the FET should have good charge control close to the pinch-off voltage. This is easier to achieve in a heterostructure device such as the GaAs *p*HEMT than in a GaAs MESFET. Figure 2.12 explains these concepts.

2.7.3.4. Gate Leakage Current

As already discussed, to compensate for the low-power at low bias voltages, the gate width of the FET has to be increased. The wider gate width also results in an

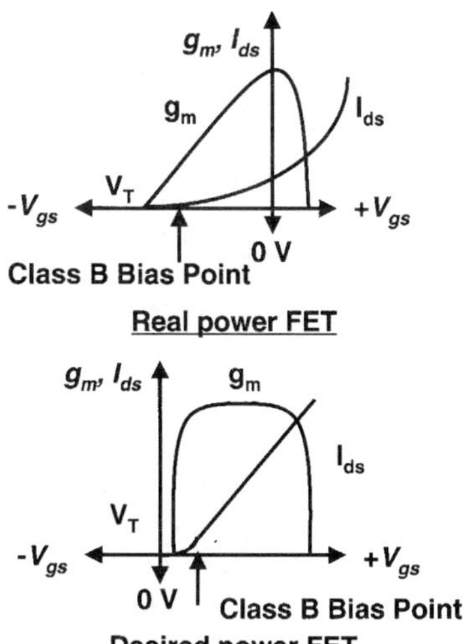

FIG. 2.12. Real and ideal transfer characteristics of FETs for wireless applications.

increase in the total gate current, resulting in an unstable setting in the quiescent gate bias point. Therefore, using a device with higher power density helps in reducing gate current as the gate width is reduced. It was demonstrated that the ACPR of a power amplifier increases with the gate leakage current (Masato *et al.* 1995). This can be achieved by increasing the Schottky barrier height at gate contact. In this area, the GaAs *p*HEMT has the advantage over the GaAs MESFET.

2.7.3.5. Power Amplifier Specifications for Wireless Applications

In a wireless hansdset, most of the baseband electronics and the low noise receiver section can be implemented using silicon bipolar or silicon MOS technologies, both of which are cost effective. However, these technologies have yet to satisfy the specifications for the high-power transmitter section. The high-power transmitter section in a wireless handset is the most crucial component of the whole system, as it decides both "talk-time" and base station spacing. The GaAs *p*HEMT, which has proven its superior power amplification characteristics at microwave and millimeter-wave frequencies, is an ideal candidate for use in power amplifiers for wireless handsets. Table 2.7 summarizes the requirements of wireless handset power amplifiers for analog and digital cellular phone systems and the corresponding device parameters. Table 2.8 summarizes the performance specifications for wireless handset power amplifiers in both the North American and Japanese cellular phone systems (Mitama 1995).

TABLE 2.7
DEVICE REQUIREMENTS FOR WIRELESS HANDSET POWER AMPLIFIERS

Cellular system	System requirement	Device requirement
Digital	High linearity	Good charge control
(CDMA–USA)	High PAE	Low knee voltage, high gain
(PDC, PHS—Japan)	High output power	High current density
(DECT—Europe)		
Analog	High power	High current density
(AMPS—USA)	High PAE	Low knee voltage, high gain
Digital and analog	Single supply voltage	Enhancement mode device
	Reduced battery size	Low-voltage operation

TABLE 2.8
PERFORMANCE SPECIFICATIONS FOR WIRELESS HANDSET POWER AMPLIFIERS

	AMPS	CDMA (J-STD-019)	Japanese PDC	Japanese PHS
Frequency	836.5 MHz	836.5 MHz	900 MHz	1.9 GHz
VDD	3.3 V	3.3 V	3.4 V	2.2 V
Output power	31 dBm	33 dBm	31.2 dBm	22 dBm
Gain		30 dB	25 dB	
PAE	55%	35	43%	>43%
ACPR	−44 dBc (2nd harmonic)	−33 dBc (885 kHz)	−45 dBc (50 kHz)	−50 dBc (600 kHz)
	−40 dBc (3rd harmonic)	−45 dBc (1.98 GHz)	−60 dBc (100 kHz)	−55 dBc (900 kHz)

2.7.3.6. Power Performance of GaAs pHEMTs for Wireless Applications

The power performance of MESFETs and GaAs *p*HEMTs for wireless applications has been compared (Ota *et al.* 1994) For the same saturation drain current density, at a frequency of 950 MHz, the saturated power output from the *p*HEMT is 2.5 W, whereas it is 1.8 W from the MESFET. The power-added efficiency of the *p*HEMT is 68%, which is 8% higher than that of the MESFET. This difference is due to the transfer characteristics of the two devices. The *p*HEMT performs as a better power amplifier than the MESFET because the input power is effectively amplified with higher gm near the pinch-off voltage. This is a direct consequence of better charge control properties of the HEMT when compared to the MESFET. The *p*HEMT also has a lower gate leakage current than the MESFET due to a higher Schottky barrier on AlGaAs.

The limitations of the AlGaAs/InGaAs material system can be overcome by using the AlGaInP/InGaAs material system. The higher band discontinuity of the AlGaInP/InGaAs material system enables higher sheet charge and consequently higher current drive. The higher bandgap of AlGaInP results in an enhanced Schottky barrier, which reduce the gate leakage current. A 1-μm gate length $In_{0.5}(Al_{0.3}Ga_{0.7})_{0.5}P/In_{0.2}Ga_{0.8}As$ double heterojunction *p*HEMT demonstrated an output power of 24 dBm and a power efficiency of 65% at a drain bias voltage of 1.2 V (Wang *et al.* 1998).

The power performance of enhancement mode GaAs *p*HEMT with a threshold voltage of 0.05 V for wireless applications has also been investigated (Kunihisa *et al.* 1997). A device with a gate width of 3.2 mm delivered an output power of 22 dBm with power-added efficiency of 41.7%. The standby current at a gate bias of 0 V was 150 μA. Enhancement mode GaAs HFET with a higher threshold voltage of 0.5 V has also been demonstrated (Glass *et al.* 1997). A 1-μm gate length device with a gate width of 12 mm delivered an output power of 31.5 dBm with a PAE of 75% at 850 MHz and a drain bias of 3.5 V. The standby current at a gate bias of 0 V was 1 μA. This eliminates the need for a switch in the drain current of the power amplifier. The device was manufactured using Motorola's CGaAs™ process, which is cost effective as it uses processes that are similar to standard silicon MOS and bipolar processes. A 0.7-μm gate length device manufactured using the same process delivered an output power of 30 dBm with a PAE of 50% at a drain bias of 3.5 V. The ACPR was −30 dBc.

Table 2.9 shows a summary of power performance of GaAs *p*HEMTs for cellular phones.

2.8. AlInAs/GaInAs/InP (InP HEMT)

Future military and commercial electronic applications will require high-performance microwave and millimeter-wave devices. Important applications include low-noise amplifiers for receiver front ends, power amplifiers for phased-array radars, ultrahigh-speed digital circuits for prescalers, and MUX/DEMUX electronics for high-speed (> 40 Gb/s) optical links.

A HEMT device capable of operating at millimeter-wave frequency requires a channel with high electron velocity, and high current density and minimal parasitics. As discussed, the $Al_{0.48}In_{0.52}As/Ga_{0.47}In_{0.53}As$ material system lattice matched to InP satisfies these criteria. The 1-μm gate length AlInAs/GaInAs HEMT with extrinsic transconductances as high as 400 mS/mm was demonstrated (Hirose *et al.* 1985). The microwave performance of 1-μm gate-length devices showed an improvement of 20–30% over the AlGaAs/GaAs HEMT (Palamateer *et al.* 1987). The superior electron properties of the AlInAs/GaInAs material system are further evident in submicron gate length (0.3 μm) HEMTs, which exhibited a current gain cut-off frequency (f_T) > 80 GHz (Mishra *et al.* 1988). The maximum drain saturation current in these devices is 700 mA/mm.

TABLE 2.9
SUMMARY OF POWER PERFORMANCE OF GaAs *p*HEMTs FOR CELLULAR PHONES

Frequency	Device width	Drain bias	Power output	PAE	ACPL	Device/Reference
850 MHz	5 mm	1.2 V	19.6 dBm	65.2%		AlGaInP/InGaAs *p*HEMT (Wang *et al.* 1998)
	10 mm	1.3 V	21.5 dBm	57.4%		InGaP/InGaAs *p*HEMT (Ren *et al.* 1997)
	12 mm	3.5 V	31.5 dBm	75%		Enhancement mode HFET (Glass *et al.* 1997)
	12 mm	3.5 V	33.1 dBm	84.8%		GaAs *p*HEMT (Martinez *et al.* 1996)
	30 mm	3.7 V	31.0 dBm	59.0%	−30.1 dBc © 30 kHz	1 μm GaAs MESFET (Masato *et al.* 1995)
900 MHz	12 mm	3.5 V	31.5 dBm	75%		Enhancement mode CGaAs^{TM} (Huang *et al.* 1997)
	21 mm	2.3 V	31.3 dBm	68%		0.8 μm MESFET (Lee *et al.* 1996)
	14 mm	3.0 V	32.3 dBm	71%		*p*HEMT (Inosako *et al.* 1994)
	40 mm	1.5 V	31.5 dBm	65%		MESFET (Tanaka *et al.* 1995)
950 MHz	28 mm	1.2 V	1.1 W	54%		*p*HEMT (Inosako *et al.* 1995)

	21 mm	2.2 V	32.7 dBm	62.8%	−50.5 dBc	*p*HEMT (Iwata *et al.* 1995)
	12 mm	3.0 V	1.4 W	60.0%		*p*HEMT (Iwata *et al.* 1993)
	7 mm	3.4 V	30.9 dBm	56.3%	−51.5 dBc @ 50 kHz	*p*HEMT (Iwata *et al.* 1996)
	16 mm	3.4 V	1.42 W	60.0%	−48.2 dBc @ 50 kHz	*p*HEMT (Bito *et al.* 1998)
	12 mm	4.7 V	2.5 W	68.0%		*p*HEMT (Ota *et al.* 1994)
	12 mm	4.7 V	1.8 W	60.0%		MESFET (Ota *et al.* 1994)
1.9 GHz	1 mm	2.0 V	20.2 dBm	45.3%	55.2 dBc @ 600 kHz	*p*HEMT (Lai *et al.* 1997)
	2.4 mm	2.0 V	21.1 dBm	54.4%	−55 dBc @ 600 kHz	MESFET (Choumei *et al.* 1998)
	3.2 mm	3.0 V	22.0 dBm	41.7%	−58.2 dBc @ 600 kHz	Enhancement mode *p*HEMT (Kunihisa *et al.* 1997)
	12 mm	3.5 V	30 dBm	50%	−30 dBc	Enhancement mode *p*HEMT (Glass *et al.* 1997)
	5 mm	2.0 V	25.0 dBm	53.0%		AlGaInP/InGaAs *p*HEMT (Wang *et al.* 1998)

Decreasing the gate length to 0.1 μm further increases the f_T to 170 GHz (Mishra *et al.* 1988). Using a T-gate to self-align the source and drain contacts results in reduction of source-gate and source drain spacing. This not only reduces the parasitic source and drain resistances but also the drain delay. Using the preceding technique, an f_T of 250 GHz was achieved in a 0.13-μm gate length self-aligned HEMT (Mishra *et al.* 1989). A 0.07 μm AlInAs/GaInAs HEMT with an f_T of 300 GHz and an f_{max} of 400 GHz was reported (Suemitsu *et al.* 1998).

The high-frequency performance of the InP HEMT can be further improved by using a pseudomorphic InGaAs channel with an indium content as high as 80%. The f_T of a 0.1-μm InP HEMT increased from 175 to 205 GHz when the Indium content in the channel was increased from 53 to 62% (Mishra *et al.* 1988). Although devices with high indium content channels have low breakdown voltages, they are ideal for low noise applications and ultrahigh-speed digital applications. An f_T of 340 GHz was achieved, a 0.05-μm gate length pseudomorphic InP HEMT with a composite $In_{0.8}Ga_{0.2}As/In_{0.53}Ga_{0.47}As$ channel (Nguyen *et al.* 1992). This is the highest reported f_T of any 3-terminal device.

Compared to the GaAs *p*HEMTs, AlInAs/GaInAs HEMTs has a higher current density that makes it suitable for ultrahigh-speed digital applications. The high current gain cutoff frequency and low parasitics makes the AlInAs/GaInAs HEMT the most suitable choice for low-noise applications extending well beyond the 100 GHz. The high current density and superior high-frequency performance can be utilized for high-performance millimeter-wave power applications provided the breakdown voltage is improved. Some state-of-the-art millimeter-wave analog circuits and ultrahigh-speed digital circuits have been implemented using InP HEMTs. A low-noise amplifier with 12 dB gain at a frequency of 155 GHz using a 0.1-μm InP HEMT with a $In_{0.65}Ga_{0.35}As$ pseudomorphic channel was demonstrated (Lai *et al.* 1997). An amplifier with 5 dB gain at 184 GHz using a 0.1 μm gate $In_{0.8}Ga_{0.2}As/InP$ composite channel HEMT was demonstrated. This is the highest frequency solid-state amplifier reported to date (Pobanz *et al.* 1998).

This section begins with a review of AlInAs/GaInAs, for low-noise applications. This will be followed by a review of AlInAs/GaInAs power HEMTs for millimeter-wave applications. A potential high-volume commercial application of AlInAs/GaInAs HEMTs is ultrahigh-speed digital circuits for use in electronic front ends of Gigabit optical communication systems. Device issues pertinent to digital circuits (e.g., nonalloyed contacts and enhancement mode devices) will be discussed. The AlInAs/GaInAs HEMT technology has been supported by high-end military and space applications where cost is not an issue. Manufacturability issues have to be taken into consideration to make the AlInAs/GaInAs HEMT technology viable for insertion in high-volume commercial applications. This section will therefore end with a discussion of these issues.

2.8.1. LOW-NOISE AlInAs/GaInAs HEMTs

AlInAs/GaInAs HEMTs have emerged as premier devices for microwave and millimeter-wave low-noise applications. The superior electronic properties of the GaInAs channel enable fabrication of extremely high f_T and f_{max} devices. The superior carrier confinement at the AlInAs/GaInAs interface results in a highly linear transfer characteristic. High transconductance is also maintained very close to pinch-off. This is essential because the noise contribution of the FET is minimized at low drain current levels. Hence high gain can be achieved at millimeter-wave frequencies under low-noise bias conditions. The high mobility at the AlInAs/GaInAs interface also results in reduced parasitic source resistance of the device. AlInAs/GaInAs HEMTs with 0.25-μm gate length exhibited a noise figure of 1.2 dB at 58 GHz (Ho *et al.* 1988). At 95 GHz a noise figure of 1.4 dB with associated gain of 6.6 dB was achieved in a 0.15-μm gate length device (Chao *et al.* 1990). At any given frequency, the InP HEMT has a noise figure that is about 1 dB lower than the GaAs *p*HEMT. Table 2.10 summarizes the low-noise performance of AlInAs/GaInAs HEMTs.

2.8.2. MILLIMETER-WAVE AlInAs/GaInAs POWER HEMTs

The millimeter-wave power capability of single heterojunction AlInAs/GaInAs HEMTs has been demonstrated (Kao *et al.* 1991; Matloubian *et al.* 1991). The requirements for power HEMTs as discussed in Section 2.5 are high gain, high current density, high breakdown voltage, low access resistance, and low knee voltage to increase power output and power-added efficiency. The AlInAs/GaInAs HEMTs satisifes all of these requirements with the exception

TABLE 2.10
SUMMARY OF LOW-NOISE PERFORMANCE OF AlInAs/GaInAs HEMTs

Gate length	Frequency	F_{min} (dB)	G_a (dB)	Comments/Reference
0.15 μm	12 GHz	0.39 dB	16.5 dB	$In_{0.7}Ga_{0.3}As$ channel (Onda *et al.* 1993)
0.25 μm	18 GHz	0.5 dB	15.2 dB	(Ho *et al.* 1988)
0.15 μm	18 GHz	0.3 dB	17.2 dB	(Chao *et al.* 1990)
0.18 μm	26 GHz	0.43 dB	8.5 dB	Passivated device (Umeda *et al.* 1992)
0.25 μm	57 GHz	1.2 dB	8.5 dB	(Ho *et al.* 1988)
0.1 μm	60 GHz	0.8 dB	8.9 dB	(Duh *et al.* 1991)
0.1 μm	63 GHz	0.8 dB	7.6 dB	Passivated (Kao *et al.* 1994)
		0.7 dB	8.6 dB	Unpassivated
0.15 μm	94 GHz	1.4 dB	6.6 dB	(Chao *et al.* 1990)
0.1 μm	94 GHz	1.2 dB	7.2 dB	(Duh *et al.* 1991)

of breakdown voltage. This limitation can be overcome by operating at a lower drain bias. In fact, the high gain and PAE characteristics of InP HEMTs at low drain bias voltages make them ideal candidates for battery-powered applications (Larson *et al.* 1993). Another advantage is the use of InP substrate that has a 40% higher thermal conductivity than GaAs. This allows higher dissipated power per unit area of the device or lower operating temperature for the same power dissipation. As low breakdown voltage is a major factor that limits the power performance of InP HEMTs, this section will discuss in detail the various approaches used to increase breakdown voltage.

Breakdown in InP HEMTs is a combination of electron injection from the gate contact and impact ionization in the channel (Bahl *et al.* 1995). The breakdown mechanism in the off-state (when the device is pinched-off) is electron injection from the gate. It is also dependent on the sheet carrier concentration in the gate-drain region. These injected hot electrons cause impact ionization in the high-field drain end of the GaInAs channel. Impact ionization is the main mechanism that determines the on-state breakdown. Some of the holes generated by impact ionization are collected by the negatively biased gate and result in increased gate leakage. The potential at the source end of the channel is modulated by holes collected by the source. This results in increased output conductance.

One of the two main reasons for low breakdown voltage is the low Schottky barrier height of AlInAs that results in increased electron injection from the gate and, consequently, higher gate leakage current. The other is the high impact ionization rate in the low bandgap GaInAs channel. Together this results in a low breakdown voltage and high output conductance, as well as degradation of the $f_{\max}$ of the device. Various approaches have been investigated to improve breakdown voltage and reduce gate leakage current and impact ionization in AlInAs/GaInAs HEMTs. These are discussed in what follows.

2.8.2.1. Reduction in Electric Field in the Gate Drain Region

The electric field in the gate drain depletion was lowered by using a double recess process, which increases the breakdown voltage from 9 to 16 V (Boos *et al.* 1991). A gate-drain breakdown voltage of 11.2 V was demonstrated for 0.15-μm gate length devices with a 0.6-μm recess width (Hur *et al.* 1995). In addition, reduction in output conductance (g_{ds}) and gate-drain feedback capacitance (C_{gd}) was observed when compared to single recessed devices. The $f_{\max}$ of a double recessed device increased from 200 to 300 GHz (Hur *et al.* 1995). Hence it is desirable for power devices.

Another approach to reduce electric field in the gate-drain region is to use an undoped GaInAs cap instead of a doped GaInAs cap (Pao *et al.* 1990). The output conductance can be reduced from 50 to 20 mS/mm for a 0.15-μm gate length device by replacing the doped GaInAs cap by an undoped cap (Ho *et al.* 1991).

This also improved the breakdown voltage from 5 to 10 V. The reduction in C_{gd} and g_{ds} resulted in an f_{max}, as high as 455 GHz.

Redistributing the dopants in the AlInAs barrier layers can also increase breakdown voltage. An increase in breakdown voltage from 4 to 9 V is achieved by reducing doping in the top AlInAs barrier layer and transferring it to the AlInAs barrier layers below the channel (Matloubian *et al.* 1991).

2.8.2.2. Reduction in Gate Leakage Current

The gate leakage current can be reduced and the breakdown voltage can be increased by using a higher bandgap strained AlInAs barrier (Matloubian *et al.* 1993). By increasing the Al composition in the barrier layers from 48 to 70%, the gate-to-drain breakdown voltage was increased from 4 to 7 V. This also results in reduction of gate leakage as the Schottky barrier height increses from 0.5 to 0.8 eV. The use of $Al_{0.25}In_{0.75}P$ as a Schottky barrier improves the breakdown voltage from −6 to −12 V (Brown *et al.* 1994).

The on-state breakdown can be improved in two ways. The first is to reduce the gate leakage current by the impact ionization generated holes by increasing the barrier height for holes. This was achieved by increasing the valence band discontinuity at the channel-barrier interface. The use of a strained 25-Å $In_{0.5}Ga_{0.5}P$ spacer instead of AlInAs increases the valence band discontinuity at the interface from 0.2 to 0.37 eV. An on-state breakdown voltage of 8 V at a drain current density of 400 mA/mm for a 0.7-μm gate length InP HEMT was achieved by using a strained InGaP barrier (Scheffer *et al.* 1994).

The various approaches to increase the breakdown voltage, as already discussed here, concentrate mainly on reducing the electron injection from the Schottky gate and reducing the gate leakage current. These approaches also have their inherent disadvantage as Al-rich barriers result in high source resistance and are more susceptible to atmospheric oxidation. Additionally these approaches to not address the problem of high-impact ionization rate in the GaInAs channel and carrier injection from contacts. In the recent past, various new approaches have been investigated to increase breakdown voltage without compromising the source resistance or atmospheric stability of the device. These include the junction-modulated AlInAs/GaInAs HEMT (JHEMT), the composite GaInAs/InP channel HEMT, and the use of regrown contacts. These approaches will be discussed in detail in section 2.8.3.

Despite the low breakdown voltage, InP HEMTs have demonstrated output power capabilities comparable to GaAs *p*HEMTs. Comparable to GaAs *p*HEMTs for a given power output, InP HEMTs have higher PAE and gain. The potential of InP HEMTs as millimeter-wave power devices is evident in the fact that comparable power performance is achieved at drain biases 2–3 V lower than those for GaAs *p*HEMTs. Table 2.11 summarizes the power performance of AlInAs/GaInAs HEMTs.

TABLE 2.11
SUMMARY OF POWER PERFORMANCE OF InP HEMTs

Frequency	Gate length	Gate width	Power density	Power output	Gain	PAE	Device/drain bias (Reference)
4 GHz	0.5 μm	2 mm	0.13 W/mm	269 mW	18 dB	66%	$V_{ds} = 2.5$ V (Larson *et al.* 1993)
	0.15 μm	0.8 mm	0.4 W/mm	320 mW	18 dB	57%	$V_{ds} = 3$ V (Larson *et al.* 1993)
12 GHz	0.22 μm	150 μm	0.78 W/mm	117 mW	8.4 dB	47%	$V_{ds} = 4$ V (Matloubian *et al.* 1991)
			0.47 W/mm	70 mW	11.3 dB	59%	$V_{ds} = 3$ V, Double HJ
18 GHz	0.15 μm	600 μm	0.74 W/mm	446 mW	13 dB	59%	Double recessed $V_{ds} = 7$ V (Hur *et al.* 1997)
20 GHz (K-Band)	0.15 μm	50 μm	0.78 W/mm	39 mW	10.2 dB	44%	$V_{ds} = 4.9$ V (Kao *et al.* 1991)
			0.41 W/mm	21 mW	10.5 dB	52%	$V_{ds} = 2.5$ V Single Heterojunction
	0.15 μm	50 μm	0.61 W/mm	30 mW	12.2 dB	44%	Single heterojunction $In_{0.69}Ga_{0.31}As$ channel $V_{ds} = 4.1$ V (Kao *et al.* 1991)
	0.15 μm	800 μm	0.65 W/mm	516 mW	7.1 dB	47%	70% AlInAs, $V_{ds} = 4$ V (Matloubian *et al.* 1993)
44 GHz (Q-Band)	0.15 μm	450 μm	0.55 W/mm	251 mW	8.5 dB	33%	$Al_{0.6}In_{0.4}As$ barrier
			0.88 W/mm	398 mW	6.7 dB	30%	Doped channel (Matloubian *et al.* 1993)

	0.2 μm	600 μm	0.37 W/mm	225 mW	5 dB	39%	$Al_{0.6}In_{0.4}As$ barrier Single heterojunction, $V_{ds} = 4$ V (Hur *et al.* 1995)
57 GHz (V-Band)	0.22 μm	450 μm	0.33 W/mm	150 mW	3.6 dB	20%	$Al_{0.6}In_{0.4}As$ barrier
			0.44 W/mm	200 mW		17%	Doped channel, $V_{ds} = 3.5$ V (Matloubian *et al.* 1993)
60 GHz (V-Band)	0.15 μm	50 μm	0.35 W/mm		7.2 dB	41%	$V_{ds} = 2.6$ V (Kao *et al.* 1991)
			0.52 W/mm	26 mW	5.9 dB	33%	$V_{ds} = 3.6$ V Single Heterojunction
	0.1 μm	50 μm	0.30 W/mm	15 mW	8.6 dB	49%	$V_{ds} = 3.35$ V
			0.41 W/mm	21 mW	8.0 dB	45%	(Ho *et al.* 1994)
	0.1 μm	400 μm	0.48 W/mm	192 mW	4.0 dB	30%	$V_{ds} = 4.12$ V, 67% In (Ho *et al.* 1994)
94 GHz (W-Band)	0.15 μm	50 μm	0.30 W/mm	15 mW	4.6 dB	21%	Single HJ, Passivated device $V_{ds} = 2.6$ V (Hwang *et al.* 1994)
	0.15 μm	640 μm	0.20 W/mm	130 mW	40 dB	13%	Double HJ, $V_{ds} = 2.7$ (Chen *et al.* 1997)
	0.1 μm	200 μm	0.29 W/mm	58 mW		33%	$In_{0.68}Ga_{0.32}As$ channel (Smith *et al.* 1995)

2.8.3. TECHNOLOGY IMPROVEMENTS IN AlInAs/GaInAs HEMTs

Significant performance improvement can be achieved by using InP HEMT technology for microwave and millimeter-wave analog and ultrahigh-speed digital circuit applications. However, due to its relative immaturity compared to the well-established GaAs *p*HEMT technology, the InP HEMT technology suffers from several limitations, including lower breakdown voltage, lower thermal stability, and device reliability. Figure 2.13 illustrates the various problem areas in InP HEMT technology.

Significant efforts are underway in many universities and industrial research laboratories to overcome these limitations. This section reviews the various improvements in InP HEMT technology. The section begins with a discussion of issues related to manufacturability and reliability of InP HEMTs. This is followed by a discussion of various approaches used to increase the breakdown voltage of InP HEMTs.

2.8.3.1. Manufacturability and Reliability of InP HEMT Technology

Gate recess uniformity

The threshold voltage of a HEMT is determined by the thickness of the higher bandgap barrier layer. This thickness is controlled by wet chemical etching or dry

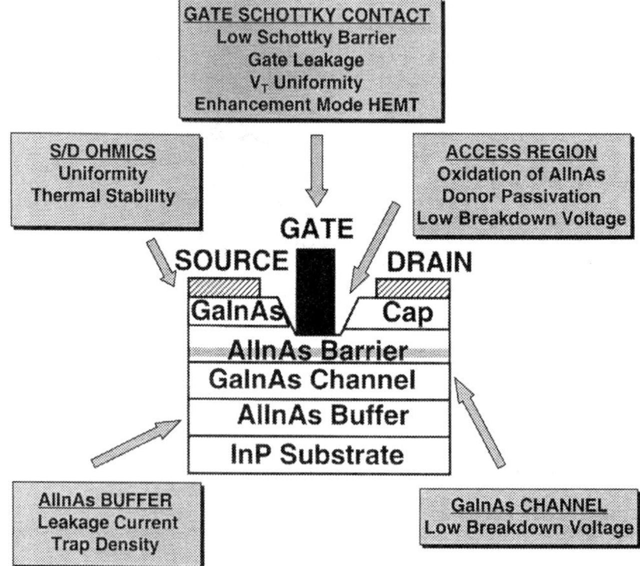

FIG. 2.13. Limitations of InP HEMT technology.

plasma/RIE etching. As the gate recess depth and its variation determines the performance and uniformity of the device characteristics, it is the most crucial step in the InP HEMT process. Various approaches have been used to improve gate recess uniformity.

The first is to use dry etch or wet etch processes, which are selective between the n^+ GaInAs cap layer and the AlInAs barrier layer. Using a citric/ NH_4OH/H_2O_2 etchant with selectivity of 30 between AlInAs and GaInAs a drain current level of 18 mA with a standard deviation of 3.2 mA was achieved (Yoshida et al. 1996). A photochemical selective dry recess etch process using CH_3Br plasma resulted in a threshold voltage uniformity of 18 mV over a 76-mm wafer (Kuroda *et al.* 1992).

Selectivity can be also improved by inserting an etch stop layer between the cap and the barrier layers. This was first demonstrated by using a strained $In_{0.75}Ga_{0.25}P$ etch stop layer (Fujita *et al.* 1993). Using a 3 : 1 : 150 $H_3PO_4 : H_2O_2 : H_2O$ etchant, a selectivity of 20 was achieved at 15 °C. The threshold voltage deviation was reduced to 80 mV from 340 mV by the use of the InGaP etch stop layer. Enoki *et al.* have demonstrated an improvement in the standard deviation of threshold voltage from 101 mV to 16.2 mV on a 50-mm wafer. It was achieved by using 60 Å thick InP as the etch stop layer (Enoki *et al.* 1995). Another advantage of using etch stop layers is that the AlInAs barrier layer, which is susceptible to oxidation due to its high reactivity, is protected from the atmosphere. This improves the thermal stability and reliability of the device. Table 2.12 summarizes the threshold voltage uniformity achieved using various approaches to recess etching.

Gate length uniformity

Process uniformity and reproducibility are essential for lowering manufacturing costs they eliminate the need for an expensive process, the high frequency testing of circuits and systems. As described, high-frequency FET devices required a T-shaped gate to reduce gate resistance. The conventional way to fabricate a T-gate

TABLE 2.12
SUMMARY OF THE THRESHOLD VOLTAGE UNIFORMITY ACHIEVED USING VARIOUS APPROACHES TO RECESS ETCHING

Process/Reference	Selectivity	Uniformity
31 : 150 $H_3PO_4 : H_2O_2 : H_2O$ (Fujita *et al.* 1993)	0	$\sigma(V_T) = 340$ mV
Citric acid (Enoki *et al.* 1994)	0	$\sigma(V_T) = 101$ mV (50-mm wafer)
InGaP Schottky layer (Fujita *et al.* 1993)	20	$\sigma(V_T) = 80$ mV
InP Recess etch stopper (Enoki *et al.* 1995)	400	$\sigma(V_T) = 16.2$ mV (50-mm wafer)
CH_3Br plasma (Kuroda *et al.* 1992)	25	$\sigma(V_T) = 18$ mV (76-mm wafer)
Citric/NH_4OH/H_2O_2 (Yoshida *et al.* 1996)	30	$\sigma(I_{ds}) = 3.2$ mA

is to use a double layer of photoresist in conjunction with electron beam lithography. The upper resist co-PMMA is highly sensitive to light while the lower layer, PMMA, is less sensitive. Hence simultaneous exposure of both layers creates a T-shaped profile. This process is convenient but creates a variation in gate length over a wafer that is not desirable.

This problem is solved by using a double exposure electron beam lithography process. The first noncritical exposure is used to define the head of the T-gate and the second critical exposure is used to define the gate footprint that defines the gate length (Nguyen et al. 1993). This process increased the DC yield from 60 to 91%.

The uniformity of a 0.1-μm T-gate was improved by replacing the conventional double-layer electron beam lithography process by a single-layer electron beam lithography process (Enoki *et al.* 1994). Silicon nitride was used to form the T-shaped gate. However, this procedure resulted in a reduction of f_T, and an increase in the uniformity of the process (which is crucial for multitransistor circuits). For a 12-cm^2 wafer area the average f_T was 201 GHz with a standard deviation of only 3.4 GHz.

Nonalloyed ohmic contacts

Lateral diffusion of AuGeNi-based alloyed contacts is a major cause of catastrophic failure in InP HEMTs. Using regrown contacts or nonalloyed contacts can solve this problem. Advantages of nonalloyed contacts are uniformity, sharp edge definition, which allows for close gate source spacing, and good surface morphology, which is necessary for gate definition. Heavily doped n^+ InGaAs/n^+ InAlAs/n^+ InAlAs/n^+ InGaAs triple capping layers were used to form a nonalloyed contact (Higuchi et al. 1995). This is achieved by reducing vertical conduction resistance between contact and 2DEG. A source resistance of 0.57 Ω mm was obtained. A source resistance of 0.2 Ω mm was achieved with a heavily doped n^+ InGaAs/n^+ InAlAs contact layer (Chen *et al.* 1995). This value compares favorably with those obtained with alloyed contacts.

Device reliability

Device reliability is a major issue of concern for insertion of InP HEMT technology in both commercial and military applications. Satellite applications require an MTTF of 10^7 h at a channel temperature of 80 °C. The principle mechanisms limiting the reliability of InP HEMTs are ohmic contact degradation, oxidation of exposed AlInAs, and passivation of donors in *n*-AlInAs by flourine.

Ohmic and Schottky contact degradation is a major failure mechanism of III–V-semiconductor FETs. The degradation of ohmic contact results increased source resistance and hence decreased transconductance of the device. The degradation of the Schottky contact results in a change in the threshold voltage of the device and also increases gate leakage current. The interdiffusion of Ti into

InGaAs and InAlAs is responsible for ohmic and Schottky contact degradation. Contact reliability is improved by insertion of a molybdenum layer because it prevents diffusion of Ti in the semiconductor (Onda et al. 1994). Another way to improve contact reliability is to use refractory contacts. Improved thermal stability was achieved by using a WSiN refractory gate contact (Enoki *et al.* 1997).

Another degradation mechanism in InP HEMTs is the reduction of drain current or 2DEG density during thermal stress. This is caused by donor compensation in *n*-AlInAs by fast diffusing impurities like fluorine (Hayafuji *et al.* 1995). Thermal degradation due to passivation of donors in AlInAs is reduced by using an AlAs/InAs superlattice to suppress fluorine diffusion. The InP recess etch stopper used to improve threshold voltage uniformity can also act as a passivant for the surface and improve the thermal stability of the device (Enoki *et al.* 1997).

Device reliability can also be improved by passivating the exposed access region. This prevents the deterioration and atmospheric oxidation of the exposed AlInAs layers. The ECR nitride deposition technique has been used for passivation of InP HEMTs. The ECR process generates a high density of ions with low ion energies. This enables the deposition of films at lower temperatures (90–100 °C) than used in conventional techniques (200–250 °C) and minimizes surface damage. Passivated InP HEMTs had a mean time to failure (MTTF) of 10^6 h at a temperature of 150 °C (Hwang *et al.* 1994). A passivated InP HEMT had a noise figure of 0.8 dB with an associated gain of 7.6 dB, whereas an unpassivated device had a noise figure of 0.7 dB with an associated gain of 8.6 dB at 62 GHz. This indicates that the passivation process has minimal effect on the device performance (Kao *et al.* 1994).

2.8.3.2. Junction High Electron Mobility Transistors (JHEMTs)

As discussed, AlInAs/GaInAs HEMTs suffer from excessive gate leakage current due to low Schottky barrier on AlInAs and low gate-drain breakdown voltage. To optimize the power performance of the AlInAs/GaInAs HEMT it is necessary to increase the gate barrier height, which then reduces gate leakage and increases breakdown voltage. As was also discussed, this can be achieved by increasing the Al content in the barrier, or by using AlInAsP barrier layers. However, this method has disadvantages that include increased susceptibility to oxidation and higher source resistance (due to higher Al content). In the case of a phosphorus-based barrier, other growth techniques, such as gas-source MBE or MOCVD, are needed. The growth of high-performance InP HEMTs using these techniques has yet to be optimized.

Another way to increase the barrier height is to replace the Schottky junction by a *pn* junction, resulting in a junction-modulated HEMT (JHEMT). This concept was first demonstrated in the AlGaAs/GaAs system (Suzuki *et al.* 1986). This concept was implemented in the AlInAs/GaInAs material system by Boos *et al.*, who used selective area Zn diffusion (Boos *et al.* 1990) and Shealy *et al.*, who used the MBE-grown AlInAs/GaInAs HEMT with a p^+ GaInAs/AlInAs gate (Shealy *et al.* 1993). A 2-terminal breakdown voltage as high as 19 V was obtained for 0.2-μm gate length AlInAs/GaInAs JHEMTs (Shealy *et al.* 1996). The improvement in the breakdown voltage is due to the reduction electron injection from the gate.

Another advantage of using a JHEMT is threshold voltage uniformity. The threshold voltage of a JHEMT depends on the work function difference between the p^+ gate and the 2DEG and the barrier layer thicknesses and is solely determined by the growth technique. This improves the threshold uniformity over the wafer as no recess etching is involved. A threshold uniformity of 13.7 mV was reported over a 1×1.5 in^2 wafer for a 0.2-μm gate length JHEMTs (Shealy *et al.* 1995). Combining this with selective etching to define the p^+ gate further improves the uniformity (Shealy *et al.* 1995). A noise figure of 0.45 dB with an associated gain of 14.5 dB was obtained for a 0.2-μm gatelength device at 12 GHz, which is comparable for Schottky gate InP HEMTs. Hence the AlInAs/GaInAs JHEMT is a manufacturable device for both microwave analog and ultrahigh-speed digital applications.

2.8.3.3. Regrown Contacts

As discussed in Section 2.8.2, the low on-state breakdown voltage of AlInAs/GaInAs HEMTs is due mainly to impact ionization in the high field drain end of the channel. Alleviating the high electric field at the drain end of the channel can enhance breakdown voltage. This is achieved by the use of selective regrown heavily doped n^+ source and drain contacts.

The breakdown mechanism AlInAs/GaInAs JHEMTs with selectively grown n^+ source and drain contacts has been investigated (Shealy *et al.* 1993). Breakdown voltages of 1-μm gate length devices with conventional alloyed contacts were compared with those with regrown contacts. Device with alloyed contacts had a 2-terminal gate to drain breakdown voltage (BV_{gd}) of 22 V, whereas devices with regrown contacts had a BV_{gd} of 31 V. The 3-terminal on-state breakdown voltage (measured at $I_{DS} = I_{DSS}/2$) was 4 V for the device with alloyed contacts and 7 V for the device with regrown contacts. As was discussed, impact ionization in the channel contributes to breakdown process in the on-state. From the observed increase in breakdown voltage for devices with heavily doped regrown contacts it can be concluded that increased doping in the drain reduces the field, thereby resulting in a lower impact ionization rate. Another advantage of using heavily doped selective regrown contacts is that a low source resistance can be achieved for the device, regardless of the layer structure.

2.8.3.4. Composite Channel HEMTs

The high-speed and power performance of InP HEMTs can be improved by the use of composite channels that are composed of two materials with complementing electronic properties.

The high-speed performance of an InP HEMT can be improved by inserting InAs layers in the InGaAs channel. The current gain cutoff frequency of a 0.15-μm gate length device increased from 179 to 209 GHz due to improved electron transport properties (Akazaki *et al.* 1992). A f_T as high as 264 GHz was achieved for a 0.08-μm gate length device.

The GaInAs channel has excellent electronic properties at a low electric field but suffers from high impact ionization at high electric fields. On the other hand, InP has excellent electronic transport properties at high field but has lower electron mobility. In a composite InGaAs/InP channel HEMT, the electrons are in the InGaAs channel at the low field source end of the channel and are in the InP channel at the high field drain end of the channel. This improves the device characteristics at high drain bias while still maintaining the advantages of the GaInAs channel at low bias voltages (Enoki *et al.* 1992). A typical submicron gatelength AlInAs/GaInAs HEMT has an off-state breakdown-voltage (BV_{dsoff}) of 7 V, and on-state breakdown voltage (BV_{dson}) of 3.5 V. Using a composite channel, (30 Å GaInAs/50 Å InP/100 Å n^+ InP), Matloubian *et al.* demonstrated a BV_{dsoff} of 10 V and BV_{dson} of 8 V for a 0.15-μm gate length device (Matloubian *et al.* 1995). A 0.25 μm GaInAs/InP composite channel HEMTs with a 2-terminal gate drain voltage of 18 V was also demonstrated (Shealy *et al.* 1996).

The increased breakdown voltage of a composite channel HEMT enables operation at a higher drain bias. This increases the drain efficiency and the PAE of the device (see Section 2.5 for discussion). An output power of 0.9 W/mm with a PAE of 76% at 7 GHz was demonstrated for a 0.15-μm GaInAs/InP composite channel HEMT at a drain bias of 5 V (Shealy *et al.* 1997). At 20 GHz, an output power density of 0.62 W/mm (280 mW), and a PAE of 46% was achieved for a 0.15-μm gate length device at a drain bias of 6 V (Matloubian *et al.* 1995). At 60 GHz, a 0.15-μm GaInAs/InP composite channel HEMT demonstrated an output power of 0.35 W/mm, a power gain of 6.2 dB with a PAE of 1290 at a drain bias of 2.5 V (Chevalier *et al.* 1998).

2.8.3.5. Buffer Layer Engineering

A low output conductance of 2.5 mS/mm was demonstrated for a 0.2-μm gate length InP HEMT with the use of low temperature grown (LTG) AlInAs buffer layers (Brown *et al.* 1989). The LTG AlInAs buffer layers are grown by MBE at a reduced temperature of 150–200 °C, whereas the growth temperature for conventional AlInAs buffer layers is approximately 530 °C. The reduction in output conductance is attributed to the reduced electron injection from the channel into the buffer. The low-frequency g_m/g_{ds} ratio was 250 for the device with the LTG AlInAs buffer. On the other hand, similar devices with conventional AlInAs buffer layers had a g_m/g_{ds} ratio of 10–20. The voltage gain, an important parameter for power devices, is proportional to the g_m/g_{ds} ratio. Reducing the leakage through the buffer can also increase the breakdown voltage of the device. An increase in the 3-terminal on-state breakdown voltage from 5 to 14 V at a current density of 300 mA/mm was achieved by using a low temperature grown AlInAs buffer (Prost *et al.* 1995).

2.8.4. AlInAs/GaInAs HEMTs FOR DIGITAL CIRCUITS

High-speed digital circuits are used in direct digital frequency synthesizers (DDS) for radar transmitters. Lightwave communication systems operating at 10 Gbit/s have been used commercially. Intensive efforts are underway to develop 40-Gbit/s systems. The decision circuit in the electronic front-end is the most critical component of the system and operates at the maximum bit rate. The maximum operating speed of a digital circuit like a D-flip flop is linearly dependent on the f_T of the device (Maeda *et al.* 1997). As the InP HEMTs has the highest reported f_T of any 3-terminal device, it is an ideal candidate for application in ultrahigh-speed digital circuits. This was first demonstrated by Mishra *et al.* who implemented 25-GHz static frequency dividers using 0.2-μm gate length AlInAs/GaInAs HEMTS (Mishra *et al.* 1988). Some of the fastest digital circuits have been implemented using InP HEMT technology. An optical repeater circuit operating at 40 Gbit/s with all the digital chips was implemented in InP HEMT technology (Yoneyama *et al.* 1997). Recently, a 2 : 1 selector type multiplexer IC operating at 80-Gbit/s was demonstrated (Otsuji *et al.* 1997).

The operating frequency and power consumption of a static frequency divider can be considered as a figure of merit for a high-speed digital circuit technology. Table 2.13 summarizes the results of static frequency dividers implemented in various device technologies.

It is clear from Table 2.13 that the highest frequency of operation and large voltage swings coupled with low power dissipation is offered by InP HEMT technology. Its performance is surpassed only by the AlInAs/GaAs HBT technology on the InP substrate.

For LSI digital circuits, speed and other parameters, including power dissipation and circuit complexity are also of importance. Table 2.14 compares the propagation delay, power dissipation, and power-delay products of various logic schemes implemented in GaAs- and InP-based FET technologies.

It is clear from Table 2.14 that for any logic scheme, the InP HEMT offers the lowest operating voltages, power consumption, and power-delay product. At a

TABLE 2.13
SUMMARY OF STATIC FREQUENCY DIVIDER PERFORMANCE IN VARIOUS BIPOLAR AND FET TECHNOLOGIES

Device technology/logic	Speed	Differential output voltage swing	P_{disp} (flip-flop)	P_{disp} (flip-flop + buffers)
Si Bipolar/ECL (Felder *et al.* 1996)	30 GHz	800 mV	230 mW	630 mW
Si Bipolar/ECL (Böck *et al.* 1996)	35 GHz	200 mV		
SiGe HBT/ECL (Wurzer *et al.* 1997)	42 GHz	400 mV	300 mW	1.15 W
AlGaAs/GaAs HBT/ECL (Amamiya *et al.* 1998)	40 GHz	600 mV	294 mW	900 mW
AlInAs/GaInAs HBT/ECL (Sokolich *et al.* 1998)	52.9 GHz	200 mV	40 mW	360 mW
AlInAs/GaInAs TS-HBT/ECL (Pullela *et al.* 1998)	48 GHz	150 mV	380 mW	
AlInAs/GaInAs TS-HBT/CML (Pullela *et al.* 1998)	48 GHz	100 mV	75 mW	
GaAs HEMT/SCFL (Lao *et al.* 1997)	35 GHz	680 mV	250 mW	
InP HEMT/SCFL (Enoki *et al.* 1995)	40.5 GHz	800 mV	228 mW	550 mW
InP HEMT/SCFL (Otsuji *et al.* 1997)	46.5 GHz	800 mV		1.1 W

*TS-HBT—Transferred substrate HBT.

TABLE 2.14
GATE DELAY, POWER DISSIPATION AND POWER DELAY PRODUCTS OF VARIOUS FET LOGIC SCHEMES

Logic	Device technology	Gate delay (ps/stage)	Power diss. (mW/stage)	Power-delay product (fJ/stage) @ supply voltage V_{DD} (volts)
BFL	0.2 μm InP HEMT	4 ps	1.5 mW	6 fJ (Brown *et al.* 1989)
SCFL	0.2 μm GaAs MESFET	13.2 ps	65 mW	858 fJ (Yamane *et al.* 1991)
	0.1 μm InP HEMT	7.0 ps	80 mW	560 fJ (Enoki *et al.* 1995)
DCFL	0.2 μm GaAs MESFET	8.1 ps	1.7 mW	13.8 fJ @ $V_{DD} = 1$ V (Tsuji *et al.* 1990)
	0.3 μm GaAs HEMT	15.0 ps	200 μW	3 fJ (Abe *et al.* 1990)
	0.2 μm GaAs *p*HEMT	6.6 ps	1.8 mW	11.9 fJ @ $V_{DD} = 2$ V (Tsuji *et al.* 1991)
	0.25 μm GaAs *p*HEMT	19.0 ps	150 μW	2.85 fJ @ $V_{DD} = 0.6$ V (Hida *et al.* 1992)
	0.25 μm InP HEMT	16.7 ps	78 μW	1.31 fJ @ $V_{DD} = 0.4$ V (Adesida *et al.* 1998)
	0.1 μm GaAs *p*HEMT	10.6 ps	50 μW	0.53 fJ @ $V_{DD} = 0.6$ V (Wada *et al.* 1998)

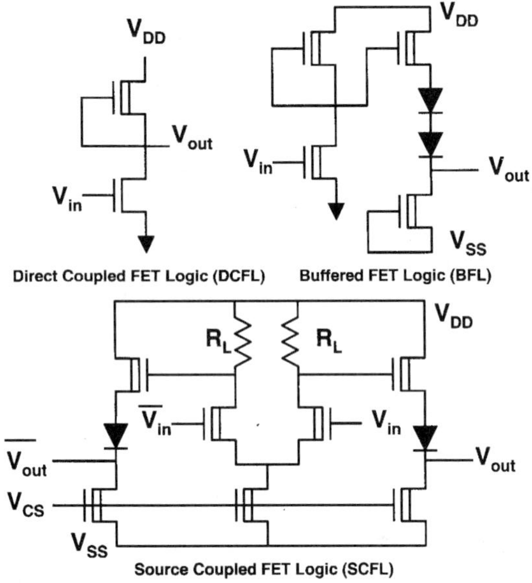

FIG. 2.14. Logic schemes for III–V FET-based Inverters.

given gate length, InP HEMT technology also offers the lowest gate delays. Figure 2.14 shows inverters implemented in various logic schemes using III–V-semiconductor MESFETs and HEMTs.

For ultrahigh-speed applications where power consumption is not an issue, source coupled FET logic (SCFL) is the ideal choice. The circuit performance of SCFL logic can be analyzed by examining the delay time for a D-latch, which is a basic component of high-speed divider circuits. The delay time for a D-latch implemented in SCFL logic is given by (Umeda *et al.* 1996),

$$\tau_{\text{tot}} = 2\frac{C_{\text{gs}} + C_{\text{gd}}}{g_{m,\max}} + R_L C_{\text{gd}}\left(1 + 2\frac{\langle g_m \rangle}{g_{m,\max}}\right) \tag{2.68}$$

To minimize delay time $(C_{\text{gs}} + C_{\text{gd}})/g_m$ has to be minimized. This implies that the f_T of the FET has to be maximized. Therefore, InP HEMT-based SCFL circuits are expected to be the fastest. Power consumption is also an important factor in decision IC, that are used in the electronic front ends of optical communication systems. A 32-Gbit/s super-dynamic decision IC fabricated using 0.13-μm GaAs MESFETs has power dissipation of 2.34 W (Otsuji *et al.* 1997). On the other hand, a 46-Gbit/s super-dynamic decision IC fabricated using 0.1-μm InP HEMTs has power dissipation of 1.7 W (Yoneyama *et al.* 1997). Hence compared with other transistor technologies, InP HEMTs enable operation at higher bit rates with lower power dissipation for SCFL logic.

The high-speed operation of SCFL logic is achieved at the cost of increased power dissipation and circuit complexity. Power consumption is also an issue for LSI digital circuits like prescalers, that operate around 20 GHz and are used in spread spectrum communication systems and microwave instrumentation. The DCFL logic, with its reduced power dissipation and circuit complexity, is an ideal choice for these circuits that typically contain 100 to 200 transistors.

The performance of DCFL logic can be analyzed by considering the propagation delay and power consumption for a DCFL inverter as shown in Fig. 2.14. It is assumed that the driver FET and the load FET operate in the saturated drain current region. The propagation delay t_{pd} or the average time required to charge/discharge the load capacitance C_L at the output of the inverter is given by (Hida *et al.* 1993),

$$t_{\mathrm{pd}} = \frac{t_{\mathrm{rise}} + t_{\mathrm{fall}}}{2} = \frac{k^2}{k-1}\frac{C_L V_{\mathrm{SW}}}{2I_D} \tag{2.69}$$

where k is the ratio of the current drive of the driver FET I_D and the current drive of the load FET I_L. V_{SW} is the voltage swing at the output. The peak power dissipation is given by

$$P_c = \frac{I_{\mathrm{DD}} V_{\mathrm{DD}}}{k} \tag{2.70}$$

Multiplying Eqs. (2.69) and (2.70) gives the expression for the power delay product:

$$P_c t_{\mathrm{pd}} = \frac{k}{2(k-1)} C_L V_{\mathrm{SW}} V_{\mathrm{DD}} \tag{2.71}$$

To reduce power consumption while maintaining high speed, it is desirable to have a device technology that can supply high current drive at low voltages with minimal parasitic resistances. The gate delay and the power delay product can also be lowered by the reduction of the output voltage swing but this compromises the noise margin of the circuit. The optimum method to reduce delay and power consumption is operation at low voltages ($V_{\mathrm{DD}} < 0.5$). This requires that the knee voltage of the FET be of the order of 0.1–0.2 V, so that the FETs operates in saturated mode during most of the output swing. Low-voltage application is also desirable for mobile applications as the battery size is reduced. Again the InP HEMT with its low operating voltage and high speed and current drive is an ideal choice for high-speed, low-power digital circuit applications. Another critical aspect of ultrahigh-speed circuit design is a compact circuit layout, that minimizes the load capacitance and signal propagation delay along interconnects.

The input/driver transistor in a DCFL gate has to be an enhancement mode device ($V_T > 0$). The typical compound semiconductor FET is a depletion mode device, that is the channel is normally on ($V_T < 0$). Hence most III–V digital logic circuits, such as BFL and SCFL, require level shifting diodes to increase power dissipation and circuit design complexity. The following sections will address this issue along with crucial requirements of device uniformity for high-speed digital circuit technologies.

2.8.4.1. Device Uniformity

A lightwave communication system, which is a commercial high-volume application, has completely different performance requirements from millimeter-wave analog applications. First and foremost is threshold voltage uniformity, because typical LSI digital circuits have more than 30 transistors. A variation of 10 Å in barrier thickness during recess etching results in up to a 60-mV variation in threshold voltage. Threshold voltage variation reduces the maximum operating frequency of a digital circuit. Therefore, for digital circuit applications having more than 30 transistors per circuit, it is necessary that the threshold voltage, gate length, and source resistance be uniform across the wafer. A detailed discussion of these issues can be found in Section 2.8.3.1 of this chapter.

2.8.4.2. Enhancement Mode Devices

As already discussed, for high-speed SSI circuits using InP HEMT, direct-coupled FET logic (DCFL) is preferred because there are no level shifting diodes and negative power supply voltages required. The driver HEMT in a DCFL circuit needs to be an enhancement mode device. Referring to Eq. (2.3), it can be seen that for an enhancement mode device the barrier thickness has to be reduced so that the depletion layer from the gate extends across the channel at zero bias. This is achieved by deep recess etching in the gate region. However, the side etching in the access region of the device results in high source resistance, which reduces the f_T of the device.

One approach to solving this problem uses an undoped AlinAs/GaInAs structure with the source and drain contacts defined by ion implantation (Feuer *et al.* 1991). The resulting device is known as a heterostructure insulated gate FET or HIGFET. However, this device suffers from lateral diffusion of the source and drain implants in the channel as the gate length of the device is reduced. A novel approach that uses a platinum-based buried gate to achieve an enhancement mode device without increasing the source resistance of the device (Harada *et al.* 1991). As seen from Eq. (2.3) to obtain a positive threshold voltage it is necessary to maximize the Schottky barrier height Φ_B. In most InP HEMT MMIC processes, Ti/Pt/Au is used for gate metallization and has a Schottky barrier height of 0.65 eV. On the other hand, the Schottky barrier height of platinum on InAlAs is 0.83 eV. Hence, in the buried gate process, the HEMT device is fabricated as a depletion mode device with Pt/Ti/Pt/Au gates. Subsequent annealing of the device in forming gas at 250 °C results in the sinking of the platinum metal into the AlInAs barrier and formation of the metallic $PtAs_2$ alloy. This effectively reduces the barrier thickness to give a threshold voltage greater than 0 V. The Ti layer inserted in the metallization acts as a barrier and controls the amount of Pt, that reacts with the AlInAs. Thus the reduction in barrier thickness is achieved without deep recess etching. This minimizes the depletion in the access region and keeps the source resistance of the device low. The concept of buried gate technology is illustrated in Figure 2.15. Using the same technique, 1-μm gate length devices with threshold voltages as high as +0.255 V have been demonstrated (Mahajan *et al.* 1998).

A 5.8-GHz frequency divider using DCFL gates based on a 1.2-μm gate InP E-HEMTs has been demonstrated (Harada *et al.* 1993). Using the forementioned technology, a 6-GHz divide-by-four prescaler with a power dissipation of 5.37 mW/stage and a 23-stage ring oscillator with a delay of 16.72 ps/stage and a power delay product of 0.322 fJ/stage at a supply voltage of 0.4 V have been demonstrated (Adesida *et al.* 1998). The propagation delay of D-HEMT-

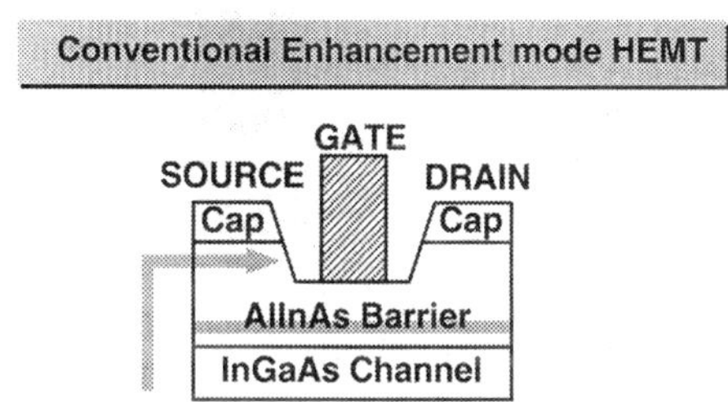

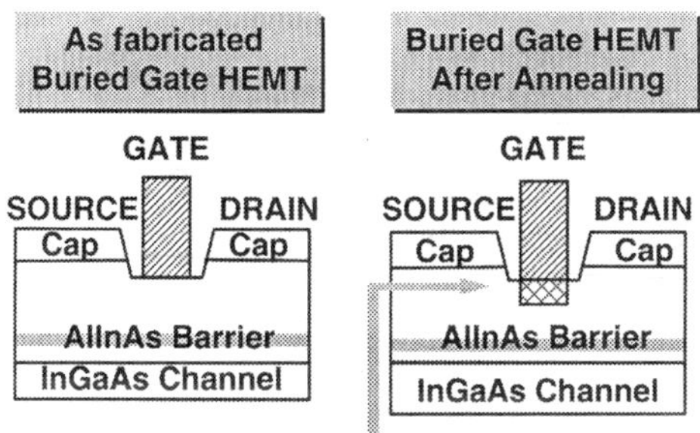

FIG. 2.15. Buried gate process for fabrication on enhancement mode HEMT.

based logic (implemented in 0.2-μm gate technology) like CEL is 6.0 ps/state with a power delay product of 138 fJ/stage (Mishra *et al.* 1988). The BFL has 9.3 ps/stage and power dissipation of 66.7 mW/gate. The SCFL ring oscillator has a delay of 7.0 ps/stage with a power dissipation of 80 mW/gate (Enoki *et al.* 1995). Hence E-HEMT technology is attractive for ultralow power applications.

Because the buried gate process relies on a temperature-activated reaction of Pt with AlInAs, it may have potential reliability problems. Recently an enhancement mode AlInAs/GaInAs HEMT using a 2-step recessed technology that used InP etch stop layer to reduce side-etching was reported (Suemitsu *et al.* 1998). The 0.1-μm gate length devices had a threshold voltage of 49 mV and the threshold voltage uniformity over a 76-mm wafer was 13.3 mV. Thermal stability, which is important for large-scale integrated circuits, was ensured by using a WSiN refractory gate.

2.9. Conclusion

GaAs- and InP-based high electron mobility transistors have emerged as premier devices for the implementation of millimeter-wave analog circuits and ultrahigh-speed digital circuits. In this chapter principles of HEMTs operation were discussed. The design aspects of HEMTs for both low-noise and high-power applications were discussed. Reduction in gate length is essential for improved performance at high frequencies. Appropriate device scaling with gate length reduction is necessary to minimize the effect of parasitics on device performance.

Millimeter-wave power modules have been demonstrated using GaAs *p*HEMT devices. The superior device performance of GaAs *p*HEMTs is being used to improve the performance of power amplifiers for wireless phone systems. The superior material characteristics of the AlInAs/GaInAs material system have been used to achieve record low-noise performance at millimeter-wave frequencies using InP HEMTs. Despite their low breakdown voltage, InP HEMTs have demonstrated superior power performance at millimeter-wave frequency. Improving the breakdown voltage using approaches that include composite channel GaInAs/InP HEMT and junction modulated HEMT will further improve power performance. The high current drive capability and high current gain cutoff frequency of the InP HEMT have enabled the realization of ultrahigh-speed circuits operating at 80 Gbit/s.

The development of the GaAs, *p*HEMT and InP HEMT technology was traditionally supported by low-volume, high-cost military and space applications. The recent emergence of high-volume commercial applications such as wireless and optical communications systems has new constraints that include manufacturability and low-voltage operation for these technologies.

References

1. Dingle R., Stromer H. L., Gossard A. C., Wiemann W. (1978). "Electron mobility in modulation doped semiconductor superlattices." *Applied Physics Letters*, **33**, 665–667.
2. Fukui H. (1979). "Optimal noise figure of microwave GaAs MESFET's." *IEEE Trans. Electron Devices*, **26**, 1032–1037.
3. Mimura T., Hiyamizu S., Hikosak S. (1981). "Enhancement mode high electron mobility transistors for logic application." *Japanese Journal of Applied Physics*, L317.
4. Delagebeaudeuf D., Linh N. T. (1982). "Metal-(n) AlGaAs-GaAs two-dimensional electron gas FET." *IEEE Trans. Electron Devices*, **29**, 955–960.
5. Vinter B. (1984). "Subbands and charge control in two-dimensional electron gas field effect transistor." *Applied Physics Letters*, **44**, 307–309.
6. Curtice W. R., Ettenberg M. (1985). "A nonlinear GaAs FET model for use in the design of output circuits for power amplifier." *IEEE Transactions on Microwave Theory and Techniques*, **33**, 1383–1394.
7. Das M. B. (1985). "A High Aspect Ratio Design Approach to Millimeter-Wave HEMT structures." *IEEE Trans. Electron Devices*, **32**, 11–17.
8. Hirose K., Ohata K., Mizutani T., Itoh T., Ogawa M. (1985). "700 mS/mm 2DEGFETs fabricated from high electron mobility MBE-grown n-AlInAs/GaInAs heterostructures." *GaAs and Related Compounds*, 529–534.
9. Ketterson, A., Moloney M., Masselink W. T., Klem J., Fischer R., Kopp, W., Morkoç H. (1985). "High transconductance InGaAs/AlGaAs pseudomorphic modulation doped field effect transistors." *IEEE Electron Device Lett.*, **6**, 628–630.
10. Blight S. R., Wallis R. H., Thomas H (1986). "Surface Influence on the conductance DLTS spectra of GaAs MESFETs." *IEEE Trans. Electron Devices*, **33**, 1447–1453.
11. Drummond T. J., Masselink W. T., Morkoç H. (1986). "Modulation Doped GaAs/(Al,Ga)As Heterojunction Field-Effect Transistors: MODFETs." *Proceedings of the IEEE*, **74**, 773–822.
12. Henderson T., Aksun M. I., Peng C. K., Morkoç H., Chao P. C., Smith P. M., Duh K.-H. G., Lester L. F. (1986). "Microwave Performance of Quarter Micrometer Gate Low Noise Pseudomorphic InGaAs/AlGaAs Modulation Doped Field Effect Transistor." *IEEE Electron Device Lett.*, **7**, 649–651.
13. Ketterson A. A., Masselink W. T., Gedymin J. S., Klem J., Peng C.-K., Kopp W. F., Morkoç H., Gleason K. R. (1986). "Characterization of InGaAs/AlGaAs Psueodomorphic Modulation-Doped Field-Effect Transistors." *IEEE Trans. Electron Devices*, **33**, 564–571.
14. Suzuki Y., Hida H., Toyoshima H., Ohata K. (1986). "High-speed ring oscillators using planar p+ gate n-AlGaAs/GaAs 2DEG FETs." *Electronics Letters*, **22**, 672–673.
15. Canfield P., Medinger J., Forbes L. (1987). "Buried-channel GaAs MESFET's with frequency-independent output conductance." *IEEE Electron Device Lett.*, **8**, 88–89.
16. Palamateer L. F., Tasker P. J., Itoh T., Brown A. S., Wicks G. W., Eastman L. F. (1987). "Microwave characterization of 1 μm gate $Al_{0.48}In_{0.52}As/Ga_{0.47}In_{0.53}As/InP$ MODFETs." *Electronics Letters*, **23**, 53–54.
17. Dambrine G., Cappy A., Heliodore F., Playez E. (1988). "A New Method for Determining the FET Small-Signal Equivalent Circuit." *IEEE Transactions on Microwave Theory and Techniques*, **36**, 1151–1159.
18. Foisy M. C., Tasker P. J., Hughes B., Eastman L. F. (1988). "The role of insufficient charge modulation in limiting the current gain cutoff frequency of the MODFET." *IEEE Trans. Electron Devices*, **35**, 871–878.
19. Ho P., Chao P. C., Duh K. H. G., Jabra A. A., Ballingall J. M., Smith P. M. (1988). "Extremely High Gain, Low Noise InAlAs/InGaAs HEMTs grown by molecular beam epitaxy." *IEDM Technical Digest*, 184–186.

20. Lester L. F., Smith P. M., Ho P., Chao P. C., Tiberio R. C., Duh K. H. G., Wolf E. D. (1988). "0.15 μm Gate Length Double Recess psuedomorphic HEMT with f_{max} of 350 GHz." *IEDM Technical Digest*, 172–175.
21. Mishra U. K., Brown A. S., Jelloian L. M., Hackett L. H., Delaney M. J. (1988). "High Performance Submicrometer AlInAs-GaInAs HEMTs." *IEEE Electron Device Lett.*, **9**, 41–43.
22. Mishra U K., Brown A. S., Rosenbaum S. E. (1988). "DC and RF Performance of 0.1 μm Gate Length $Al_{0.48}In_{0.52}As - Ga_{0.38}In_{0.62}As$ Pseudomorphic HEMTs." *Technical Digest*, 180–183.
23. Mishra U. K., Brown A. S., Rosenbaum S. E., Hooper C. E., Pierce M. W., Delaney M. J., Vaughn S., White K. (1988). "Microwave performance of AlInAs-GaInAs HEMT's with 0.2 μm and 0.1 μm Gate Length." *IEEE Electron Device Lett.*, **9**, 647–649.
24. Mishra U. K., Jensen J. F., Brown A. S., Thompson M. A., Jelloian L. M., Beaubien R. S. (1988). "Ultra High Speed Digital Circuit performance in 0.2 μm Gate length AlInAs/GaInAs HEMT Technology." *IEEE Electron Device Lett.*, **9**, 482–484.
25. Moll N., Hueschen M. R., Fischer-Colbrie A. (1988). "Pulse Doped AlGaAs/InGaAs Pseudomorphic MODFETs." *IEEE Trans. Electron Devices*, **35**, 879–885.
26. Nguyen L. D., Radulescu D. C., Tasker P. J., Schaff W. J., Eastman L. F., (1988). "0.2 μm Gate-Length Atomic-Planar Doped Pseudomorphic $Al_{0.3}Ga_{0.7}As/In_{0.25}Ga_{0.75}As$ MODFET's with f_τ over 120 GHz." *IEEE Electron Device Lett.*, **9**, 374–376.
27. Saunier P., Matyi R. J., Bradshaw K. (1988). "A Double-Heterojunction Doped-Channel Pseudomorphic Power HEMT with a Power Density of 0.85 W/mm at 55 GHz." *IEEE Electron Device Lett.*, **9**, 397–398.
28. Brown A. S., Chou C. S., Delaney M. J., Hooper C. E., Jensen J. F., Larson L. E., Mishra U. K., Nguyen L. D., Thompson M. S. (1989). "Low-Temperature Buffer AlInAs/GaInAs on InP HEMT Technology for Ultra-High-Speed Integrated Circuits." *Proceedings of GaAs IC Symposium*, 143–146.
29. Chao P. C., Duh K. H. G., Ho P., Smith P. M., Ballingall J. M., Jabra A. A., Tiberio R. C. (1989). "94 GHz Low Noise HEMT." *Electronics Letters*, **25**, 504–505.
30. Chao P. C., Shur M. S., Tiberio R. C., Duh K. H. G., Smith P. M., Ballingall J. M., P. Ho, Jabra A. A. (1989). "DC and Microwave Characteristics of Sub 0.1 μm Gate Length Planar doped Pseudomorphic HEMT's." *IEEE Trans. Electron. Devices*, **36**, 461–473.
31. Ferguson D. W., Smith P. M., Chao P. C., Lester L. F., Smith R. P., Ho P., Jabra A., Ballingall J. M. (1989). "44 GHz hybrid HEMT Power Amplifiers". *IEEE MTT-S Int. Microwave Symp. Dig.*, 987–990.
32. Kao M.-Y., Smith P. M., Ho P., Chao P.-C., Duh K. H. G., Jabra A. A., Ballingall J. M. (1989). "Very High Power-Added Efficiency and Low Noise 0.15 μm Gate-Length Pseudomorphic HEMTs." *IEEE Electron Device Lett.*, **10**, 580–582.
33. Kushner L. J. (1989). "Output performance of idealized microwave power amplifiers." *Microwave Journal*, 103–116.
34. Mishra U. K,. Brown A. S., Jeelloian L. M., Thompson M., Nguyen L. D., Rosenbaum S. E. (1989). "Novel High Performance Self Aligned 0.15 μm long T-gate AlInAs-GaInAs HEMTs." *IEDM Technical Digest*, 101–104.
35. Nguyen L. D., Brown A., Delaney M., Mishra U., Larson L., Jelloian L., Melendes M., Hooper C., Thompson M. (1989). "Vertical Scaling of Ultra-High-Speed AlInAs-GaInAs HEMTs." *IEDM Technical Digest*, 105–108.
36. Nguyen L. D., Tasker P. J., Radulescu D. C., Eastman L. F. (1989). "Characterization of Ultra-High Speed Pseudomorphic AlGaAs/InGaAs (on GaAs) MODFET's." *IEEE Trans. Electron Devices*, **36**, 2243–2247.
37. Pospieszalski M. W. (1989). "Modeling of Noise Parameters of MESFET's and MODFET's and Their Frequency and Temperature Dependence." *IEEE Transactions on Microwave Theory and Techniques*, **37**, 1340–1350.

38. Smith P. M., Lester L. F., Chao P.-C., Ho P., Smith R. P., Ballingall J. M., Kao M.-Y. (1989). "A 0.25 μm Gate length pseudomorphic HFET with 32 mW Output power at 94 GHz." *IEEE Electron Device Lett.*, **10**, 437–439.

39. Tasker P. J., Hughes B. (1989). "Importance of Source and Drain Resistance to the Maximum f_T of Millimeter-Wave MODFETs." *IEEE Electron Device Lett.*, **10**, 291–293.

40. Abe M., Mimura T. (1990). "Ultrahigh-speed HEMT LSI Technology." *Proceedings of GaAs IC Symposium*, 127–130.

41. Berroth M., Bosch R. (1990). "Broad-Band Determination of the FET Small-Signal Equivalent Circuit." *IEEE Transactions on Microwave Theory and Techniques*, **38**, 891–895.

42. Boos J. B., Binari S. C., Kruppa W., Hier H. (1990). "InAlAs/InGaAs/InP Junction HEMTs." *Electronics Letters*, **26**, 1172–1173.

43. Chao P. C., Tessmer A. J., Duh K. H. G., Ho P., Kao M.-Y., Smith P. M., Balligall J. M., Liu S.-M. J., Jabra A. A. (1990). "W-band Low-Noise InAlAs/IUnGaAs Lattice matched HEMT's. *IEEE Electron Device Lett.*, **11**, 59–61.

44. Maas S. A., Neilson D. (1990). "Modeling MESFET's for Intermodulation Analysis of Mixers and Amplifiers." *IEEE Transactions on Microwave Theory and Techniques*, **38**, 1964–1971.

45. Nguyen L. D., Tasker P. J. (1990). "Scaling issue of ultra-high speed HEMTs." *SPIE Conf. on High Speed Electronics and Device Scaling*, **1288**, 251–257.

46. Pao Y. C., Nishimoto C. K., Majidi-Ahy R., Archer J., Betchel N. G., Harris J. S. (1990). "Characterization of surface undoped $In_{0.52}Al_{0.48}As/In_{0.53}Ga_{0.47}As$ High Electron Mobility Transistors." *IEEE Trans. Electron Devices*, **37**, 2165–2170.

47. Smith P. M., Chao P. C., Ballingall J. M., Swanson A. W. (1990). "Microwave and mm-Wave Power Amplification Using Pseudomorphic HEMT's." *Microwave Journal*, **33**, 71–86.

48. Tan K. L., Dia R. M., Streit D. C., Lin T., Trinh T. Q., Han A. C., Liu P. H., Chow P. D., Yen H. C., (1990). "94 GHz 0.1 μm T-Gate low noise Pseudomorphic InGaAs HEMTs." *IEEE Electron Device Lett.*, **11**, 585–587.

49. Tsuji H., Fujishiro H. I., Nakamura H., Nishi S. (1990). "A Sub-10 ps/gate Direct-Coupled FET Logic Circuit with 0.2 μm-Gate GaAs MESFET." *Japanese Journal of Applied Physics*, **29**, L2438–L2441.

50. Boos J. B., Kruppa W. (1991). "InAlAs/InGaAs/InP HEMTs with High Breakdown voltages using double recess gate process." *Electronics Letters*, **27**, 1909–1910.

51. Duh K. H. G., Chao P. C., Liu S. M. J., Ho P., Kao M. Y., Ballingall J. M. (1991). "A Super Low Noise 0.1 μm T-gate InAlAs-InGaAs-InP HEMT." *IEEE Microwave and Guided Wave Letters*, **1**, 114–116.

52. Feuer M. D., He Y,. Shunk S. C., Huang J.-H., Vang T. A., Brown-Goebeler K. F., Chang T.-Y. (1991). "Direct-Coupled FET Logic Circuits on InP." *IEEE Electron Device Lett.*, **12**, 98–100.

53. Harada N., Kuroda S., Katakami T., Hikosaka K., Mimura T., Abe M. (1991). "Pt-Based Gate Enhancement Mode InAlAs/InGaAs HEMTs For Large-Scale Integration." *Proceedings of International Conference on InP and Related Materials*, 377–380.

54. Helms D., Komiak J. J,. Kopp W. F., Ho P., Smith P. M., Smith R. P., Hogue D. (1991). "Ku Band Power Amplifier using pseudomorphic HEMT Devices for improved efficiency." *IEEE MTT-S Int. Microwave Symp. Dig.*, 819–821.

55. Ho P., Kao M. Y., Chao P. C., Duh K. H. G., Ballingall J. M., Allen S. T., Tessmer A. J., Smith P. M. (1991). "Extremely high gain 0.15 μm Gate-length InAlAs/InGaAs/InP HEMTs." *Electronics Letters*, **27**, 325–327.

56. Kao M. Y., Smith P. M., Chao P. C., Ho P. (1991). "Millimeter wave power performance of InAlAs/InGaAs/InP HEMTs." *IEEE/Cornell Conference on Advanced Concepts in High Speed Semiconductor Devices and Circuits*, 469–477.

57. Matloubian M., Nguyen L. C., Brown A. S., Larson L. E., Melendes M. A., M. A. Thompson (1991). "High Power and High Efficiency AlInAs/GaInAs on InP HEMTs." *IEEE MTT-S Int. Microwave Symp. Dig.*, 721–724.
58. Streit D. C., Tan K. L., Dia R. M., Liu J. K., Han A. C., Velebir J. R., Wang S. K., Trinh T. Q., Chow P. D., Liu P. H., *et al.* (1991). "High Gain W-band Pseudomorphic InGaAs Power HEMTs." *IEEE Electron Device Lett.*, **12**, 149–150.
59. Tan K. L., Dia R. M., Streit D. C., Shaw L. K., Han A. C., Sholley M. D., Liu P. H., Trinh T. Q., Lin T., Yen H. C. (1991). "60 GHz Pseudomorphic $Al_{0.25}Ga_{0.75}As/In_{0.28}Ga_{0.72}As$ Low-Noise HEMTs." *IEEE Electron Device Lett.*, **12**, 23–25.
60. Tan K. L., Streit D. C., Dia R. M., Wang S. K., Han A. C., Chow P. D., Trinh T. Q., Liu P. H., Velebir J. R., Yen H. C. (1991). "High Power V-band Pseudomorphic InGaAs HEMT." *IEEE Electron Device Lett.*, **12**, 213–214.
61. Tokue T., Nashimoto Y., Hirokawa T., Mese A., Ichikawa S,. Negishi H., Toda T., Kimura T., Fujita M., Nagtasako I., *et al.* (1991). "Ku Band Super Low Noise Pseudomorphic Heterojunction Field Effect Transistor (HJFET) with High Producibility and High Reliability." *IEEE MTT-S Int. Microwave Symp. Dig.*, 705–708.
62. Tsuji H., Fujishiro H. I., Shikata M., Tanaka K., Nishi S. (1991). "0.2 μm gate pseudomorphic inverted HEMT for high speed digital ICs". *Proceedings of GaAs IC Symposium*, 113–116.
63. Yamane Y., Ohhata M,. Kikuchi H., Asai K., Imai Y. (1991). "A 0.2 μm GaAs MESFET Technology for 10 Gb/s Digital and Analog ICs". *IEEE MTT-S Int. Microwave Symp. Dig.*, 513–516.
64. Akazaki T., Enoki T., Arai K., Umeda Y., Ishii Y. (1992). "High Frequency performance for sub 0.1 μm Gate InAs-inserted-channel InAlAs/InGaAs HEMT." *Electronics Letters*, **28**, 1230–1231.
65. Angelov I,. Zirath H., Rorsman N. (1992). "A New Empirical Nonlinear Model for HEMT and MESFET Devices." *IEEE Transactions on Microwave Theory and Techniques*, **40**, 2258–2266.
66. Danzilio D., White P., Hanes L. K., Lauterwasser B., Ostrowski B., Rose F. (1992). "A High Efficiency 0.25 μm pseudomorphic HEMT power process", *Proceedings of GaAs IC Symposium*, 255–258.
67. Dow G. S., Tan K., Ton N., Abell J., Siddiqui M., Gorospe B., Streit D., Liu P., Sholley M. (1992). "Ka-band High Efficiency 1 Watt Power Amplifier". *IEEE MTT-S Int. Microwave Symp. Dig.*, 579–582.
68. Enoki T., Arai K., Kohzen A., Ishii Y. (1992). "InGaAs/InP Double channel HEMT on InP". *Proceedings of International Conference on InP and Related Materials*, 14–17.
69. Hida H., Tokushima M., Fukaishi M., Maeda T., Ohno Y. (1992). "A 0.25 μm Inner Sidewall-assisted Super Self-Aligned Gate Heterojunction FET Fabricated Using All Dry-Etching Technology for Low Voltage Controlled LSIs". *IEDM Technical Digest*, 982–983.
70. Hughes B. (1992). "A Temperature Noise Model for Extrinsic FETs." *IEEE Transactions on Microwave Theory and Techniques*, **40**, 1821–1832.
71. Kuroda S., Imanishi K., Harada N., Hikosaka K., Abe M. (1992). "Highly Uniform n-AllnAs/InGaAs HEMTs on a 3 in InP Substrate Using Photochemical Selective Dry Recess Etching." *IEEE Electron Device Lett.*, **13**, 105–107.
72. Nguyen L. D., Brown A. S., Thompson M. A., Jelloian L. M. (1992). "50-nm Self-Aligned-Gate Pseudomorphic AlInAs/GaInAs High Electron Mobility Transistors." *IEEE Trans. Electron Devices*, **39**, 2007–2014.
73. Nguyen L. D., Larson L. E., Mishra U. K. (1992). "Ultra-high-speed modulation-doped field-effect transistors: A Tutorial Review." *Proceedings of the IEEE*, **80**, 494–518.
74. Umeda Y,. Enoki T., Arai K., Ishii Y. (1992). "Silicon Nitride Passivated Ultra Low Noise InAlAs/InGaAs HEMT's with n^+ InGaAssn$^+$-InAlAs Cap Layer." *IEICE Transactions on Electronics*, **E75-C**, 649–655.

75. Aucoin L., Bouthillette S., Platzker A., Shanfield S., Bertrand A., Hoke W., Lyman P. (1993). "Large Periphery, High Power Pseudomorphic HEMTs". *Proceedings of GaAs IC Symposium*, 351–354.
76. Fujita S., Noda T., Nozaki C., Ashizawa Y. (1993). "InGaAs/InAlAs HEMT with a strained InGaP Schottky Contact layer." *IEEE Electron Device Lett.*, **14**, 259–261.
77. Harada N., Kuroda S., Watanbe Y., Hikosaka K. (1993). "Frequency divider using InAlAs/InGaAs HEMT DCFL-NOR gates." *Electronics Letters*, **29**, 2100–2101.
78. Hida H., Tokushima M., Maeda T., Ishikawa M., Fukaishi M., Numata K., Ohno Y. (1993). "0.6 V Supply Voltage 0.25 μm E/D HFJET (IS^3T) LSI Technology For Low Power Consumption and High Speed LSIs." *Proceedings of GaAs IC Symposium*, 197–200.
79. Huang J. C., Boulais W., Platzker A., Kazior T., Aucoin L., Shanfield S., Bertrand A., Vafiades M., Niedzwiecki M. (1993). "The Effect of Channel dimensions on the millimeter Wave performance of Pseudomorphic HEMT." *Proceedings of GaAs IC Symposium*, 177–180.
80. Iwata N., Inosako K., Kuzuhara M. (1993). "3V Operation L-Band Power Double-Doped Heterojunction FETs." *IEEE MTT-S Int. Microwave Symp. Dig.*, 1465–1468.
81. Lai R., Wojtowicz M., Chen C. H., Giedenbender M., Yen H. C., Streit D. C., Tan K. L., Liu P. H. (1993). "High Power 0.15 μm V-band Pseudomorphic InGaAs-AlGaAs-GaAs HEMT." *IEEE Microwave and Guided Wave Letters*, **3**, 363–365.
82. Larson L. E., Matloubian M., Brown J. J,. Brown A. S., Rhodes R., Crampton D., Thompson M. (1993). "AlInAs/GaInAs on InP HEMTs for Low Power Supply Voltage Operation of High Power-Added Efficiency Microwave Amplifiers." *Electronics Letters*, **29**, 1324–1325.
83. Matloubian M., Brown A. S., Nguyen L. D., Melendes M. A., Larson L. E., Delaney M. J., Pence J. E., Rhodes R. A., Thompson M. A., Henige J. A. (1993). "High Power V-Band AlInAs/GaInAs on InP HEMT's." *IEEE Electron Device Lett.*, **14**, 188–189.
84. Matloubian M., Brown A. S., Nguyen L. D., Melendes M. A., Larson, L. E., Delaney M. J., Thompson M. A., Rhodes R. A., Pence J. A. (1993). "20 GHz High Efficiency AlInAs-GaInAs on InP Poer HEMT." *IEEE Microwave and Guided Wave Letters*, **3**, 142–144.
85. Matloubian M., Larson L, Brown A., Jelloian L., Nguyen L., Lui M., Liu T., Brown J., Thompson M., Lam W., *et al.* (1993). "InP based HEMTs for the realization of Ultra High Efficiency Millimeter Wave Power Amplifiers". *IEEE/Cornell Conference on Advanced Concepts in High Speed Semiconductors Devices and Circuits*, 520–527.
86. Nguyen L., Le M., Delaney M., Lui M., Liu T., Brown J., Rhodes R., Thompson M., Hooper C. (1993). "Manufacturability of 0.1 μm Millimeter Wave Low noise HEMTs". *IEEE MTT-S Int. Microwave Symp. Dig., 345–348.*
87. Onda K., Fujijara A., Miyamoto H., Nakayama T., Mizuki E., Samoto N., Kuzuhara M. (1993). "Low noise and high gain InAlAs/InGaAs heterojunction FETs with high indium composition channels". *GaAs and Related Compounds*, 139–144.
88. Shealy, J. B., Hashemi M. M., DenBaars S. P., Mishra U. K. (1993). "Breakdown Characterization of AlInAs/GaInAs Junction Modulated HEMTs (JHEMTs) with Regrown Ohmic Contacts by MOCVD". *Proc. of IEEE/Cornell Conference on High Speed Semiconductor Devices and Circuits*, 548–556.
89. Shealy J. B., Hashemi M. M., Kiziloglu K., Denbaars S. P., Mishra U. K., Liu T. K., Brown J. J., Lui M. (1993). "High Breakdown Voltage AlInAs/GaInAs Junction Modulated HEMT's (JHEMTs) with Regrown Ohmic Contacts by MOCVD." *IEEE Electron Device Lett.*, **14**, 545–547.
90. Boulais W., Donahue R. S., Platzker A., Huang, J., Aucoin L., Shanfield S., Vafiades M. (1994). "A High Power Q-Band GaAs Pseudomorphic HEMT Monolithic Amplifier". *IEEE MTT-S Int. Microwave Symp. Dig.*, 649–652.
91. Bouthillette S., Platzker A., Aucoin L. (1994). "High Efficiency 40 Watt PsHEMT S-Band MIC Power Amplifiers". *IEEE MTT-S Int. Microwave Symp. Dig.*, 667–670.

92. Brown J. J., Matloubian M., Liu T. K., Jelloian L. M., Schmitz A. E., Wilson R., G., Liu M., Larson L., Melendes M. A., Thompson M. A. (1994). "InP Based HEMTs with $Al_xIn_{1-x}P$ Schottky Layers grown by Gas Source MBE". *Proceedings of International Conference on InP and Related Materials*, 419–422.
93. Chen C. H., Saito Y., Yen H. C., Tan K., Onak G., Mancini J. (1994). "Reliability Study on Pseudomorphic InGaAs Power HEMT Devices at 60 GHz". *IEEE MTT-S Int. Microwave Symp. Dig.*, 817–820.
94. Enoki T., Kobayashi T., Ishii Y. (1994). "Device Technologies for InP-Based HEMTs and Their Application to ICs". *Proceedings of GaAs IC Symposium*, 337–340.
95. Enoki T., Tomizawa M., Umeda Y., Ishii Y. (1994). "0.05 μm Gate InAlAs/InGaAs High Electron Mobility Transistor and Reduction of Its Short Channel Effects." *Japanese Journal of Applied Physics*, **33**, 798–803.
96. Fu S. T., Lester L. F., Rogers T. (1994). "Ku Band High Power High Efficiency Pseudomorphic HEMT". *IEEE MTT-S Int. Microwave Symp. Dig.*, 793–796.
97. Ho. P., Smith P. M., Hwang K. C., Wang S. C., Kao M. Y., Chao P. C., Liu S. M. J. (1994). "60 GHz power performance of 0.1 μm gate length InAlAs/GaInAs HEMTs". *Proceedings of International Conference on InP and Related Materials*, 411–413.
98. Hwang K. C., Ho P., Kao M. Y., Fu S. T., Liu J., Chao P. C., Smith P. M., Swanson A. W. (1994). "W band High Power Passivated 0.15t μm InAlAs/InGaAs HEMT device". *Proceedings of International Conference on InP and Related Materials*, 18–20.
99. Hwang K. C., Reisinger A. R., Duh K. H. G., Kao M. Y., Chao P. C., Ho P., Swanson A. W. (1994). "A Reliable ECR Passivation Technique on the 0.1 μm InAlAs/InGaAs HEMT Device", *Proceedings of International Conference on InP and Related Materials*, 624–627.
100. Inosako K., Matsunaga K., Okamato Y., Kuzuhara M. (1994). "Highly Efficient Double Doped Heterojunction FET's for Battery Operated Portable Power Applications." *IEEE Electron Device Lett.*, **15**, 248–250.
101. Kao, M.-Y., Duh K. H. G., Ho P., Chao P.-C. (1994). "An Extremely Low Noise InP-Based HEMT with Silicon Nitride Passivation". *IEDM Technical Digest*, 907–910.
102. Kraemer B., Basset R., Baughman C., Chye P., Day D. Wei J. (1994). "Power PHEMT module delivers 4 Watts, 38% PAE over the 18.0 to 21.2 GHz Band". *IEEE MTT-S Int. Microwave Symp. Dig.*, 801–804.
103. Kraemer B., Basset R,. Chye P., Day D., Wei J. (1994). "Power PHEMT Module Delivers 12 Watts, 40% PAE over the 8.5 to 10.5 GHz Band". *IEEE MTT-S Int. Microwave Symp. Dig.*, 683–686.
104. Onda K., Fujihara A., Mizuki E., Hori Y., Samoto M., Kuzuhar M. (1994). "Highly Reliable InAlAs/InGaAs Heterojunction FETs Fabricated Using Completely Molybdenum-Based Electrode Technology (COMET)". *IEEE MTT-S Int. Microwave Symp. Dig.*, 261–264.
105. Ota Y., Adachi C., Takehara H., Yangihara M., Fujimoto H., Hasato H., Inoue K. (1994). "Application of heterojunction FET to power amplifier for cellular telephone." *Electronics Letters*, **30**, 906–907.
106. Scheffer F., Heedt C., Reuter R., Lindner A., Liu Q., Prost W., Tegude F. J., (1994). "High Breakdown voltage InGaAs/InAlAs HFET Using $In_{0.5}Ga_{0.5}P$ spacer layer." *Electronics Letters*, **30**, 169–170.
107. Smith P. M., Creamer C. T., Kopp W. F., Ferguson D. W., Ho P., Willhite J. R., (1994). "A High Power Q-Band HEMT for Communication Terminal Appications". *IEEE MTT-S Int. Microwave Symp. Dig.*, 809–812.
108. Yarborough R., Heston D., Saunier P., Tserng H. Q., Salzman K., B. Smith (1994). "Four Watt, Kt-Band MMIC Amplifier". *IEEE MTT-S Int. Microwave Symp. Dig.*, 797–800.
109. Actis R., Nichols K. B., Kopp W. F., Rogers T. J., Smith F. W. (1995). "High Performance 0.15 μm Gate length *p*HEMTs Enhanced with a Low Temperature Grown GaAs Buffer". *IEEE MTT-S Int. Microwave Symp. Dig.*, 445–448.

110. Bahl S. R., Azzam W. J., delAlamo J. A., Dickmann, J., Schildberg S. (1995). "Off-State Breakdown in InAlAs/InGaAs MODFET's." *IEEE Trans. Electron Devices*, **42**, 15–22.
111. Chen K., Enoki T., Maezawa K., Arai K., Yamamoto M. (1995). "High performance Enhancement mode InAlAs/InGaAs HEMT's Using non-alloyed Ohmic Contact and Pt-Based Buried Gate". *Proceedings of International Conference on InP and Related Materials*, 428–431.
112. Enoki T., Ito H., Ikuta K., Ishii Y. (1995). "0.1 μm InAlAs/InGaAs HEMTs with an InP recess etch stopper grown by MOCVD". *Proceedings of International Conference on InP and Related Materials*, 81–84.
113. Enoiki, T., Umeda Y., Osafune K., Ito H., Ishii Y. (1995). "Ultrahigh-Speed InAlAs/InGaAs HEMT Ics Using pn-Level Shift Diodes". *IEDM Technical Digest*, 193–196.
114. Hayafuji N., Yamamoto Y., Yoshida N., Sonoda T., Takamiya S., Mitsui S. (1995). "Thermal stability of AlInAs/GaInAs/InP heterostructure." *Applied Physics Letters*, **66**, 863–865.
115. Higuchi K., Mori M., Kudo M., Mishima T. (1995). "High Gm MBE grown InP Based HEMTs with a Very Low Contact Resistance Triple Capping Layer". *Proceedings of International Conference on InP and Related Materials*, 741–744.
116. Hori Y, Onda K., Funabashi M., Mizutani H., Maruhashi K., Fujihara A., Hosoya K., Inoue T., Kuzuhara M. (1995). "Manufacturable and Reliable Millimeter Wave HJFET MMIC Technology Using Novel 0.15 μm MoTiPtAu Gates". *IEEE MTT-S Int. Microwave Symp. Dig.*, 431–434.
117. Hur K. Y., McTaggart R. A., LeBlanc B. W., Hoke W. E., Lemonias P. J., Miller A. B., Kazior T. E., Aucoin L. M. (1995). "Double Released AlInAs/GaInAs/InP HEMTs with High Breakdown Voltages". *Proceedings of GaAs IC Symposium*, 101–104.
118. Hur K. Y., McTaggart R. A., Miller A. B., Hoke W. E., Lemonias P. J., Aucoin L. M. (1995). "DC and RF characteristics of double recessed and double pulse doped AlInAs/GaInAs/InP HEMTs." *Electronics Letters*, **31**, 135–136.
119. Hur K. Y., McTaggart R. A., Ventresca M. P., Wohlert R., Aucoin L. M. Kazior T. E. (1995). "High Gain AlInAs/GaInAs/InP HEMT's with Individually Grounded Source Finger Vias." *IEEE Electron Device Lett.*, **16**.
120. Hur K. Y., McTaggart R. A., Ventresca M. P., Wohlert R., Hoke W. E., Lemonais P. J., Kazior T. E., Aucoin L. M. (1995). "High efficiency single pulse doped $Al_{0.60}In_{0.40}As$/GaInAs HEMTs for Q band power applications." *Electronics Letters*, **31**, 585–586.
121. Inosako K., Iwata N., Kuzuhara M. (1995). "1.2 V Operation 1.1 W Heterojunction FET for Portable Radio Applications". *IEEE Electron Devices Meeting*, 185–188.
122. Inosako K., Iwata N., Kuzuhara M. (1995). "Power Heterojunction FETs for Low-Voltage Digital Cellular Applications." *IEICE Transactions on Electronics*, **E78-C**, 1241–1245.
123. Iwata N., Inosako K., Kuzuhara M. (1995). "2.2 V Operation power heterojunction FET for personal digital cellular telephones." *Electronics Letters*, **31**, 2213–2215.
124. Kasody R., Wang H., Biedenbender M., Callejo L,. Dow G. S., Allen B. R. (1995). "Q Band high efficiency monolithic HEMT power prematch structures." *Electronics Letters*, **31**, 505–506.
125. Lee J.-H., Yoon H.-S., Park C.-S., Park H.-M. (1995). "Ultra Low Noise Characteristics of AlGaAs/InGaAs/GaAs Pseudomorphic HEMT's with Wide Head T-Shaped Gate." *IEEE Electron Device Lett.*, **16**, 271–273.
126. Masato H., Maeda M., Fujimoto H., Morimoto S., Nakamura M. Yoshikawa Y., Ikeda H., Kosugi H., Ota Y. (1995). "Analogue/Digital Dual Power Module Using Ion-Implanted GaAs MESFETs". *IEEE MTT-S Int. Microwave Symp. Dig.*, 567–570.
127. Matloubian M., Liu T., Jelloian L. M., Thompson M. A., Rhodes R. A. (1995). "K-Band GaInAs/InP channel power HEMTs." *Electronics Letters*, **31**, 761–762.
128. Matsunaga K., Okamoto Y., Kuzuhara, M. (1995). "A 12-GHz, 12-W HJFET Amplifier with 48% Peak Power Added-Efficiency." *IEEE Microwave and Guided Wave Letters*, **5**, 402–404.
129. Mitama M. (1995). "Mobile Communications Systems Trend in Japan and Device Requirements". *Proceedings of GaAs IC Symposium*, 6–9.

130. Prost W., Tegude F. J. (1995). "High speed, high gain InP-based heterostructure FETs with high breakdown voltage and low leakage". *Proceedings of International Conference on InP and Related Materials*,729–732.
131. Shealy J. B., Hafizi M., Thompson M. A., Sun H. C., Hooper C. E., Mishra U. K., Nguyen L. D. (1995). "High Uniformity 75 GHz Junction HEMTs (JHEMTs) using a Dry-Etch Gate Technology". *IEEE/Cornell Conference on Advanced Concepts in High Speed Semiconductor Devices and Circuits*, 532–541.
132. Shealy J. B., Liu T. Y., Thompson M. A., Wilson R. G., Nguyen L. D., Mishra U. K. (1995). "High Threshold Uniformity, Millimeter-Wave p^+-GaInAs/n-AlInAs/GaInAs JHEMT's." *IEEE Electron Device Lett.*, **16**, 560–562.
133. Smith, P. M., Liu S. M. J., Kao M. Y., Ho P., Wang S. C., Duh K. H. G., Fu S. T., Chao P. C. (1995). "W-Band High Efficiency InP-Based Power HEMT with 600 GHz f_{max}." *IEEE Microwave and Guided Wave Letters*, **5**, 230–232.
134. Staudinger J. (1995). "Modeling GaAs MESFETs for RF Power Amplifiers." *Microwave Journal*, **38**, 20–34.
135. Tanaka T., Furukawa H., Takenaka H., Ueda T., Noma A., Fukui T., Tateoka K., Ueda D. (1995). "1.5 V Operation GaAs Spike-Gate Power FET with 65% Power-Added Efficiency". *IEDM Technical Digest*, 181–184.
136. Teeter D., bouthillette S., Aucoin L., Platzker A., Alfaro C., Bradford D. (1995). "High Power, High Efficiency PHEMTs for Use at 8 GHz". *IEEE MTT-S Int. Microwave Symp. Dig.*, 323–326.
137. Böck, J., Felder A., Meister T. F., Franosch M., Aufinger K., Wurzer M., Schreiter R., Boguth S., Treitinger L. (1996). "A 50 GHz Implanted Base Silicon Bipolar Technology with 35 GHz Static Frequency Divider". *Symposium on VLSI Technology Digest of Technical Papers*, 108–109.
138. Felder A., Möller M., Popp J., Böck J., Rein H.-M. (1996). "46 Gb/s DEMUX, 50 Gb/s MUX, and 30 GHz Static Frequency Divider in Silicon Bipolar Technology." *IEEE Journal of Solid State Circuits*, **31**, 481–486.
139. Fujimara A., Onda K., Nakayama T., Miyamoto H., Ando Y., Wakejima A., Mizuki E., Kuzuhara M. (1996). "Thermally stable InAlAs/InGaAs heterojunction FET with AlAs/InAs superlattice insertion layer." *Electronics Letters*, **32**, 1039–1041.
140. Hiokawa T., Negishi H., Nishimura Y., Ichikawa S., Tanaka J., Kimura T., Watanbe K., Nashimoto Y. (1996). "A Ku-band Ultra Super Low-noise Pseudomorphic Heterojunction FET in a Hollow Plastic PKG". *IEEE MTT-S Int. Microwave Symp. Dig.*, 1603–1606.
141. Hwang T., Kao T. M., Glajchen D., Chye P. (1996). "Pseudomorphic AlGaAs/InGaAs/GaAs HEMTs in low-cost plastic packaging for DBS application." *Electronics Letters*, **32**, 141–143.
142. Iwata N., Tomita M., Yamaguchi K., Oikawa H., Kuzuhara M. (1996). "7 mm Gate Width Power Heterojunction FETs for Li-Ion Battery Operated Personal Digital Cellular Phones". *Proceedings of GaAs IC Symposium*, 119–122.
143. Kao M.-Y., Saunier P., Ketterson A. A., Yarborough R., Tserng H. Q. (1996). "20 GHz Power PHEMTs with Power-Added Efficiency of 68% at 2 Volts". *IEDM Technical Digest*, 931–933.
144. Lee J.-L., Mun, J. K., Kim H., J.-J., Park H.-M. (1966). "A 68% PAE, GaAs Power MESFET Operating at 2.3 V Drain Bias for Low Distortion Power Applications." *IEEE Trans. Electron Devices*, **43**, 519–526.
145. Martinez M. J., Schirmann E., Durlam M., Halchin D., Burton R., Huang J.-H., Tehrani S., Reyes A., Green D., Cody N. (1996). "P-HEMTs for Low-Voltage Portable Applications Using Filled Gate Fabrication Process". *Proceedings of GaAs IC Symposium*, 241–244.
146. Matsunaga K., Okamoto Y., Miura I., Kuzuhara M. (1996). "Ku-Band 15 W Single Chip HJFET Power Aplifier". *IEEE MTT-S Int. Microwave Symp. Dig.*, 697–700.
147. Pereiaslavets B., Bachem K. H., Braunstein J., Eastman L. F. (1996). "GaInP/InGaAs/GaAs Graded Barrier MODFET Grown by OMPVE : Design, Fabrication, and Device Results." *IEEE Trans. Electron Devices*, **43**, 1659–1664.

148. Shealy J. B., Matloubian M., Liu T. Y., Thompson M. A., Hashemi M. M., DenBaars S. P., Mishra U. K., (1996). "High-Performance Submicrometer Gatelength GaInAs/InP Composite Channel HEMT's with Regrown Contacts." *IEEE Electron Device Lett.*, **16**, 540–542.

149. Shealy J. B., Mishra U. K. (1996). "0.2 μm gate length, non-alloyed P^+-AlInAs/n-AlInAs/GaInAs JHEMTs with $f_t = 62$ GHz." *Electronics Letters*, **32**, 2180–2181.

150. Umeda Y., Enoki T., Osafune K., Ito H., Ishii Y. (1996). "High-Yield Design Technologies for InAlAs/InGaAs/InP-HEMT Analog-Digital ICs." *IEEE Transactions on Microwave Theory and Techniques*, **44**, 2361–2368.

151. Yoshida N., Kitano T., Yamamoto Y., Katoh T., Minami H., Kashiwa T., Sonoda T., Takamiya S., Mitsui S. (1996). "A Super Low Noise AlInAs/GaInAs HEMT Fabricated by Selective Gate Recess Etching." *IEEE Trans. Electron Devices*, **43**, 178–180.

152. Chen Y. C., Lai R., Lin E., Wang H., Block Y., Yen H. C., Streit D., Jones W., Liu P. H., Dia R. M., *et al.* (1997). "A 94-GHz 130-mW InGaAs/InAlAs/InP HEMT High-Power MMIC Amplifier." *IEEE Microwave and Guided Wave Letters*, **7**, 133–135.

153. Enoki T., Ito H., Ishii Y. (1997). "Reliability Study on InAlAs/InGaAs HEMTs with an InP recess etch stopper and refractory gate metal." *Solid State Electronics*, **41**, 1651–1656.

154. Glass E., Huang J.-H., Abrokwah J., Bernhardt B., Majerus M., Spears E., Droopad R., Ooms B. (1997). "A True Enhancement Mode Single Supply Power HFET for Portable Applications". *IEEE MTT-S Int. Microwave Symp. Dig.*, 1399–1402.

155. Halchin D., Golio M. (1997). "Trends for portable wireless applications." *Microwave Journal*, **40**, 62–78.

156. Huang J. H., Glass E. J. A., Bernhardt B., Majerus M., Spears E., Parsey Jr. J. M., Scheitlin D., Droopad R., Mills L. A., Hawthorne K., Blaugh J., (1997). "Device and Process Optimization for a Low Voltage Enhancement Mode Power Heterojunction FET for Portable Applications". *Proceedings of GaAs IC Symposium*, 55–58.

157. Hur K. Y., McTaggart R. A., Lemonias P. J., Hoke W. E. (1997). "Development of Double Recessed AlInAs/GaInAs/InP HEMTs for Millimeter Wave Power Applications." *Solid State Electronics*, **41**, 1581–1585.

158. Kunihisa T., Yokoyama T., Nishijima M., Yamamoto S., Nishitsuji M., K. Nishii, Nakayama, Ishikawa O. (1997). "A High-Efficiency Normally-Off MODFET Power MMIC for PHS Operating under 3.0 V Single-Supply Conditions". *Proceedings of GaAS IC Symposium*, 37–40.

159. Lai R, Wang H., Chen Y. C., Block T., Liu P. H., Streit D. C., Tran D., Barsky M., Jones W., Siegel P., *et al.* (1997). "155 GHz MMIC LNAs with 12 dB Gain Fabricated Using a High Yield InP HEMT MMIC Process." *Microwave Journal*, **40**, 166–171.

160. Lai Y.-L., Chang E. Y., Chang C.-Y., Liu T. H. Wang S. P., Hsu H. T. (1997). "2-V-Operation δ-Doped Power HEMT's for Personal Handy-Phone Systems." *IEEE Microwave and Guided Wave Letters*, **7**, 219–221.

161. Lao Z., Bronner W., Thiede A., Schlechtweg M., Hülsmann A., Rieger-Motzer M., Kaufel G., Raynor B., Sedler M. (1997). "35-GHz Static and 48-GHz Dynamic Frequency Divider IC's Using 0.2-μm AlGaAs/GaAs-HEMT's." *IEEE Journal of Solid State Circuits*, **32**, 1556–1562.

162. Maeda T., Fujii M. (1997). "Analytical Expression for operating speed of GaAs SCFL D-Type Flip-Flops," *Solid State Electronics*, **41**, 1687–1691.

163. Mesquida-Kusters A., Heime K. (1997). "Al-Free InP-Based High Electron Mobility Transistors: Design, Fabrication and Performance." *Solid State Electronics*, **41**, 1159–1170.

164. Otsuji, T., Murata K., Enoki T., Umeda Y. (1997). "An 80 Gbit/s Multiplexer IC Using InAlAs/InGaAs/InP HEMTs". *Proceedings of GaAs IC Symposium*, 183–186.

165. Otsuji, T., Murata K., Tokumitsu M., Sutitani S. (1997). "32 Gbit/s super-dynamic decision IC using 0.13 μm GaAs MESFET's with multilayer-interconnection structure." *Electronics Letters*, **33**, 480–482.

166. Otsuji T., Yoneyama M., Murata K., Imai Y., Enoki T., Umeda Y. (1997). "2-46.5 GHz quasi-static 2:1 frequency divider IC using InAlAs/InGaAs/InP HEMTs." *Electronic Letters*, **33**, 1376–1377.

167. Ren F., Lothian J. R., Tasi H. S., Kuo J. M., Lin J., Weiner J. S., Ryan R. W., Tate A., Chen Y. K. (1997). "High Performance Pseudomorphic InGaP/InGaAs Power HEMTs." *Solid State Electronics*, **41**, 1913–1915.

168. Shealy, J. B., Matloubian M., Liu T. Y., Lam W., Ngo C. (1997). "0.9 W/mm, 76% PAE (7 GHz) GaInAs/InP Composite Channel HEMTs". *Proceedings of International Conference on InP and Related Materials*, 20–23.

169. Wurzer M., Meister T. F., Schäfer H., Knapp H., Böck J., Stengl R., Aufinger K., Franosch M., Rest M., Möller M., *et al.* (1997). "42 GHz Static Frequency Divider in a Si/SiGe Bipolar Technology". *IEEE International Solid-State Circuits Conference*, 122–123.

170. Yoneyama M., Otsuji, T., Imai Y., Yamaguchi S., Enoki T., Umeda Y., Hagimoto K. (1997). "46 Gbit/s super-dynamic decision circuits module using InAlAs/InGaAs HEMTs." *Electronic Letters*, **33**, 1472–1474.

171. Yoneyama M., Sano A., Hagimoto S., Otsuji, T., Murata K., Imai Y., Yamaguchi S., Enoki T., Sano E. (1997). "Optical Repeater Circuit Design Based on InAlAs/InGaAs HEMT Digital IC Technology." *IEEE Transactions on Microwave Theory and Techniques*, **45**, 2274-2282.

172. Adesida I., Mahajan A., G. Cueva (1998). "Enhancement-Mode InP-Based HEMT Devices and Applications". *Proceedings of International Conference on InP and Related Materials*, 493–497.

173. Amamiya Y,. Niwa T., Nagano N., Mamada M., Suzuki Y, Shimawaki H. (1998). "40 GHz frequency dividers with reduced power dissipation fabricated using high-speed small-emitter areea AlGaAs/InGaAs HBTs". *Proceedings of GaAs IC Symposium*, 121–124.

174. Bito Y., ıwata N., Tomita M. (1998). "Single 3.4 V operation power heterojunction FET with 60% efficiency for personal digital cellular phones." *Electronics Letters*, **34**, 600–601.

175. Chevalier P., Wallart X., Bonte B., Farquembergue R. (1998). "V-band high-power/low-voltage InGaAs/InP composite channel HEMTs." *Electronics Letters*, **34**, 409–411.

176. Choumei K., Yamamotot K., Kasai N., Moriwaki T., Y. Y., Fujii T., Otsuji J., Miyazaki Y., Tanino N., Sato K. (1998). "A High Efficiency, 2 V Single-Supply Voltage Operation RF front-end MMIC for 1.9 GHz Personal Hand Phone Systeems". *Proceedings of GaAs IC Symposium*, 73–76.

177. Mahajan A., Fay P., Arafa M., Adesida I. (1998). "Integration of InAlAs/InGaAs/InP Enhancement and Depletion-Mode High Electron Mobility Transistors for High-Speed Circuit Applications." *IEEE Trans. Electron Devices*, **45**, 338–340.

178. Pobanz C., Matloubian M., Lui M., Sun H.-C., Case M., Ngo C., Janke P., Gaier T., Samoska L. (1998). "A High-Gain Monolithic D-Band InP HEMT Amplifier". *Proceedings of GaAs IC Symposium*, 41–44.

179. Pullela R., Mensa D., Lee Q., Agarwal B., Guthrie J., Jagannathan S., Rodwell M. J. W. (1998). "48 GHz static frequency dividers in transferred-substrate HBT Technology." *Electronics Letters*, **34**, 1580–1581.

180. Sokokich M., Docter D. P., Brown Y. K., Kramer A. R., Jensen J. F., Stanchina W. E., III S. T., Fields C. H., Ahmar D. A., Lui M., *et al.* (1998). "A Low Power 52.9 GHz Static Divider Implemented in a Manufacturable 180 GHz AlInAs/GaInAs HBT IC Technology". *Proceedings of GaAs IC Symposium*, 117–120.

181. Suemitsu T., Enoki T., Yokoyama H,. Umeda Y., Ishii Y. (1998). "Impact of two-step-recessed gate structure on RF performance of InP-based HEMTs." *Electronics Letters*, **34**, 220–222.

182. Suemitsu T., Yokoyama H., Umeda Y., Enoki T., Ishii Y. (1998). "High-Performance 0.1-μm-Gate Enhancement-Mode InAlAs/InGaAs HEMTs Using Two-Step-Recessed Gate Technology". *Proceedings of International Conference on InP and Related Materials*, 497–500.

183. Wada S., Maeda T., Tokushima M., Yamazaki J. Ishikawa M., Fujii M. (1998). "A 27 GHz/151 mW GaAs 256/258 Dual-Modulus Prescaler IC with 0.1 μm Double-Deck-Shaped (DDS) Gate E/D-HJFETs". *Proceedings of GaAs IC Symposium*, 125–128.

184. Wang Y. C., J. M., Kuo J. R. L., Ren F., Tsai H. S., Weiner J. S., Lin J., Tate A., Chen Y. K., Mayo W. F. (1998). "An $In_{0.5}(Al_{0.3}Ga_{0.7})_{0.5}P/In_{0.2}Ga_{0.8}As$ power HEMT with 65.2% power-added efficiency under 1.2 V operation." *Electronics Letters*, **34**, 594–595.

Antimony-Based Infrared Materials and Devices

C.E.A. GRIGORESCU[1] AND R.A. STRADLING

Blackett Laboratory, Imperial College of Science, Technology and Medicine, London, United Kingdom

3.1. Introduction

Much of the research involving infrared optoelectronics has focused mostly on military sector needs, with a particular emphasis on the development of high-performance detectors for the 10-μm wavelength band. Here the preferred materials system has been HgCdTe despite its rather poor mechanical properties and thermal instability. Civilian needs have become more dominant and a major area of development, for infrared LED and lasers, has become important due to an urgent need to provide low cost, sensitive pollution monitoring systems that detect trace gases by their fundamental vibrational-rotational absorption bands [1]. Other applications include landfill gas monitoring, flue gas analysis, personal safety, sports medicine, heating ventilation and air conditioning, and a variety of horticultural uses that include total organic carbon dioxide measurements, incubators, fruit storage, livestock husbandry, and mushroom farms.

These III–V materials have stronger chemical bonds and are therefore more attractive than II–VI compounds provided that the same range of bandgaps can be covered. The introduction of epitaxial growth methods has widened the range of material combinations available.

---

[1] Permanent address: Institute of Optoelectronics PO Box MG-5 Bucharest Romania

Vol. 28
ISBN 0-12-533028-6/$35.00
ISSN 1079-4050

The most developed semiconductor heterostructures are the lattice-matched systems consisting of GaAs (lattice constant = 5.654 Å) with $Ga_{1-x}Al_xAs$ and InP (lattice constant = 5.868 Å) with $Ga_{1-x}In_xAs$. However, neither of these systems provides a bandgap suitable for either long-wavelength optical sources or mid-infrared detectors. There are also potential applications for narrow gap materials for low power and fast electronic devices. These developments have produced demands for new materials based on InAs, InSb, AlSb, and GaSb and alloy combinations of these binaries. The lattice constants of InAs (6.058 Å at room temperature), GaSb (6.095 Å) and AlSb (6.135 Å) are quite similar but the lattice constant of InSb (6.479 Å) is much greater and all these materials are poorly matched to GaAs or InP. Reasonable quality substrates are available for InAs, GaSb, and InSb but all of these materials are quite conductive at room temperature even without intentional doping. These GaAs substrates are frequently employed for both cost- and electrical isolation-related reasons. In this case an $InAs/In_{0.7}Ga_{0.3}As$ superlattice can be useful as a buffer to prevent threading dislocations from reaching the surface [2].

Two other methods are now emerging that offer great promise for antimony-based laser devices and other structures requiring low dislocation densities with lattice constants different from 6.1 Å and 6.5 Å (i.e., when conventional substrates are unavailable). Compliant substrate technology is currently undergoing rapid development. The growth of dislocation-free InSb was demonstrated on a compliant GaAs substrate formed by wafer bonding a 3-nm GaAs layer with a large angular misalignment to a (011)-bulk GaAs crystal [3]. Reasonable quality ternary substrates of $In_{1-x}Ga_xSb$ ($x \leq 0.11$) have been grown [4]. Both approaches will allow alloys such as $In_{1-x}Al_xSb$ and $InAs_{1-x}Sb_x$ with greater ranges of x to be incorporated into heterostructures without the penalty of increased dislocation density.

The band structure of InAs and InSb is characterized by the small direct bandgap at the centre of the Brillouin zone and the large separation in energy between the conduction band minima at the $\Gamma - X$ and $\Gamma - L$ points. The primary electronic properties arise from the band structure, which results in high mobilities and saturation drift velocities for the electrons. In the case of InSb the room temperature electron mobility is nearly 10 times that of GaAs and the saturation drift velocity exceeds that of silicon by a factor of 5. The low effective masses of InAs and InSb give high quantum confinement energies and large optical nonlinearities. The latter point was vividly demonstrated by the first observation of optical bistability, which was found because nonlinear refraction was already substantial at long wavelengths with milliwatt power levels [5].

The heterostructure combinations of particular interest are InAs/GaSb, InAs/AlSb, $InSb/Al_{1-x}Sb_x$, $InAs_{1-x}Ga_xSb$, $InAs/InAs_{1-x}Sb_x$, and InAs/In(As,Sb,P).

3.2. Overview of Materials and Electronic Properties

3.2.1. NARROW GAP III–V BINARY COMPOUNDS

Indium arsenide has a particularly low lying conduction band that leads to the formation of type II band alignments at heterojunctions and to very large conduction band offsets that can be exploited in such devices as tunnel diodes. The type II alignment can drastically modify the electronic properties, for example, by leading to the suppression of Auger recombination. Another special property is that the deep lying conduction band causes native defect levels to lie about 200 meV above the conduction band edge rather than in the middle of the forbidden gap [6–7]. Consequently electron accumulation layers form naturally at the surfaces of bulk layers and the Fermi energy at a metal semiconductor contact is pinned within the conduction band at a similar energy. Thus a Schottky barrier is not formed and contacts that are extremely transparent to electron flow are readily fabricated.

The defect levels that cause surface pinning also act to stabilize the Fermi level, thereby providing an electronic reference level for the defect annihilation energies. In the case of amphoteric impurities this determines the maximum free carrier concentration that can be obtained from doping [6]. Consequently, InAs can be doped very heavily with Si donors (the preferred dopant in Molecular Beam Epitaxy (MBE) growth) where concentrations as high as $5 \times 10^{19}\,\mathrm{cm}^{-3}$ can be achieved. With GaAs, where the Fermi level is pinned midgap, the donor doping limit is $10^{19}\,\mathrm{cm}^{-3}$. With InSb the pinning energy is near to the valence band so Si acts amphoterically, however, almost complete activation of the silicon as a donor up to concentrations of $\sim 3 \times 10^{18}\,\mathrm{cm}^{-3}$ can be obtained by reducing the temperature to 350 °C . With GaSb and AlSb silicon acts only as an acceptor. These trends can be understood qualitatively in terms of the amphoteric native defect model introduced in Reference [6], where the defects act to stabilize the Fermi level. The position of the defect levels with respect to the band edges therefore determines the maximum free carrier concentration that can be obtained by silicon doping. The defect levels lie in the conduction band of InAs but close to the valence band edge in GaSb.

Doping with elemental tellurium (and other group VI atoms) is discouraged in MBE as Te has a very high vapor pressure and severe long-term memory effects are found. Congruent evaporation using PbTe or GaTe works well with little contamination [9, 10] but ties up an additional cell in the MBE chamber as Si will also be required for *n*-type doping of InAs or InSb. Surface segregation appears to be a problem in AlSb [10]. The untreated GaSb surface is known to produce donor-like levels, which act to pin the Fermi energy at the surface about 0.2 eV above the valence band edge [11–13]. The surface donors provide an additional source of electrons for the InAs quantum well above the intrinsic concentration arising from the semimetallic band alignment with the concentration of extra electrons varying approximately inversely with the thickness of the GaSb cap [11–12]. Apart from the question of segregation, the use of group VI elements for doping presents less of a problem with Metallorganic Vapor Phase Epitaxy (MOVPE).

Biefeld *et al.* [14] have studied the doping of InSb and InSb/$InAs_{1-x}Sb_x$ strained-layer superlattices (SLS) grown by MOVPE, using Se and Sn as dopants for InSb, instead of the more usual (MOVPE) donor dopant Te. Their results show that control of the doping levels obtained with these dopants is better than for the standard MOVPE dopant Te. From their results on the current-voltage characteristics obtained with *p–n* junctions it appears that Sn could be the preferred doping source for InSb.

Gallium antimonide grown without deliberate doping always turns out to be *p*-type with a hole concentration at room temperature of $\sim 10^{16}\,cm^{-3}$. The native defect responsible for the residual *p*-type conductivity is thought to be a double acceptor with levels 33 and 80 meV about the valence band edge [15, 16]. The defect has been variously identified including the possibilities that an antisite point defect formed by a Ga atom on the Sb site or a gallium vacancy complexed with a Ga atom on the Sb site may be responsible. The lower the growth temperature the lower is the density of the native defects and the higher the hole mobility; room temperature mobilities of $900\,cm^2/Vs$ together with peak mobilities at 50 K of 15,000 cm/Vs have been reported.

Aluminum antimonide grown by MBE is generally *p*-type with a hole concentration of $\sim 10^{16}\,cm^{-3}$ at room temperature. Material grown by MOVPE generally shows very large carbon acceptor contamination although the use of new precursors such as tritertiarybutylaluminum [17], trimethylamine alane and ethyldimethylamine alane [18] have dropped the carbon level down from $10^{19}\,cm^{-3}$ to $10^{18}\,cm^{-3}$.

A particular problem arises with AlSb-containing structures because of the instability in air of AlSb. In order to prevent corrosion, the final AlSb layer has to be capped by either a thin GaSb or InAs layer. When GaSb is employed the Fermi energy is pinned by the surface donors [11–13]. Alternatively, a $Ga_{1-x}xAl_xSb$ alloy with $x < 0.5$ can be used as an air-stable cap.

A favorable feature of AlSb is its relatively low refractive index compared with its near lattice-matched partner GaSb. The ratio of the two refractive indices at 2-μm wavelength is 1.24, which is considerably larger than corresponding value for GaAs/AlAs. Thus a simple 10-period AlSb/GaSb-distributed Bragg reflector (DBR) has been shown to have a reflectance of over 98% at 1.92-μm wavelength (see Fig. 3.1) and a 12-period $Al_{0.2}Ga_{0.8}Sb$/AlSb DBR had a reflectivity of 99% at 1.38-μm wavelength [19].

3.2.2. III–V ALLOYS

The $InAs_{1-x}Sb_x$ alloy system with $x \sim 0.65$ has the narrowest direct bandgap of any thick-film III–V material. Remarkably, when InSb is mixed together with the

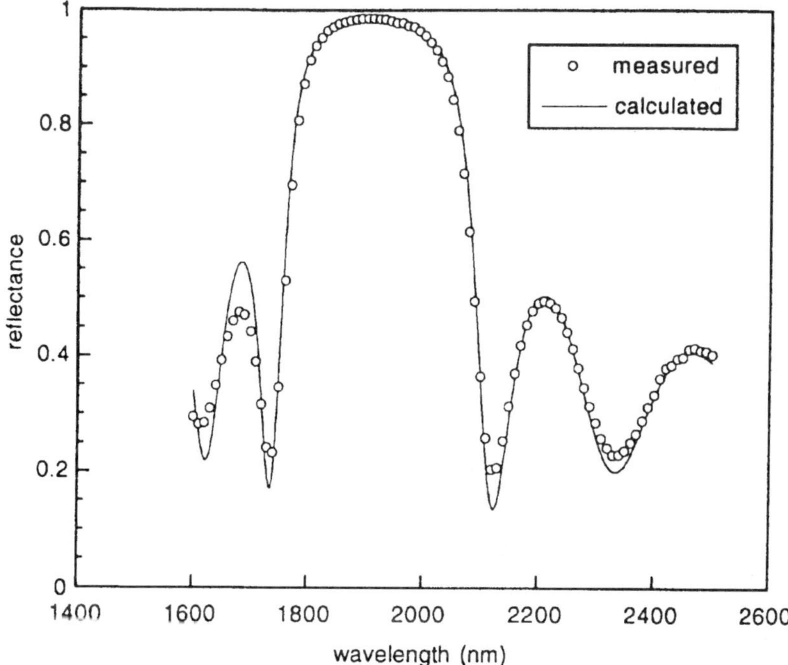

FIG. 3.1. Shows the measured and calculated reflectance for a 10-period AlSb/GaSb distributed Bragg reflector [19].

wider bandgap material InAs at low compositions, the bandgap decreases, reaching a minimum value of 0.13 eV at 10 K and a value of 80.9 ± 1.5 meV at room temperature for $x = 0.63$ [20] (see Fig. 3.2. here).

Together with $InAs_{1-x}Sb_x$, other ternaries such as $In_{1-x}Ga_xAs$, $In_{1-x}Ga_xSb$ [21], $InP_{1-x}Sb_x$ $In_{1-x}Al_xSb$, $AlAs_{1-x}Sb_x$ and various quaternary combinations are frequently used with the binaries to form quantum wells and superlattices. $InP_{1-x}Sb_x$ [22, 23] or the quaternary In(As, Sb, P) [24, 25] is commonly employed for the barriers when the structure is grown by MOVPE. Unfortunately, all the alloys of interest are prone to metallurgical problems such as ordering [26, 27] and phase separation [28, 29] in the mid-alloy range and even "natural superlattices" can be grown when material is supplied at constant composition [28].

3.2.3. HETEROSTRUCTURES

The low-lying conduction band for InAs gives rise to an exceptionally high conduction band offset (1.36 eV at room temperature) for the InAs/AlSb system

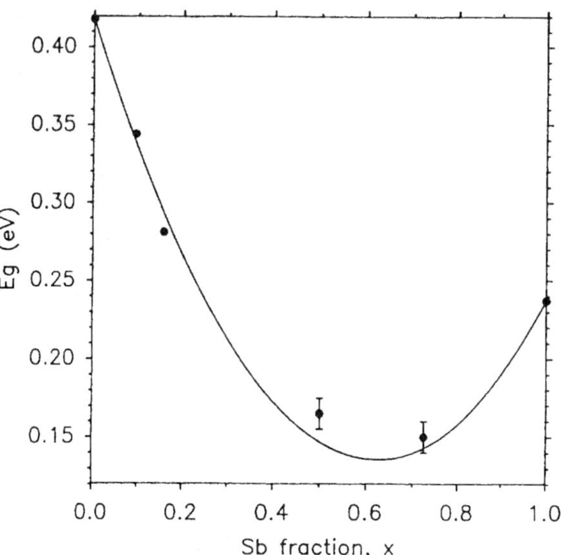

FIG. 3.2. Shows the variation of bandgap (E_g) with alloy composition (x) deduced from interband magnetooptics for the $InAs_{1-x}Sb_x$ alloy system [20]. The full curve corresponds to the quadratic expression $E_g = 0.418 - 0.903x + 0.722x^2$ (eV).

and to a semimetallic band alignment with InAs/GaSb heterostructures [30–32]. The large conduction band offsets are particularly suited to tunnel and other microelectronic devices and also provide the possibility of short wavelength ($\sim 2\,\mu m$) operation of sources and detectors utilizing intersub-band transitions. The semimetallic alignment gives the possibility of interband tunneling processes. The InAs/GaSb/AlSb combinations are therefore attractive for a wide range of infrared and electronic applications

The band offsets for the InAs/GaSb/AlSb system have been determined quite precisely. In the case of InAs/GaSb, the InAs conduction band minimum lies about 150 meV lower than the top of the GaSb valence band for the [001] orientation (type III alignment) with differences of about 20 meV [31] depending on whether the bonding at the interfaces is InSb-like or GaAs-like [32].

For the [111]A orientation the overlap is found to be 200 meV and the difference in energy compared with [001] is attributed to the dipole expected at the interface [32]. Thus the system is naturally semimetallic.

There has been considerable speculation for some time that the simultaneous presence of electrons and holes in undoped structures of InAs/GaSb could lead to the formation of stable excitonic states at low temperatures. Experimental evidence for such states has been lacking until recently when Cheng *et al.* [33]

reported the presence of an additional line approximately 2 meV above the cyclotron resonance line in a number of InAs/(Ga,Al)Sb quantum wells which disappeared with increasing temperature. In this experiment the persistent photoconductive effect was used to control the carrier concentration and the line was only present when the electron and hole concentrations were approximately equal.

A metal insulator transition occurs with single InAs/GaSb quantum wells when the well width is reduced below about 10 nm [30]. When short-period superlattices are grown the material can remain closely intrinsic down to the lowest temperatures provided that a thick surface cap is grown to reduce the influence of the surface [34, 35]. The InAs/AlSb quantum well and superlattices on the other hand appear much more *n*-type and it has been speculated that antisite donors in the bulk consisting of As substituting on the Al site and similar native defects such as As on the Al site at AlAs interfaces can be responsible for these residual donors [36–38]. Donor planes can be formed in AlSb by an interrupted growth technique where the growth surface is first made aluminum rich and then "soaked" with As for 60 s, after which further aluminum is supplied. By this means As antisite defects are formed [38].

There is no common anion or cation across the interface between InAs and GaSb (or AlSb). It is therefore possible to induce two different types of bonding, InSb-like or AlAs (or GaAs)-like at the interface by careful control of the shutter sequences [37, 38]. Without such control the bonding at the interfaces will be random. The band offsets, the local vibrational properties and the electronic mobilities will depend on the nature of the interfaces.

Samples grown with AlAs (GaAs)-like interfaces generally have inferior structural quality and much lower mobility than those with InSb interfaces [37]. The problem appears to be the roughening or intermixing of the surface during the growth of the first AlAs (GaAs) interface due to the exposure of the AlSb (GaSb) surface to excess As [39, 40]. Atomic force microscopy has demonstrated a severe roughening of the first AlSb/InAs interface for structures grown between 490 and 500 °C [41, 42]. The optimum temperature in MBE growth for obtaining high mobility with InSb-interfaces is about 450 °C [43] for InAs/GaSb quantum wells. However, InAs/GaSb quantum wells with even higher structural quality and mobility can be grown at 400 °C by the use of minimum As overpressure to reduce the interface roughening [43].

The barrier and well width dependence of the electrical properties of InAs/AlSb quantum wells has been studied in an extensive series of experiments by the Santa Barbara group [11–13, 40–45]. For studies of the well width dependence all samples had a top barrier consisting of 50 nm of AlSb capped by a 5-nm GaSb layer and InSb interfaces. The bottom barrier consisted of 20-nm of AlSb grown on top of a ten period (2.5 nm + 2.5 nm) GaSb/AlSb smoothing superlattice. The growth temperature was 500 °C. The optimum mobility was found for a well thickness of 12.5 nm [44]. A decrease in mobility with decreasing thickness was found for well widths of below 10 nm, which is thought to be due to interface roughness [45]. In contrast to the rapid decrease of carrier concentration with decreasing well thickness found with InAs/GaSb wells and explained by the metal-insulator transition [30], the carrier concentration in the InAs/AlSb wells increased rapidly with decreasing well thicknesses below 7 nm [44]. This behavior has yet to be explained.

Similar structures to those employed in Reference [45] were employed in References [11–13] for a study of the effect of the barrier thickness on the mobility except that the well width was kept the same for all samples at 15 nm and the GaSb surface cap was increased to 10 nm. On decreasing the thickness of the AlSb barrier, the carrier concentration was found to increase rapidly with decreasing distance (L) between the quantum well and the surface in the manner expected on the surface pinning model. An increase in mobility was also observed with decreasing L. This was assumed to be correlated with the increasing carrier concentration as would be expected with a fixed concentration of scattering centers and transfer of carriers from the surface. For very small values of L (or large values of carrier concentration), however, the mobility fell sharply, either because of surface roughness or because of the onset of intersubband scattering.

Remote doping of InAs/GaSb or InAs/AlSb quantum wells presents a problem with MBE. Use of a group VI dopant risks long-term memory effects. Silicon, which is the preferred MBE dopant for most other Ill-V systems, acts as an acceptor with GaSb and in AlSb.

An alternative technology has been developed [46–49] for remote doping of the InAs quantum wells where a double well structure is employed. The second InAs well is thin (~2 nm) and doped with silicon where concentrations as high as 5×10^{19} cm^{-3} can be employed (see Section 3.1). Because of high confinement energy this well acts as a source of electrons for the first well. Superlattices formed by heavily *n*-doped InAs regions separated by undoped GaSb or AlSb spacers are also used to provide current transport through the *n*-type contact forming an optical cladding layer surrounding the active regions in InAs/GaSb laser or detector structures.

In contrast to the accurate band offsets established for InAs/GaSb and InAs/AlSb, the band alignments at the InSb/$InAl_{1-x}Sb_x$ interface are not known accurately. In a study of the electrical properties of InSb/$InAl_{1-x}Sb_x$ quantum wells [50], the value of x was kept small ($x = 0.09$) to limit to 0.5% the lattice mismatch to InSb. This restricted the concentration of electrons achievable in the wells by remote doping to 3×10^{11} cm^{-2}. The corresponding mobilities were 45,000 cm^2/Vs at room temperature and 300,000 cm^2/Vs at low temperatures.

The low-temperature mobilities are the highest yet achieved for epitaxial InSb but are still substantially less than those routinely achieved with undoped *n*-type substrate material.

Strain-layer superlattices of both InAs/InAs $_{1-x}Sb_x$ and InSb/InAs $_{1-x}Sb_x$ have been grown with reasonable mobility despite the large mismatch. Interband magneto-optics and luminescence [51–54]; show that the photon energy emitted falls extremely rapidly with increasing x with the bandgap decreasing at 10 K from 0.44 eV for $x = 0$ to 0.10 eV for $x = 0.4$.

The reason for this rapid narrowing of the superlattice energy gap has been controversial with different groups suggesting conflicting schemes [55]: (i) "type II" band alignments with the valence band offsets changing rapidly with x (in which case the photon emission is spatially indirect) with the conduction band minimum being either in the InAs [52] or in the alloy [54]; or alternatively (ii) a spatially direct bandgap in which the alloy bandgap is anomalously narrowed by microstructural effects such as atomic ordering [51, 52].

Strained-layer superlattices (SLS) are grown from lattice mismatched epilayers, with layer thickness below a critical value. Above the critical thickness strain relaxation occurs by generation of dislocations. The SLS layers with larger bulk lattice constants are under bi-axial compression and those with smaller bulk lattice constants are under bi-axial tension. For the layers under bi-axial compression the bulk conduction-band minimum energy is increased and a splitting of the bulk light- and heavy-hole bands occurs. In the case of the layers under bi-axial tension, a reduction of the bulk conduction-band minimum energy is found to be accompanied by a splitting (of opposite sign from the compressive case) of the light- and heavy-hole bands. Both effects tend to decrease the bandgap of the structure with respect to the unstrained case.

3.2.4. EXOTIC Sb-BASED INFRARED MATERIALS

The compatibility of two apparently disparate materials has been demonstrated with the growth of high-quality heterostructures between metallic Sb and GaSb [56, 57]. It has proved possible to perform multilayer growth on [111] surfaces using MBE and migration-enhanced epitaxy despite the fact that the crystal structure is different, with GaSb being zinc blende while Sb is rhombohedral. Regrowth of one on the other is possible because on a [111] surface there is near perfect lattice match (to 0.06%) and the interface atomic nets are both hexagonal.

Two other alloys of InSb ($InSb_{1-x}Bi_x$ and $In_{1-x}Tl_xSb$) are known to result in a reduction in the bandgap. Thin layers of $InSb_{1-x}Bi_x$ ($x = 0.29$) were grown by MBE and these were found to be strongly n-type from electrical measurements [58–60]. At 77 K bismuth substitution appeared to optimize the electron-concentration ($1.6 \times 10^{18}\,cm^{-3}$) and Hall mobility ($480\,cm^2/V\,s$). This alloy could provide an infrared material with a cutoff wavelength beyond 8 μm, at 77 K.

The ternary compound $In_{1-x}Tl_xSb$ has been grown by liquid phase epitaxy onto semi-insulating GaAs substrates [61]. The test structure contained an InSb buffer layer. Infrared transmission and photoresponse measurements showed

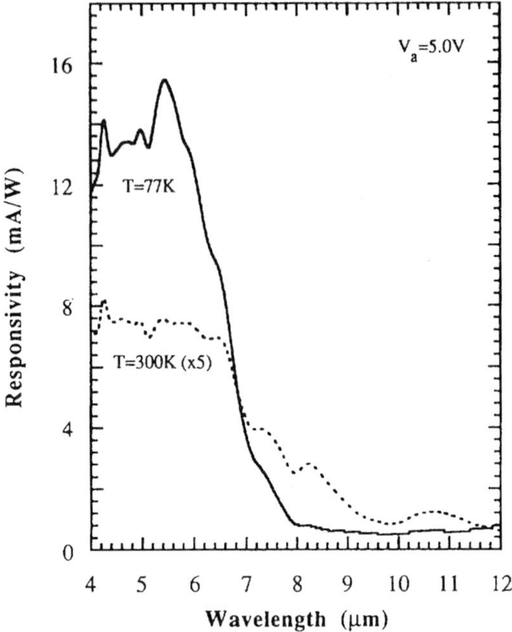

FIG. 3.3. Shows the spectral response measured at 300 and 77 K for an $In_{1-x}Tl_xSb$ alloy photoconductor.

evidence of an increase in the absorption edge with increasing thallium concentration [62, 63]. The cutoff wavelength increased linearly with increasing lattice mismatch to InSb, reaching a value of 9.0 μm at 77 K at a lattice mismatch of -1.3% (see Fig. 3.3). The wavelength for the peak photoresponse also increases with the thallium content. The specific detectivity D^* at 77 K and 7 μm wavelength reached a value of 3×10^8 cm $Hz^{1/2}W^{-1}$.

The quaternary compound, GaInSbBi, has been grown by MBE for infrared applications [64].

3.3. Mechanisms Limiting the Performance of Sources and Detectors

Generation–recombination processes play the main role in limiting the performances of infrared optoelectronic devices [65–78].

3.3.1. SHOCKLEY–READ RECOMBINATION

The Shockley–Read (S–R) mechanism occurs via defect levels in the forbidden energy band and is therefore not a fundamental limitation [70]. Frequently the S–R generation–recombination process can be controlled by reducing the concentrations of native defects and foreign impurities during growth of the material.

3.3.2. RADIATIVE MECHANISM

The fundamental barrier to the improvement in detector performance is the background limit [66]. The radiative lifetime as defined by Roosbroeck and Shockley [79] can be strongly modified because of reabsorption effects. Reabsorption can also modify the performance of infrared sources; in this case however the effect is to reduce the efficiency. Equating the internal radiative lifetime to the overall lifetime can be incorrect by a factor of at least 25. This implies that very high radiative efficiencies are necessary to attain the background limit and Humphreys' work [66] actually refers to an idealized photodetector.

The same author [67] also considered less ideal materials with realistic surface models and operation at relatively high temperatures. Remarkably, the modeling is not limited to small signal approximation and applies to photodetectors of any thickness up to a diffusion length. It is shown within a good approximation that the decay rate of an uniform excess carrier population in a semiconductor sample is given by the sum of the non-radiative decay rate and the net rate of photon emission from the sample. The correct expression for the radiative lifetime is

$$\tau = \frac{dn_i^2}{2\eta_B \phi_B}(n+p) \tag{3.1}$$

with n_i the intrinsic carrier concentration, η_B the quantum efficiency for background radiation at the temperature of the sample, Φ_B the photon flux from the background and n, p, the electron and hole concentrations, respectively.

The conclusions are that radiative recombination followed by reabsorption is essentially noise–free in a semiconductor infrared photodetector and that internal radiative recombination can be neglected for all practical purposes. Moreover, the fundamental process involved is photon re-emission, whose characteristic rate is given by the correct expression for the radiative lifetime (Eq. (3.1.)) and not by the conventional one. For this quantity to be a limiting one, the radiative efficiency should approach 100%.

3.3.3. AUGER RECOMBINATION

Auger recombination provides a nonradiative process that increases strongly with decreasing bandgap and with increasing carrier concentration. Auger generation-recombination processes therefore determine the performance of most infrared devices operated close to room temperature. In the case of interband lasers, Auger recombination invariably contributes to the high threshold currents and can require low temperatures. There are 10 Auger processes possible in materials with InSb-like band structure but only the two having the lowest threshold energy are of practical importance [80] — these are Auger-1 and Auger-7.

The net Auger–generation rate due to these two processes has the form:

$$g_A - r_A = \frac{n_i^2 - np}{2n_i^2}\left[\frac{n}{(1+an)\tau_{A1}^i} + \frac{p}{\tau_{A7}^i}\right] \tag{3.2}$$

where n_i is the intrinsic carrier concentration, a is a parameter that differs from unity only in the case of degenerate statistics [69] and τ_{Aj}^i are the intrinsic Auger lifetimes [76].

Auger recombination sets a fundamental limit to the radiative recombination efficiency in any optoelectronic device. In the case of infrared photodetectors, it results in high generation-recombination diode currents that actually lead to low zero–bias resistance-area products and low detectivities unless the detector is cooled and/or the carrier concentrations are reduced by other means [81].

Auger suppression was previously observed in diode or multi-element structures of narrow bandgap materials operated close to room temperature [70].

If one of the components of the structure is very lightly doped, the material can be considered intrinsic and the diodes will exhibit different properties in comparison with conventional bipolar diodes in the following aspects: (a) diffusion and drift are dominated by ambipolar effects, since the electrons and holes are present in equal proportions and are therefore charge-coupled; (b) perturbations of carrier distributions, especially near junctions, can only be described by a large signal theory as no distinction exists between minority and majority carriers; and (c) the photoexcited carrier concentrations in low bandgap materials at near room temperature are dominated by carrier-dependent Auger processes and therefore the characteristic lifetimes cannot be regarded as independent of position, in contrast to the case with the usual diode structures.

The analysis of Auger processes in the *i* zone of *p-i-n* structures, developed by White [70], shows that Auger suppression is accompanied by strong negative resistance effects as increasing bias forces the carrier concentration to lower values. It is demonstrated that total Auger suppression in a *p-i-n* diode is dependent on device length, which should not be greater than a critical value defined in relation to the ambipolar diffusion length

$$L = \sqrt{D_a \tau} \tag{3.3}$$

D_a is the ambipolar diffusion coefficient having the form

$$D_a = \frac{2D_n D_p}{D_n + D_p} \tag{3.4}$$

with D_n and D_p the diffusion coefficients for electrons and holes, respectively.

To achieve Auger suppression the device should be shorter than $\pi L/\sqrt{2}$. Longer devices exhibit a partial effect only.

Due to their properties, low-gap *p-i-n* diodes are best suited to the requirements for low-noise infrared detectors operating under moderate temperature conditions.

In quantum well (QW) and strained layer superlattices (SLS) devices the Auger mechanism can also be controlled via band-structure engineering [82, 83].

Due to their special importance, Auger processes have been extensively studied both experimentally and theoretically. A simple analytic approximation for electron-initiated Auger transitions valid over a wide range of electron and hole Fermi levels and temperature was developed by Beattie and White [84] using time-dependent perturbation theory. Since in narrow-gap semiconductors the ratio of heavy-hole effective mass to conduction band effective mass is large, the valence band can be considered flat, with Auger transitions involving states close to the zone center. Flat valence band, Fermi-Dirac, statistics and the assumption of constant overlap functions are used to derive the Auger rates. The method is applied to InSb and cadmium mercury telluride, the results being compared with accurate evaluations [85].

Shockley-Read and Auger lifetimes for MBE-grown InAs and $InAs_{0.91}Sb_{0.09}$ thin films and $InAs_{0.85}Sb_{0.15}$-InAlAsSb multiple quantum wells have been determined from the intensity-dependent photoconductive response to 2.6-μm excitation [86] The Auger coefficient in the alloy decreases with decreasing temperature, whereas that in the quantum well increases.

The Auger recombination rate has been studied as a function of carrier density in InSb and InAs MBE layers by means of the more accurate technique involving time-resolved pump-probe saturation transmission using a novel infrared parametric oscillator as source [77].

Murdin *et al.* [78] performed similar pump-probe measurements with a picosecond free-electron laser with InSb epilayers and arsenic-rich InAs/$InAs_{1-x}Sb_x$ ($x = 0.68$)-strained layer superlattices (type II structure). This experiment shows that Auger processes are substantially suppressed in InAs/$InAs_{1-x}Sb_x$ strained layer superlattices at room temperature. Moreover, this provides the strongest example to date of Auger recombination suppression at long wavelengths, which has important implications for both infrared sources and detectors based on this system.

Similar effects have been predicted theoretically for another type II strained layer superlattices system—InAs/$In_{0.25}Ga_{0.75}Sb$ [87]. In this case, by increasing the indium composition the lattice mismatch is also increased, which leads to a strain splitting of the highest two valence bands. This result together with the quantum confinement limits the available phase space for Auger transitions.

The subject of suppression of non-radiative processes has been the subject of a review by Pidgeon *et al.* [88].

3.4. Infrared Emitters

3.4.1. INTRODUCTION

There has been great progress in the development of III–V antimony-based lasers operating between 2- and 4-μm wavelength. The first report of operation with III–V antimony materials system was with bulk InSb where lasing was observed at a wavelength of 5.2 μm but only at a temperature of 10 K [89]. The threshold currents required were extremely high (60 kA/cm^2) even though a magnetic field was employed to increase the density-of states close to the conduction band edge. Until 1995 the longest wavelength for the room-temperature operation of an antimony-based III–V interband laser was 2.78 μm [90]. Table 3.1 lists the characteristics of the newer mid-IR laser systems that have been developed with III–V materials and compares the results with the newer quantum well cascade lasers and LEDs operating with intersub-band transitions that employ the AlInAs/GaInAs- [91–97] or GaAs/GaAlAs- [98] materials systems and with II–VI ($Hg_{1-x}Cd_xTe$)- [99]; and IV–VI (PbSe/PbSrSe)-laser systems [100]. Column 1 gives the institute and first author; column 2 gives the materials system involved; column 3 gives the structure; columns 4–8 list the operating parameters at particular temperatures; and column 9 gives the mode of excitation and the maximum operating temperature.

3.4.2. TERNARY AND QUATERNARY LASERS (MAINLY InAs/$InAs_{1-x}Sb_x$ DEVICES)

The quarternary (Ga,In)(As,Sb) has been used as the basis for lasers operating between 2 and 3 μm [90, 101–103]. The ternary In(As,Sb) is a popular choice [103–115] as a component in the active region of the laser structure at wavelengths of 3.4 and 4.5 μm. Intense luminescence is seen from "strained

TABLE 3.1

PERFORMANCE OF INFRARED LASER SYSTEMS

Group	Material	Structure	λ (micron)	T	Max. power	Threshold	T_σ	Max temperature mode of operation
			Antimonide-based Infrared Lasers					
MIT								pulsed
Melngailis [89]	Insb	diode	5	10 K		60 kA/cm^2		in B-field (0.7 to 1.1 T)
Choi and Eglash [101, 102]	(GainAl)(AsSb)	broad stripe	2	300 K	1.3 W(cw)	140 A/cm^2	110 K	
		ridge waveguide	2	300 K	100 mW	20 mA		
		tapered	2	300 K	200 mW			
		DH[a]	3	100 K	90 mW	9 A/cm^2 at 40 K		255 K pulsed/cw to 170 K
Lee *et al.* [103]	InAsSb/GaSb		3.9	85 K	0.8 W			210 K optically pumped
Eglash and Choi [104]	InAsSb/AllnAs	DH	3.9	70 K	30 mW			170 k pulsed/105 k cw
	InAsSb/AllnAs		4.5					85 K pulsed
	InAsSb/AlAsSb	DH/diode pumped	4	80 K	200 mW (mean)			155 K pulsed/80 K cw
Choi and others [105–107]	InAsSb/InAlAsSb	SLS[b]	3.9	80 K	60 mW	78 A/cm^2	30 K	165 K pulsed/128 K cw
Le *et al.* [108]	InAsSb/InAlAsSb	broad stripe SLS	3.3	80 K	5 mW/F cw	30 A/cm^2	30–40 4K	225 K pulsed/175 K cw
	InAs/GaInSb	MQW[d]	4	80 K	350 mW av			optically pumped
Sarnoff Research Center								
Lee *et al.* [90]	AlGaAsSb/InGaAsSb	MQW	2.78	288 K	30 mW	10 kA/cm^2	8 K	333 KL pulsed
Sandia								
Kurtz *et al.* [109]	InAsSb/InGaAs	SLS	3.9	<100 K				optically pumped
Allerman *et al.* [110]	InAs/InAsSb/InPSb	SLS diode/	3.8–3.9				30–40 K	210 K pulsed
Kurtz *et al.* [111]	InAsSb/InPbSb	SL[e] MQW cascade	3.8–3 9	80 K	>100 mW pk[f]	1 kA/cm^2		180 K pulsed
Kurtz *et al.* [23]	InAsSb/InPSb	SLS diode	3.86	80 K	6 mW/F pk	4 kW/cm^2	33 K	240 K optically pumped
Ioffe		DH LPE	3.2	80 K	8 mW	40 mA	30 K	80 K cw
Baranov *et al.* [112]	InAsSbP/InAsSb			180 K		6 A		180 K pulsed
Aidaraliev [113]	InAsSbP/InGaAsSb	DH LPE	3.55	80 K	130 mW pk	87 A/cm^2		
Illinois								
Diaz *et al.* [25]	InAs/InAsSb/InAsSbP	DH stirpe	3.2	77 K	260 mW	40 A/cm^2		pulsed
Rybaltowski *et al.* [24]	InAs/InAsSb/InAsSbP	DH stripe3.290 K	3 W			good far field props		
Lane *et al.* [114]	InAs/InAsSb/InAsSbP	MQW stripe	3.65	90 K	1 W	35 A/cm^2		pulsed up to 200 K
Hughes								
Zhang and Zhang *et al.* [53, 115]	InAs/InAsSb	diode pumped SLS	3.4	95 K		56 A/cm^2	32 K	cw optically pumped
Miles *et al.* [119]	InAs/GaInSb	Type II SL	3.3	40 K		350 A/cm^2	63 K	170 K pulsed
Chow *et al.* [120]	InAs/GaInSb		3.8					
NRL/Houston								
Felix *et al.* [121]	InAs/GaSb/AlSb	VCSEL Type II SL	2.9	260 K	2 W pk	100 kW/cm^2	30 K	Optically pumped to 280 K

(*continued*)

TABLE 3.1 (*continued*)

Group	Material	Structure	λ (micron)	T	Max. power	Threshold	T_σ	Max temperature mode of operation
Bewley *et al.* [122]	InAs/GaInSb/AlSb	type II SL	2.9	200 K	200 mW/F pk	1.1 kA/cm^2	33 K	Pulsed to 260 K
Malin *et al.* [123]	InAs/GaInSb/AlSb	type II SL	3.1	100 K	6.5 W/F pk	340 W/cm^2	27 K	Optically pumped to 260 K
Lin *et al.* [124]	InAs/GaInSb/AlSb	type II SL	3.22	300 K	270 mW			Optically pumped to 350 K
Bewley *et al* [125]	InAs/GaInSb/AlSb	type II SL	4.25	220 K		3.25 kW/cm^2	30 K	Optically pumped to 300 K/
			3.83	80 K	250 mW/F pk	150 W/cm^2		CW to 100 K (14.7 mW at 74 K)
NRL/Houston								
Felix *et al.* [126]	InAs/AlSb/GaInSb	interband cascade	3.0	100 K	532 mW	170 A/cm^{-2}		pulsed to 225 K
Oliafsen *et al.* [127]	InAs/AlSb/GaInSb/	interband cascade	3.6	196 K	160 mW/F	3.3 kA/cm^2	53 K	pulsed to 286 K
Iowa								
Flatte *et al.* [100]	InAs/GaInSb/	type II SL	4.95	80 K		62 kW/cm^2	37 K	optically pumped
	AlGaInAsSb		5.2 μ at 185 K					
DRA Malvern	InSb/AlInSb	DH	5.1	80 K	56 mW	1.48 kA/cm^2	17 K	90 K pulsed
Ashley *et al.* [117]								
			Other Infrared Laser Systems					
ATT				0.2 Wpk				
Faist *et al.* [96]	AllnAs/GainAs	unipolar cascade	5.2	300 K	6 mW av	10 kA/cm^2	114 K	320 K pulsed
Sirtori *et al.* [97]	AllnAs/GaInAs	unipolar cascade	11.2	110 K	50 mW pk	3.3 kA/cm^2	132 K	
				10 K	7 mW (cw)	3.2 kA/cm^2		
Paris-Sud								
Gauthier-Lafaye *et al.* [98]	GaAs/AlGaAs	unipolar cascade	15.5	80 K	0.4 W/F pk	0.5 MW/cm^2		optically pumped (CO^2) to 110 K
MIT	CMT/CdZnTe	II-VI QW	3.2	88 K	1.3 W peak			154 K diode pumped
Le *et al.* [99]				105 mW mean				
Fraunhofer	PbSe/PbSrSe	IV-VI MQW	4.2	282 K		330 Ma		pulsed
Shi *et al.* [100]			5	120 K	300 μW	800 mA	cw	
			7.3	30 K			cw	

[a] DH = double heterostructure, [b] SLS = strained layer superlattice, [c] SL = superlattice, [d] MQW = multiple quantum well, [e] F = facet, [f] pk = peak

PERFORMANCE OF INFRARED LED SYSTEMS

Kurtz/Alleman	InAsSb/InPSb	4	80 μW	280 K	SLS LED	cw
ICSTM	InAs/lnAsSb	3–10		290 K	SLS LED	cw
Golding et al [57]		4.5	~1 μW			
Wood et al [58]		7	~0.2 μW			
ATT	AllnAs/GaInAs	8 to 13	6 nW	10–200 K	cascade LED	cw
Capasso [51–54]					(Blue Stark shift)	

layer superlattices (SLS)" formed by $InAs/InAs_{1-x}Sb_x$ or $In_{1-x}Ga_xAs/InAs_{1-x}Sb_x$. The MIT, Sandia, Ioffe and Illinois groups have investigated lasers based on $InAs_xSb_{1-x}$ [23–25, 101–119]. With these SLS it is thought that band structure effects can be used to quench nonradiative Auger recombination either by the strain splitting of the valence band [83, 84] or by the type II band structure [78, 82]. Operation up to 240 K at a wavelength of 3.9 μm has been reported for an In(As,Sb)/In(P,Sb)-strained-layer superlattice diode laser [23]. Peak powers of 100 mW have been measured at 3.9 μm and 1 W at 3.65 μm for low-temperature operation with In(As,Sb)-based lasers.

3.4.3. InAs/(GaAlIn)Sb LASERS

Hughes, NRL, Houston, and Iowa [101–127] have developed type II superlattice lasers based on InAs/GaSb and InAs/AlSb. The structural quality of these heterostructures is generally superior to $InAs/InAs_{1-x}Sb_x$ structures because of the greater strain generally present with the latter system and the metallurgical problems discussed in Section 3.2.2. The $InAs/Ga_{1-x}In_xSb$ combination is often used to introduce strain as a design parameter to modify the superlattice band structure. Room temperature operation at 4.3 μm wavelength has been achieved with an optically pumped InAs/(Ga,In)Sb multiple quantum well laser but the peak powers reported are lower at comparable wavelengths and temperatures compared with those quoted in the preceding for the In(As,Sb) systems.

The longest wavelength operation yet reported for a III–V interband laser near to room temperature is at 5.2 μm where pulsed operation was achieved at 185 K by optically pumping a $InAs/Ga_{0.6}In_{0.4}Sb$ multiquantum well structure [116].

3.4.4. InSb BASED LASERS

Despite the early reports of the magnetically tuned bulk InSb laser [89], binary InSb was not featured as a component for interband MIR lasers until liquid nitrogen temperature operation was achieved at 5.2 μm wavelength with an InSb [117] diode laser that used a relatively simple pseudo double heterostructure involving an $In_{1-x}Al_xSb$ barrier with $x = 0.14$. The $InSb/InAs_{1-x}Sb_x$ superlattice system is found to give low luminescence efficiency in contrast to the $InAs/InAs_{1-x}Sb_x$ combination.

3.4.5. INTERSUBBAND QUANTUM WELL CASCADE LASERS

A different approach, not involving band-to-band radiation, uses transitions between the sub-bands in a multiple quantum well structure [91–99]. A sophisticated superlattice structure is employed to obtain laser action where the lower state of the laser transition is separated by an optical phonon energy from the ground state to ensure population inversion (see Fig. 3.4). The distinctive feature of cascade lasers is that each injected electron is reused with the

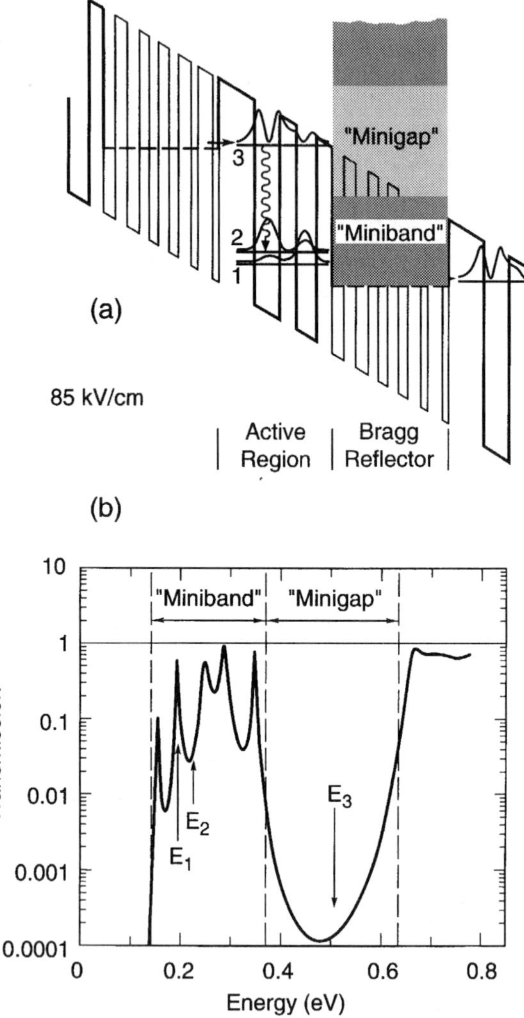

FIG. 3.4. Gives the schematic band structure for a portion of an intraband quantum cascade laser structure consisting of $Ga_{0.47}In_{0.53}As/Al_{0.48}In_{0.52}As$ quantum wells. The lasing transition is between level 3 and 2, which is in turn emptied rapidly by a non-radiative transition into level 1 accompanied by the emission of an LO phonon.

possibility of extra photons being generated each time it cascades down a step in an energy staircase. The active region is designed so that tunneling out of the higher of the two levels involved in the lasing transition is impeded by means of a minigap in the downstream superlattice band structure as a Bragg reflector for the electrons in the excited laser state. Build-up of electrons in the lower of the two states involved in the lasing action is avoided by designing the thicknesses of the active layers so that a third level is located exactly an optical phonon energy below the final lasing state, which is therefore emptied very rapidly. This fast non-radiative relaxation creates substantial local heating and limits the performance of intersub-band cascade lasers. Typically the electrons are recycled through 25 stages.

The system is generally described as a "quantum cascade" laser. A similar structure can be used for LED devices [91]. Pulsed operation at room temperature has been achieved at a wavelength of 5.2 μm [96] and laser emission at a wavelength as long 15.5 μm has been reported at temperatures of up to 110 K [98]. Quantum cascade lasers are characterized by both high values of T_0 and high threshold currents. They are unipolar devices and thus are not restricted by the requirement for diode lasers of having highly doped *n*- and *p*-type contacts. Current intersub-band cascade lasers based on AlInAs/GaInAs do not work well at wavelengths < 5 μm because of the limited conduction band offset available for this materials combination.

3.4.6. INTERBAND CASCADE LASERS

A different type of cascade laser structure has been developed in which the optical transition involved is between the conduction band and valence band in a type II quantum well structure (see Fig. 3.5) [126–128]. By using this spatially indirect type of transition the fast phonon scattering step can be avoided, thereby reducing the local heating and leading to high output powers; simultaneously the advantages of cascade operation and wavelength tunability are still retained. Higher thresholds have already been shown for an interband laser based on the "W" type II combination of InAs/GaSb than for any other "W" laser structure. Further improvements can be expected with better optimized designs.

3.4.7. LED AND NEGATIVE LUMINESCENCE

The reverse biasing of a *p-n* junction can reduce the emission of photons below the density expected from a blackbody of the same temperature. This negative luminescence [135] can be used to cool the surroundings or as part of LED

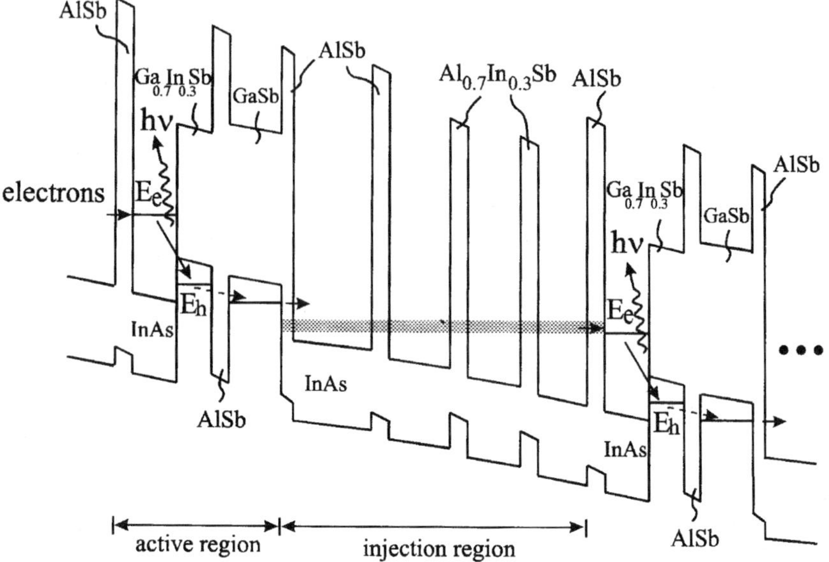

FIG. 3.5. Gives the schematic band structure for a portion of an interband cascade laser consisting of InAs/(Ga, Al)Sb type II quantum wells.

sources where the enhanced dynamic range of modulation of the emission can be used to good effect in modulated-source gas sensing.

3.4.8. II–VI and IV–VI Lasers

In comparison, pulsed operation very near to room temperature (282 K) was reported at a 4.2 μm wavelength with lead salt multiple quantum well lasers ($PbSe/Pb_{0.9785}Sr_{0.0215}Se$) [100]. These authors suggest that IV–VI laser systems may be preferable to III–Vs for long wavelengths but that III–Vs are likely to be superior for wavelengths 3 μm.

The HgCdTe/ZnCdTe lasers have been designed to pulse at a wavelength of 5.3 μm up to 60 K and up to 154 K at 3.2 μm wavelength using optical pumping [99].

3.4.9. Summary

In only a few years the maximum wavelength for III–V laser operation near to room temperature has increased from 2.78 μm [90], to beyond 5 μm [96, 108]. For interband lasers the main competition has been between the $InAs/InAs_{1-x}Sb_x$

and InAs/(GaAlIn)Sb- materials combinations. In terms of longest wavelength operation, from Table 3.1 the latter appears to have the edge. However, care should be taken in comparing the performance of diode and optically pumped lasers as the latter are free of the compromises in design that determine the doping levels for the optical and electrical confinement layers in diode lasers. Consequently, although it is not yet clear which will be the preferred materials system for MIR interband lasers, the III–Vs generally look to be more capable of development as compared with II–VIs and IV–VIs. Intersub-band cascade lasers have made spectacular advances but threshold currents remain high. Interband cascade lasers have shown encouraging initial results and have the potential of operation with much reduced thresholds.

3.5. Infrared Detectors

3.5.1. INTRODUCTION

Infrared photodetectors can be classified into three categories: *(i) photon detectors*—in which absorbed radiation excites electronic transitions in the detector material and an electric output signal is observed due to the change in the distribution function; *(ii) thermal detectors*—in which one of the temperature-dependent properties of the material changes follows the radiation absorption; and *(iii) radiation field detectors*—which respond directly to the radiation field, that is, they do not depend on thermal or carrier generation effects (for example parametric up-conversion in a non-linear optical material).

Early work on infrared photodetectors involved mainly IV–VI compounds, such as PbS, PbSe, PbTe, all of which cover the wavelength range up to 5 μm. A period then followed in which InSb won the top position in the infrared detection competition. Both photoconductive and photovoltaic InSb detectors were fabricated [61, 65]. For photoconductive detection *p*-type InSb cooled below the intrinsic region was preferred because of the high electron-to-hole mobility ratio.

The need to develop infrared imaging systems for infrared astronomy, environmental and defense reasons gave rise to the emergence of new semiconductor materials, such as $Hg_{1-x}Cd_xTe$, whose bandgap can be tailored to match any application within the spectral range from about 2-μm to beyond 10 -μm wavelengths [75].

Cooling is still required for HgCdTe-(intrinsic material) detectors to achieve background limited performance (BLIP) at long wavelengths. Although cheaper and easier to operate, near-room temperature detectors are generally far from achieving BLIP and are hence inadequate for many applications [132–134].

Cooling requirements have been dramatically reduced however by the use of non-equilibrium modes of operation based on device structures where the carrier densities are held below their equilibrium values. A non-equilibrium state can be achieved in photoconductive devices by means of minority carrier exclusion or in photodiodes by means of minority carrier extraction [81, 135, 136] to suppress the noise associated with Auger generation processes. This implies that detectors can operate at higher temperatures with noise figures that are appropriate to lower temperatures (see Fig. 3.6).

The technologies for InSb- and HgCdTe-infrared detectors are now extremely well developed in comparison with other midinfrared detectors, with 2D photovoltaic arrays consisting of up to 100,000 pixels available.

3.5.2. The Design Parameters of Infrared Photodetectors

3.5.2.1. Responsivity

The theory underlying infrared photodetectors has been developed by many authors [65, 73–75, 137].

An interesting point of view was advanced by Williams [138, 139], who relates the characteristic detector parameters involved, that is, responsivity and

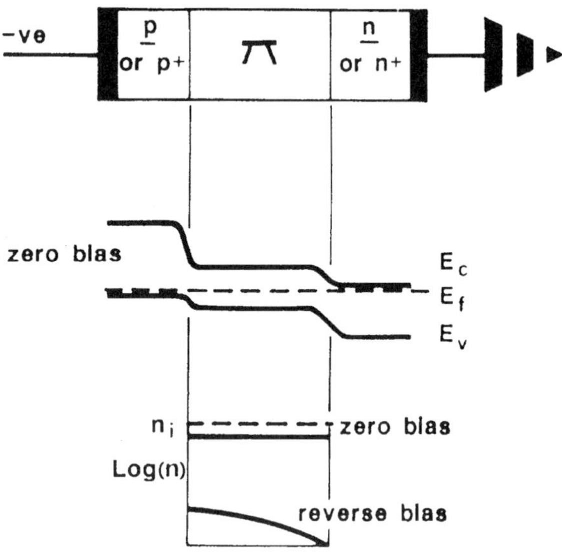

FIG. 3.6. Shows the principle of a p π n heterojunction diode extraction structure showing energy levels and electron concentrations [81].

detectivity, to the parameters of interest to the designer and device user, which are *signal current, rms noise current and signal- to-noise ratio (SNR)*. The photodetector responsivity is examined in the small-signal range, and the magnitude of the responsivity, together with detector frequency response and linearity, are discussed in relation to the photogain.

The standard definition for responsivity relates the signal current (i_s) developed per unit signal power (E_s) incident to the detector surface to its area:

$$R_i(T) = \frac{dI}{AdE} = \frac{i_s}{AE_s(T)} \tag{3.5}$$

where A is the detector area and T the effective blackbody temperature of the irradiant target. For a linear detector, the slope dI/dE of the current-irradiance curve (which is measured at the background irradiance value) is constant.

The magnitude of the photocurrent involves the convolution of several processes, the carrier generation, the decay in the excess energy of the photoexcited carriers towards the band edge, the magnitudes of carrier mobilities, and finally the recombination kinetics. The recombination processes in particular can severely limit quantum efficiency and hence the photogain of the device.

In the case of long-wavelength, room-temperature operation, the thermal generation rates are high, producing currents that short the photocurrent and reduce the performance of the devices. Thus, to achieve high performance the thermal generation rate should be drastically decreased.

Voltage responsivity is directly obtained from the preceding equation:

$$R_v(T) = rR_i(T) \tag{3.6}$$

where $r = dV/dI$ is the dynamic resistance.

The spectral (or monochromatic) responsivity is obtained in the form:

$$R_i(\lambda) = \frac{eG\lambda\eta(\lambda)}{hc} \tag{3.7}$$

with G the photogain, $\eta(\lambda)$ the spectral quantum efficiency, λ the wavelength value where the current responsivity is determined, e the electronic charge, h the Planck constant, and c the velocity of light. The photogain G is defined as the number of carriers passing through the contacts for each photogenerated electron-hole pair. It embodies the magnitude of the responsivity, the detector linearity, and the frequency response. For photodiodes at midband frequencies, where minority carriers are swept rapidly across the junction and no gain process occurs, $G = 1$, and in photoconductors, where the excess carrier motion is governed by the ambipolar transport equations [140], the photogain can become of the order 10^2. The photogain is dependent also on the recombination lifetime, which depends on the magnitude of the irradiance through the excess photocarrier density. Therefore, G is dependent also on the irradiance.

3.5.2.2. *Current Noise*

Accurate expressions for the noise in infrared photodetectors have been derived in many references [64, 75, 137, 138, 140–142].

In the case of photoconductors, noise arises from generation-recombination processes involving both photo- and thermally generated carriers, Johnson-Niquist noise, and $1/f$ (flicker) noise. The later contribution is believed to be caused by the surface properties of the photoconductor, and suitable design or passivation steps can limit its significance to low frequencies. The main properties of $1/f$ noise in homogeneous semiconductors are described by Hooge's empirical formula [143]. A detailed theory of flicker noise, extended to include photodiodes, has been developed by Arutyunyan *et al.* [144]. This assumes fluctuations in the mobility of the free carriers [145, 146] arising from fluctuations in the scattering cross section. The conclusion arises that the electron-phonon interaction plays the main role in the generation of $1/f$ noise in homogeneous, non-degenerate n-type semiconductors. However, with narrow gap semiconductors, degenerate statistics frequently apply even at room temperature.

3.5.2.3. *Detectivity*

Specific detectivity D^* is the main parameter that characterizes the normalized SNR of photodetectors:

$$D^* = \frac{(A\Delta f)^{1/2}}{\mathrm{NEP}} \tag{3.8}$$

where noise-equivalent power (NEP) is defined by the ratio between the rms current noise and the current responsivity R_i, previously defined.

$$\mathrm{NEP} = \frac{I_n}{R_i} \tag{3.9}$$

A is the area of the detector, Δf is the frequency band where the responsivity is measured, and I_n is the current noise.

In order to provide an accurate calibration of photodetectors, separate measurements are carried out to determine the blackbody signal response and the rms current noise. The latter quantity (Eq. (3.9)) is measured when the photodetector is exposed to blackbody radiation, in the absence of signal, at a well-defined ambient temperature (usually 290–300 K). The results of the two different measurements are used to evaluate the specific detectivity.

3.5.2.4. *Cutoff Wavelength*

It is important to define carefully the value of wavelength known as "cutoff wavelength" (λ_c) where the photodetector quantum efficiency and, consequently, the responsivity fall to zero. Often other characteristic wavelengths close to λ_c appear in the literature, for example, the value of the wavelength where responsivity falls to half of its maximum value [147, 148].

The true cutoff wavelength is defined by

$$\lambda_c = \frac{hc}{E_g} \tag{3.10}$$

where E_g is the gap energy of the photodetector.

3.5.2.5. *Specification of Infrared Photodetectors and Arrays*

The change from single-element detectors to focal plane arrays requires the definition of a new figure of merit, to cover the widest possible range of devices. As in a focal plane array the signals from individual detectors are not directly accessible, so the "classic" figures of merit cannot apply. Therefore a new quantity should be defined to quantify the performance of the focal plane as a unit. An attempt in this direction was made in the very detailed work by Humphreys [68].

It is shown that a good figure of merit should fulfil the following requirements: *(a) the definition should give a good measure of the device performance; (b) the test conditions should be simple and reproducible; (c) the expression should apply to all detection mechanisms; (d) all parameters involved should be measurable; and (e) the definition should be clear enough not to give rise to misinterpretation*.

The preferred figure of merit as results from the study has the following form:

$$D_a^* = \frac{SB^{1/2}}{N\frac{P}{A}A_D^{1/2}} \tag{3.11}$$

D_a^* signifies the specific detectivity of the array, S is the signal response to a 500 K blackbody, which illuminates the detector array at a power density of P/A W cm^{-2}, N is the standard deviation of successive readings of an individual pixel output, B is the bandwidth, and represents half of the pixel rate, and A_D is the ratio between the area A of the focal plane and the number of its individual detectors. The detector area A is either the physical area (as in the case of a masked-off photoconductor) or the full width half maximum (FWHM) of a point spread function used to define mathematically the focal plane area.

The main use of a figure of merit is to compare different types of detectors.

3.5.3. Bulk InSb- and $InAs_{1-x}Sb_x$-Infrared Photodetectors

3.5.3.1. InSb Infrared Photodetectors

The properties of indium antimonide as a material for infrared photodetectors have been extensively discussed for more than 40 years. However, improvements and consequently new results have arisen from advances in InSb technology.

At room temperature the intrinsic carrier density in narrow gap materials has values in the range 10^{15}–$10^{17}\,cm^{-3}$ and the thermal generation rate is of the order 10^{23}–$10^{24}\,cm^{-3}s^{-1}$, which results in a high noise level in photodetectors at room temperature. A method of obtaining better performance at RT was found by Ashley *et al.* [81] and Ashley [149] through the use of "non-equilibrium" operation.

The structure designed to achieve "non-equilibrium" (or "HOT": *Higher Operating Temperature*) conditions is of the type $p{+}p^{+}\pi n^{+}$ or $p^{+}p^{+}\nu n^{+}$, where π and ν refer to near intrinsic material (either p- or n-type), which forms an "active" region, and the underlined p^{+} refers to wider bandgap material (see Fig. 3.6). The active region has a low doping level and therefore is intrinsic or near-intrinsic at room temperature. The p- and n-type contacts are made to the active region via regions with high doping level, or larger energy gap, or both, so that under appropriate bias conditions minimal transport of minority carriers is ensured through the active region. The thickness of the near-intrinsic region ranges from 0.3–5 μm, depending on the type of device, and the n^{+} and p^{+} regions are usually about 1-μm thick. The n^{+} InSb is heavily degenerate and therefore has a very low hole density, resulting in a very large step in the valence band at the πn^{+} or νn^{+} junction in comparison with p^{+} InSb, which is not highly degenerate. To achieve a larger step in the conduction band at the junction a region of $In_{1-x}Al_xSb$ ($x = 0.15$) is used with p^{+} InSb. The p^{+}InSb forms a low resistance contact with relatively low recombination velocity to the $In_{1-x}Al_xSb$.

The epitaxial layers optimize the active part of the device to achieve high quantum efficiency over a large temperature range within the minimum material volume. At zero bias the band structure of the device ensures little transport of the minority carriers from the contact regions so that additional noise is minimized. An increase in the temperature of operation is obtained in the vicinity of 200 K. At 294 K the detectivity D^* of these devices is 2.5×10^{9} cm $Hz^{1/2}W^{-1}$, calculated from measured values of R_0A. This latter product is an order of magnitude higher than that in commercially available single- element thermal detectors.

For certain applications, particularly those requiring low-temperature operation, it is often desirable to narrow the spectral responsivity, thereby increasing detectivity by reducing the influence of the background radiation. This situation can be achieved either by an external filter or by embodying the filter in the photodetector structure. A remarkable method [150] of self-filtering involves the application of the Moss-Burstein effect, which is particularly strong in n^+-p-InSb photodiodes.

The structure described by Djuric *et al.* [150] consists of a lightly doped ($5 \times 10^{14}\,\mathrm{cm}^{-3}$ at 77 K) p-type InSb substrate on which an n^+ layer (20 μm thick) is grown by liquid phase epitaxy. The dopant for the n^+ layer is tellurium and an electron concentration of $(4–8) \times 10^{18}\,\mathrm{cm}^{-3}$ at 77 K is achieved. Analysis of capacitance-voltage characteristics shows an abrupt junction. The quantum efficiency of the InSb detectors using the Moss-Burstein effect decreases almost linearly with the wavelength. This allows approximately constant sensitivity over a wide range of wavelengths when choosing appropriate material parameters.

Later work on InSb, with the aim of preparing high-performance infrared photodetectors, has been directed towards reducing the surface recombination rate and increasing the carrier lifetime in near-bulk-like InSb grown by MBE.

Surface recombination has an important role in limiting the performance of InSb-infrared photodetectors in general and in focal plane arrays in particular. There are two approaches that can lead to significant improvements: *a) the minimization of surface recombination; and b) its uniformity across the surface of the array.*

Studies of surface passivation have proved for InSb detectors that passivated (111) surfaces provide lower recombination rates and better electrical isolation than (001) ones [151]. Michel *et al.* [152] investigated the MBE growth of complete InSb photovoltaic structures on (111)B-GaAs and compared the results with those obtained for (001) surfaces. It was shown that the structural, electrical and optical properties of the (111) B material are very close to those of (001) devices, but provide lower surface recombination (see Fig. 3.7). The same group also performed a mobility spectrum analysis of Hall data and demonstrated the presence of a low mobility group of carriers that were believed to be located at the InSb/GaAs interface.

Michel *et al.* [152] also demonstrated a near-bulk value for the carrier lifetime in InSb grown by MBE on GaAs substrates despite the high dislocation densities close to the interface. The carrier lifetime was derived from measurements of photoresponse, performed on a 4.4 μm thick epilayer, at 80 K. A value of 240 ns resulted, which is the highest carrier lifetime in heteroepitaxial InSb reported to date.

3.5.4. QUANTUM WELLS AND STRAINED LAYER SUPERLATTICE INFRARED PHOTODETECTORS

3.5.4.1. Quantum Wells - Intersub-band Infrared Photodetectors

An alternative approach for fast and efficient infrared detectors has emerged involving intersub-band transitions in quantum well where the well width and the

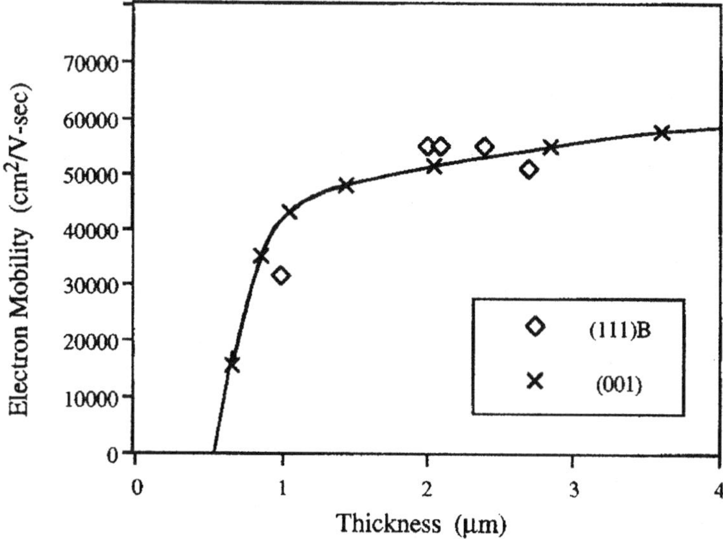

FIG. 3.7. Compares the results for the variation of mobility with epilayer thickness for InSb grown on (111)B GaAs substrates with the mobilities obtained for (001) surfaces.

height of the potential barrier determine the operating wavelength [153, 154] (i.e., Quantum Well Intersub-band Photodetectors or QWIP structures). In comparison with the classical photoconductors, photons are only absorbed in the wells. Most of the studies [86, 155–157] of QWIP detectors have involved GaAs/$Ga_{1-x}Al_xAs$ heterostructures, but the much higher band offsets for the InAs/AlSb system may offer advantages for short wavelength operation.

Liu [156] has developed a specific model, which explains the mechanism of the photoconductive gain and its dependence on the number of wells. The photoconductive gain is defined by the number of electrons flowing through the external circuit for each mobile carrier generated in the sample. The model assumes that: (a) *the intensity of the incident light is low, and therefore photocurrent is at most a small fraction of the total current through the device*; (b) *the dark current is limited by thermal effects, neglecting the interwell contributions*; (c) *the contacts are ideal* (not rigorously true, as emphasized by the author); and (d) *the well is assumed to hold only one bound state.*

There is a direct photoemission of electrons from the well that contributes to the photocurrent. The photoconductive gain (G) results from the extra current injection (from the assumed perfect contacts), which balances the loss of electrons from the well caused by photoemission.

In Section 3.5.2 a general definition of G was given (for both photoconductor and photodiode cases), and it was pointed out that G is dependent on the recombination lifetime; for the QW device the photoconductive gain is defined as a function of the carrier lifetime, τ_{life}, and τ_{transit}, which is the transit time across the detector active region; τ_{life} is associated only with the trapping processes (scattering of an electron into a bound state in the well) and is therefore assumed to be equal to the intersub-band relaxation time, τ_{relax}.

A more precise definition of the photoconductive gain for the QW-infrared photodetector involves the *capture* (trapping) probability for an excited electron crossing the well:

$$p = \frac{\tau_{\text{escape}}}{\tau_{\text{relax}} + \tau_{\text{escape}}} \tag{3.12}$$

τ_{escape} signifies the time spent by the electron in the region of the well while passing by. Trapping in the well originates in scattering processes by a variety of mechanisms, such as impurities and electrons in the well, phonons, and interface roughness.

By accounting for the trapping probability and for the number of wells N, the photogain may be written as

$$G = \frac{1-p}{Np} \tag{3.13}$$

The current responsivity may be independent of the number of wells, but detector performance still depends on it, from noise consideration [155].

A further developed model for photoconductive gain and generation-recombination noise in multiple quantum wells (MQW)-infrared photodetectors is given by Beck [157].

Under negligible tunneling conditions Eq. (3.13) is written as

$$G = \frac{1}{Np} \tag{3.14}$$

and the noise power is found

$$I_n^2 = 4e\bar{I}GB\left(1 - \frac{p}{2}\right) \tag{3.15}$$

where $\bar{I}$ is the mean current, G is the photogain, and B is the measurement frequency bandwidth. From measurements of *g-r* noise power and mean current, respectively, the photogain is inferred by solving Eqs. (3.14) and (3.15).

When $p \ll 1$, the *g-r* noise contains equal contributions from carrier generation and decay, as in a homogeneous photoconductor. For $p \rightarrow 1$ the recombination noise decreases to zero, which makes the total noise equal to the shot noise of N independent junctions connected in series.

Low bias or tunneling dominated current, when $p \approx 1$, leads to an overestimation of the noise by a factor of 2 when using the noise equation for a homogeneous photoconductor. For a QW photodetector under the conditions specified here, shot noise is half that in a homogeneous photoconductor.

For QW photodetectors with tunneling barriers the current responsivity depends on the number of wells, and the detectors have a higher dark current.

Because the capture probability arises from scattering phenomena, the reproducibility of the detector performance cannot be very high, having in view at least the variations in the interface roughness caused by different growth processes. On the other hand, by tailoring the well width and the height of the tunneling barrier, longer cutoff wavelengths can be achieved.

3.5.4.2. Superlattice Based Detectors

Antimony-bearing superlattice structures are often highly strained. Strained layer superlattices (SLS) employ transitions across the fundamental gap [75, 158–161], and therefore carrier lifetime is limited by radiative and Auger recombination processes.

These systems often have type II interfaces [162–164] where the conduction band of one material is lower than the valence band of the other one, and hence the opportunity occurs for the SLS to have a bandgap narrower than either of the constituents. The effective bandgap of the SLS depends on the composition, well width, and strain within the SL [165]. The absorption of infrared radiation is spatially indirect between the states in the valence band of the first material and the conduction band minimum of the second one. Because the optical absorption mechanism involves interband transitions, SLS are intrinsic devices.

Carrier recombination occurs primarily by a relatively slow Auger process and therefore carrier lifetimes are longer than those in extrinsic MQW photoconductors, thus providing better performance at higher temperatures.

3.5.4.3. InAs/(GaIn)Sb SLS Photodetectors

A drawback of type-II superlattices is that they offer a low optical-absorption coefficient due to the separation of electrons and holes in real space. The strength of the band edge transitions is determined by the overlapping of the wavefunctions and can be increased by using very thin layers. In addition this can enhance the quantum size contributions to the band gap.

The first III–V superlattice exhibiting a longer cutoff wavelength than the binary constituents was formed by InAs/GaSb, which has a semimetallic band alignment [75]. With these superlattices a finite bandgap only occurs as a result of the size effect. The bandgap is therefore a strong function of the superlattice period. Recent results with InAs/GaSb superlattices [166] have shown a 14-μm absorption edge at 300 K in good agreement with the theoretical value calculated using an envelope function approximation that contains strain [167]. The

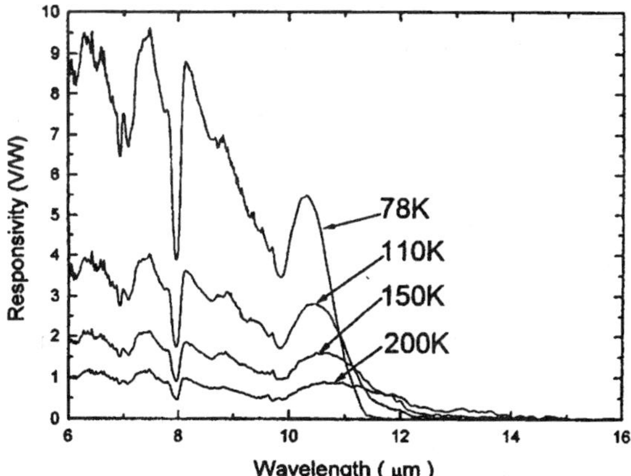

FIG. 3.8. Shows the responsivity of an InAs/GaSb type II quantum well photodetector at various temperatures. The structure has a cut-off wavelength of 12 µm and peak in the response of about 9.5 V/W at a wavelength of about 8 µm at 78 K [166].

responsivity measurements (see Fig. 3.8) show a peak of about 9.5 V/W at a wavelength of about 8 µm, at 78 K. Although maximum responsivity decreases in strength with increasing temperature, it is remarkable that at 150 K responsivity is still about 2 V/W. The detectivity was 1.3×10^9 cm $Hz^{1/2}W^{-1}$ ($\lambda = 10.3$ µm, 78 K) with the limiting mechanism Johnson noise.

An alternative infrared system based on $InAs/In_xGa_{1-x}Sb$ SLS with strain increasing with In content (x) also shows promise for the long wavelength infrared range [82, 168–173]. The SL is strained, with the In_x $Ga_{1-x}Sb$ layers under bi-axial compression. The strain lifts the degeneracy of the light- and heavy-hole edges, which are shifted in energy downward and upward, respectively. When the resulting splitting exceeds the energy gap, Auger recombination is suppressed in p-type material.

Grein *et al.* [170] calculated the theoretical performance of 11–19 µm $InAs/In_xGa_{1-x}Sb$ SL-infrared detectors for n-on-p photodiodes with a doping level of 10^{15} cm^{-3} on the p-type side, as a function of temperature. The result has been compared with the variation of the detectivity of bulk HgCdTe photodetectors over the same temperature- and wavelength range. Detectivity was calculated as a function of the quantum efficiency η [174] and, implicitly, of the absorption coefficient α and the minority carrier diffusion length L_n.

This latter parameter is proportional to $\tau_n^{1/2}$, the minority carrier lifetime, so that $D^* \sim \tau_n^{1/4}$. The minority carrier lifetime embodies the lifetimes due to radiative, electron-electron band-to-band Auger, hole-hole band-to-band Auger, and acceptor-assisted Auger recombination mechanisms.

An acceptor-assisted Auger mechanism limits the performance of the 13–19-μm SL below temperatures ranging from 130 to 100 K and band-to-band Auger recombination dominates at higher temperatures. This parameter is proportional to $\tau_n^{1/2}$, the minority carrier lifetime, so that $D^* \sim \tau_n^{1/4}$. The minority carrier lifetime includes the lifetimes due to radiative, electron-electron band-to-band Auger, hole-hole band-to-band Auger and acceptor-assisted Auger recombination mechanisms. The acceptor-assisted Auger mechanism limits the performance of the 13–19-μm SLs below temperatures ranging from 130–100 K and band-to-band Auger recombination dominates at higher temperatures.

For the long-wavelength infrared detectors considered, background-limited temperatures, T_{BLIP}, of 94, 86, and 73 K are predicted for 13-, 15- and 19 μm, respectively, with an RT background and a field of view of 60°. The suppression of band-to-band Auger recombination in the SL is most effective in increasing T_{BLIP} for lower values of D^*_{BLIP}; for higher D^*_{BLIP} the performance enhancement is reduced due to the rapidly decreasing importance of band-to-band Auger recombination in comparison with that of radiative recombination when temperature decreases.

A detailed study on InAs/(GaIn)Sb diodes grown on GaSb, by MBE, has been performed by Fuchs *et al.* [172]. Superlattices consisting of 12 monolayers (ML) of InAs alternated with 10 ML $Ga_{0.8}In_{0.2}Sb$ were grown on undoped GaSb substrates with a bi-axial tensile mismatch of between 2 and 2.6×10^{-3}. In-plane electron mobilities of approximately $10^4\ \mathrm{cm^2/Vs}$ at RT and hole mobilities of slightly above $10^3\ \mathrm{cm^2/Vs}$ were found by magnetotransport measurements. A broad spectral photoresponse over the wavelength range 5–8 μm was obtained at 77 K, most likely extending to shorter wavelengths at values higher than 1.5 A/W. The peak responsivity is around 2 A/W, which depends very weakly on the reverse bias voltage and is almost independent of temperature up to 180 K.

The Johnson-noise limited detectivity of a photodiode operated at a temperature T depends on the current responsivity R_i and the dynamic impedance (r_0A) at zero voltage, with A the photodiode area. At temperatures above 77 K the diodes are limited by diffusion currents. The high r_0A products at 77 K lead to Johnson-noise limited detectivity, greater than $1 \times 10^{12}\ \mathrm{cm\ Hz^{1/2}W^{-1}}$. The background-limited performance is found at 110 K, with a cutoff wavelength of 8 μm, at a field of view of 45°.

The tunneling contributions in the InAs/(GaIn)SbSL system are found to be less important than with $Hg_{1-x}Cd_xTe$ [172]. In comparison with $Hg_{1-x}Cd_xTe$, the higher electron effective mass of InAs/(GaIn)SbSLs leads to a significant reduction of leakage currents caused by band-to-band tunneling. Therefore, a higher background doping level, associated with a smaller depletion width of the *p-n* junction is acceptable. Moreover, the effective mass of heavy holes for the in-plane motion becomes comparable to the electron effective mass, leading to a higher symmetry in the band structures and hence to an important reduction of the Auger rates for electrons in the *p*-type region of the diode [82, 170, 171]. The high electron effective mass in the InAs/(GaIn)Sb SL is associated with a higher density of states, which can explain the strong absorbency and increased current responsivities observed in this system [159, 166, 172].

Excellent performance at long wavelength has been demonstrated for $InAs/Ga_{1-x}In_xSb$ SLSs grown by MBE in Reference [168]. The composition x for the $Ga_{1-x}In_xSb$ alternating layers was chosen so that the superlattice was symmetrically strained with respect to the GaSb substrate. Both photoconductive and photovoltaic detectors were fabricated. For photoconductors with $x = 0.25$, both the photoresponse at 5 K and the cutoff wavelength are increased with increasing thickness of the $Ga_{1-x}In_xSb$ layer. The longest cutoff wavelength is 12 μm for this composition.

For the photovoltaic detectors, which are designed to have a 12-μm energy gap, the dynamic impedance is $10\,\Omega\,cm^2$, corresponding to a detectivity of 1.5×10^{11} cm $Hz^{1/2}W^{-1}$ at 80–85 K. These values are comparable to the figures of merit for high-performance HgCdTe-photovoltaic detectors that have a similar energy gap.

For infrared imaging, gas sensing and spectroscopy, an important wavelength range is 2–2.6 μm. $In_xGa_{1-x}Sb/GaSb$ photodiodes with $x = 0.4$, grown by MOVPE and operating at these wavelengths at room temperature with a cutoff wavelength of 2.7 μm have been reported in Reference [175]. The 2.5% mismatch of the alloy with the GaSb substrate leads to a large number of dislocations in the *p-n* junction. Intermediate buffer layer structures were investigated involving either steps in the concentration of indium or a combination of ramp and strained layer superlattices and a maximum quantum efficiency of 42% was achieved.

An improvement in the sensitivity over a limited wavelength range can be achieved by placing the absorbing region between two Bragg mirrors to form a resonant cavity-enhanced (RCE) photodetector [176].

Mansoor *et al.* [176] demonstrated for the first time RCE photodetectors based on bulk Ga(In)Sb- or GaInSb/Ga(In)Sb-quantum wells with a detectivity D^* (1.7 μm) = 7.7×10^9 cm $Hz^{1/2}W^{-1}$ at room temperature. The resonant devices show a fourfold increase in the photocurrent in comparison with non-resonant structures. The quantum efficiency and detectivity of the RCE detectors are limited by the high dark current observed. Cooling to 77 K significantly reduces the dark current, so that detectivity increases by at least one order of magnitude.

3.5.4.4. *The System $InAs_{1-x}$.Sb_xSLS*

Alone among the bulk III–V semiconductor alloys available as alternatives to $Hg_{1-x}Cd_xTe$ for the detection of long-wavelength infrared radiation, $InAs_{1-x}Sb_x$ is the one capable of covering the entire wavelength range 3–12 μm at room temperature [177]. The intrinsic carrier concentration, a key parameter in determining the performance of IR detectors, has been calculated for In(As,Sb) in Reference [178].

Photovoltaic devices are more favorable for array applications. Double heterostructure p-InSb/p$^-$ $InAs_{0.15}Sb_{0.85}$/n-InSb photovoltaic detectors have been grown on semi-insulating GaAs [61]. The top and bottom InSb layers were doped at the level of 10^{18} cm^{-3} and a 3×10^{16} cm^{-3} doping level was achieved for the p-type $InAs_{0.15}Sb_{0.85}$ layer. The room-temperature responsivity of the device is comparable to that of HgCdTe detectors operated under the same conditions.

Various photovoltaic detectors based on SLS—$InAs_{1-x}Sb_x$/(InAs; InSb; GaAs; InAsSb)—have been proposed where the wavelength range is extended because of the type II band alignments [158]. Typical structures contain a strain-balanced buffer layer corresponding to the lattice constant of the substrate. The first high-performance In(As,Sb)-infrared photodetectors were reported by Kurtz *et al.* [179]. These devices were $InAs_{0.15}Sb_{0.85}$/InSb SLS photodiodes, exhibiting a detectivity of 1×10^{10} cm Hz$^{1/2}$ W^{-1} at a wavelength value of about 10 μm at 77 K The dynamic impedance at this temperature had a value of 9 Ω cm^2

Some InSb/$InAs_{1-x}Sb_x$/InSb ($p^+\pi n^+$) photovoltaic devices, grown by low-pressure MOVPE and operated at near-to-room temperature in the 8–13 μm wavelength region have been reported by Kim *et al.* [180]. A value of $R_V A$ of 3×10^5 V cm^2 W^{-1} close to the theoretical limit calculated assuming Auger thermal generation [181] was measured at room temperature for a device optimized for $\lambda = 10.6$ μm.

Photodetectors have poor performance at high temperature due to the strong thermal carrier generation-recombination processes involved [183]. Advanced solutions to suppress thermal generation noise include optimization of the doping level and the thickness of the structure.

With this aim, p-$InAs_{0.23}Sb_{0.77}$/p-InSb photoconductors were grown on GaAs substrates by LP-MOVPE [183]. The InSb-like band structure with p-type doping assures a better compromise between optical and thermal generation than n-type doping [181]. The optimized hole concentration for this device is 3×10^{16} cm^{-3}, with a hole mobility of 923 cm^2/V s at 77 K. The 300 K peak responsivity is about 4 mV/W at 8 μm and the cutoff wavelength is as long as 14 μm. The corresponding Johnson noise-limited detectivity is estimated to be approximately 3.27×10^7 cm Hz$^{1/2}$ W^{-1} at room temperature, which is below the theoretical limit set by Auger generation-recombination processes. The photoconductive lifetime at 300 K is close to the limit of Auger recombination, suggesting that the p-$InAs_{0.23}Sb_{0.77}$/p-InSb could achieve higher performance than reported to date.

Band offsets, ordering effects on the band structure of the superlattices and Auger processes in (In,As)Sb SLS grown by MBE were investigated in References [182, 184]. Free electron laser measurements showed that this system has the longest room-temperature Auger lifetime yet measured for a III–V semiconductor, either in bulk or heterostructure form.

The performance of the various types of midinfrared photodetectors is summarized in Table 3.2.

TABLE 2.2
PERFORMANCE OF INFRARED DETECTORS

Group	Material	Structure	Detector type	Performance	T operation [K]	λ_{cutoff} [μ]
Kim *et al.* [180]	p^+-InSb/π-$InAs_{1-x}Sb_x$/ n^+-InSb	DH[a]	photodiode	$R_vA = 3 \times 10^5$ Vcm^2W^{-1}	300	13
				D* (10.16 μm) = 3.27×10^7 $cmHz^{1/2}W^{-1}$	300	
	p-$InAs_{0.23}Sb_{0.77}$/p-InSb	SL	photo-conductor	R(8 μm) = 4 mV/W	300	14
				D* (10.16 μm) = 3.27×10^7 $cmHz^{1/2}W^{-1}$	300	
Young *et al.* [168, 169]	$Ga_{1-x}In_xSb$/InAs	Type II SLS[b]	photo-conductor	D* = 1.5×10^{11} $cmHz^{1/2}W^{-1}$	80–85	12
				D*(10.6 μm) = 1×10^8 $cmHz^{1/2}W^{-1}$		
Grein *et al.* [170, 171]	InAs/$In_xGa_{1-x}Sb$	Type II SL[c]	photodiode	D* = 10^{11} cm $Hz^{1/2}W^{-1}$ (x = 0.4)	100	11
				D* = 10^{10} cm $Hz^{1/2}W^{-1}$ (x = 0.25)	100	19
Mosheni *et al.* [166]:	InAs/GaSb	Type II SL	photodiode	R = 0.5 V W^{-1}	78	8.5
Michel *et al.* [152]				R = 2 V W^{-1}	150	
				D* = 1.3×10^9 cm $Hz^{1/2}W^{-1}$	78	
Fuchs *et al.* [172]	InAs/(Ga, In)Sb	Type II SL	photodide	R = 2 A W^{-1}	77	8
Djuric *et al.* [150]	InSb (Moss-Burnstein)	Bulk	photodiode	R_{rel} = 100%	77	5.6
Ashley [135]	InSb	ML[d]	photodiode (HOT)	D* = 2.5×10^9 $cmHz^{1/2}$ W^{-1}	294	5
Ashley and Elliot [149]						
Pascal-Delannoy *et al.* [175]	InGaSb/GaSb	SL	photodiode	η (2.4 μm) = 42%	300	2.9
Kurtz *et al.* [179]	InAsSb/InSb	Type II SLS	photodiode	D* = 1×10^{10} $Hz^{1/2}W^{-1}$	77	⩽10
Razeghi *et al.* [61, 62]	p-InSb/$p^{-1}$$InAs_{0.15}Sb_{0.85}$/ n-InSb	DH	photodiode	comparable with MCT	77	
	InTlSb	SL	photo-conductor	D*(7 μm) = 3×10^8 $cmHz^{1/2}W^{-1}$	77	8
Levine *et al.* [153]	GaAs/AlGaAs		QWIP	D* = 10^{10} $cmHz^{1/2}W^{-1}$	77	10.7
				D* = 10^{13} $cmHz^{1/2}W^{-1}$	40	10.7

[a]DH double heterostructure. [b]SLS strained layer superlattice. [c]SL superlattice. [d]ML multilayer.

3.6. Conclusions

When reviewing the work and results on infrared devices based on antimony compounds, our main purpose had been to emphasize the new achievements in this field. Comparison with other materials systems highlights the recent advance of III–V antimonide structures. The advances have been dramatic in the case of infrared lasers, where improved performance is linked with suppression of nonradiative Auger recombination by band structure engineering associated with the introduction of thin InAs layers into the devices and associated type II band alignments at the interfaces. The energy band structure can be "adjusted" by strain and the transition energy can be controlled by varying the thickness of the layer, which may result in better uniformity over a larger area.

Because commercial infrared detectors have yet to incorporate the design features that have created spectacular advances of sources, $Hg_{1-x}Cd_xTe$ still retains its place as the established material for long wavelength (~10 μm) detection despite its thermal instability and other materials problems. However, the technology for bulk-InSb infrared detectors is extremely well developed for somewhat shorter wavelengths (~5 μm) with 2D photovoltaic arrays consisting of over 65,000 pixels commercially available. These detectors cooled to between 3 to 30 K have revolutionized infrared astronomy [185] and are competitive with all other detectors at wavelengths as short as 2 μm. It remains to be seen whether the introduction of thin film heterostructure technology will enable III–Vs to overcome commercial inertia and to supplant $Hg_{1-x}Cd_xTe$ for long wavelength infrared detection.

References

1. Wang, C. H., Crowder, J. G., Mannheim, V., Ashley, T., Dutton, D. T., Johnson, A. D., Pryce, G. J., and Smith, S. D. (1998). Detection of nitrogen dioxide using a room temperature operation mid-infrared InSb light emitting diode. *Electronics Letters* **34**: 300–331
2. Brown, B. R., Eglash, S. J., Turner, G. W., Parker, C. D., Pantano, J. V., and Calawa, D. R. (1994). Effect of lattice mismatched growth on InAs/AlSb resonant tunneling diodes. *IEEE Trans. Electron Device* **41**: 879–882.
3. Ejackam, F. E., Seaford, M. L., Lo, Y. H., Hou, H. Q., and Hammons, B. E. (1997). Dislocation-free InSb grown on GaAs compliant universal substrates. *Appl. Phys. Lett.* 776–778
4. Ashley, T., Beswick, J. A., Cockayne, B., and Elliott, C. T., (1996). The growth of ternary substrates of indium gallium antimonide by the double crucible Czochralski technique. *Proc. 22nd Int. Symposium on Compound Semiconductors. IOP Conf. Ser.* **145**: 209–213.
5. Miller, D. A. B., Mozolowski, M. H., Miller, A., and Smith, S. D. (1978). Nonlinear optical effects in InSb with a cw CO laser. *Optics* Comm. **27**: 133.
6. Walukiewicz, W. (1988). *Mat Res Soc Symp. Proc.* **104**: 483; (1988). Mechanism of Fermi-level stabiliza-tion in semiconductors. *Phys. Rev. B. - Condensed Matter.* **37**: 4760–4763.

7. Wang, P. D., Holmes, S. N., Stradling, R. A., Droopad, R., Ferguson, I. T., d'Oliveira, A. G., Parker, S. D., and Williams, R. L. (1992). Electrical 86
8. Egdell, R. G., Evans, S. D., Li, Y. B., Stradling, R. A., and Parker, S. D. (1992). Observation of spatial dispersion of plasmon modes in HREELS of heavily-Doped InAs(001). *Surface Science* **262**: 444–450.
9. Newstead, S. M., Kerr, T. M., and Wood, C. B. C. (1988). Properties of GaAs and $Al_xGa_{1-x}Sb$ grown by molecular-beam epitaxy. *J. Appl. Phys.* **66**: 4184–4187.
10. Subbana, S., Tuttle, G., and Kroemer, H. (1988). N-type doping of GaSb and A1Sb grown by MBE using PbS as a dopant source. *J. Electron. Mater.* **17**: 297.
11. Nguyen, C., Brar, B., Kroemer, H., and English, J. H. (1992). Effects of barrier thickness on the electron concentration in not-intentionally doped InAs-AlSb quantum wells. *J. Vac. Sci. Techn.* **B10**: 898–900.
12. Nguyen, C., Brar, B., Kroemer, H., and English, J. H. (1992). Surface donor contribution to electron sheet concentrations in not-intentionally doped InAs-AlSb quantum wells. *Appl. Phys. Lett.* **60**: 1854–1856.
13. Nguyen, C., Brar, B., and Kroemer, H. (1993). Surface layer modulation of electron concentration in InAs-AlSb quantum wells. *J. Vac. Sci. Techn.* B **11**: 1706–1709.
14. Biefeld, R. M., Wendt, J. R., and Kurtz, S. R. (1991). Improving the performance of $InAs_{1-x}Sb_x$/InSb Infrared detectors grown by metalorganic chemical vapor deposition. *J. Crystal Growth.* **107**: 836–839.
15. Nakashima, K. (1981). Electrical and optical studies in gallium antimonide. *Jap. J. Appl. Phys.* **20**: 1085–1094.
16. Poole, I., Lee, M. B., Cleeverly, I. R., Peaker, A. R., and Singer, K. (1990). Deep donors in GaSb grown by molecular beam epitaxy. *Appl. Phys. Lett.* **57**: 1645–1647.
17. Wang, C. A., Finn, M. C., Salim, S., Jensen, K. F., and Jones, A. C. (1995). Tritertiarybutylaluminum as an organometallic source for epitaxial growth of AlGaSb. *Appl. Phys. Lett.* **67**: 1384–1386.
18. Biefeld, R. M., Kurtz, S. R., and Allerman, A. A. (1997). Metalorganic chemical vapor deposition growth of A1AsSb and InAsSb/InAs using novel source materials for infrared emitters. *J. Electronic. Materials.* **26**: 903–909.
19. Tuttle, G., Kavanaugh, J., and McCalmont, S. (1993). (A1,Ga)Sb long-wavelength distributed Bragg reflectors. *IEEE Photonics Technology Letters* **5**: 1376.
20. Stradling, R. A., Yuen, W. T., Miura, N., and Arimoto, H. (1998). *Proc. Narrow Gap Semiconductor Conference* (Shanghai), pp. 320–4 (Singapore: World Scientific)
21. Yang, M. J., Moore, W. J., Bennett, B. R., and Shanabrook, B. V. (1998) Growth and characterisation of InAs/InGaSb/InAs/AlSb infrared laser structures. *Electronic Letters* **34**: 270–272.
22. Drews D., Schneider, A., Werninghaus, T., Behres, A., Heuken, M., Heime, K., and Zahn, D. R. T. (1998). Characterization of MOVPE grown InPSb/InAs heterostructures. *Applied Surface Science* **123**: 746–750.
23. Kurtz, S. R., Allermanm, A. A., and Biefeld R. M. (1997). Midinfrared lasers and light-emitting diodes with InAsSb/InAsP strained-layers superlattice active regions. *Appl. Phys. Lett.* **70**: 3188
24. Rybaltowski,. A., Xiao, Y., Bu, D., Lane, B., Yi, H., Feng, Diaz, J., and Razeghi, M. (1997). High power InAsSb/InPAsSb/InAs mid-infared lasers. *Appl. Phys. Lett.* **71**: 2430.
25. Diaz, J., Yi, H., Rybaltowksi, A., Lane, B., Lukas, G., Wu, D., Kim, S., Erdtmann, M., Kaas, E., and Razeghi, M. (1997). InAsSbP/InAsSb/InAs laser diodes ($\lambda = 32\,\mu m$) grown by low-pressure metal-organic chemical-vapor deposition. *Appl. Phys. Lett.* **70**: 40.
26. Jen, H. R., Ma, K. Y., and Stringfellow ,G. B. (1989) Long range order in InAsSb. *Appl. Phys. Lett.* **54**: 1154.

27. Cheong, H. M., Ahrenkiel, S. P., Hanna, C., and Mascarenhas, A. (1998). Phonon signatures of spontaneous CuPt ordering in $Ga_{0.47}In_{0.53}As$. *Appl. Phys. Lett.* **73**: 2648.

28. Ferguson, I. T., Norman, A. G., Seong, T. Y., Thomas, R. H., Phillips, C. C., Zhang, X. M., Stradling, R. A., Joyce, B. A., and Booker, R. (1991). Molecular beam epitaxial growth of $InAs_{1-x}Sb_x$ strained layer superlattices. Can nature do it better? *Appl. Phys. Lett.* **59**: 3324.

29. Dobretsov, V. Y., Vaks, V. G., and Martin, G. (1996). Kinetic features in phase separation under alloy ordering. *Phys. Rev.* B **54**: 3227.

30. Chang, L. L., Kwai, Sai-Halasz, G. A., Ludeke, R., and Esaki, L. (1979). Observation of semiconductor-semimetal transition in InAs-GaSb superlattices *Appl. Phys. Lett.* **35**: 939

31. Symons, D. M., Lakrimi, M., van der Burgt, M., Vaughan, T. A., Nicholas, R. J., Mason, N. J., and Walker, P. J. (1995). Temperature dependence of the band overlap in InAs/GaSb structures. *Phys Rev.* **B51**, 1729–1734.

32. Symons, D. M., Lakrimi, M., Warburton, R. J., Nicholas, R. J., Mason, N. J., Walker, P. J., and Eremets, M. I. (1994). Orientation and pressure dependence of the band overlap in InAs/GaSb structures. *Semicond. Sci. Techn.* **9**: 118–122; (1994). [001]-oriented and piezo-electric [111] oriented InAs/GaSb structures under hydrostatic pressure. *Phys. Rev.* **B49**, 16614–16621.

33. Cheng, J. P., Kono, J., McCombe, B. D., Lo, I., Mitchel, W. C., and Stutz, C. E. (1995). Evidence for a stable excitonic ground state in a spatially separated electron-hole system. *Phys Rev. Lett.* **74**: 450–453; (1995). *Proc. Int. Conf. on Physics of Semiconductors* (Vancouver) p. 751, (Singapore: World Scientific).

34. Booker, G. R., Klipstein, P. C., Lakrimi, M., Lyapin, S., Mason, N. J., Murgatroyd, I. J., Nicholas, R. J., Seong, T. Y., Symons, D. M., and Walker, P. J. (1995). Growth of InAs/GaSb strained layer superlattices. *J. Crystal Growth* **146**: 495–502.

35. Nicholas, R. J., van der Burg, M., Daly, M. S., Dalton, K. S. H., Lakrimi, M., Mason, N. J., Symons, D. M., Walker, P. J., Warburton, R. J., Barnes, D. I, and Miura, N. (1994). Optical and magnetotransport properties of semimetallic InAs/(In,Ga)Sb superlattices. *Physica* B **201**: 271–279.

36. Shen, J., Goronkin, H., Dow, J. D., and Ren, S. Y. (1995). Explanation of the origin of electrons in the unintentionally doped InAs/A1Sb system. *J. Vac. Sci Techn.* B**13**: 1736–1739.

37. Tuttle, G., Kroemer, H., and English, J. H. (1989). Electron concentration and mobilities in AlSb/InAs/ AlSb quantum wells. *J. Appl. Phys.* **65**: 5239–5242.

38. Tuttle, U., Kroemer, H., and English, J. H. (1990). Effects of interface layer sequencing on the transport properties of InAs/A1Sb quantum wells — evidence for antisite donors at the InAs/A1Sb interface. *J. Appl. Phys.* **67**: 3032–3037.

39. Feenstra, R. M., Collins, D. A., Ting, D. Z-Y., Wang, M. W., and McGill, T. C. (1994) Interface roughness and asymmetry in InAs/GaSb superlattices studied by scanning – tunneling micro-scopy. *Phys. Rev. Lett.* **72**: 2749–2752.

40. Spitzer, J., Hopner, A., Kuball, M., Cardona, M., Jenichen, B., Neoroth, H., Brar, B., and Kroemer, H. (1995). Influence of the interface composition of InAs/A1Sb superlattices on their optical and structural properties. *J. Appl. Phys.* **77**: 811–820.

41. Blank, H. R., Thomas, M., Wong, K. C., and Kroemer, H. (1996). The influence of the buffer layers on the morphology and transport properties of InAs/(Al,Ga)Sb quantum wells grown by molecular beam epitaxy *Appl. Phys. Lett.* **69**: 2080–2082.

42. Wong, K. C., Yang, C., Thomas, M., and Blank, H. R. (1997). Study of the morphology of the InAs-on-A1Sb interface. *Appl. Phys. Lett.* **69**: 2080–2082.

43. Chung, S. J., Norman, A. C, Yuen, W. T., Malik, T. and Stradling, R. A. (1996). The dependence on growth temperatures of the electrical and structural properties of GaSb/InAs Single Quantum Well Structures grown by MBE. *Proc. 22nd Int. Symposium on Compound Semiconductors. IOP Conf. Ser.* **145**: 45–50.

44. Bolognisi, C. R., Kroemer, H., and English, J. H. (1992). Well width dependence of electron transport in molecular beam epitaxially grown InAs/AlSb quantum wells. *J. Vac. Sci. Techn.* B**10**: 877–879.
45. Bolognisi, C. R., Kroemer, H., and English, J. H. (1992). Interface roughness scattering in InAs/AlSb quantum wells. *Appl. Phys. Lett.* **61**: 213–215.
46. Malik, T., Chung, S., Harris, J. J., Norman, A. G., Stradling, R. A., and Yuen, T. (1995). Remote doping of InAs/GaSb quantum wells by means of a second InAs well doped with silicon. Int. Conf. on Narrow Gap Semiconductors (Santa Fe). *IOP Conference Series* **144**: 229–233.
47. Bolognisi, C. R., Bryce, J. B., and Chow, D. H. (1996). InAs channel heterostructure-field effect transistors with InAs/AlSb short-period superlattice barriers *Appl. Phys. Lett.* **69**: 3531.
48. Sasa, S., Yamamoto, Y., Izumiya, S., Yano, M., Iwai, Y., and Inoue, M. (1997). Increased electron concentration in InAs/AlGaSb heterostructures using a Si planar doped ultrathin InAs quantum well. *Japanese Jour. Appli. Phys.* Part 1, **36**: 1869–1871.
49. Boos, J. B., Bennett, B. R., Kruppa, W., Park, D., Yang, M. J., and Shanabrook, B. V. (1998). AlSb/InAs HEMTs using modulation InAs(Si)-doping. *Electronics Letters*. **34**, 403–404.
50. Goldammer, K.J., Liu, W. K., Khodaparast, U. A., Lindstrom, S. C., Johnson, M. B., Doezema, R. E., and Antos, M. B, (1998). Electrical properties of InSb quantum wells remotely doped with Si. *J. Vac. Sci. Technol* B **16**: 1367–7 1 See also Goldammer *et al.* to be published in *J. Crystal Growth*.
51. Kurtz, S. R., Biefeld, R. M., Dawson, L. R., Baucom, K. C., and Howard, A. J. (1994). Midwave (4 μm) infrared lasers and light-emitting diodes with biaxially compressed InAsSb active regions. *Appl. Phys. Lett.*. **64**: 812.
52. Kurtz, S. R. and Biefeld, R. M. (1995). Magnetophotoluminescence of biaxially compressed InAsSb quantum wells. *Appl. Phys. Lett.*. **66**: 364–366.
53. Zhang, Y. H., Miles, R. H., and Chow, D. H. (1995). InAs-$InAs_xSb_{1-x}$ type-II superlattice midwave infrared lasers grown on InAs substrates. *IEEE Jour. Selected Topics in Quantum Electronics*. **1**: 749.
54. Li, Y. B., Bain, D. J., Hart, L., Livingstone, M., Ciesla, C. M., Pullin, M. J., Tang, P. J. P., Yuen, W. T, Galbraith, I., Phillips, C. C., Pidgeon, C. R., and Stradling, R. A. (1997). Band alignments and offsets in In(As,Sb)/InAs. *Phys. Rev.* B **55**(7): 4589–4595.
55. Wei, S-H. and Zunger, A. (1995). InAsSb/InAs: A type-I or a type-II band alignment. *Phys. Rev.* B **52**: 12039.
56. Golding, T. D., Dura, J. A., Wang, W. C., Vigliante, A., Moss, S. C., Chen, H. C., Miller, J. H., Hoffman, C. A., and Meyer, J. R. (1993). Sb/GaSb heterostructures and multilayers. *Appl. Phys. Lett.* **63**: 1098–1100.
57. Golding, T. D., Meyer, J. R., Huang, J., Hoffman, C. A., Ang, E. G., Xu, J. H., and Zborowski, J. T. (1995). Electronic devices based on III-V/V heterostructures. IOP Conference Series **144**: 359
58. Wood, C. E. C., Noreika, A., and Francombe, M. (1986). Thallium incorporation in molecular-beam-epitaxial InSb. *J. Appl. Phys.* **59**: 3610–3612.
59. Noreika, A. J., Takei, W. J., and Francombe, M. H. (1982). Structure and electrical properties of MBE-grown In-Sb based composition. MBE-CST-2, 1982, Tokyo. A-9-2. 161–164.
60. Noreika, A. J., Greggi, J. Jr., Takei, W. J., and Francombe, M. H. (1983). Properties of MBE-grown InSb and $InSb_{1-x}Bi_x$. *J. Vac. Sci. Technol.* A**1**(2): 558–561.
61. Razeghi, M., Kim, J. D., Park, S. J., Choi, Y. H., Wu, D., Michel, B., Xu, J., and Bigan, B. (1996). New infrared materials and detectors. *Inst. Phys. Conf. Ser.* No. 145, 1085–1090.
62. Razeghi, M., Choi, Y. H., He, X., and Sun, C. J. (1995). *Mat. Sci. Technol.* **11**: 3.
63. Staveteig, P. T., Choi, Y. H., Bigan, B., and Razeghi, M. (1994). Photoconductance measurements on InTlSb/InSb/GaAs grown by low-pressure metalorganic chemical vapor deposition. *Appl. Phys. Lett.* **64**: 460–462.

64. Du, Q, Alperin J., and Wang, W. I. (1997). Molecular beam epitaxial growth of GaInSbBi for infrared detector applications *J. Crystal Growth*. **175**: 849–852.
65. Kruse, P. W., McGlauchlin, L. D., and McQuistan, R. B. (1963). *Elements of Infrared Technology*, NewYork-London: John Wiley & Sons, Inc.
66. Humphreys, R. G. (1983). Radiative lifetime in semiconductors for infrared detection. *Infrared Phys*. **23**: 171–175.
67. Humphreys, R. G. (1986). Radiative lifetime in semiconductors for infrared detection. *Infrared Phys*. **26**: 337–342.
68. Humphreys, R. G. (1988). Specification of infrared detectors and arrays. *Infrared Phys*. **28**: 29–35.
69. White, A. M. (1985). The characteristic of minority carrier – exclusion in narrow gap semiconductors. *Infrared Phys*. **25**: 729–741.
70. White, A. M. (1986). Auger suppression and negative resistance in low gap pin diode structures. *Infrared Phys*. **26**: 317–324.
71. White, A. M. (1987). Negative resistance with Auger suppression in near-intrinsic, low-band gap photodiodes structures. *Infrared Phys*. **27**: 361–369.
72. Davis, A. P. and White, A. M. (1990). Effects of residual Shockley-read traps on the efficiency of Auger suppressed infrared detectors diodes. *Semicond. Sci. Technol*. **5**: S38–S40.
73. Elliott, C. T., Day, D., and Wilson, D. J. (1982). An integrated detector for serial scan thermal imaging. *Infrared Phys*. **22**: 31–42.
74. Elliott, C. T. (1990). *IEE Proc. 4th Int. Conference on Advanced Detectors and Systems*. **61**.
75. Elliott, C. T. and Gordon, N. T. (1993). Infrared detectors, in *Handbook on Semiconductors*, vol. 4, T. S. Moss, C. Hilsum, eds., North Holland, 841–936.
76. Piotrowski, J., Gawron, W., and Djuric, Z. (1994). New generation of near-room-temperature photodetectors. *Opt. Eng*. **5**: 1413–1421.
77. Chazapis, V., Blom, H. A., Vodopyanov, K. L., Norman, A. G., and Philipps, C. C. (1995). Midinfrared picosecond spectroscopy studies of Auger recombination in InSb. *Phys. Rev*. B. **52**: 2516–2521.
78. Murdin, B. N., Pidgeon, C. R., Stradling, R. A., Phillips, C. C., Ciesla, C. M., Livingstone, M, Galbraith, I., Jaroszinski, D. A., and Langerak, C. J. G. M. (1996). Suppression of Auger recombination in arsenic-rich $InAs_{1-x}Sb_x$ strained layer superlattices at 300 K. *J Appl. Phys*. **80**: 2994–7.
79. van Roesbroeck, W. and Shockley, W. (1954). Photon radiative recombination of electrons and holes in germanium. *Phys. Rev*. **94**: 1558.
80. Antoncik, E. and Tauc, J. (1966). Quantum efficiency of the internal photoelectric effect in InSb, in *Semiconductors and Semimetals*, **2**, R. K. Willardson and A. C. Beer, eds., New York and London: Academic Press: pp. 245–262.
81. Ashley, T., Elliott, C. T., and Harker, A. T. (1986). Non-equilibrium modes of operation for infrared detectors. *Infrared Phys*. **26**: 303–315.
82. Youngdale, E. R., Meyer, J. R., Hoffman, C. A., Bartoli, F. J., Grein, C. H., Young, P. M., Ehrenreich, H., Miles, R. H., and Chow, D. H. (1994). *Appl. Phys. Lett*. **64**: 3160.
83. Kurtz, S. R., Biefeld, R. M., and Dawson, L. R. (1995). Modification of valence-band symmetry and Auger threshold energy in biaxially compressed $InAs_{1-x}Sb_x$. *Phys. Rev*. B **51**: 7310–7313.
84. Beattie, A. R. and White, A. M. (1996). An analytic approximation with a wide range of applicability for electron initiated Auger transitions in narrow-gap semiconductors. *J. Appl. Phys*. **79**: 802–813.
85. Beattie, A. R., Abram, R. A., and Scharoch, P. (1990). Realistic evaluation of impact ionization and Auger recombination rates for the CCCH transition in InSb and InGaAsP. *Semicond. Sci. Technol*. **5**: 738–744.

86. Lindle, J. R., Meyer, J. R., Hoffman, C. A., Bartoli, J. F., Turner, G. W., and Choi, H. K. (1995). Auger lifetime in InAs, InAsSb, and InAsSb-InAlAsSb quantum wells. *Appl. Phys. Lett.* **67**: 3153–3155.
87. Grein, C. H., Flatte, M. E., Ebrenreich, H., and Miles, R. H. (1995). Long wavelength InAs/InGaSb infrared detectors: Optimization of carrier lifetimes. *J. Appl. Phys.* **78**: 7143–7152.
88. Pidgeon, C. R., Ciesla, C. M., and Murdin, B. N. (1997). Suppression of nonradiative processes in semiconductor mid-infrared emitters and detectors. *Progress in Quantum Electronics.* **21**(5): 361–419.
89. Melngailis, I. (1965). Longitudinal injection plasma laser of InSb. *Appl. Phys. Lett..* **6**: 59–60; also Phelan, R. J. and Rediker, R. H. (1965). Optically pumped semiconductor laser. *Appl. Phys. Lett.* **6**: 70–71.
90. Lee, H., York, P. K., Menna, P. J., Martinelli, R. U., Garbuzov, D. Z., Narayan, S. Y., and Connolly, J. C. (1995). Room-temperature 2.78 μm AlGaAsSb/InGaAsSb quantum-well lasers. *Appl. Phys. Lett.* **66**: 1942.
91. Sirtori, C., Capasso, F., Faist, J., Sivco, D. L., Hutchison, A. L., and Cho, A. Y. (1995). Quantum cascade unipolar intersubband light emitting diodes in the 8–13 μm wavelength region. *App. Phys. Lett.* **66**: 4.
92. Faist, J., Capasso, F., Sirtori, C., Sivco, D. L., Hutchison, A. L., and Cho, A. Y. (1995). Vertical transition quantum cascade laser with Bragg confined excited state. *Appl. Phys. Lett..* **66**: 538.
93. Sirtori, C., Capasso, F., Faist, J., Sivco, D. L., Hutchison, A L., and Cho, A. Y. (1995). Quantum cascade laser with plasmon-enhanced waveguide operating at 8.4 μm wavelength. *Appl. Phys. Lett..* **66**: 3242.
94. Faist, J., Capasso, F., Sirtori, C., Sivco, D. L., Hutchison, A. L., and Cho, A. Y. (1995). Continuous wave operation of a vertical transition quantum cascade laser above T = 80 K. *Appl. Phys. Lett.* **67**: 3057.
95. Sirtori, C., Faist, J., Capasso, F., Sivco, D. L., Hutchison, A. L., Chu, S. N. U., and Cho, A. Y. (1996). Continuous wave operation of midinfared (7.4–8.6 μm) quantum cascade lasers up to 110 K temperature. *Appl. Phys. Lett.* **68**: 1745.
96. Faist, J., Capasso, F., Sirtori, C., Sivco, D. L., Balliargeon J. N., Hutchison, A. L., Chu, S. N. U., and Cho, A. Y. (1996). High power mid-infrared ($\lambda \sim 5$ μm) quantum cascade lasers operating above room temperature. *Appl. Phys. Lett.* **68**: 3680.
97. Sirtori, C., Faist, J., Capasso, F., Sivco, D. L., Hutchison, X. L., and Cho, A. Y. (1996). Long wavelength infrared ($\lambda \sim 11$ μm) quantum cascade lasers. *Appl. Phys. Lett.* **69**: 2810.
98. Gauthier-Lafaye, O., Boucaud, P., Julien, F. H., Sauvage, S., Cabaret, S., Thiery-Mieg, V., and Planel, R. (1997). Long-wavelength (~ 15.5 μm) unipolar semiconductor laser in GaAs quantum wells. *Appl. Phys. Lett.* **71**: 3619–21.
99. Le, H. Q., Sanchez, A., Arias, J. M., Zandian, M., Zucca, R. R., and Liu, Y. Z. (1995). High-power diode-laser pumped midwave infrared HgCdTe/CdZnTe quantum well lasers. Int Conf on Narrow Gap Semiconductors (Santa Fe). *IOP Conference Series* **144**: 24.
100. Shi, Z., Tacke, M., Lambrecht, A., and Bottner, H. (1995). Midinfrared lead salt multi-quantum-well diode lasers with 282 K operation. *Appl. Phys. Lett.* **66**: 2537.
101. Choi, H. K. and Eglash, S. J. (1991). Room temperature cw operation at 2.2 μm of GaInAsSb/AlGaAsSb diode lasers grown by MBE. *Appl. Phys Lett.* **59**: 1165.
102. Choi, H. K. and Eglash, S. J. (1992). High-power multiple quantum well GaInAsSb/AlGaAsSb diode lasers emitting at 2.2 μm with low threshold current densities. *Appl. Phys Lett.* **61**: 1154.
103. Le, H. Q, Turner, U. W., Eglash, S. J., Choi, H. K., and Coppeta, D. A. (1994). Highpower diode-laser-pumped in InAsSb/GaASb and GaInAsSb/GaSb lasers emitting from 3 to 4 μm. *Appl. Phys. Lett.* **64**: 152.
104. Eglash, S. J. and Choi, H, K. (1994). InAsSb/AlAsSb Double heterostructure diode lasers emitting at 4 μm. *Appl. Phys. Lett.* **64**: 833–835.

105. Choi, H. K,. and Eglash, S. J. (1994). InAsSb/AlAsSb double-heterostructure diode lasers emitting at 4 μm. *Appl. Phys. Lett.* **64**: 833–835.
106. Choi, H. K. and Turner, G. W. (1995). InAsSb/InAlAsSb strained quantum-well diode lasers emitting at 3.9 μm *Appl. Phys. Lett.* **67**: 332.
107. Choi, H. K, Turner, G. W., Manfra, M. J., and Connors, M. K. (1996). 175 K continuous wave operation of InAsSb/InAlAsSb quantum well diode lasers emitting at 3.5 μm. *Appl. Phys. Lett.* **68**: 2936–2938.
108. Le, H. Q., Lin, C. H., and Pei, S. S. (1998). Low-loss high-efficiency and high-power diode-pumped mid-infrared GaInSb/InAs quantum well lasers. *Appl. Phys. Lett.* **72**: 3434–3436.
109. Kurtz, S. R., Biefeld, R. M., Dawson, L. R., Baucom, K. C., and Howard, A. J. (1994). Midwave (4 μm) infared lasers and light-emitting diodes with biaxially compressed InAsSb active regions. *Appl. Phys. Lett.* **64**: 812.
110. Allerman, A. A, Biefeld, R. M., and Kurtz, S. R. (1996). InAsSb based mid-infrared lasers (3.8–3.9 μm) and light emitting diodes with AlAsSb claddings and semimetal electron injection, grown by metalorganic vapor deposition. *Appl. Phys. Lett.* **69**: 465–467.
111. Kurtz, S. R, Allerman, A. A., Biefeld, R. M., and Baucom, K. C. (1998). High slope efficiency, cascaded mid-infrared lasers with type I InAsSb quantum wells. *Appl. Phys. Lett.* **72** :2093–2095.
112. Baranov, A. N., Imentkov, A. N., Sherstnev, V. V., and Yakovlev, Y. P. (1994). 2.7–3.9 μm InAsSb(P)/ InAsSbP low threshold diode lasers. *Appl. Phys. Lett..* **64**: 2480.
113. Aidaraliev, M., Zotovam N, V., Karandashov, S. A., Matveev, B. A., Stua, N. M., and Talaalakin, U. N. (1996). Midwave (3–4 μm) InASsSbP/InGaAsSb infrared diode lasers as a source for gas sensors. *Infrared Phys. & Techn.* **37**: 83–6; (1993). Low threshold long wave lasers ($\lambda = 3.o$ to 3.6 μm) based on III–V alloys. *Semicond. Sci. Techn.* **9**: 1575–1580.
114. Lane, B., Wu, D., Rybaltowski, A., Yi, H., Diaz, J., and Razeghi, M. (1997). Compressively strained multiple quantum well InAsSb lasers emitting at 3.6 mm grown by metal-organic chemical vapor deposition. *Appl. Phys. Lett.* **70**: 443–445.
115. Zhang, Y. H. (1995). Continuous wave operation of $InAs/InAs_xSb_{1-x}$ Midinfared lasers. *Appl. Phys. Lett.* **66**: 118; (1996). Int. Conf. on Narrow Gap Semiconductors (Santa Fe). *IOP Conference Series* **144**: 36.
116. Flatte, M. E., Hasenberg, T. C., Olseberg, J. T., Anson, S. A., Boggess, T. F., Chi Yan, and McDaniel, D. L. (1997). III–V Interband 5.2 μm laser operating at 185 K. *Appl. Phys. Lett.* **71**: 3764–3766.
117. Ashley, T., Elliott, C. T., Jefferies, R., Johnson, A. D., Pryce, U. J. , White, A. M., and Carroll, M. (1997). Mid-infared $In_{1-x}Al_xSb/InSb$ heterostructure diode lasers. *Appl. Phys. Lett.* **70**: 931.
118. Vurgafiman, I. H., Meyer, J. R., Julien, F. H., and RamMohan, L. R. (1998). Design and simulation of low-threshold antimonide intersubband lasers. *Appl. Phys. Lett.* **73**: 711–713.
119. Miles, R. H., Chow, D. K., Zhang, Y. H., Brewer, P. D., and Wilson, R. U. (1995). Midwave infrared stimulated emission from a GaInSb/InAs superlattice. *Appl. Phys. Lett.* **66**: 1921.
120. Chow, D. K., Miles, R. H., Hasenberg, T. C., Kost, A. R., Zhang, Y. H., Dunlap, H. L., and West, L. *Appl. Phys. Lett.* **67**: 3700.
121. Felix, C. L., Bewley, W. W., Vurgaftman, I., Meyer, J. R., Goldberg, L., Chow, D. H., and Selvig, E. (1997). Midinfrared vertical-cavity surface emitting laser. *Appl. Phys. Lett.* **71**: 3483–3485.
122. Bewley, W. W., Aifer, E., Felix, C. L., Vurgaftman, I., Meyer, J. R., Lin, C. H., Murry, S. J., Zhang, D., Pei, S. S. (1997). High-temperature type II superlattice diode laser at 2.9 μm. *Appl. Phys. Lett.* **71**: 3607–3609
123. Malin, J. I., Felix, C. L., Meyer, J. R., Hoffman, C. A., Pinto, J. F., Lin, C.-H., Chang, P. C., Murry S. J., and Pei, S.-S. (1996).Type II mid-IR lasers operating above room temperature. *Electronics Lett.* **32**: 1593.

124. Lin, C. H., Pei, S. S., Le, H. Q., Meyer, J. R., and Felix, C. L. (1997). Low-threshold quasi-cw type-II quantum well lasers at wavelengths beyond 4 μm. *Appl. Phys. Lett.* **71**: 3281–3283.
125. Bewley, W. W., Felix, C. L, Aifer, B. H., Vurgaftman, I. H., Olafsen, L. J., Meyer, J. R., Lee, H., Martinelli, R. U., and Connolly, J. C. (1998). Above-room-temperature optically pumped midinfrared W lasers. *Appl. Phys. Lett.* **73**: 3833–3835.; Bewley, W. W., Vurgaftman, I. H., Felix, C. L., Meyer, J. R., Lin, C. H., Zhang, D., Murry, S. J., Pei, S. S., RamMohan, L. R. (1998). Role of internal loss in limiting type-II mid-IR laser performance. *J. Appl. Phys.* **83**: 2384–2391.
126. Felix, C. L, Bewley, W. W., Aifer, E. H., Vurgaftman, I., Meyer, J. R., Lin, C. H., Zhang, D., Murry, S. J., Yang, R. Q., and Pei, S. S. (1997). Midinfrared vertical-cavity surface emitting laser. *J. Electronic Mat.* **27**: 77–80.
127. Olafsen, L. J., Aifer, E., Vurgaftman, I., Bewley, W. W., Felix, C. L., Meyer, J. R., Zhang, D., Lin, C. H., and Pei, S. S. (1998). Near-room temperature mid-infrared interband cascade laser. *Appl. Phys. Lett.* **72**: 2370.
128. Yang, R. Q., Yang, B. H., Zhang, D., Lin, C.-H., Murry, S. J., Wu, H., and Pel, S. S. (1997). High power mid-infrared interband cascade lasers based on type II quantum wells. *Appl. Phys. Lett.* **71**: 2409.
129. Krier, A., Chubb, D., and Hopkinson, M. (1998). 2.7 μm LEDs for water vapour detection grown by MBE on InP. *Electronics Lett.* **16**: 1606–1607.
130. Tang, P. J. P., Pullin, M. J., Chung, S. J., Phillips, C. C., Stradling, R. A., Norman, A. U., Li, Y. B., and Hart, L. (1995). 4–11 μm infrared emission and 300 K light emitting diodes from arsenic rich InAs/In(As,Sb) strained layer superlattices. *Semiconductor Sci. Techn.* **10**: 1177; Tang, P. J. P., Pullin, M. J., Phillips, C. C., Li, Y. B, Norman, A. U., and Stradling, R. A. (1995). Efficient 300 K light emitting diodes for the 3–10 μm band from arsenic rich InAs/In(As,Sb) strained layer superlattices. Int. Conf. on Narrow Gap Semiconductors (Santa Fe). *IOP Conference Series* **144**: 8–12.
131. Tang, P. J. P., Hardaway, H., Heber J., Phillips, C. C., Pullin, M. J., and Stradling, R. A. (1998). Efficient 300 K light emitting diodes at $\lambda = 5\,\mu m$ and $\sim 8\,\mu m$ from InAs/In(As,Sb) single qauntum wells. *Appl. Phys. Lett.* **72**: 3473–3475.
132. Piotrowski, J., Galus, W., and Grudzien, M. (1991). Near-room temperature infrared photodetectors. *Infrared Phys.* **31**: 1–48.
133. Rogalski, A. and Piotrowski, J. (1988). Intrinsic infrared detectors. *Prog. Quantum Electron.* **12**: 87–289.
134. Rogalski, A. (1994). New trends in infrared detector technology. *Infrared Physics and Technology.* **35**: 1–21.
135. Ashley, T. and Elliot, C. T. (1991). Operation and properties of narrow-gap semiconductor devices near room temperature using non-equilibrium techniques. *Semicond Sci. Technol.* **6**: C99–C 105; see also Ashley, T. and Elliott, C. T. (1985). Nonequilibrium devices for infrared detection. *Electron. Lett.* **21**: 451.
136. Ashley, T., Dean, A. B., Elliott, C. T., Johnson, A. D., Pryce, U. J., White, A. M., and Whitehouse, C. R. (1993). A heterojunction minority carrier barrier for InSb devices. *Semicond Sci. Technol.* **8**; S386–S389.
137. Long, D. (1977). *Optical and Infrared Detectors*, R. J. Keyes, ed., Berlin: Springer-Verlag, p. 101.
138. Williams, O. M. (1986). A critique on the application of infrared photodetector theory. *Infrared Phys.* **26**: 141–153.
139. Williams, O. M. (1987). Infrared photodetector photon formalism: extension and application. *Infrared Phys.* **27**: 167–179.
140. van der Ziel, A. (1959). *Fluctuation Phenomena in Semiconductors*, London: Butterworths.
141. Hill, J. E. and Vliet, K. M. (1958). *Physica* **24**: 709.

142. Pruett, G. R. and Petritz, R. L. (1959). *Proc. Inst. Radio Engrs.* 47: 1524.

143. Hooge, F. N. (1974). *Physica* **83B**: 14.

144. Arutyunyan, V. M., Gasparyan, F. V., and Melkonyan, S. V. (1989).Theory of $1/f$ noise in medium and far infrared photodetectors. *Infrared Phys.* **29**: 243–250.

145. Weissman, M. B. (1980). *J. Appl. Phys.* **51**: 5872.

146. Hooge, F. N., Kleinpenning, T. G. M., and Vandame, L. K. J. (1981). Experimental studies of $1/F$ noise. *Rep. Prog. Phys.* **44**: 479–532.

147. Limperis, T. and Mudar, J. (1978). *The Infrared Handbook*, Chap. 11, W. L. Wolfe and U. J. Zissis eds., Washington.

148. Chiari, J. A. and Morten, F. D. (1982). *Electronic Components and Applications*, vol. 4, Mullard Techn. Publ., p. 242.

149. Ashley, T. (1995). Electronic and optoelectronic devices in narrow-gap semiconductors. *IOP Conference Series.* **144**: 345–352.

150. Djuric, Z., Livada, B., Jovic, V., Smiljanic, M., Matic, M., and Lazic, Z. (1989). Quantum efficiency and responsivity of InSb photodiodes utilizing the Moss-Burustein effect. *Infrared Phys.* **29**: 1–7.

151. Bloom, I. and Nemirowsky, Y. (1993). Surface passivation of backside illuminated indium antimonide focal plane array. *IEEE Trans. Electron. Devices* **40**: 309–314.

152. Michel, B., Mosheni, H., Kim, J. D., Wojkowski, J., Sandven, J., Xu, J., Razeghi, M., Bredthauer, R., Vu, P., Mitchel W., and. Ahoujja, M. (1997). *Appl. Phys. Lett.* **71**: 1071–1073.

153. Levine, B. F., Bethea, C. U., Hasnain, U., Shen, V. O., Pelve, B., Abbott, R. R., and Hsieh, S. J. (1990). High sensitivity low dark current 10 pm GaAs quantum well infrared photodetectors. *Appl. Phys. Lett.* **56**: 851–853.

154. Steele, A. U., Liu, H. C., Buchanan, M., and Wasilewski, Z. R. (1991). Importance of the upper state position in the performance of quantum well intersubband infrared detectors. *Appl. Phys. Lett.* **59**: 3625–3627.

155. Kane, M. J., Millidge, S., Emeny, M. T., Lee, D., Guy, D. R. P., and Whitehouse, C. R. (1992). Intersubband transitions in Quantum Wells, B. Rosencher, B. Vinter, and B. F. Levine, eds., London: Plenum.

156. Liu, H. C. (1992). Photoconductive gain mechanism of quantum-well intersubband infrared detectors. *Appl. Phys. Lett.* **60**: 1507–1509.

157. Beck, W. A. (1993). Photoconductive gain and generation-recombination noise in multiple-quantum-well infrared detectors. *Appl. Phys. Lett.* **63**: 3589–3591.

158. Osbourn, U. C. (1984). InAsSb strained layer superlattices for long wavelength detector applications. *J. Vac. Sci. Technol.* B**2**: 176–178.

159. Smith, D. L. and Mailhiot C. (1987). Proposal for strained type II superlattice infrared detectors. *J. Appl. Phys.* **62**: 2545–2548.

160. Mailhiot, C. and Smith, D. L. (1989). Long wavelength infrared detectors based on strained InAs-$Ga_{1-x}In_xSb$ type II superlattices. *J. Vac. Sci. Technol.* A **7**: 445–449.

161. Manasreh, M. O. (ed). (1993). *Semiconductor Quantum Wells and Superlanices for Long-Wavelength Infrared Detectors*, Norwood, MA: Artech House.

162. Tang, P. J. P., Pullin, M. J., Chung, S. J., Phillips, C. C., and Stradling, R. A (1995). Photo- and electroluminescence studies of uncooled arsenic rich In(As,Sb) strained layer superlattice light emitting diodes for the 4–12 μpm band. Conference Procs. of "Photonics West" – San Jose. *SPIE Conf. Procs.* **2397**: 389–396.

163. Li, Y. B., Stradling, R. A., Norman, A. G., Tang, P. J. P., Chung, S. J., and Phillips, C. C. (1995). Band offsets for $InAs_{1-x}Sb_x/InAs_{1-y}Sb_y$ strained layer superlattices derived from interband magneto-optical studies. ICPS – Vancouver, 496–499.

164. Stradling, R. A., Chung, S. J., Ciesla, C. M., Langerak, C. J. M., Li, Y. B., Malik, T. A., Murdin, B. N., Norman, A. G., Phillips, C. C., Pidgeon, C. R., Pullin, M. J., Tang, P. J. P., and Yuen, W. T. (1997) *Mat. Sci. Eng.* B **22**: 260–265.

165. Rogalski, A. (1997). Comparison of the performance of quantum wells and conventional bulk infrared photodetectors. *Infrared Phys. Technol.* **38**: 295–310.
166. Mohseni, H., Michel, B., Sandoen, J., Razeghi, M., Mitchel, W., and Brown, U. (1997). Growth and characterization of InAs/GaSb photoconductors for long wavelength infrared range. *Appl. Phys. Lett.* **71**: 1403–1405.
167. Marzin, J. Y. (1985). *Heterostructures and Semiconductor Superlattices*, G. Allan, ed., Berlin: Springer-Verlagp. 161.
168. Young, P. M., Grein, C. H., Ehrenreich, H., and Miles, R. H. (1993). Temperature limits on infrared detectivities of $InAs/In_xGa_{1-x}Sb$ superlattices and bulk $Hg_xCd_{1-x}Te$. *J. Appl. Phys.* **74**: 4774–4776.
169. Young, M. H., Chow, D. H., Hunter, A. T., and Miles, R. H. (1998). Recent advances in $Ga_{1-x}In_xSb/InAs$ superlattice IR detector materials. *Appl. Surf. Sci.* **123/124**: 395–399.
170. Grein, C. H., Cruz, H., Flatté, M. B., and Ehrenreich, H. (1994). Theoretical performance of very long wavelength $InAs/In_xGa_{1-x}Sb$ superlattice based infrared detectors. *Appl. Phys. Lett.* **65**: 2530–2532.
171. Grein, C. H., Young, P. M., Flatté, M. B., Ehrenreich, H., Miles, R. H., and Cruz, H. (1995). Theoretical performance limits of 2.1–4.1 μm InAs/InGaSb, HgCdTe, and InGaAsSb lasers. *J. Appl. Phys.* **78**: 4552–4559.
172. Fuchs, F., Weimer, U., Pletchen, W., Schmitz, J., Ahlswede, B., Walther, M., and Wagner, J. (1997). High performance $InAs/Ga_{1-x}In_xSb$ superlattice infrared photodiodes. *Appl. Phys. Lett.* **71**: 3251–3253.
173. Fuchs, F., Weimer, U., Pletchen, W., Schmitz, J., Ahlswede, B., Walther, M., and Wagner, J. (1998). SPIB Conference – Materials and Devices III, San Jose, CA., Jan.
174. Kinch, M. A. and Borrello, S. R., (1974). *Infrared Phys.* **15**: 111.
175. Pascal-Delannoy, F., Mason, N. J., Bougnot, U., Walker, P. J., Bougnot, J., Giani, A., and Allogho, U. G. (1992). InGaSb/Gasb photodiodes grown by MOVPE. *J. Crystal Growth* **124**: 409–414.
176. Mansoor, F., Haywood, S. K., Mason, N. J., Nicholas, R. J., Walker, P. J., Grey, R., and Hill, U. (1995). Resonant cavity-enhanced (RCB) photodetector based on Ga(In)Sb for gas-sensing applications. *Semicond. Sci. Technol.* **10**:1017–1021.
177. Bethea, C. G., Levine, B. F., Yen, M. Y., and Cho, A. Y. (1988). Photoconductive measurements on $InAs_{0.22}Sb_{0.78}/GaAs$ grown using molecular beam epitaxy. *Appl. Phys. Lett.* **53**: 291.
178. Orman, Z. and Rogalski, A.(1986). Calculation of the intrinsic carrier concentration in $InAs_{1-x}Sb_x$. *Phys. Stat Sol.* **135b**: K85–K88.
179. Kurtz, S. R., Dawson, L. R., Zipperian, T. B., and Whaley, R. D. (1990). High detectivity ($10^{10}\,cmHz^{1/2}\,W^{-1}$) InAsSb strained-layer superlattice photovoltaic infrared detector. *IEEE Electron Device Lett.* **11**: 54–56.
180. Kim, J. D., Kim, S., Wu, D., Wojkowski, J., Xu, J., Piotrowski, J., Bigan, B., and Razeghi, M. (1995). 8–13 μm InAsSb heterojunction photodiode operating at near room temperature. *Appl. Phys. Lett.* 67: 2645–2647.
181. Piotrowski, J. and Razeghi, M. (1995). *Proc. SPIE* **2397**: 180.
182. Stradling, R. A. (1996). The electronic properties and applications of quantum wells and superlattices of III–V narrow gap semiconductors. *Brazilian Jour. of Physics* **26**: 7–19.
183. Kim, J. D., Wu, D., Wojkowski, J., Piotrowski, J., Xu, J., and Razeghi, M. (1996). Long-wavelength InAsSb photoconductors operated at near room temperatures (200–300 K). *Appl. Phys. Lett.* **68**: 99–101.
184. Stradling, R. A. (1991). InSb-based materials for detectors. *Semicond Sci. Technol.* **6**: C52–C58.
185. Fazio, G. G. (1994). Infrared array detectors in Astrophysics. *Infrared Phy.* **35**:107–117.

HgCdTe Infrared Detectors

ARVIND I. D'SOUZA

Boeing Sensor and Electronic Products, Anaheim, California, USA

P.S. WIJEWARNASURIYA

Rockwell Science Center, Thousand Oaks, California, USA

JOHN G. POKSHEVA

Analysis Associates, Whittier, California, USA

4.1. Introduction

The $Hg_{1-x}Cd_xTe$ ($0 \leq x \leq 1$) alloy system has a forbidden energy gap that falls between the end points for HgTe ($E_g = -0.3$ eV) and CdTe ($E_g = 1.6$ eV). Because of its bandgap tunability with x, $Hg_{1-x}Cd_xTe$ has evolved to become the most important/versatile material for detector applications over the entire infrared range. As the Cd composition increases, the energy gap for $Hg_{1-x}Cd_xTe$ gradually increases from the negative value for HgTe to the positive value for CdTe. The bandgap tunability results in infrared detector applications that span the Short Wavelength InfraRed (SWIR: 1–3 μm), middle wavelength (MWIR: 3–5 μm), mid–long wavelength (MLWIR: 6–8 μm), long wavelength (LWIR: 8–14 μm), and very long wavelength (VLWIR: 14 μm) ranges. Primary applications have been in the military and astronomy fields. The significant military applications of $Hg_{1-x}Cd_xTe$ technology have resulted in a significant portion of the

ISSN 1079-4050
Vol. 28
ISBN 0-12-533028-6/$35.00

research being performed at government laboratories in various countries and within a few industrial concerns and universities in conjunction with the national laboratories.

4.2. HgCdTe Material Properties and Background

4.2.1. Crystal Structure

Semiconducting properties of the $Hg_{1-x}Cd_xTe$ ($0 \leq x \leq 1$) alloy system were reported in 1959 by Lawson *et al.* [1]. This system has since been investigated extensively and because of its tunability with x has evolved to become the most important/versatile material for detector applications over the entire infrared range. Semiconducting $Hg_{1-x}Cd_xTe$ is formed by alloying II-VI compound semiconductors composed of semimetal mercury telluride (HgTe with lattice constant 6.42 Å) and wide gap semiconductor cadmium telluride (CdTe with lattice constant 6.48 Å). HgTe, CdTe, and the alloy $Hg_{1-x}Cd_xTe$ all have the zincblende crystalline structure. The lattice constant for $Hg_{1-x}Cd_xTe$ falls between the end points for HgTe and CdTe but is not linear with Cd composition x [1–5]. Similarly, the forbidden energy gap for $Hg_{1-x}Cd_xTe$ falls between the end points for HgTe ($E_g = -0.3$ eV) and CdTe ($E_g = 1.6$ eV). As the Cd composition increases, the energy gap for $Hg_{1-x}Cd_xTe$ gradually increases from the negative value for HgTe to the positive value for CdTe. A gapless state occurs when $x = 0.14$ at 77 K [1]. In the semimetal region, the conduction and light-hole valence bands are highly nonparabolic but as the x value increases they approach more parabolic behavior.

The growth of HgCdTe prior to 1981 had been effected by a variety of earlier growth techniques involving bulk methods [6–8], liquid phase epitaxy (LPE) [9–14] and vapor phase epitaxy (VPE) [15–18]. In 1981 Faurie and Million [19] achieved the growth of HgCdTe by means of molecular beam epitaxy (MBE).

4.2.2. Energy Gap—E_G

The energy gap for $Hg_{1-x}Cd_xTe$ falls between the end points for HgTe and CdTe and rather approximately has linear dependence on Cd composition x. This behavior has been established from a variety of independent methods involving optical absorption [20], photoluminescence [21], electroluminescence [22], and interband magnetoreflection experiments [23]. For CdTe, the temperature dependence of the energy gap is similar to that of the usual semiconductors. At low temperatures it varies as the square of the temperature [24], but 100 K it behaves approximately linearly with temperature with a slope of $dE_g/dT = -5.6 \times 10^{-4}$ eV/K. On the other hand, HgTe has a positive slope of $dE_g/dT = 5.0 \times 10^{-4}$ eV/K. A gapless state occurs when $x = 0.14$ at a temperature of 77 K. In the semimetal region, the conduction and light-hole valence bands are highly nonparabolic and as the x value increases they approach a more parabolic behavior.

Several expressions [25–28] can be found in the literature relating the energy gap E_g for $Hg_{1-x}Cd_xTe$ to Cd composition x and temperature T. In particular, Hansen *et al.* [28] have combined optical absorption and magneto–optical experimental data to develop a new empirical expression for the energy gap E_g covering a wide range of x and T values. From their work, E_g (in eV) as a function of x and T is given by,

$$E_g(x, T) = -0.302 + 1.93x + 5.35 \times 10^{-4} T(1 - 2x) - 0.810x^2 + 0.832x^3 \tag{4.1}$$

where, $0 \leq x \leq 0.6$ along with $x = 1$, and $4.2\,K \leq T \leq 300$ K. Special significance attaches to the wavelength equivalent for E_g defined by

$$\lambda_c = \frac{hc}{E_g} \tag{4.2}$$

where h is Planck's constant, c multiplying h is the speed of light, and λ_c on the left-hand side denotes the so-called cutoff wavelength associated with the bandedge value. Calculated cutoff wavelength versus temperature behavior for $Hg_{1-x}Cd_xTe$ is shown in Fig. 4.1 for different Cd composition x values.

4.2.3. ABSORPTION COEFFICIENT—α

$Hg_{1-x}Cd_xTe$ is a direct bandgap semiconductor. As such, the absorption coefficient for photon energies $E = h\nu$ greater than the bandgap E_g is often calculated from a Kane model approach. Results given by Anderson [29] express Kane model results in the form

$$\alpha(h\nu) = \beta(E - E_g)^{1/2} \tag{4.3}$$

where [30, 31]

$$\beta = \frac{(2.109x10^5)(1 + x)}{(T + T_0)} \tag{4.4}$$

with energy values expressed in eV, temperature values in K, and α in cm^{-1}. Near the bandedge, direct bandgap semiconductors normally follow the Urbach rule

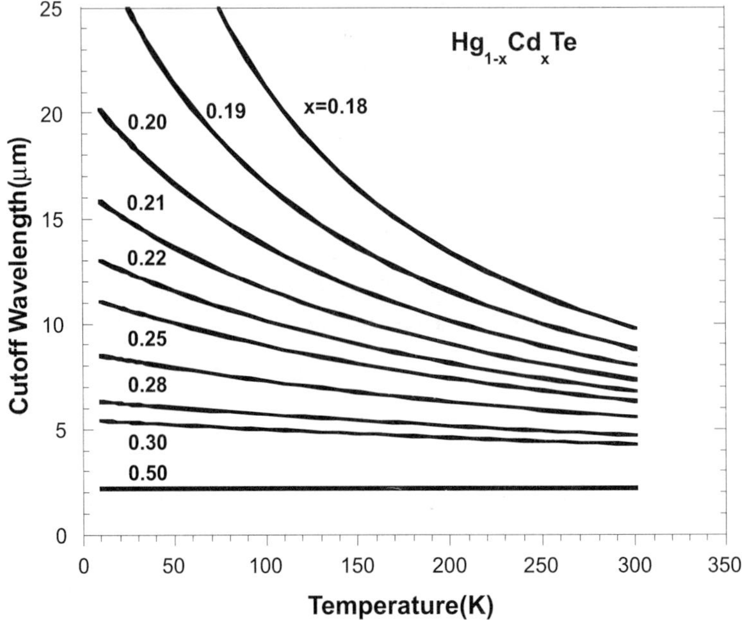

FIG. 4.1. Cutoff wavelength of $Hg_{1-x}Cd_xTe$ versus temperature calculated from Eq. 1.1 for different Cd composition.

[31]. In $Hg_{1-x}Cd_xTe$, however, $\alpha(E)$ near and below the band edge follows a modified Urbach rule described by [32],

$$\alpha(E) = \alpha_0 e^{[\sigma_0(E-E_0)/T+T_0]}, \tag{4.5}$$

where α_0, σ_0, E_0 and w are fitting parameters characterized by

$$\begin{aligned} \alpha_0 &= e^{(53.61x-18.88)} \\ E_0 &= 1.838x - 0.3424 \\ \sigma_0 &= 3.267x10^4(1+x) \\ T_0 &= 81.9 \end{aligned} \tag{4.6}$$

The transition region from Kane's square root law to the modified Urbach rule tail occurs at [33]

$$\alpha(E_g) = 100 + 5000x \tag{4.7}$$

again with α expressed in cm^{-1}. An alternative means for characterizing β in Eq. (4.3) is invoked by Gopal *et al.* [35]. A review of optical constants for $Hg_{1-x}Cd_xTe$ is given by Amirtharaj [34]. Variations in $\alpha(E)$ below the bandedge can occur due to the presence of defects, compositional grading, and spatial nonuniformity in the material.

4.2.4. INTRINSIC CARRIER CONCENTRATION—n_i

As with parameter E_g, intrinsic carrier concentration n_i is also a fundamental quantity of interest. It varies strongly with temperature, Cd composition, and energy gap. Several authors have tried to calculate n_i from the different empirical equations available for E_g [36–38]. Schmit [36], for example, computes n_i using the energy gap E_g given by Schmit and Stelzer [25]. Nemirovsky and Finkman [37] develop another expression for n_i but restrict the x range to $0.2 \leq x \leq 0.4$. More recently, Hansen and Schmit [38], assuming nonparabolic bands and the $\vec{k} \cdot \vec{p}$ approach [39] with degenerate statistics, derive expressions for n_i in the form

$$n_i = (5.585 - 3.820x + 1.75310^{-3}T - 1.364 \times 10^{-3}xT) \times 10^{14} E_g^{3/2} T^{3/2} \exp\left(\frac{E_g}{2kT}\right) \text{cm}^{-3} \qquad (4.8)$$

with $E_g$being described in terms of the x and T range encompassed in Eq. (4.1).

4.2.5. RECOMBINATION MECHANISMS

Minority carrier lifetime in n-type HgCdTe is determined by several different recombination mechanisms, discussed extensively in the literature [40–53]. In particular, Auger, radiative and Shockley-Read-Hall (SR) processes are important recombination mechanisms in HgCdTe. When the lifetime is not dominated by a single process it is necessary to calculate the lifetime accordingly. Thus if all three recombination mechanisms are important, the overall effective lifetime τ is then calculated from

$$\frac{1}{\tau} = \frac{1}{\tau_A} + \frac{1}{\tau_R} + \frac{1}{\tau_{\text{SR}}} \qquad (4.9)$$

where, τ_{A}, τ_{R} and τ_{SR} are the Auger, radiative and Shockley-Read-Hall lifetimes, respectively. The Auger-1 process [50] is the dominant recombination mechanism in n-type material. In this mechanism an excited electron in the conduction band recombines with a hole in the valence band and the loss in energy is transferred to a second conduction-band electron by the electron–electron coulombic interaction. Under low injection conditions, lifetime for Auger-1 limited recombination has been described in terms of [40],

$$\tau_{A1} = \frac{2n_i^2\tau_{\mathrm{Ai}}}{(n_0 + p_0)n_0}, \tag{4.10}$$

where n_i is the intrinsic carrier concentration, n_0 is the equilibrium electron concentration, p_0 is the equilibrium hole concentration, and τ_{Ai} is the Auger lifetime for intrinsic HgCdTe, given by [40–42, 45]

$$\tau_{\mathrm{Ai}} = \frac{3.8 \times 10^{-18}\varepsilon^2(1+\eta)^{1/2}(1+\eta)}{(m_e^*/m_0)|F_1F_2|^2(kT/E_g)^{3/2}} \times \exp\left(\frac{(1+2\eta)}{(1+\eta)}\frac{E_g}{kT}\right)(\text{seconds}) \tag{4.11}$$

giving τ_{Ai} in sec. In this last expression, k is Boltzmann's constant and T is the absolute temperature, E_g is the gap energy [28], and ε is the dielectric constant, given in Reference [41] to be 20–9.4x. Also, m_0 is the electron rest mass, m_e^* and m_h^* are electron and hole effective masses, and $\eta = m_e^*/m_h^*$. The value for $|F_1F_2|$ typically varies between 0.1 to 0.3 [41–49].

In p-type material in addition to the Auger-1 process, the Auger-7 process [50, 54] is also important. In this process, an excited electron in the conduction band recombines with a hole in the valence band and the loss in energy is transferred to an electron in the light hole band. Under low injection conditions, Auger-7 lifetime in p-type material is given by [45, 48],

$$\tau_{A7} = \frac{2n_i^2\gamma\tau_{A1}}{(n_0 + p_0)p_0} \tag{4.12}$$

where the ratio $\gamma = \tau_{\mathrm{A7}}/\tau_{\mathrm{A1}}$ has been calculated in Reference [54].

Next, radiative recombination lifetime has been described in terms of (see References [41, 42, 45]):

$$\tau_R = \frac{1}{B(n_0 + p_0)} \tag{4.13}$$

with B being given by

$$B = 5.8 \times 10^{-13}\varepsilon^{1/2}\left(\frac{m_0}{m_e^* + m_h^*}\right)\left(1 + \frac{m_0}{m_e^*} + \frac{m_0}{m_h^*}\right) \times \left(\frac{300}{T}\right)^{3/2}(E_g^2 + 3kTE_g + 3.75k^2T^2) \tag{4.14}$$

giving τ_R in seconds.

Lopes *et al.* [45] have calculated how the lifetime varies with the Cd composition. For a doping level of $\sim 2 \times 10^{15}\,\mathrm{cm}^{-3}$ they claimed that the Auger lifetime dominates for Cd composition 0.24, while the radiative lifetime dominates when x is closer to 0.30 in n-type material. Calculations of Auger-1 and radiative lifetime as a function of temperature for different doping levels are

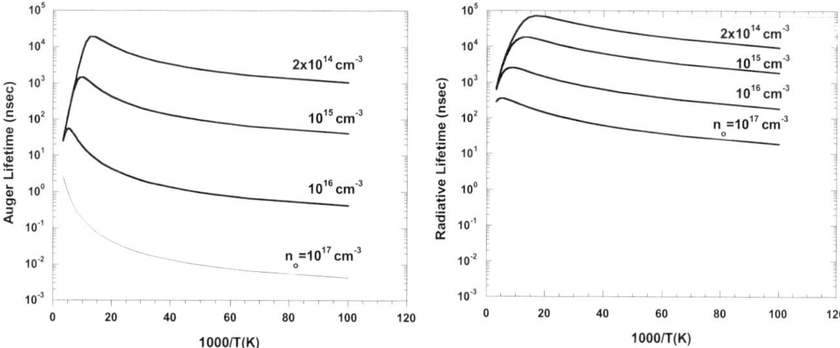

FIG. 4.2. Calculated Auger-1 and radiative lifetime as a function of temperature for $Hg_{0.80}Cd_{0.2}Te$ versus doping concentration.

given here in Fig. 4.2. In these calculations $m_e^*/m_0 = 0.55$ and $|F_1F_2| = 0.22$ were used. Finally, Shockley-Read lifetime, assuming a single recombination level E_t below the conduction bandedge, and low injection levels, has been described in terms of [40, 45, 53],

$$\tau_{SR} = \frac{\tau_{p0}(n_0 + n_1)}{n_0 + p_0} + \frac{\tau_{n0}(p_0 + p_1)}{n_0 + p_0} \tag{4.15}$$

where τ_{n0} and τ_{p0} denote shortest-time constant values for electron and hole capture, respectively, $n_1 = n_0 \exp{(E_t - E_F)/kT}$, $p_1 = p_0 \exp{(E_F - E_t)/kT}$, and E_F denotes the Fermi energy.

4.3. HgCdTe Growth

Bulk and epitaxial growth techniques are available for producing high-quality HgCdTe layers [6–8]. Among epitaxial techniques, liquid phase epitaxy (LPE) [9–14], molecular beam epitaxy (MBE) [15–18] and metallic-organic vapor phase epitaxy (MOVPE) [55–61] are promising. The LPE technique for HgCdTe has been in development since the late 1970s and has been reviewed by several authors [6, 9, 13, 62, 63, 64]. The LPE HgCdTe can result from either Hg- or Te-rich solutions of the phase diagram. The basic principle in the Te-rich case is that a saturated solution of Hg and Cd is dissolved in liquid Te; in the Hg-rich case, a saturated solution of Cd and Te is dissolved in liquid Hg. Tipping, dipping, and horizontal sliding techniques for HgCdTe LPE growth are described in the literature [13]. Outstanding progress has been made in the last decade regarding LPE growth in producing device-quality HgCdTe material.

The MBE technique, employed in HgCdTe growth since 1981, has emerged as a viable growth technique for manufacturing high-performance InfraRed Focal Plane Arrays (IRFPA). Due to its relative processing simplicity and its precision control over growth parameters [17, 65-67], MBE is now one of the leading growth techniques for developing new HgCdTe detector structures [68, 69]. In MBE for HgCdTe, the growth temperature never exceeds 200 °C. As the resulting MBE growth is carried out at relatively low temperature, flexibility accrues in allowing the synthesis of multilayer structures with abrupt interfaces. Inherent to the MBE technology is its amenability for modifying the bandgap during a growth run or from run-to-run; a change in the bandgap is easily and readily achieved by adjusting the beam fluxes [65]. Excellent run-to-run reproducibility has been demonstrated in Cd composition x, etch-pit-density (EPD), void defect density, and electron concentration and mobility on MBE HgCdTe material [65–67]. Spatial uniformity of Cd composition x and layer thickness have been achieved over large areas. For typical 4 cm × 4 cm MBE-LWIR HgCdTe growth on CdZnTe, statistical behavior (mean, standard deviation) of 16 measurements from wafer center to edge shows (0.224, 0.001325) and (7.4 μm, 0.1835 μm) results for composition and layer thickness, respectively [65]. Developments in MBE HgCdTe growth and doping control have led to the demonstration of high-performance IRFPA for imaging applications spanning the VSWIR to VLWIR spectral range [68–88].

CdZnTe (Zn ~ 3.5%) is the lattice-matched substrate of choice for HgCdTe material growth [89, 90, 77–79]. Diodes with the highest performance are fabricated in HgCdTe grown on lattice-matched CdZnTe [80–82]. Alternate substrates such as sapphire [83], silicon [72, 84, 85] or germanium [86] are also used for the growth of HgCdTe. Alternate substrates may be used because they are cheaper than CdZnTe, are available in larger sizes, are more rugged-making them easier to handle and process, or have less thermal mismatch with silicon wafer readout circuitry used to extract signals generated by the diodes. When alternate substrates are used, a CdTe buffer layer [87, 88, 88a] is grown on the substrate prior to HgCdTe growth. In general, diodes fabricated on alternate substrates are not comparable in performance to diodes fabricated in HgCdTe grown on lattice-matched CdZnTe [72, 85].

4.4. Native Defects and Impurity Doping Behavior

4.4.1. Unintentional Doping

It is difficult to determine what controls the electronic properties in HgCdTe material—native defects (Hg vacancies, interstitial Te, etc.) or background impurities introduced during growth and/or diffusing from the substrates [91, 92]. Intrinsic doping by stoichiometric deviation is unavoidable in II-VI materials. There is abundant evidence in the literature that native point defects appear in the form of vacancies and interstitials. Annealing of this material under Hg- or Te-rich vapor usually gives *n*- or *p*-type behavior, respectively [2, 9, 13]. Hg vacancies generally behave as acceptor states [2]; similarly, there is some evidence that *n*-type behavior is due to Te vacancies rather than to Hg interstitials. Defects of this type are electrically active so that when the concentration of residual electrically active impurities is low, the conduction is determined mainly by these active point and extended defects. In the case of LPE, layers grown under Te-rich conditions show *p*-type conduction (due to Hg vacancies or to electrically active impurities) whereas layers grown under Hg-rich conditions are usually *n*-type (due to residual impurities). In MBE material, several factors such as substrate orientation, choice of substrate, source materials, growth temperature, Hg flux, and alloy composition appear to influence the electrical properties. A complex picture of the influence of these factors has been described [91]. Early investigations have shed some light on understanding the influence of native defects (Hg vacancies, interstitial/antisite Te, etc.) or background impurities introduced during the growth by source materials and/or by the diffusion of impurities from the substrates [92]. For several years, erratic behavior in the electrical properties of as-grown as well as annealed layers was observed in different laboratories [92–95]. It has been found that this was not due to a lack of control in the growth but to diffusion of impurities from the substrate. Recently, there has been an increased emphasis on replacing native acceptor doping (Hg vacancies) with external acceptor dopants [96, 97]. Data on the transport properties (mobility, carrier concentration, lifetime, etc.) are described in the literature extensively [2, 98].

4.4.2. IMPURITY DOPING

Optimal photodiode operation ultimately requires intentionally doped material with known and controllable concentrations of both donor and acceptor elements on both sides of the junction. Hence, a solid understanding of doping in $Hg_{1-x}Cd_xTe$ layers is a prerequisite to fabricate high-performance devices. In the following two sections we will present results of the impurity doping of $Hg_{1-x}Cd_xTe$.

4.4.2.1. P-Type Doping

Column I and V elements could act as acceptor dopants in HgCdTe if they substitute in metallic (column II) or nonmetallic (column V) sites, respectively. In earlier work, group I elements were thought to be favorable for p-type doping as they could easily be incorporated in the metallic sites with any growth technique including MBE. Among the group I elements, incorporation of Cu, Ag, Au, and Li as p-type dopants in HgCdTe layers has been studied [91, 99–105]. Doping of Ag and Li has successfully been achieved in MBE growth with (100) and (111)B orientations [99,100]. With Li, carrier concentrations of up to $8 \times 10^{18}\,cm^{-3}$ have been achieved by MBE, and the doping levels could be controlled by varying the Li cell temperature during growth. Hole mobilities of between 300–400 cm^2/vs were obtained at low temperatures. Both the incorporation coefficient and electrical activity were found to be close to 100%. Wijewarasuriya *et al.* claimed [99] from SIMS profiling that Li diffuses uniformly in unintentionally doped layers grown under or above the Li-doped regions. Further they demonstrated that Li can diffuse out of the crystals during isothermal annealing because the annealed layers became n-type. Similar behavior also has been reported in LPE materials [102, 103].

Similarly, Wroge *et al.* [100] have studied the potential use of Ag for p-type doping of HgCdTe grown by MBE along (100) and (111)B orientations. Doping levels of between 1×10^{16} and $1 \times 10^{18}\,cm^{-3}$ were observed and the layers exhibited good electrical properties. However, the average diffusion coefficient, calculated to be $7 \times 10^{-14}\,cm^2/s$ at 165 °C, is too high to produce controlled junctions.

Due to the high diffusion of group I elements in $Hg_{1-x}Cd_xTe$, use of group V elements as p-type dopants is being studied [62–64, 89, 90, 102–104, 106–132]. The larger atom size of the group V elements and hence their lower diffusivity in HgCdTe is expected to result in stable and well-controlled p-n junctions with sharp transitions. With LPE from Te-rich solutions, group V elements were tried and the resulting electrical activity, solubility, and mode of incorporation were investigated [63, 106]. Depending upon the growth conditions (either Te-rich or Hg-rich), the mode of incorporation of group V elements in HgCdTe is different. In LPE growth, with both Te-rich [62, 63] and Hg-rich [106] conditions, contrasting behavior in incorporation of group V elements is seen. A full account of the growth and the dopant behavior has been reported by Kalisher [107] and Tung [64] in Hg-rich LPE. The P, As, Sb and Bi were observed mostly to be electrically inactive in as-grown conditions and when annealed at 200 °C in Te-rich LPE HgCdTe. However, when a pre-anneal at 500 °C (in Hg) was used before the 200 °C anneal, p-type conduction was obtained for P, As and Sb, but not for Bi [64]. This behavior was explained by the movement of dopants occupying metal sites in the as-grown state to active Te sites during the 500 °C anneal. It is well established that As is a 100% active p-type dopant in LPE-HgCdTe grown under Hg-rich conditions [62] for doping concentrations of between 10^{15}–$10^{18}\,cm^{-3}$.

In MBE growth (usually occuring under Te-saturated conditions), it is not surprising that As and Sb incorporated during the MBE process have been found to act primarily as donors in as-grown materials [101, 108, 109]. This "intrinsic" difficulty in MBE explains why many approaches have been tried in attempting to achieve *p*-type doping in the MBE growth of HgCdTe using arsenic. *In situ* doping using cracked elemental As_4 [108, 109] or arsenic compounds [108, 116, 123], arsenic diffusion [90, 114], arsenic implantation [110, 111], arsenic doping in multilayers [90, 119] or superlattices [121, 122], and delta doping [126–128] have all been tried in doping of HgCdTe material. However, most of these approaches demand a high-temperature anneal (300 °C) in order to fully activate the incorporated As.

Direct incorporation of arsenic during the MBE growth process using As_4 was achieved several years ago [108, 109, 128, 129] with (111)B- and (211)B-oriented epilayers. Resulting layers exhibited poor *n*-type characteristics. After isothermal annealing at temperatures of between 200 and 250 °C, the samples remained *n*-type. It was concluded that these impurities interact preferentially with the metal site. It was further noted that if during the MBE growth the substrate were illuminated with ultraviolet light or a green Nd-YAG pulsed laser (10-kHz rep rate), the resulting HgCdTe material remained *n*-type with no substantial change in carrier concentration character.

The *p*-type doping of MBE-HgCdTe using arsenic [123, 125] was done at Hughes Research Laboratory in 1990. In this work, *p*-type doping is achieved *in situ* on an As-doped HgCdTe layer grown at 170–180 °C without any photo-assistance. The *p*-type conduction is reported to have been observed in the as-grown layers with $0.28 < x < 0.35$ after annealing under a mercury ambient. This result is attributed to the use of atomic tellurium, produced by an undescribed proprietary cracker cell found to enhance Te reactivity on the surface [124]. The same research group is also using Cd_3As_2 in the effusion cell. In a U.S. patent [125] it is claimed that cation-rich conditions can be achieved for $Hg_{1-x}Cd_xTe$ ($x = 0.22$) with cadmium overpressure in MBE. A striking similarity in As activation as a function of temperature has been observed for As diffusion and implantation in MBE HgCdTe.

Diffusion of As at 380–450 °C from a Hg solution containing As was used [90, 116, 118] to produce $p+/n$ homojunction diodes in HgCdTe grown by MBE on (211)B-CdZnTe substrates. Further, MBE-based planar HgCdTe heterostructure photodiodes with the *p*-on-*n* configuration have been formed by selective pocket diffusion of arsenic deposited on the wide gap surface by ion implantation [118]. In order to fabricate high-performance photodiodes, the structure has to experience two Hg-overpressure anneals after the implantation process, one at 435 °C for 20 min followed by another at 250 °C for 24 h. High-temperature annealing is also required to achieve *p*-type conversion in this technology.

Using the photoassisted MBE growth technique, Harper *et al.* [119] have successfully grown *p*-type doped (100) CdTe epilayers with arsenic at 180 °C, which is a suitable temperature to grow high-quality HgCdTe. Han and coworkers [120] have grown *p*-type modulation-doped HgCdTe samples. These samples were obtained by using thin CdTe layers of 50 Å heavily doped with arsenic and located periodically between $Hg_{1-x}Cd_xTe$ layers 1000–1100-Å thick with x values of 0.18–0.26. The layers were grown on (100)-CdZnTe substrate at 170 °C, and p-type doping levels of 5×10^{16} to 1×10^{18} cm^{-3} were achieved. The authors proposed to consider these structures as a new quantum alloy. The direct doping of HgCdTe with As was overcome by utilizing the fact that CdTe can be doped directly as a p-type. Neither hole freeze-out nor hole mobility enhancement due to modulation doping has been observed in these layers.

This method was used later to fabricate p-on-n homojunction superlattice (SL) detectors operating in the 3–5-μm region [121,122]. In this work, it was reported that the As-doped superlattice structures can be used for SL detector fabrication or interdiffused in a standard n-type anneal (the compositionally modulated structure technique) [121], resulting in p-type homogeneous HgCdTe. The same approach has been tested by Arias *et al.* [112], who have reported the p-type As-doping of CdTe and HgTe/CdTe superlattices (period of 50–180 Å) by both photoassisted and conventional MBE in the (100) orientation. They have shown that at low growth temperatures and under cation-stabilized growth conditions, p-type doping due to arsenic can be observed on as-grown material. The CdTe epilayers grown at low temperature (180 °C) with photoassisted MBE exhibit superior structural, optical and electrical properties to those grown by conventional MBE. In this work, the importance of establishing cation-stabilized conditions to incorporate arsenic in tellurium sites is stressed. The extension of this CdTe doping process to the doping of HgTe/CdTe superlattice structures at a growth temperature 155 °C has resulted in *in situ* growth of p-type modulated doped superlattices with a cut-off response of up to 12.6 μm at 300 K. All p-type HgTe/CdTe samples interdiffused at 250 °C for 20 h under a saturated Hg environment have produced p-type HgCdTe alloys with higher hole concentrations. This indicates that even a low-temperature annealing process is sufficient to increase the amount of electrically active arsenic acceptor atoms in the lattice. These experiments have been utilized to fabricate *in situ* arsenic doped p-on-n HgCdTe heterojunctions using the interdiffused superlattice process (ISP) doping approach described in the preceding [95].

An alternative approach was proposed recently to achieve arsenic activation at a temperature close to MBE-growth temperature. The University of Illinois implemented a planar doping technique with arsenic as a p-type dopant during the MBE growth of HgCdTe [126–128]. In these experiments $Hg_{1-x}Cd_xTe$ alloy is grown at 185 °C by MBE on (211)B-CdZnTe substrates. Periodically, CdTe and Te effusion cell shutters are closed and the arsenic effusion cell shutter is opened. The periodicity interval ranges from 30–200 Å and the duration of the As shutter cell opening is a few seconds for an As flux in the range of 10^{15} atm^{-2}s^{-1}, while the Hg flux is still impinging on the surface, which is expected to enhance the Hg-As bond formation. Results showed that some planar As-doped layers exhibit p-

type character without *ex situ* annealing. In as-grown layers, *p*-type doping in the high 10^{16} cm^{-3} levels and a minimum 40 arcsec x-ray full width at half maximum (FWHM) has been achieved in these layers by MBE.

Figure 4.3 is a plot of acceptor concentration as measured by Hall versus SIMS arsenic concentration. The solid line indicates 100% activation efficiency. Data includes MBE [126, 129], LPE [69] and MOVPE [55]. In all cases, the highest arsenic electrical activation occurred when the total arsenic concentration is less than the 10^{18} cm^{-3} range. In MBE regardless of the doping mechanisms and annealing temperature (above 300 °C), close to 100% arsenic electrical activity is obtained when the total arsenic is no more than 2.0×10^{18} cm^{-3}.

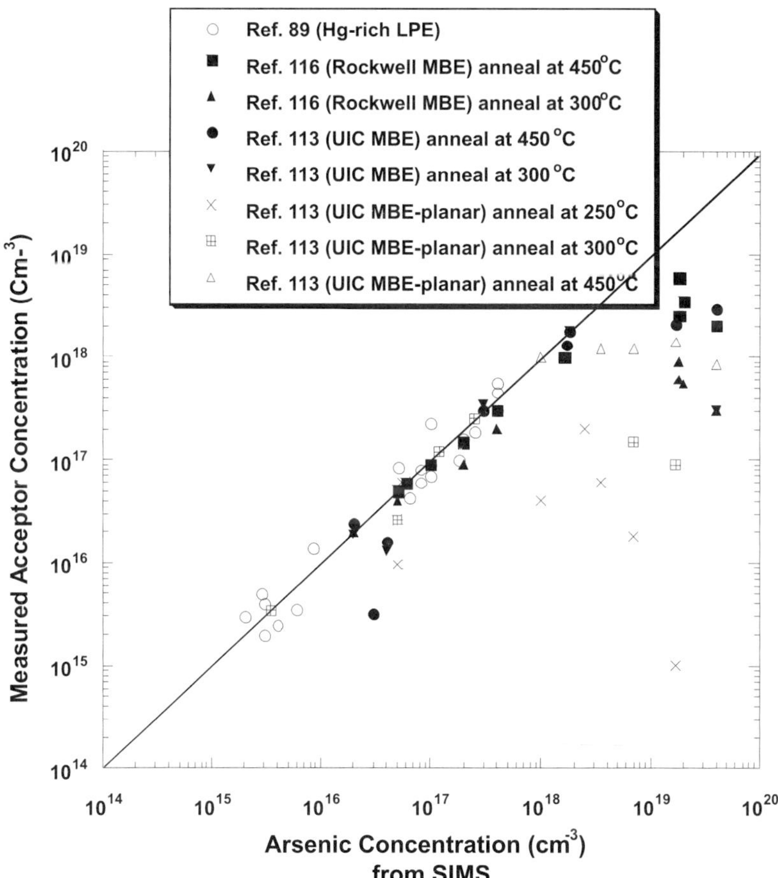

FIG. 4.3. Measured doping concentrations versus SIMS arsenic concentration in arsenic-doped HgCdTe layers grown by MBE, Hg-rich LPE, and MOCVD.

Some authors have claimed [126] that at higher total arsenic concentrations. electrical activity falls off drastically and they suggest that at high concentrations much of the incorporated arsenic occupies neutral or donor sites in the metal sublattice, or forms neutral compounds. For Cd composition of about 30% and doping levels in the $10^{17}\,cm^{-3}$ range, measured minority carrier lifetime follows an inverse dependence on the p-type doping concentration [64, 97, 129].

4.4.2.2. Extrinsic n-Type Doping

Indium-doped n-type HgCdTe is being used extensively for MBE [46, 47, 105, 90, 52]. In MBE material several factors such as substrate orientation, substrate supplier, growth temperature, and Hg flux appear to influence the electrical properties. Small concentrations of unintentionally added elements can, if electrically active, affect the transport properties of the material. The first successful layer with indium doping during MBE growth was reported in the (111)B growth orientation by Boukerche, *et al.* at the Microphysics Laboratory [105]. Carrier concentrations two orders of magnitude higher than can be achieved by stoichiometry deviation have been reached for MBE layers. A maximum doping level of $2 \times 10^{18}\,cm^{-3}$ with indium was obtained. Electron efficiencies close to 100% were achieved for indium concentrations of up to $2 \times 10^{17}\,cm^{-3}$. Transport and SIMS data have shown that, in this growth orientation, indium is incorporated as a singly ionized donor occupying metal sublattices. Unlike stoichiometry-deviation doping, a high Cd composition layer with $x = 0.55$ could also be doped in the $10^{15}\,cm^{-3}$ range.

The incorporation of indium, also has been carried out with different indium fluxes during MBE growth in (211)B-HgCdTe layers at the Microphysics Laboratory at the University of Illinois at Chicago [46, 47] and at Rockwell [65]. Grown layers show excellent electrical properties after n-type annealing at 250 °C. The Cd composition of these layers is ≈ 0.22. Electron mobilities ranging from $(2 - 3) \times 10^5\,cm^2/vs$ at 23 K were obtained with doping levels down to $1 \times 10^{15}\,cm^{-3}$. These values are comparable with the best mobilities reported for HgCdTe for this Cd composition and this doping level.

Measured minority carrier lifetime data fits very well with intrinsic band-to-band recombination mechanisms down to $\approx 1 \times 10^{15}\,cm^{-3}$. As can be seen, MBE-grown layers compare very well with the LPE layers with comparable doping densities. The Auger lifetime dominates throughout the entire temperature region from 300–80 K.

When the net doping level falls below $\approx 2 \times 10^{15}\,cm^{-3}$, the measured lifetime data of the layers can be explained by assuming SR recombination centers (at $\approx 3/4\,E_g$) in addition to band-to-band recombination processes. The combined Auger and radiative lifetimes are in good agreement with the experimental data in the intrinsic region. They believe that these SR levels are somehow related to the Hg vacancies.

4.5. Photovoltaic Detectors

Features and fabrication of photodiode architectures in prevalent use are described in Section 4.5.1. Basic equations governing photodiode behavior are then described in Section 4.5.2. Parameters used to describe and assess photodiode performance are discussed and compared to recent experimental results and theoretical calculations.

4.5.1. Architecture and Fabrication

Planar or mesa schemes are the most frequently used photodiode architectures. Representative examples of $p+$ on-n-double-layer planar and mesa diodes are depicted in Figs. 4.4 and 4.5. The base active layer in these cases is doped n-type with a concentration $\sim 1 \times 10^{15}\,\mathrm{cm}^{-3}$. The cap layer is doped p-type $\sim 5 \times 10^{17}\,\mathrm{cm}^{-3}$ and has a wider bandgap than the base layer. Growth of a single type (n-type or p-type)-HgCdTe layer permits the fabrication of planar diodes. If the base layer is grown n-type, a group V element such as arsenic is then used to form the p-type region. Arsenic is implanted in selected regions and

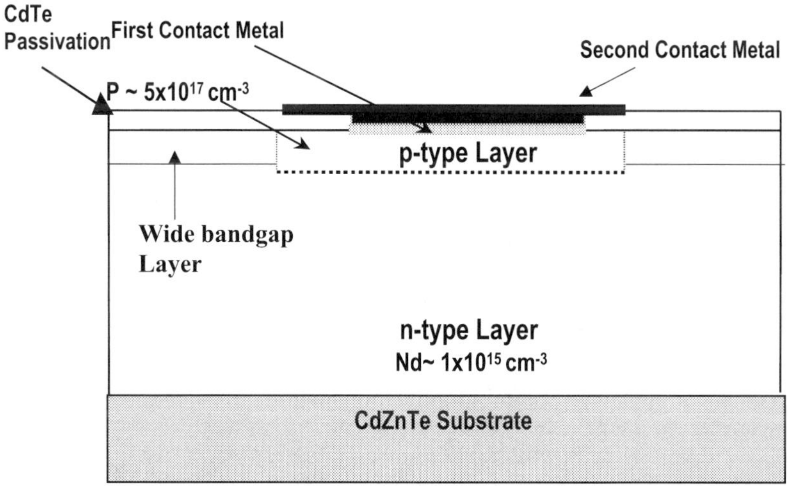

FIG. 4.4. Representative $p+$ on n-double-layer planar diode.

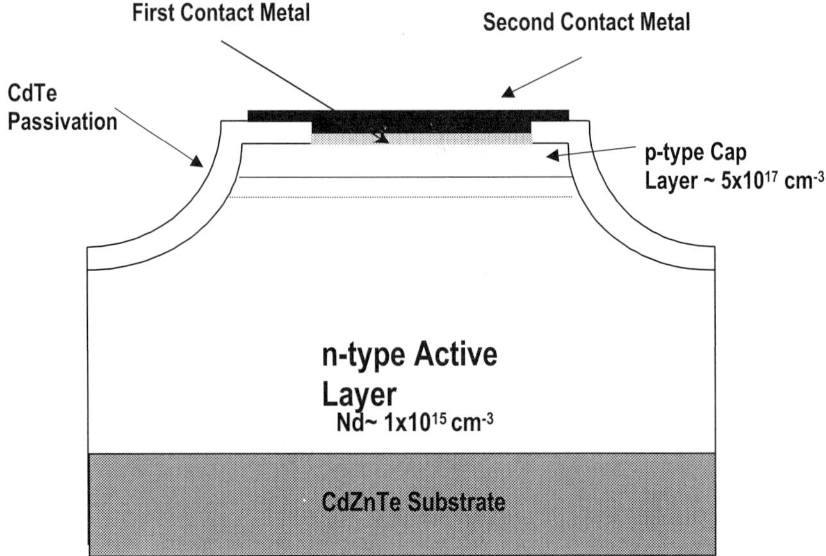

FIG. 4.5. Representative $p+$ on n mesa diode.

annealed to form a planar $p+$ on n diode architecture [133–135]. When both n- and p-type layers are grown prior to diode delineation, the only diode architecture permitted is a mesa scheme. Speaking briefly to n-on-p cases when the base layer is grown p-type, boron implantation is typically employed in selected regions to form n-on-p planar diodes [136, 137]. A variety of different detector architectures has been described previously [98, 140–142, 152, 152a].

4.5.2. Basic Equations and Results

The current-voltage relationship for photovoltaic detector operation is conventionally expressed in the form [143]

$$I = I_0\left(e^{qV_d/nkT} - 1\right) - I_L \tag{4.16}$$

where q is the electron charge, k is Boltzmann's constant, T is absolute temperature, n is an empirical slope factor $\sim 1 - 2$, V_d is the bias voltage across the diode, I_0 is the so-called reverse saturation current, and I_L is the photocurrent generated by impinging photons. Ideally, $n = 1$ for diffusion current components and 2 for generation-recombination components [144].

For flood illumination conditions,

$$I_L = q\eta\phi A_{\text{opt}} \tag{4.17}$$

where η is the detector quantum efficiency, ϕ is the incident flux in photons/cm^2/s, and A_{opt} is the photodiode optical area in cm^2. Detector quantum efficiency is defined as the ratio of the number of minority carriers collected by the detector to the number of photons incident on the detector.

The choice of detector architecture determines the type of dark currents that impact detector performance. A widely used architecture now aggressively pursued is the double layer p-on-n heterostructure described in Section 4.5.1. The dominant current mechanisms for this specific case are diffusion current from the field-free bulk n-side, generation-recombination current in the depletion region, along with band-to-band or trap-assisted tunneling currents. The double layer heterostructure architecture coupled with recent improvements in passivation techniques have reduced the impact of surface tunneling currents.

4.5.2.1. Diffusion Current

Diffusion current contributions to reverse saturation current I_0 consist of components from p- and n-sides of the junction, expressed here in current density superposition form as [143–146]

$$J_{0\text{diff}} = qn_i^2 \left[\frac{L_p}{N_d \tau_p} + \frac{L_n}{N_a \tau_n} \right] \tag{4.18}$$

where $J_{0\text{diff}}$ denotes diffusion current density, $L_{p,n}$ and $t_{p,n}$ denote minority diffusion lengths and lifetimes on respective n and p sides of the junction , and $N_{d,a}$ denote donor and acceptor concentration levels on the n- and p-sides of the junction. For the case of p-on-n double heterostructures where the p-side has a wider bandgap and high doping N_a, the second term in $J_{0_{\text{diff}}}$ is negligible compared to the first and $J_{0_{\text{diff}}}$ reduces to

$$J_{0\text{diff}} = \frac{qn_i^2 L_p}{N_d \tau_p} \tag{4.19}$$

When the n-type active (absorber) layer thickness d is smaller than the diffusion length L_p, dimension d effectively replaces L_p to give

$$J_{0\text{diff}} = \frac{qn_i^2 d}{N_d \tau_p} \tag{4.20}$$

Diffusion current is seen to be proportional to n_i^2 and tends to become the dominant current mechanism at higher temperatures. Another performance parameter of paramount interest is the so-called R_0A_{diff} product defined by

$$\frac{1}{R_0A_{\text{diff}}} = \left.\frac{dJ_{\text{diff}}}{dV}\right|_{V=0} = \frac{qJ_{0\text{diff}}}{kT} \tag{4.21}$$

where R_0 denotes the diode zero-bias impedance value and A_{diff} denotes the effective area associated with absorber layer diffusion processes. Substituting results from Eq. (4.20) into Eq. (4.21) and inverting gives

$$R_0A_{\text{diff}} = \frac{kT}{qJ_{0\text{diff}}} = \left(\frac{kT}{q}\right)\frac{N_d\tau_p}{qn_i^2d} \tag{4.22}$$

with higher R_0A_{diff} values characterizing a higher figure-of-merit.

The doping level N_d is generally in the mid-10^{14} cm^{-3}-low-10^{15} cm^{-3} range. The minority carrier lifetime τ_p has been described earlier. Material growth should be tailored to maximize the $N_d\tau_p$ product. In addition, active (absorber) layer thickness d should be grown as thin as possible to minimize diffusion volume and associated dark diffusion current. However, a lower constraint on thickness d, is that it be thick enough to absorb all of the incident photons within the spectral band of interest, that is, there should be no appreciable loss in quantum efficiency.

4.5.2.2. Generation-Recombination Current

Early treatments developed the seminal formulation for generation-recombination phenomena in semiconductor junctions [138]. Phenomenologically, electron-hole pairs generated thermally in the depletion region give rise to a current that is recombination behaving under forward bias and generation behaving under reverse bias. Derivations for this current historically take the form

$$J_{g-r} = \frac{qn_iW}{\sqrt{\tau_{n0}\tau_{p0}}}\frac{\sinh(qV/2\text{kT})}{q[(V_{bi}-V)/2\text{kT}]}f(b) \tag{4.23}$$

where

$$f(b) = \int_0^\infty \frac{du}{u^2+2bu+1}, \qquad b = e^{-qV/2\text{kT}}\cosh\left[\frac{E_i-E_i}{kT}+\frac{1}{2}\ln\left(\frac{\tau_{p0}}{\tau_{n0}}\right)\right] \tag{4.24}$$

Trap levels due to defects are most effective when they exist close to the intrinsic Fermi level, that is, when $E_t = E_i$. For bias conditions where $V \approx 0$, along with assumptions that $E_t = E_i$ and $\tau_{p0} = \tau_{n0} = \tau_0$, it follows that $b \approx 1$ and $f(b) \approx 1$. Under these provisos, J_{g-r} reduces to

$$J_{g-r} = \frac{qn_i W}{\tau_0} \frac{2\mathrm{kT}}{q(V_{bi} - V)} \sinh(qV/2\mathrm{kT}) \tag{4.25}$$

Generation-recombination current J_{g-r} current is seen to be proportional to n_i. As such, generation-recombination J_{g-r} current will tend to dominate over diffusion J_{diff} current at low temperatures. Again, as with diffusion current it is useful to introduce performance relations defined by

$$\frac{1}{R_0 A_{g-r}} = \left.\frac{dJ_{g-r}}{dV}\right|_{V=0} = \frac{qn_i W}{\tau_0 V_{bi}} \tag{4.26}$$

giving

$$R_0 A_{g-r} = \frac{\tau_0 V_{bi}}{qn_i W} \tag{4.27}$$

with higher $R_0 A_{g-r}$ values characterizing a higher figure-of-merit. Generation-recombination currents in HgCdTe photodiodes have been discussed by several authors [140, 152].

4.5.2.3. Tunneling Currents

Equations for tunneling current can be found in the following references [147, 148, 152, 152a, 153a, 154a, 155a, 156a]. Tunneling currents can dominate at low temperatures where thermally generated currents are low. Currents tunneling through depletion regions are of two types, namely: i) Zener or band-to-band tunneling composed of a single-step process; and ii) trap-assisted tunneling where forbidden gap energy states promote and participate in a 2-step process.

Band-to-band-single-step tunneling is an energy conserving process that requires an occupied energy state on one side of a depletion region to be matched by an unoccupied state of the same energy on the other side of the depletion region. The probability for this type of tunneling characteristically has an exponential dependence on depletion width. Detectors with low doping on at least one side of the metallurgical junction are not apt to have a high probability for band-to-band tunneling because the depletion width will generally be too large, with this size promoted by requisite deeper penetration into the low doped side. In general, due to the exponential dependence with depletion width, together with other influences involving depletion region shape and electric field magnitudes, tunneling currents estimated from theory can differ widely from values measured.

Defects in the material give rise to energy states in the forbidden gap. These states act as traps and can serve as intermediate levels for a 2-step tunneling process. The latter can take on such characters as: i) an energy conserving valence band-to-trap level tunneling step followed by thermal activation to the conduction band; and ii) two energy-conserving steps composed of a valence band-to-trap level step followed by a trap level to conduction band step, etc.

Tunneling is often the dominant current mechanism in narrowgap LWIR devices at low (50 K) temperatures. Example $I_d - V_d$ and $R_d - V_d$ behavior for a detector dominated by tunneling currents is shown in Fig. 4.6. Results here show R_d to be a maximum at zero bias in tunneling-current dominated detectors as compared to diffusion-limited detectors where reverse bias R_d values exceed zero-bias R_0 by two orders of magnitude or more.

In keeping with R_0A definitions for diffusion and generation-recombination currents, R_0A_{tunn} for tunneling currents is defined by

$$\frac{1}{R_0A_{\text{tunn}}} = \left.\frac{dJ_{\text{tunn}}}{dV}\right|_{V=0} \tag{4.28}$$

where J_{tunn} denotes tunneling current density.

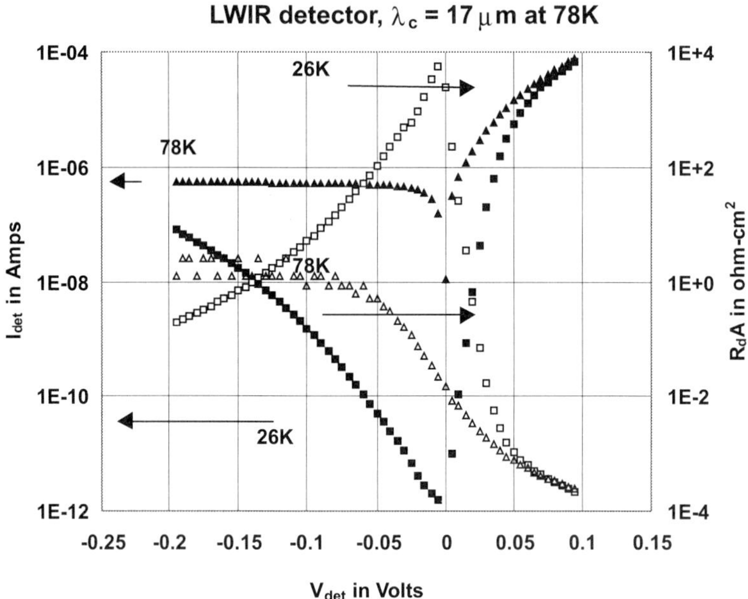

FIG. 4.6. $I_d - V_d$ and $R_dA - V_d$ curves for a LWIR diode at 26 K and 78 K dominated by tunneling current and diffusion current, respectively.

Combining these considerations relative to diffusion, generation-recombination and tunneling current mechanisms [149], detector overall R_0A [150] is defined by

$$\frac{1}{R_0A} = \frac{1}{R_0A_{\text{diff}}} + \frac{1}{R_0A_{g-r}} + \frac{1}{R_0A_{\text{tunn}}} \tag{4.29}$$

4.5.3. $I_d - V_d$ AND $R_d - V_d$ CURVES

Figure 4.6 gives $I_d - V_d$ and $R_dA - V_d$ curves for an 8 μm diode measured at 26 K and 78 K. The dominant current mechanism at 26 K is tunneling current, as can be seen by the shape of the $I_d - V_d$ curve in reverse bias and R_dA peaks near zero bias. The $I_d - V_d$ characteristic at 78 K is limited by diffusion current.

R_0A_{opt} experimental data at 78 K is plotted versus λ_c in Figure 4.7 for detectors grown on PACE-1(sapphire), CdZnTe and CdTe/Si substrates. The dashed line displayed in the figure is based on a one-dimensional theoretical model that assumes diffusion currents are dominated by the narrower bandgap n-side, and minority carrier recombination is via Auger and radiative processes [151]. Parameters used in the calculation of the theoretical R_0A_{opt} values are an n-side

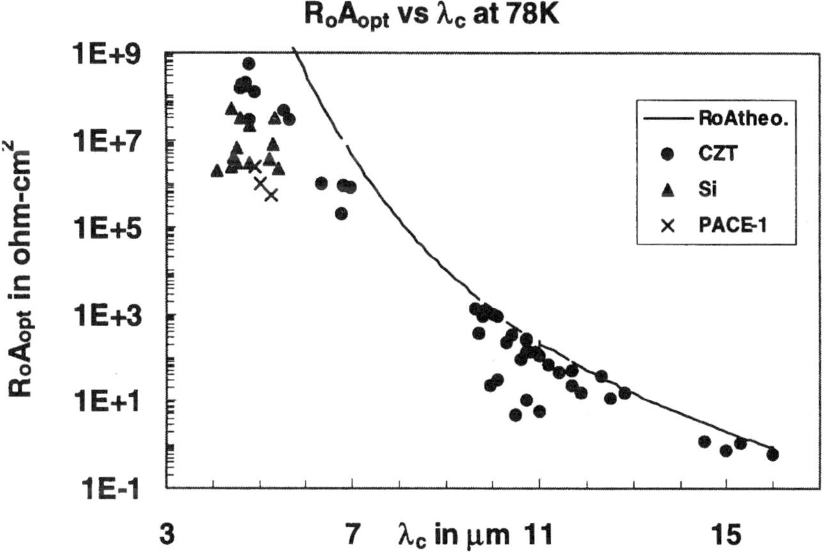

FIG. 4.7. R_0A_{opt} product vs cut-off wavelength at 78 K showing the performance of HgCdTe photovoltaic detectors grown by different growth techniques.

donor concentration of $N_d = 1 \times 10^{15}\ \text{cm}^{-3}$ and a narrow bandgap active layer thickness of 16 μm.

4.5.4. Photoresponse, Quantum Efficiency and Cross Talk

Elements bearing upon detector photoresponse behavior are summarized in synopsis form in Fig. 4.8. Internal mechanisms cited in this figure refer in general to a complicated combination of processes involving: 1) optical absorption giving rise to electron-hole pair carrier generation; 2) consequent minority carrier photocurrent flow toward collecting junctions; 3) loss mechanisms where minority carriers are lost to various recombination processes; and 4) carriers lost to neighboring detectors in the form of crosstalk. Photocurrent flow toward junctions tend to be dominated by diffusion transport in the HgCdTe technology, with drift component contributions considered to be negligible or insignificant in many cases.

In analyzing small-pixel, square-like detector photoresponse behavior, three-dimensional (3D) treatments must generally be employed. Example results for a 40 μm pitch, 100% fill factor, two-dimensional (2D) square array design are illustrated in Figs. 4.9. The absorption coefficient treatment summarized by Gopal *et al.* [35] is used in generating these results. Quantum efficiency behavior as a function of active region thickness is explored in Fig. 4.9a, with related crosstalk behavior being explored in Fig. 4.9b. Symbols d and L_D in the figures

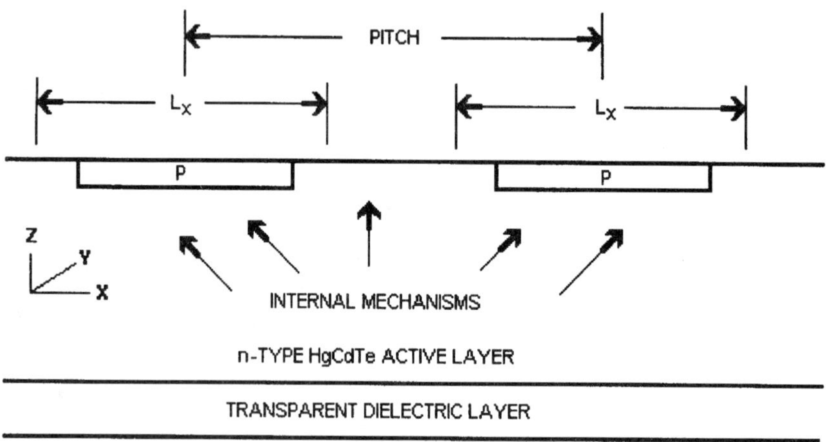

FIG. 4.8. Overview of factors determining detector photoresponse.

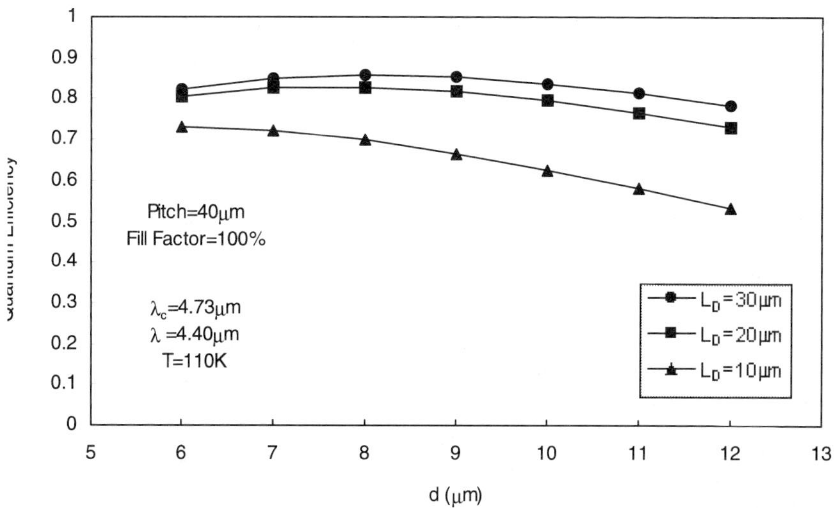

(a) Quantum Efficiency as a function of active region thickness, d.

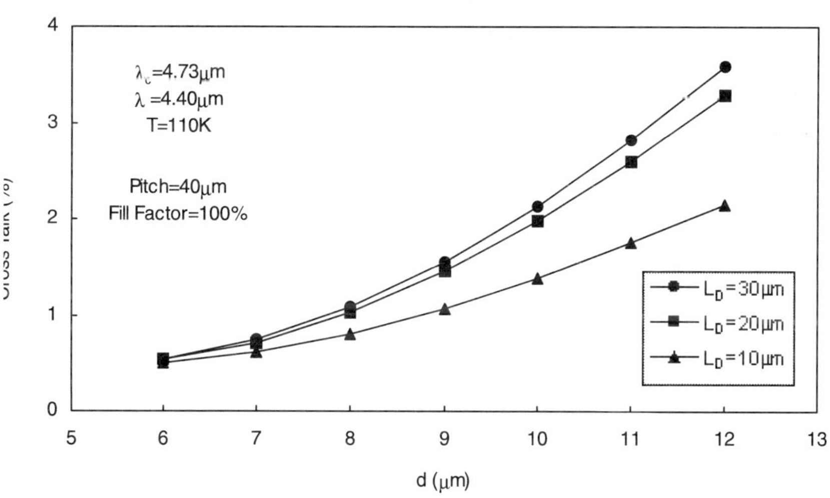

(b) Cross Talk as a function of active region thickness, d.

FIG. 4.9. Quantum efficiency and Cross Talk behavior as a function of active region thickness d for various diffusion length values L_D.

denote active region thickness and diffusion length values, respectively. Crosstalk values here represents relative crosstalk defined by

$$\text{Crosstalk} = 100^* \frac{r_{i+1,j}}{r_{i,j}} \tag{4.30}$$

where $r_{i,j}, r_{i+1,j}$ denote responses observed in adjacent detectors with a photon point source centered on detector i, j.

Results in Fig. 4.9 show quantum efficiency and crosstalk performance to be compromised concomitantly when active region thicknesses are made too large. Smaller diffusion lengths and smaller solid angle collection attributes both diminish performance in the larger thickness cases. At the other extreme, as active layer thicknesses are reduced to improve photocurrent collection, the active layer itself becomes more transmissive. To the extent that photons transmitted through the active layer are lost, absorption is rendered less than complete with some negative impact upon quantum efficiency. These considerations may arise in modern state-of the-art designs implemented with top-surface capping layers employed to reduce diode quiescent leakage. Photons transmitted into these ~0.3μm-thick, higher bandgap, lesser absorbing layers are apt to be lost unless somehow reflected back into the active region.

Finally, quantum efficiency behavior as a function of wavelength is explored in Figure 4.10, showing the asymptotic approach to flat-like behavior for wave-

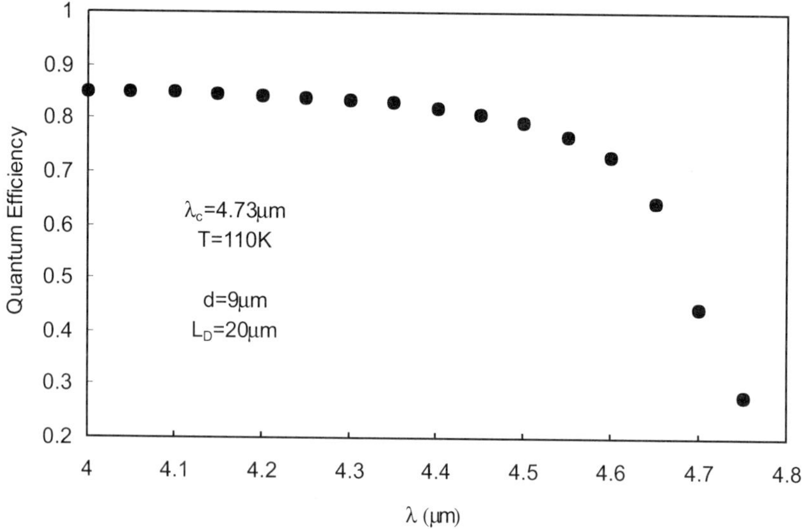

FIG. 4.10. Quantum efficiency behavior as a function of wavelength showing the asymptotic approach to flat-like behavior suitably below the cutoff wavelength.

lengths suitably below cutoff. In the latter range, active region absorption becomes virtually complete and devoid of transmitted components.

4.5.5. SPECIFIC DETECTIVITY PARAMETER D^*

Specific detectivity parameter D^* is a performance measure used to compare photodetectors. D^* is defined by [152]:

$$\frac{D^* = \Re\sqrt{A_{\text{det}}}}{N} \tag{4.31}$$

where $\Re$ denotes detector responsivity expressed in Amps/Watt, N denotes detector noise expressed in Amps/Hz$^{1/2}$, with D^* then being expressed in units of cmHz$^{1/2}$/Watt. Detector responsivity is defined in terms of [152]

$$\Re = \frac{\eta\lambda q}{hc} \tag{4.32}$$

with all factors here having been described earlier. Photovoltaic detector noise associated with diffusion and photocurrent flow is given by [152]:

$$N = \sqrt{2q(I_0 + I_0 e^{qV/kT} + I_L) + i^2_{\text{excess}}} \tag{4.33}$$

with possible $1/f$ noise contributions [153, 154, 162a, 163a] being incorporated in the i_{excess} term. Substituting for photocurrent from Eq. (4.12) and assuming zero-bias operation, the expression for noise takes the form

$$N = \sqrt{\frac{4\text{kT}}{R_0} + 2q^2\eta\Phi A_{\text{opt}} + i^2_{\text{excess}}} \tag{4.34}$$

High-performance photodiodes have low excess noise in comparison to the other contributions. Moreover, photodiode applications where the following conditions prevail,

$$\frac{4\text{kT}}{R_0} \ll 2\eta A_{\text{det}} q^2 \Phi \tag{4.35}$$

are said to be background-noise limited. Background-noise limited performance is the highest attainable performance for photodetectors.

4.6. Recent Progress in Focal Plane Arrays (FPAs)

The hybrid focal plane array is made up of two separate components: a detector array and a readout integrated circuit (ROIC). The HgCdTe detector array consists of photovoltaic diodes processed in epitaxially grown material on a suitable substrate described previously. The ROIC are custom designed to convert the photocurrent into a voltage for each element in the detector array. Indium bumps are deposited on each element of the detector array and a corresponding bump is deposited on the ROIC. The detector array and the ROIC are cold-welded together to form the hybrid FPA. Large format FPA reliability issues are mitigated by the use of a shim that constrains the detector array and ROIC to expand and contract at the same rate as they are cycled from room to cryogenic temperatures. Figure 4.11 shows a cross section of a hybrid focal plane array. The FPA is backside illuminated through the detector substrate.

Input circuits are optimized based on the specific applications [155, 156]. For tactical and remote sensing applications, where the backgrounds are high and detector impedance values are moderate, direct injection (DI) is a commonly used input circuit [157]. Strategic applications usually have low backgrounds and require low noise ROIC interfaced to high-impedance detectors. A commonly used input circuit for strategic applications is the capacitative transimpedance amplifier (CTIA) input circuit [158]. There are two kinds of CTIA inputs, inverter-and differential-amplifier-based. The former is preferred for high-density applications.

A typical complementary-metal-oxide-semiconductor (CMOS) ROIC architecture consists of fast (column) and slow (row) shift registers at the edges of the active area. Pixels are addressed one-by-one through the selection of a slow register, while the fast register scans through a column, and so on. The advantages of CMOS are that existing commercial foundries that fabricate

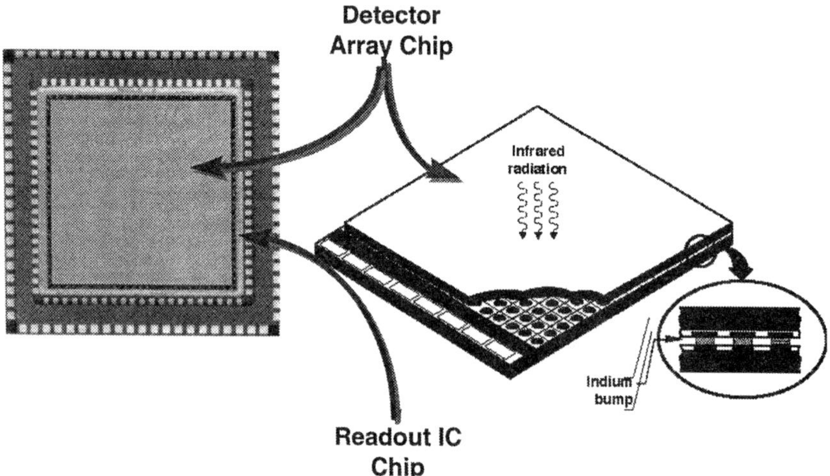

FIG. 4.11. Cross section of a hybrid focal plane array, backside illuminated through the substrate.

application specific integrated circuits (ASIC) can be readily used to fabricate the custom ROIC needed by adapting their design rules. The impact of evolving foundry capability on the 640 × 480 ROIC, for example, is evident from the fact that the number of die per wafer has increased from 11 to 70 in going from 4 to 8 in (10- to 20-cm) wafers. In addition, as foundry design rules get smaller, input cell circuitry can be fit into a smaller area, permitting the fabrication of arrays that have smaller pitch. This in turn leads to the fabrication of larger configuration arrays. Wafer diameters of up to 20 cm are now available, potentially increasing to 30 cm. Due to the size of the commercial silicon market, significant advances are expected in silicon technology; thus more on-chip signal processing capabilities will be available.

Detector and FPA performance continues to improve in all infrared spectral bands. Consequently, the literature is the best source of information for FPA performance in terms of the latest configuration, pitch, D^*, operating temperature, background conditions, integration and frame times, data rate, etc. A plethora of FPA data is published in the IRIS Detector Specialty Conference Proceedings. However, this information is restricted in its availability. The IEEE and SPIE conferences and their proceedings are also extremely good sources of FPA results. Some of these references are listed for the reader [159–181].

4.7. Conclusions

Significant advances in the areas of mercury cadmium telluride HgCdTe-material growth, detector array fabrication, and ROIC design and fabrication have been accumulated over the past 20 years. These advances have led to the demonstration of high-resolution-, low-noise- and large-format reliable focal plane arrays (FPA). In the 1–3 μm SWIR region, high-performance, reliable 1024 × 1024 FPA for astronomy applications have been developed and 2048 × 2048 arrays are in development. In the 3–5-μm MWIR spectral region, high-performance 640 × 480 and 1024 × 1024 FPA have been developed, while 256 × 256 FPA are in continuous production to support systems applications. In the 8–12-μm- and beyond long-wavelength-infrared (LWIR) region, high-performance, 128 × 128 FPA have been developed on CdZnTe substrates.

The MBE HgCdTe-based technique has emerged as a viable technology for manufacturing high-performance IRFPA. Device-quality LPE, MOCVD and MBE material is being grown routinely for applications in the 1.5–16.0-μm spectral region. Excellent control of composition, growth rate, layer thickness, doping concentration, dislocation density, and transport characteristics has been demonstrated. Excellent material quality has made possible high-performance VLWIR 128 × 128 focal plane arrays operating between 40 and 60 K, low background LWIR detectors operating at 40 K, and MWIR FPA for strategic surveillance applications. The future will see further exploitation of the advances made to further fabricate multicolor FPA and "smart" FPA with increasing on-chip processing.

Acknowledgments

The authors thank their colleagues at Boeing, the Rockwell Science Center, and at the University of Illinois at Chicago who have encouraged and supported the authors through the years. Thanks are due to everyone who has worked in the HgCdTe material, detector, and infrared FPA field for helping achieve the rewards that have accrued so far.

References

1. Lawson, W. D., Nielson, S., Putley, E. H., and Young, A. S. (1959). *J. Phys. Chem. Solids* **9**: 325.
2. Dornhaus, R. and Nimtze, G., (1974). *Narrow Gap Semiconductors*, vol. 98, in *Springer Tracts in Modern Physics*, (Berlin: Springer, pp. 202–221. R. Dornhaus, H. Happ, K. H. Muller, G. Nimtze, W. Schlobitz, P.Zaplinski, and G. Baner, in *Proc. XIIth Int. Conf. Phys. Semicond.* Stuttgart, 1974, M. H. Pilkuhm, ed., p.1157.
3. Wooley, J. C. and Ray, B. (1969). *J. Phys. Chem. Sol.* **13**: 151.
4. Blair, J. and Newnham, R. (1967) in *Metallurgy of Elemental and Compound Semiconductors* New York: Wiley, Interscience, vol. 12:, p. 393.
5. Alper, T. and Saunders, G. A. (1967). *J. Phys. Chem. Sol.* **28**: 1637.
6. Capper, P. (1991). *J. Vac. Sci. Technol.* **B9**: 1667.
7. Nimtz, G., Schlicht, B., and Dornhaus, R. (1973). *Appl. Phys. Lett.* **34**: 490.
8. Micklethwaite, W. F.H. and Redden, R. F. (1980). *Appl. Phys. Lett.* **36**: 379.
9. Tung, T. (1988). *J. Crys. Growth* **86**: 161.
10. Astles, M. G. (1994). *Properties of Narrow Gap Cadmium-Based Compounds*, EMIS Data Review Series No. 10, P. Capper ed., U. K.:INSPEC, p.13.
11. Tung, T., De Armond, L. D., Herald, R. F., Herning, P. E., Kalisher, M. H., Olson, D. A., Risser, R. F., Stevens, A. P., and Tighe, S. J. (1992). *Proc. SPIE* (USA) **1735**: 109.
12. Gertner, E. R. (1985). *Ann. Rev. Mat. Sci.* **15**: 303.
13. Vydyanath, H. R. (1995). *J. Elec. Mat.* **24**: 1275.
14. Wang, C. C., Shin, S., Chu, M., Lanir, M. and Vanderwyck, A. H. B., (1980). *J. Electrochem. Soc.* **127**: 175.
15. Arias, J. M. (1993). in *II-VI Semiconductor Compounds*, M. Pain, ed., Singapore: World Scientific Publishing Co. Pte. Ltd., p. 509–536.
16. Ghandi, S. K. and Bhat, I. (1984). *Appl. Phys. Lett.* **44**: 779.
17. Faurie, J. P. (1994). *Prg. Crystal Growth and Charact.* **29**: 85.
18. Farrow, R. F.C. (1985). *J. Vac. Sci. Technol.* **A3**: 60.
19. Faurie, J. P. and Million, A. (1981). *J. Cryst. Growth* **54**: 582.
20. Blue, M. D., (1964). *Phys. Rev.* **134**: A226.
21. Melingails, I. and Strauss, A. J., (1966). *Appl. Phys. Lett.* **8**: 179.

22. Verie, C. and Grager, G. (1965). *Compt. Rend.* **261**: 3349.
23. Harman, T. C., Strauss, A. J., Dickey, D. H., Dresselhauss, M. S., Wright, G. B. and Mavroides, J. G., (1961). *Phys. Rev. Lett.* **7**: 403.
24. Mahan, G. D. (1965). *J. Phys. Chem. Solids.* **26**: 751.
25. Schmit, J. L. and Stelzer, E. L. (1969). *J. Appl. Phys.* **40**: 4865.
26. Scott, M. W. (1969). *J. Appl. Phys.* **40**: 4077.
27. Finkman, E. and Nemirovsky, Y. (1979). *J. Appl. Phys.* **50**: 4356.
28. Hansen, G. L., Schmit, J. L. and Casselman, T. N. (1982). *J. Appl. Phys.***53**: 7099.
29. Anderson, W. W. (1980). *Infrared Phys.* **20**: 363.
30. Finkman, E. and Nemirovsky, Y. (1979). *J. Appl. Phys.* **50**: 4356.
31. Urbach, F. (1953). *Phys. Rev.* **92**: 1324.
32. Finkman, E. and Schacham, S. E. (1984). *J. Appl. Phys.* **56**: 2896.
33. Hougen, C. A. (1989). *J. Appl. Phys.* **66**: 3763.
34. Amirtharaj, P. M. (1991). *Handbook of Optical Constants in Solids II*, E. P. Palik, ed., New York: (Academic Press Inc., p. 655–689).
35. Gopal, V., Ashokan, R. and Dhar, V., (1992). *Infrared Phys.* **33**: 39.
36. Schmit, J. L. (1970). *J. Appl. Phys.* **41**: 2876.
37. Nemirovsky, Y. and Finkman, E. (1979). *J. Appl. Phys.* **50**: 8107.
38. Hansen, G. L. and Schmit, J. L. (1983). *J. Appl. Phys.* **54**: 1639.
39. Kane, E. O. (1959). *J. Phys.Chem. Solids.* **8**: 38.
40. Blakemore, J. S. (1962). *Semiconductor Statistics* New York: Pergamon.
41. Pratt, R. G., Hewett, J., Capper, P., Jones, C. L. and Quelch, M. J. (1983). *J. Appl. Phys.* **54**: 5122.
42. Kinch, M. A., Brau, M. J., and Simmons, A. (1973). *J. Appl. Phys.* **44**: 1649.
43. Casselman, T. N. (1982). *Proc. Fourth Intern. Conf. on Physics of Narrow Gap Semiconductors*, Linz, Austria, 1981, Berlin: Springer Verlag, p. 147, *Lecture Notes on Physics*, vol. 152.
44. Peterson, P. E. (1970). *J. Appl. Phys.* **41**: 3465.
45. Lopes, V. C., Syllaios, A. J., and Chen, M. C. (1993). *Semicond. Sci. Technol.* **8**: 824.
46. Wijewarnasuriya, P. S., Lange, M. D., Sivananthan, S., and Faurie, J. P. (1994). *J. Appl. Phys.* **75**: 1005.
47. Wijewarnasuriya, P. S., Lange, M. D., Sivananthan, S., and Faurie, J. P. (1995). *J. Elect. Mat.* **24**: 545.
48. de Souza, M. E., Boukerche, M., and Faurie, J. P. (1990). *J. Appl. Phys.* **68**: 5195.
49. Edwall, D. D., Zandian, M., Chen, A. C., and Arias, J. M. (1987). *J. Elect. Mat.* **26**: 493.
50. Beattie, A. R. (1987). *Semicond. Sci. Technol.* **2**: 281.
51. Capper, P. (1982). *J. Crystal Growth* **57**: 280.
52. Jones, C. L. (1987). *Properties of Mercury Cadmium Telluride*, J. Brice and P. Capper eds., New York: INSPEC, p. 137.
53. Baryshew, N. S., Gelmont, B. L., and Ibragimova, M. J. (1990). *Sov. Phys. Semocond.* **24**: 127.
54. Casselman, T. N. and Petersen, P. E. (1980). *Sold State Communications*, Vol. **33**: 615 ; *J. Appl. Phys.* **52**: 848.
55. Mitra, P., Tyan, Y. L., Case, F. C., Starr, R., and Reine, M. B. (1996). *J. Elec. Mat.* **25**: 1328.
56. Mitra, P., Tyan, Y. L., Schimert, T. R., and Case, F. C. (1994). *Appl. Phys. Lett.* **65**: 195.
57. Edwall, D. D., Chen, J. S., and Bubulac, L. O. (1991). *J. Vac. Sci. Technol.* **B9**: 1691.
58. Edwall, D. D., Bubulac, L. O., and Gertner, E. R. (1992). *J. Vac. Sci. Technol.* **B10**: 1423.
59. Elliot, J. and Kreismanis, V. G. (1992). *J. Vac. Sci. Technol.* **B10**: 1428.
60. Rao, V., Ehsani, H., Bhat, I. B., Kestigian, M., Starr, R., Weiler, M. H., and Reine, M. B. (1995). *J. Electron. Mater.* **24**: 437.
61. Reine, M. B., Norton, P. W., Starr, R., Weiler, M. H., Kestigian, M., Musicant, B. L., Mitra, P., Schimert, T., Case, F. C., Bhat, I. B., Ehsani, H., and Rao, V. (1995). *J. Electron. Mater.* **24**: 669.

62. Vydyanath, H. R., Ellisworth, J. A., and Devancy, M. (1987). *J. Electron. Mat.*, **16**: 13.
63. Harman, T. C. (1979). *J. Electron. Mat.* **8**: 191.
64. Tung, T. (1988). *J. Cryst. Growth* **86**: 161.
65. Edwall, D. D., Zandian, M., Chen, A. C., and Arias, J. M. (1997). *J. Elect. Mat.* Vol. **26**: 493.
66. Babaj, J., To be published in SPIE.
67. Bajaj, J., Arias, J. M., Zandian, M., J. G. Pasko, Kozlowski, L. J., DeWames, R. E., and Tennant, W. E. (1996). *J. Electron. Mater.* **25**: 1394.
68. Arias, J. M., Pasko, J. G., Zandian, M., Kozlowski, L. J., and DeWames, R. E. (1994). *Optical Eng.* **33**: 1422.
69. Blazejewski, E. R., Arias, J. M., Williams, G. M., Zandian, M., Pasko, J. G., and McLevige, W. (1992). *J. Vac. Sci. Technol.* **B 10**: 1794.
70. Arias, J. M., Pasko, J. G., Zandian, M., Babaj, J., Kozlowski, L. J., DeWames, R. E., and Tennant, W. E. (1994). *Proceeding of SPIE Symposia on Producibility of II-VI Materials and Devices* **2228**: 210.
71. Arias, J. M. (1994). *Properties of Narrow Gap Cadmium Based Compounds*, EMIS Data Review Series No. 10, Peter Capper, ed., U. K.: INSPEC, p.30.
72. Wijewarnasuriya, P. S., Zandian, M., Edwall, D. D., McLevige, W. V., Chen, C. A., Pasko, J. G., Hildebrandt, G., Chen, A. C., Arias, J. M., D'Souza, A. I., Rujirawat, S., and Sivananthan, S. (1998). *J. Electron. Mater.* **27**: 546.
73. Wu, O. K. *et al.* (1990). *J. Vac. Sci. Technol.* **A8**: 1034.
74. Rajavel, R. D., Jamba, D. M., Jensen, J. E., Wu, O. K., Le Beau, C., Wilson, J. A., Patten, E., Kosai, K., Johnson, J., Rosbeck, J., Goetz, P. and Johnson, S. M. (1997). *J. Electron. Mater.* **26**: 476.
75. Zandian, M. *et al.* (1991). *Appl. Phys. Lett.* **59**: 1022.
76. DeWames, R. E., Arias, J. M., Kozlowski, L. J., and Williams, G. M. (1992). *Proc. SPIE* **1735**: 2.
77. Price, S. L., Hettich, H. L., Sen, S., Currie, M. C., Rhiger, D. R., McLean, E. O. (1998). *J. Elect. Mat.* **V 27**: 564.
78. Rhiger, D. R., Peterson, J. M., Emerson, R. M., Gordon, E. E., Sen, S., Chen, Y. and Dudley, M., (1998). *J.Elect. Mat.* **27**: 615.
79. Sen, S., Liang, C. S., Rhiger, D. R., Stannard, J. E., and Arlinghaus, H. F. (1996). *J. Elect. Mat.* **V 25**: 1188.
80. Bubulac, L. O., Tennant, W. E., Pasko, J. G., Kozlowski, L. J., Zandian, M., Motamedi, M. E., DeWames, R. E., Bajaj, J., Nayar, N., McLevige, W. V., Gluck, N. S., Melendes, R., Cooper, D. E., Edwall, D. D., Arias, J. M., Hall, R., and D'Souza, A. I. (1997). *J. Elect. Mat.* **V 26**: 649.
81. D'Souza, A. I., Dawson, L. C., Anderson, E. J., Markum, A. D., Tennant, W. E., Bubulac, L. O., Zandian, M., Pasko, J. G., McLevige, W. V., Edwall, D. D., Derr, J. W., Jandik, J. E. (1997). *J. Elect. Mat.* **26**: 656.
82. Mitra, P., Case, F. C., and Reine, M. B. (1998). *J. Elect. Mat.* **27**: 510.
83. D'Souza, A. I., Bajaj, J., De Wames, R. E., Edwall, D. D., Wijewarnasuriya, P. S., and Nayar, N. (1998). *J. Elect. Mat.* **27**: 727.
84. de Lyon, T. J., Rajavel, R. D., Vigil, J. A., Jensen, J. E., Wu, O. K., Cockrum, C. A., Johnson, S. M., Venzor, G. M. Bailey, S. L. Kasai, I. Ahlgren, W. L., and Smith, M. S. (1998). *J. Elect. Mat.* **27**: 550.
85. de Lyon, T. J., Rajavel, R. D., Jensen, J. E., Wu, O. K., Johnson, S. M., Cockrum, C. A., and Venzor, G. M. (1996). *J. Elect. Mat.* **25**: 1341.
86. Zanatta, J. P., Ferret, P., Theret, G., Million, A., Wolny, M., Chamonal, J. P., and Destefanis, G. (1998). *J. Elect. Mat.* **27**: 542.
87. Ebe, H., Okamoto, T. Nishino, H., Saito, T., Nishijima, Y., Uchikowski, M., Nagoshima, M., and Wada, H. (1996). *J. Elect. Mat.* **25**: 1358.

88. Almeida, L. A., Chen, Y. P., Faurie, J. P., Sivananthan, S., Smith, D. J., and Tsen, S.-C. Y. (1996). *J. Elect. Mat.* **25**: 1402.

88a. Wang, W-S., and Bhat, I. B. (1995). *J. Elect. Mat.* **24**: 1047.

89. Arias, J. M., Zandian, M., Shin, S. H., DeWames, R. E. and Gertner E. R. (1991). *J. Appl. Phys.* **69**: 2143.

90. Arias, J. M., Shin, S. H., Pasko, J. G., DeWames, R. E., and Gertner, E. R. (1989). *J. Appl. Phys*, **65**: 1747.

91. Faurie, J. P., Sivananthan, S., and Wijewarnasuriya, P. S. (1992). *SPIE Proc.* **1735**: 141.

92. Elliot, C. T., Melngoils, I., Harman, T. C., and Foyt, A. G. (1972). *J. Phys. Chem. Solids* **33**: 252, 1527 (1972).

93. Faurie, J. P., Sivananthan, S., and Wijewarnasuriya, P. S. (1992). *SPIE Proc.* **1735**: 141.

94. Sivananthan, S., Wijewarnasuriya, P. S., and Faurie, J. P. (1995). *SPIE* **2554**: 55.

95. Myers, T. H., Harris, K. A., Yanka, R. W., Mohnkern, L. M., Williams, R. J., and Dudoff, G. K. (1992). *J. Vac. Sci. Tech.* **B10**: 1438.

96. Jones, C. E., James, K., Merz, J., Bravnstein, R., Burd, M., Eetemadi, M., Hutton, S., and Drumheller; J. (1995). *J. Vac. Sci. Technol.* **A3**: 131.

97. Chen, M. C., Colombo, L., Dodge, J. A., and Tregilgas, J. H. (1993). U. S. Workshop on the Physics and Chemistry of HgCdTe (extended abstracts, p. 71).

98. Rogalski A., *et al.* (1995). *Infrared Photon Detectors*, Bellingham, WA: SPIE Press.

99. Wijewarnasuriya, P. S., Sou, I. K., Kim, J. Mahavadi, K. K., Sivananthan, S., Boukerche, M., and Faurie, J. P. (1987). *Appl. Phys. Lett.* **51**: 2045.

100. Wroge, M. L., Peterman, D. J., Morris, B. J., Leopold, D. J., Broerman, J. G., and Feldman, B. J. (1988). *J. Vac. Sci. Technol.* **A6**: 2826.

101. Capper, P. (1991). *J. Vac. Sci. Technol.* **B9**: 1667.

102. Bubulac, L. O., Tennant, W. E., Riedel, R. A., Bajaj, J., and Edwall, D. (1983). *J. Vac. Sci. Technol.* **A1**: 1646.

103. Astles, M., Hill, H., Blackmore, G., Courtney, S., and Shaw, N. (1988). *J. Crystal. Growth.* **91**: 1.

104. Edwall, D. D., Gertner, E. R., and Tennant, W. E. (1985). *J. Elec. Mat*er. **14**: 245.

105. Boukerche, M., Wijewarnasuriya, P. S., Sivananthan, S., Sou, I. K., Kim, J. Y., Mahavadi, K. K. and Faurie, J. P. (1988). *J. Vac. Sci. Tech.* **A6**: 2830.

106. Herning, P. (1984). *J. Electron. Mat.* **13**: 1.

107. Kalisher, M. H. (1984). *J. Cryst. Growth* **70**: 365.

108. Boukerche, M., Sivananthan, S., Wijewarnasuriya, P. S., Sou, I. K., and Faurie, J. P. (1989). *J. Vac. Sci. Technol.*, **A7**: 311.

109. Vydyanath, H. R., Lichtman, L. S., Sivananthan, S., Wijewarnasuriya, P. S., and Faurie, J. P. (1995). *J. Electron. Mater.* **Vol. 24**: 625.

110. Shin, S. H., Arias, J. M., Zandian, M., Pasko, J. G., Bubulac, L. O., and DeWames, R. E. (1993). *J. Electron. Mater.* **22**: 1039.

111. Bubulac, L. O., Irvine, S. J.C., Gertner, E. R., Bajaj, J., Lin, W. P., and Zucca, R. (1993). *Semicond. Sci. Technol.* **8**: 270.

112. Arias, J. M., Zandian, M., Pasko, J. G., Shin, S. H., Bubulac, L. O., and DeWames, R. E. (1991). *J. Appl. Phys.* **69**: 2141.

113. Faurie, J. P. (1992). 2nd Annual Report - DARPA contract # F49620-91-C-0007 (Nov. 1992).

114. Sivananthan, S., Wijewarnasuriya, P. S., and Faurie, J. P. (1993). U. S. Workshop on the Physics and Chemistry of HgCdTe (extended abstracts p. 7).

115. Wijewarnasuriya, P. S., Faurie, J. P., and Sivananthan, S. (1996). *J. Crystal Growth.* **159**: 1136.

116. Bubulac, L. O., Edwall, D. D., Irvine, S. J.C., Gertner, E. R., and Shin, S. H. (1995). *J. Electron. Mater.* **24**: 617.

117. Johnson, E. S. and Schmit, J. L. *J. Electron. Mater.* **6**: 25 (1977).

118. Arias, J. M., Pasko, J. G., Zandian, M., Shin, S. H., Williams, G. M., Bubulac, L. O., DeWames, R. E., and Tennant, W. E. (1993). *J. Electron. Mater.* **22**: 1019.

119. Harper, R. L. Jr., Hwang, S., Giles, N. C., Schetzina, J. F., Dreifus, D. L., and Myers, T. H. (1989). *Appl. Phys. Lett.* **54**: 170.

120. Han, J. W., Hwang, S., Lansari, Y., Harper, R. L. Jr., Yang, Z., Giles, N. C., Cook, J. W. Jr., Schetzina, J. F., and Sen, S. (1989). *Appl. Phys. Lett.* **54**: 63 .

121. Harris, K. A., Myers, T. H., Yanka, K. W., Mohnkern, L. M., and Otsuka, N. (1991). *J. Vac. Sci. Technol.*, **B9**: 1752 (1991).

122. Harris, K. A., Myers, T. H., Yanka, K. W., Mohnkern, L. M., Green, R. W., and Otsuka, N. (1990). *J. Vac. Sci. Tech.* **A8**: 1013.

123. Wu, O. K., Kamath, G. S., Radford, W. A., Bratt, P. R., and Patten, E. A. (1990). *J. Vac. Sci. Technol.* **A8**: 1034.

124. Cheung, J. T. (1987). *Appl. Phys. Lett.* 1940.

125. Kamath, G. S. and Wu, O. K. U. S. Patent no. 5,028,561.

126. Wijewarnasuriya, P. S. and Sivananthan, S. (1998). *Appl. Phys. Lett.* **72**: 1694.

127. Aqariden, F., Wijewarnasuriya, P. S., Rujirawat, S., and Sivananthan, S. (1997). *Mat. Sci. Soc. Symp., Proc. Vol.* **450**: 251.

128. Sivananthan, S., Wijewarnasuriya, P. S., Aquriden, F., Vydyanath, H. R., Zandian, M., Edwall, D. D., and Arias, J. M. (1997). *J. Elec. Mater.*, Vol. **26**: 621.

129. Chen, A. C. (1998). *J. Electron. Mater.* **27**: 595.

130. Berding, M. A. *et al.* (1998). *J. Electron. Mater.* **27**: 605.

131. Berding, M. A. *et al.* (1998). *J. Electron. Mater.* **27**: 573.

132. Berding, M. A. and Sher, A. (1999). *Appl. Phys. Lett.* **74**: 685.

133. Arias, J. M., Pasko, J. G., Zandian, M., Shin, S. H., Williams, G. M., Bubulac, L. O., DeWames, R. E., and Tennant, W. E. (1993). *J. Elect. Mat.* **V 22**: 1049.

134. Shin, S. H., Arias, J. M., Zandian, M., Pasko, J. G., Bubulac, L. O., and DeWames, R. E. (1993). *J. Elect. Mat.* **22**: 1039.

135. Bubulac, L. O., Bajaj, J., Tennant, W. E., Zandian, M., Pasko, J., and McLevige, W. V. (1996). *J. Elect. Mat.* **25**: 1312.

136. Nemirovsky, Y. and Rosenfeld, D. (1990). *J. Vac. Sci. Tech.* A **8**: No.2, 1159.

137. Robinson, H. G., Mao, D. H., Williams, B. L., Holander-Gleixner, S., Yu, J. E., and Helms, C. R. (1996). *J.Elect. Mat.* **25**: 1336.

138. Sah, C. T., Noyce, R. N., and Shockley, W. (1957). *Proc IRE* **45**: 1228.

139. Schoolar, R., Price, S., and Rosbeck, J. (1992). *J. Vac. Sci. Tech.* Bio(4): 1507.

140. Rogalski, A. and Piotrowski, J. (1988). *Prog. Quantum Electronics*, **12**: 87.

141. Rajavel, R. D., Jamba, D. M., Jensen, J. E., Wu, O. K., LeBeau, C., Wilson, J. A., Patten, E., Kosai, K., Johnson, J., Rosbeck, J., Goetz, P., and Johnson, S. M. (1997). *J. Elect. Mat.* **26**: 476.

142. Mitra, P., Barnes, S. L., Case, F. C., Reine, M. B., O'Dette, P., Starr, R., Hairston, A., Kuhler, K., Weiler,M. H., and Musicant, B. L. (1997). *J. Elect. Mat.* **26**: 482.

143. Streetman, B. G. (1980). *Solid State Electronic Devices*, Englewood Cliffs, New Jersey: Prentice-Hall.

144. Sze, S. M. (1981). *Physics of Semiconductor Devices*, 2nd ed., New York: John Wiley & Sons, p. 92.

145. Shockley, W. (1950). *Electrons and Holes in Semiconductors*, Princeton, NJ: D. Van Nostrand.

146. Nemirovsky, Y., Rosenfeld, D., Adar, R., and Kornfeld, A. (1989). *J. Vac. Sci. Tech.* **A 7**: 528.

147. Anderson, W. W. (1980). *Infrared Phys.* **80**: 353.

148. Unikovsky, A. and Nemirovsky, Y. (1992). *Appl. Phys. Lett.* 61, 330.

148a. Anderson, W. W. (1977). *Infrared Phys.* **17**: 147.

149. Rosbeck, J., Starr, R. E., Price, S. L., and Riley, K. J. (1982). *J. Appl. Phys.* **53**: 6430.

149a. Wong, J. Y. (1980). *IEEE Trans. Electron Devices* **ED-27**: 48.

150. Rogalski, A. (1988). *Infrared Phys.* **28**: 139.
151. Williams, G. M. and DeWames, R. E. (1994). *J. Electron. Mater.* **24**: 1239.
152. Reine, M. B., Sood, A. K., and Tredwell, T. J. (1981): "Photovoltaic Infrared Detectors" in *Semiconductors and Semimetals*, vol. 18: R. K. Willardson and A. C. Beer, eds,. New York: Academic Press, 201.
152a. Kinch, M. A. (1981). "Metal-Insulator-Semiconductor Infrared Detectors,", in *Semiconductors and Semimetals*, vol. 18: R. K. Willardson and A. C. Beer, eds. (New York: Academic Press, 201.
153. Tobin, S. P., Iwasa, S. and Tredwell, T. J. (1980). *IEEE Trans. Elect. Dev.* **ED-27**: 43.
153a. Anderson, W. W. (1982). *Appl. Phys. Lett.* **41**: 1081.
154. Bajaj, J., Williams, G. M., Sheng, N. H., Hinnrichs, M., Cheung, D. T., Rode, J. P. and Tennant, W. E. (1985). *J.Vac. Sci. Tech.* **A 3**(1), 192.
154a. Nemirovsky, Y. and Bloom, I. (1987). *Infrared Phys.* **27**: 143.
155. Kozlowski, L. J. (1995). Conference on Lasers and Electro-Optics (CLEO), May 25–27.
155a. Rosenfeld, D. and Bahir, G. (1992). *IEEE Trans. Electron Devices*, **ED-39**,1683.
156. Kozlowski, L. J. (1989). Electrochemical Society 175_th_ Meeting, May.
156a. Nemirovsky, Y. and Unikovsky, A. (1992). *J. Vac. Sci. Tech.* **B10**: 1602.
157. Kozlowski, L. J., Bailey, R. B., Cabelli, S. A., Cooper, D. E., Gergis, I. S., Chen, C. A., McLevige, W. A., Bostrup, G. L., Vural, K., and Tennant, W. E. (1994). *Optical Engineering* 33, 549.
158. Kozlowski, L. J., Cabelli, S. A., Cooper, D. E., and Vural, K. (1993). *SPIE* **1946**: 199–213.
159. Vural, K., Kozlowski, L. J., Cooper, D. E., Chen, C. A., Bostrup, G., Cabelli, C., Arias, J. M., Bajaj, J., Hodapp, K., Hall, D. N., and Kleinhans, W. E. 2048 $\times$ 2048 SWIR and MWIR HgCdTe focal plane arrays for astronomy applications, presented at the AereSense Conference, April 5–9, Orlando, FL. To be published in *SPIE Proceedings* 3698.
160. D'Souza, A. I., Dawson, L. C., Berger, D. J., Markum, A. D., Bajaj, J., Tennant, W. E., Arias, J. M., Kozlowski, L., Vural, K., and Wijewarnasuriya, P. S. (1998). HgCdTe multi-spectral infrared FPAs for remote sensing applications", *SPIE* **3498**, 192–202.
161. Mackay, C. D., Beckett, M. G., McMahon, R. G., Parry, I. R., Piche, F., Ennico, K. A., Kenworthy, M., Ellis, R. S., and Aragon-Salamanca, A. (1998). *SPIE* **3354**: 14–23.
162. Kozlowski, L. J., Vural, K., Cabelli, S. A., Chen, A., Cooper, D. E., Bostrup, G., Cabelli, C., Hodapp, K., Hall, D., and Kleinhans, W. E. (1998). *SPIE* **3354**: 66–76.
162a. Radford, W. A. and Jones, C. E. (1985). *J. Vac. Sci. Tech.* **A 3(1),** 183.
163. Rajavel, R. D., Jamba, D. M., Jensen, J. E., Wu, O. K., Brewer, P. D.,Wilson, J. A., Johnson, J. L., Patten, E. A., Kosai, K., Caulfield, J. T., and Goetz, P. M. (1998). *J. Electron. Mater.* **27**: 747.
163a. Chung, H. K., Rosenberg, M. A., and Zimmerman, P. H. (1985). *J. Vac. Sci. Tech.* **A 3**(1), 189.
164. Bailey, R., Arias, J., McLevige, W., Pasko, J., Chen, A., Cabelli, C., Kozlowski, L., Vural, K., Wu, J., Forrest, W., and Pipher, J. (1998). *SPIE* **3354**: 77–86.
165. Finger, G., Biereichel, P., Mehrgan, H., Meyer, M., Moorwood, A. F.M., Nicolini, G., and Stegmeier, J. (11998). *SPIE* **3354**: 87–98.
166. Chamonal, J. P., Mottin, E., Audebert, P., Medina, P., Ravetto, M., Deschamps, J., Girard, M., Chatard, J-P. (1997). 1500-element linear MWIR and LWIR HgCdTe arrays for high resolution imaging, *SPIE* **3221**: 384–394.
167. Kozlowski, L. J., Vural, K., Arias, J. M., Tennant, W. E., and DeWames, R. E. (1997). *SPIE* **3182**: pp 2–13.
168. Mitra, P., Barnes, S. L., Case, F. C., Reine, M. B., O'Dette, P., Starr, R., Hairston, A., Kuhler, K., Weiler, M. H., and Musicant, B. L. (1997). *J. Electron. Mater.* **26**: 482.

169. Bubulac, L. O., Tennant, W. E., Pasko, J. G., Kozlowski, L. J., Zandian, M., Motamedi, M. E., DeWames, R. E., Bajaj, J., Nayar, N., McLevige, W. V., Gluck, N. S., Melendes, R., Cooper, D. E., Edwall, D. D., Arias, J. M., Hall, R., and D'Souza, A. I. (1997). *J. Electron. Mater.* **26**: 649.
170. Cockrum, C. A. (1996). *SPIE* **2685**: 2–15.
171. Wu, O. K., Rajavel, R. D., DeLyon, T. J., Jensen, J. E., Cockrum, C. A., Johnson, S. M., Venzor, G. M., Chapman, G. R., Wilson, J. A., Patten, E. A., and Radford, W. A. (1996). *SPIE* **2685**: 16–27.
172. Kozlowski, L. J., Vural, K., Cooper, D. E., Chen, C. A., Stephenson, D. M., and Cabelli, S. A. (1996). *SPIE* **2817**: 150–159.
173. Kozlowski, L. J., Tennant, W. E., Zandian, M., Arias, J. M., and Pasko, J. G. (1996). *SPIE* **2746**: pp 93–100.
174. Murakami, S., Nishino, H., Ebe, H., and Nishjima, Y. (1994). *J. Electron. Mater.* **24**: 1143 (1995).
175. Wilson, J. A., Patten, E. A., Chapman, G. R., Kosai, K., Baumgratz, B., Goetz, P., Tighe, S., Risser, R., Herald, R., Radford, W. A., Tung, T., and Terre, W. A. (1994). *SPIE* **2274**: 17–125.
176. Kozlowski, L. J., Arias, J. M., Williams, G. M., Vural, K., Cooper, D. E., Cabelli, S. A., and Bruce, C. (1994). *SPIE* **2274**: 93–116.
177. Piotrowski, J. (1994). *SPIE* **2225**: 166–173.
178. Fossum, E. R., Pain, B., SPIE Vol. 2020, pp 262–2851(1993).
179. Scribner, D. A., Kruer, M. R., and Kiliany, J. M. (1991). *Proc. IEEE* **79**: 66.
180. Norton, P. R. (1991). *Opt. Eng.* **30**: 1649.
181. Jensen, A. S. (1990). *SPIE* **1308**: 284–292.

Synthesis and Characterization of Superconducting Thin Films

CHANG-BEOM EOM

Department of Mechanical Engineering and Materials Science, Duke University, Durham, North Carolina, USA

JAMES M. MURDUCK

TRW, Space and Electronics Group, Redondo Beach, California, USA

Since the discovery of the high-T_c superconductors [1], much effort has been devoted to experiments exploring their superconductivity and device applications of these new materials. However, the success of this work depends on the availability of high-quality thin-film samples and careful characterizations of the structural and superconducting properties.

Many physical properties of epitaxial thin films are comparable to or even better than those of available single crystals in spite of the presence of many defects in the samples. Thin films have very well defined and controllable dimensions. For example, the thickness can be controlled down to one unit cell, which permits the deposition of ultrathin films or superlattices. The width of the bridges can be made easily by photolithographic techniques with chemical etching or ion milling down to 1 μm, and down to 0.2 μm by electron beam lithography or focused ion beam etching. On the other hand, the deposition conditions can be maintained very far from equilibrium where it is conceivable that metastable phases exist and can be obtained by quenching. It thus seems likely that there is potential for the discovery of new phases by using vapor phase

Vol. 28
ISBN 0-12-533028-6/$35.00
ISSN 1079-4050

deposition methods in creative ways. Furthermore, it is possible to grow artificially layered multilayer structures and to control film orientations. Therefore, thin films are the most useful form of the material for both model experiments and superconducting device applications.

Central to these device applications are the characteristics and quality of the superconducting film. The techniques involved in fabricating these films have followed the path of, and greatly benefited from, monolithic semiconductor processing. These techniques have advanced to the point that nearly any bulk superconducting material can be transformed into a thin film and quite often with enhanced or tailored characteristics. In this chapter, the thin-film deposition processes and characterizations for both low-T_c and high-T_c superconducting thin films are reviewed. Specifically, with respect to low-T_c films, niobium and niobium nitride will be the focus of consideration as these are the material systems of practical interest for circuit fabrication.

5.1. Synthesis

5.1.1. Growth Techniques for Low-T_c Superconductors

The low-T_c superconductors (LTS), which are mainly metals and alloys, are typically deposited by conventional physical vapor deposition processes such as evaporation and sputtering. The objective of these techniques is to obtain a controllable transfer of atoms from a source to a substrate where film formation and growth can proceed atomistically. In order to prevent contamination of the film and have a large mean-free path for the atomic species, these deposition processes are carried out in a high-vacuum environment.

In evaporation, the atoms are removed from the source by thermal heating. A typical evaporation system consists of a high-vacuum chamber with pressure monitoring and gas flow control systems. The high-purity source material is placed in its elemental form in a crucible or water-cooled copper hearth for evaporation. The material in these crucibles is heated thermally or by electron beam. The atomic species emitted from the source can travel relatively longer distances in the absence of gas-phase scattering due to the low operating pressures (ranging from 1 mtorr to 10^{-6} torr) and deposit on the heated substrate placed directed towards the source material. In the deposition of alloys, there is preferential vaporization of one component, resulting in composition changes of the melt. Therefore, for multicomponent film deposition, the constituent elements are evaporated from individual sources maintained at different temperatures. The deposition rate and film stoichiometry are controlled by quartz crystal monitors, which are used in a feedback loop with the power applied to evaporate the source materials [2].

In MBE, the main growth chamber is connected to a load-lock chamber and an *in situ* analysis chamber separated by gate valves. Sample manipulators and transfer arms are installed to transport the samples between the various chambers. The UHV deposition chamber is typically pumped down to 10^{-10}–10^{-11} torr. The single crystal substrates are heated to high temperatures by a resistive contact heater or radiation lamps to clean them and also attain sufficient surface mobility of the film adatoms during growth.

Deposition of each material is carried out from resistively heated effusion cells or by *e*-beam heating. The deposition rate and stoichiometry is controlled with a rate controller (quartz crystal monitors, atomic absorption spectroscopy, or electron impact emission spectroscopy) connected in a feedback loop with the power applied to the source. For Nb thin films, good rate control has been obtained using a cross-beam mass spectrometer [3].

Sputtering is among the most well-developed and mature thin-film processes. One of the main reasons for this development has been the apparent ease of extending results obtained empirically on a small research size sputter chamber to a higher-reliability production process. Sputtering is essentially a kinetic process involving momentum exchange rather than a chemical or thermal process and, therefore, virtually any material can be introduced into a gas discharge or sputtered from the solid. Typically, the target (material to be sputtered) is connected to a negative voltage supply (dc or rf). A gas (usually argon) is introduced, which provides a medium for glow discharge. The operating pressure ranges from a few to 100 mtorr. When the glow discharge is started, positive ions strike the target plate and remove target atoms by momentum transfer, which condense into thin films on the substrates. The substrate, oriented in line with the target, can be ground or floated or held at a positive potential with respect to the target. The most popular sputter source is the planar magnetron source, because of its high efficiency of sputtered species yield and convenient geometry for target fabrication.

In contrast to the fractionation of alloy melts during evaporation, with corresponding changes in film stoichiometry, sputtering allows a stoichiometric transfer of flux from the target to the substrate and is hence more attractive for the deposition of multicomponent thin films such as Nb_3Sn etc. This occurs even for elements with differing sputtering yields. This is due to an initial preferential sputtering of the highest sputtering yield element from the target. The target surface will then become slightly deficient of this element, reducing its relative sputtering rate and compensating for its higher sputtering yield. This proceeds until a steady-state deposition with the composition of the parent material is reached [4].

An exception to this can occur when there is enough oxygen in the system to oxidize the target surface even after the initial contamination layer is presputtered away and affect the sputtering rate of the target. A second mechanism that can cause a resultant preferential sputtering is due to backsputtering of the deposited film. This can occur when a bias voltage, either intentionally applied or as a consequence of the sputtering plasma, is placed on the substrate and the highest sputtering yield element is preferentially ejected from the deposited film.

5.1.1.1. Nb Film Overview

With the material problems that were faced with lead-based Josephson junctions in IBM's superconducting computer program in the 1970s, a need for more robust materials was evident [5]. Niobium, with a T_c of 9.2 K, large coherence length, and relatively small penetration depth was an attractive alternate material. However, there were a number of difficulties that had to be resolved before its use as a junction electrode. Niobium as a refractory metal has a melting point above 2400 °C, making it impossible to thermally evaporate as lead alloys had been deposited. The T_c of niobium is strongly dependent of impurity concentration and readily oxidizes. With every 1% oxygen included in a Nb film, film T_c is reduced by approximately 1 K. Typical vacuum systems of the time had insufficient base pressures to prevent degradation of the critical temperature of the deposited niobium films. The resolution of both these issues is found in what is today's standard deposition equipment. Niobium is sputtered, usually by dc magnetron, in vacuum systems capable of ultrahigh vacuum.

With niobium, researchers had a material that was mechanically hard, could be thermally cycled, and had T_c significantly above liquid helium. One advantage is niobium's capability of forming an anodic oxide by simply exposing a photolithographically defined sample into an electrolytic solution of ammonia pentaborate with a dc voltage applied to the niobium layer. Junction barrier deposition on niobium base electrodes was still problematic. Although niobium readily oxidizes, it can form various oxides and was in practice difficult to control as a barrier material. It was left to the invention of Gurvitch to solve this problem by applying an overlayer of aluminum onto the niobium (see Chapter 2, Fabrication of Superconducting Devices and Circuits).

Nb film and junction deposition techniques. There are numerous methods of depositing niobium ranging from electron beam evaporation to pulsed laser deposition and a variety of techniques within a given deposition capability (such as rf, dc and bias sputtering). This being said, it is still useful to illustrate the considerations that go into depositing a Nb film in order to understand how a given process is developed. The following is intended as an example rather than as the sole or even optimum approach. As is often true in fabrication, the quantitative characterization must be done on a system-by-system basis due to the unknown effect of the particular geometry of the deposition on the film fabrication process.

1) A typical manner of depositing niobium films due to both the ease of fabrication and successful reproducibility is by dc magnetron sputtering.
2) The system preferably has a chamber for the Nb sputter gun and is load-locked to allow rapid cycle time. For junction fabrication it is also preferable to have an aluminum sputter gun and a separate chamber for thermal oxidation (residue water vapor in the load-lock can alter targeting of junction I_c and oxidation within the Nb deposition chamber can affect the Nb target surface).
3) A target-to-substrate distance of 10 cm and a sputter target diameter of 15 cm will allow reasonable across-wafer film uniformities.
4) A base vacuum of 2×10^{-7} torr is a reasonable base pressure. This is often reduced further by predeposition of Nb that efficiently getters residue gases in the chamber.
5) Due to the sensitivity of the Nb deposition to oxygen impurity, stainless steel gas lines baked out over 100 °C are to be recommended. Note that even small sections of polyurethane tubing can account for film degradation.
6) Sputtering Nb (99.9% purity) in a 3 to 4 mtorr argon sputtering pressure (99.999% purity) is a reasonable starting point. The majority of impurity atoms arriving at the substrate are evaporated off by high energy particles during the sputter deposition process. Lower sputtering pressures reduce both partial pressures of impurity gases and the amount of thermalization of the high-energy particles in the sputtering process.
7) Varying this pressure along with the deposition rate as controlled by the power going to the sputter gun is one method of readily optimizing film parameters. Characterizing the Nb film T_c, resistivity ratio, and film stress as a function of pressure and deposition rate is important at this point. Niobium film T_c of over 9 K should be obtained before proceeding with further device fabrication.
8) It should also be noted that there is often a threshold deposition rate at which Nb will grow with $T_c \sim 9$ K, presumably due to the incorporation of impurities and defects into the Nb film at slower deposition rates.
9) A more sensitive measure of film morphology is found by measuring a film's resistivity ratio ($R_{293\mathrm{K}}/R_{10\mathrm{K}}$). Values of 3 to 6 are typical and usable for insertion into a circuit process. Variation in this value can often be observed before significant change in film critical temperature and can be an 'early-warning' of possible process drift.

Stress in films can occur during the sputtering process, for example, by the inclusion of the sputtering gas into the film during the deposition process [6]. The results of film stress range from degraded transport properties to film de-adhesion from the substrate or underlying layer [7]. When Nb films are used as electrodes

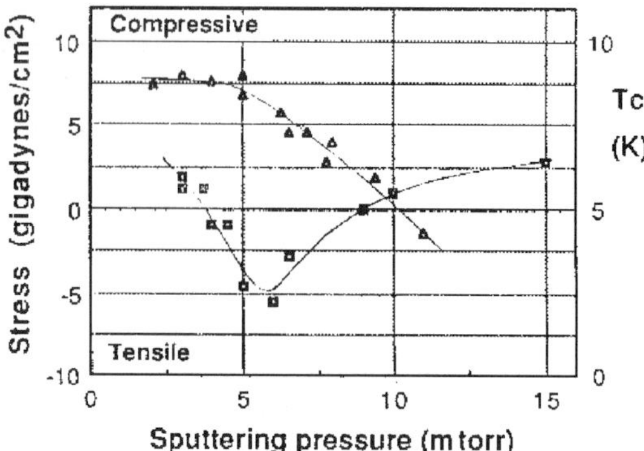

FIG. 5.1. Niobium film stress (dark squares) and superconducting critical temperature (triangles) as a function of sputtering pressure at a constant deposition rate of 8.3 A/s.

of a Josephson junction, film stress can effectively degrade device performance. This yield-reducing failure mechanism becomes increasingly critical as junction areas decrease. As such, efforts are typically made early in developing a niobium deposition process to operate in a regime that results in near-zero film stress. Near-zero film stress can typically be achieved at multiple pressures (Fig. 5.1). Of course, degraded film T_c is not desired and in this case, increasing the sputtering pressure depressed the Nb film T_c, possibly as a result of the inclusion of impurities in the sputtering gas. For the film process illustrated in Fig. 5.1, the sputtering pressure was chosen at approximately 4 mtorr in order to allow both high- and low-stress films.

Due to the difference in the coefficient of thermal contraction between the Nb film and the silicon substrate, Nb films will be under tensile stress at 4 K. In order to achieve near-zero stress at low temperature, Nb films can be targeted to be under compressive stress at room temperature. It is also not always sufficient simply to characterize the Nb film stress on a single surface. It has been noted [8] that film stress of Nb films can differ significantly from film stress in Nb/Al/AlOx/Nb junctions. The Al layer is so thin as to be insufficient to account for this variation in film stress. One difference that could account for this is that the base electrode is typically deposited on SiO_2 whereas the counter-electrode is deposited on oxidized aluminum. Thus, film stress should be characterized for the entire planar junction deposition in order to be sure to avoid the deleterious effects of film stress on junction quality.

5.1.1.2. NbN Film Overview

Bulk NbN had been determined to be superconducting with 16 K critical temperature as early as 1941 [9] but it was not until 1971 that high-quality films were fabricated by reactive sputtering [10]. Reactive sputtering soon became the deposition method of choice for NbN as attempts at sputtering from targets of stoichiometric NbN produced degraded superconducting properties. This was presumably due to the presence of oxide formation within the grain boundaries formed within the target material. More recent attempts using deposition by pulsed laser deposition similarly have had greater success in reactive deposition than from pulsed laser deposition from a NbN target [11].

The inclusion of nitrogen is most commonly accomplished with a reactive gas of nitrogen; however, ammonia and other gases have also been used. Controlling the partial pressure of the reactive gas is crucial to reproducible fabrication of NbN with critical temperatures of 16 K. This process window can be widened by heating the substrate during deposition (on the order of 300 °C). In practice, however, this may not always be an option, especially if there are temperature-sensitive components already existing on the wafer. One such component in superconducting circuit fabrication is the Josephson junction. Temperatures greater than 300 °C induces impurity diffusion and interaction at the junction interface degrading junction characteristics.

Regardless, precise control of the nitrogen partial pressure is always advantageous and techniques have been developed to ensure proper NbN formation by monitoring the gun voltage. As the reactive gas is introduced into the deposition chamber, the voltage developed on the sputter gun will be a sensitive measure of both the gas within the chamber and the condition of the nitrided target surface. This can be typically determined with greater precision and if used as a feedback control to the gas controller, affords greater process control [12].

In reactive sputtering of NbN, there are two schools of thought as to where the compound is formed, on the sputtering target or at the substrate (NbN formation in the plasma is ruled out due to both dissipating the heat of reaction and conserving momentum). The preponderance of evidence is that Nb reacts with nitrogen on the substrate surface [13]. Clearly, as evidenced by the change in voltage of the sputter gun with time as the reactive gas is introduced, the target surface does react. For this reason, presputtering should be done initially in argon and then in nitrogen prior to deposition on the substrate. The presputtering in argon has the added benefit of coating surfaces with "fresh" niobium, which serves to getter oxygen and other contaminants that could otherwise end up in the deposited film.

Film morphology is a considerable issue for NbN. Typical NbN films used in practice are polycrystalline with textured (111) growth in a rock-salt structure. Grain growth is typically on the order of 5–10 nm and is a slightly tilted columnar growth that coalesces with increasing film thickness into larger grains. As the columnar grains are slightly tilted the film roughness increases with grain size. The observation that NbN film roughness increases with increasing film thickness has been attributed to this same mechanism. Single-crystal NbN films have been fabricated by sputter deposition onto heated substrates of temperatures from 350 to 1200 °C. Use of well latticed-matched substrates, such as (100) single-crystal, MgO substrates, allows growth of single-crystal NbN at temperatures below 100 °C [14].

The resistivity of superconducting NbN films as a function of decreasing temperature is a competition of electron transport between the reduced resistivity material within the grains and increasing resistivity of the intergranular material. The NbN films have been successfully modeled as superconducting grains of film interconnected by grain boundary Josephson junctions [15]. Film resistivities at room temperature of 160 μΩ-cm and resistivity ratios (R_{293K}/R_{20K}) of 0.7–0.9 are typical.

As a compound, NbN is quite stable and does not dissociate in oxygen until 1400 °C. Only a thin layer of oxides and suboxides are on the NbN surface such that, unlike niobium, it can be readily analyzed by scanning tunneling microscopy (STM). However, the greater reactivity of oxygen as compared to nitrogen with respect to niobium means that even small amounts of oxygen contamination during the deposition process can be problematic for formation of high-T_c NbN as it is with Nb. This issue can be readily resolved in practice by first determining conditions such as base pressure, deposition rate, and target-to-substrate distance that can produce high-quality Nb. This then can serve as the starting point for developing NbN with reasonable assurance that gas impurities will have limited effect.

Like other nitrides, niobium nitride has been a target of research in fields other than superconductivity due to its extreme hardness. This feature along with its relatively low reactivity to atmospheric oxygen makes for mechanically robust integrated circuits that are unaffected by aging even without passivation.

NbN film deposition techniques. Compared to Nb, as discussed previously, NbN has the relative disadvantages of both a reduced coherence length and longer penetration depth. In addition, NbN has the added complexity of an additional element. As such, proper stoichiometry is vital to maximize critical temperature in NbN films (Fig. 5.2). The variation of nitrogen in the NbN films has the effect of changing the lattice constant of the NbN which in turn impacts the value of the film critical temperature.

The following is once again one example of assembling a NbN process rather than the only or optimum method.

1) The Nb films are deposited with ambient substrate temperature as was already described here. A Nb-film critical temperature of greater than or equal to 9 K is a reasonable assurance that target quality and system base pressure are not problematic.

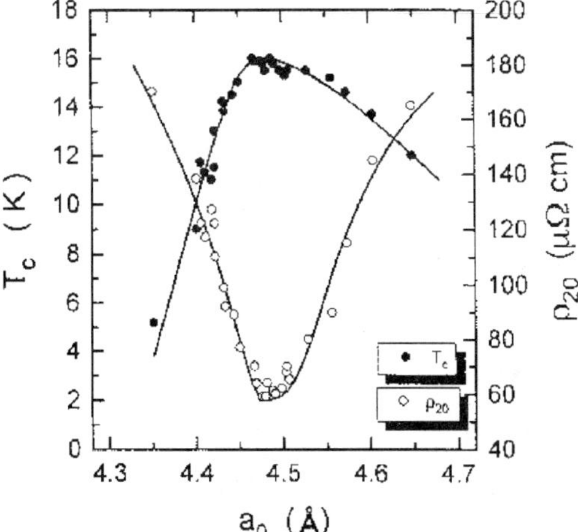

FIG. 5.2. The transition temperature T_c and resistivity of ρ_{20} as a function of lattice constant of NbN films from Reference [14].

2) The voltage of the dc sputter gun is noted, nitrogen is introduced into the system progressively, and the gun voltage is noted. As nitrogen is introduced the gun voltage will increase greater than linearly.
3) Sputter gun voltage is then graphed as a function of nitrogen partial pressure, and greatest variation of voltage with pressure is determined.
4) A series of film depositions is done using that partial pressure of N_2 and varying the nitrogen partial pressure by ± 10 and 20%.
5) Measure the T_c of these films and plot nitrogen partial pressure vs T_c; if the T_c does not go through an optimum, continue increasing or decreasing partial pressure of N_2 until an optimum is determined.
6) Film roughness of this film deposition process should be less than or equal to 2-nm roughness to be useful as a base electrode. Generally, NbN films become increasingly rough with increasing film thickness, presumably due to the coalescing of NbN grains.
7) The increasing roughness of the base electrode must be balanced with the necessary thickness in order to obtain near-bulk values of the film parameters. One can decrease this thickness and obtain near-bulk values with thinner films by heating the substrate temperature above ambient. This can be done by external heating or depositing at higher power. The resulting film is also rougher due to the increased substrate temperature and therefore this approach is not a panacea.
8) The x-ray diffraction of these samples will typically indicate that maximum T_c films are found to have predominantly a (111) orientation of B1 rock-salt

crystal structure with the highest T_c films corresponding to a lattice constant of 0.442–0.446 nm, in agreement with Moodera *et al.* [16] and others [14, 17].

Related compounds. An increase in critical temperature and a reduction of film resistivity can be obtained by incorporating carbon along with nitrogen during the niobium deposition to form NbCN. This can be accomplished by reactively sputtering niobium in partial pressures of diatomic nitrogen and methane. The choice of methane will nearly certainly result in not only carbon but hydrogen being trapped within the film unless deposited at elevated substrate temperatures. This can be avoided, with proper care taken to ensure that gases do not escape into the laboratory environment, by use of a partial pressure of cyanide, CN. In this way, carbon and nitrogen can be reactively included in the NbCN without the potential ill effects of trapped hydrogen.

As a candidate material for superconducting integrated circuits, NbCN holds some advantages over undoped-NbN. The NbCN has a critical temperature 1–2° higher and a lower resistivity, and forms a smoother surface in which to grow subsequent layers. All three of these attributes will facilitate lower inductance per square interconnects, desirable for SFQ circuit operation. The temperature dependence of the penetration depth from the two-fluid model along with theoretical for superconductors in the "dirty" limit gives:

$$\text{penetration depth, } \lambda \propto [(\rho/T_c)/(1-(T/T_c)^4)]^{1/2}$$

Thus, increased T_c and reduced resistivity reduces penetration depth, which for a given film thickness will reduce circuit inductance.

Film smoothness is also highly desirable with respect to reducing circuit inductance in a less direct but perhaps more significant way. Ideally, one would prefer film thicknesses to be greater than their penetration depth in superconducting circuitry so that both the inductance will be limited and the inductance per square will not be highly sensitive to inevitable variation of film thicknesses. However, due to its granular film growth mode, surface roughness of NbN films has been observed to scale with film thickness. Currently, base electrodes of NbN junctions are purposefully on the order of only 200 nm or less in order to be sufficiently smooth to allow complete coverage of the thin insulating barrier in SIS junctions. This film thickness is on the order of half the penetration depth of typical NbN films. An inherently smoother film could allow thicker base electrode formation with resulting lower and more consistent circuit inductance.

Another differentiating aspect of NbCN is that, unlike NbN which makes a minimal surface oxide, NbCN can be readily anodized and has been used to define junction geometries in a manner similar to the selective niobium anodization process (SNAP) [18]. The primary issue with directly inserting NbCN in place of NbN is the increased substrate temperature (>500 °C) required to gain full advantage of the improvement in characteristics. Josephson junction devices can typically only withstand temperatures on the order of 300 °C with little degradation of properties. This implies that NbCN fabricated at temperatures in excess of 300 °C can be used for base electrodes but not for counterelectrodes. As the temperature dependence of the Josephson junction parameters correlates with the temperature dependence of the lower-T_c electrode, fabricating NbCN base electrodes is of limited significance for raising the operating temperature of fabricated devices.

Another related compound that has been explored as a candidate material for electronic applications due to its lower resistivity and surface smoothness is NbTiN [19]. These films can be made by reactive sputtering from alloy targets of 70% Nb and 30% Ti in an argon and nitrogen atmosphere. Control and reduction of unwanted film stress in these films while maintaining superconducting film properties has been done by purposefully unbalancing the magnetic field configuration of the dc magnetron sputtering [20]. This serves as a good example of the flexibility that sputtering affords in the hands of a researcher well-versed in their art.

5.1.2. GROWTH TECHNIQUES FOR HIGH T_c SUPERCONDUCTORS

The epitaxial growth of high-T_c compounds was found to be a challenge because the high-T_c superconductors have characteristic properties that require growth conditions and techniques different from those used to deposit metals or semiconductors like low T_c superconductors or Si. Compared to these materials, the HTS compounds have a far more complex unit cell, which involves several different types of atoms, and requires growth temperatures of 600–900 °C which is 0.7–0.85 times the melting temperature. Furthermore, most physical vapor deposition techniques require high vacuum; therefore these techniques are not normally compatible with the synthesis of some oxide high-T_c superconductors, which require high background oxygen pressure. In this section, both conventional and new techniques currently used for high-T_c superconductors will be discussed along with their advantages and disadvantages.

The most intensive research and interest in thin film applications has so far concentrated on the $YBa_2C_3O_7$ (YBCO) [21, 22], due to the high transition temperature ($T_c = 90$ K) and good material properties. These trends remain true in spite of the discovery of the higher T_c superconductors such as the Bi-Ca-Sr-Cu-O system [23] ($T_c = 110$ K), Tl-Ba-Ca-Ba-Cu-O system [24] ($T_c = 126$ K), and the Hg-Ba-Ca-Cu-O system ($T_c = 134$ K at ambient pressure [25] $T_c = 164$ K at high pressure [26]) due to their intrinsic and extrinsic material problems. In addition to these material problems of the Bi, Tl and Hg compounds, they are difficult to synthesize. Therefore, most successful thin-film growth has been reported in the YBCO system.

An *in situ* thin-film deposition process is desirable for device fabrications. For *in situ* growth the desired structure forms at the surface during the actual film deposition at elevated substrate temperature and at relatively high oxygen pressure. The film has the superconducting phase when removed from the deposition chamber; no further annealing is needed. The high-quality YBCO films grown by these methods enable one to study many important structural, normal state, and dc and rf superconducting properties of the films.

Mobility of the surface species is a key factor. The freedom to choose the substrate and its orientation makes it possible to introduce surface energy as an extra variable to control the initial formation of the layer, which in turn serves as the template for the rest of the growth, a process known as heteroepitaxial growth. Experimentally, epitaxial growth reduces interfacial diffusion and permits deposition at an elevated temperature without running into substrate contamination. In general, any successful *in situ* deposition requires a favorable equilibrium in the surface layer. This can be achieved by controlling deposition parameters, such as the substrate and its orientation, the temperature, the pressure, the composition, and the rate.

This *in situ* process allows us to grow artificially layered multilayer structures, which are very important for model experiments to study superconductivity and superconducting device applications and is promising for integration with conventional electronics due to the low growth temperature. In addition, we can find new metastable phases using the *in situ* process. Films sometimes form many defects if they are grown at low temperatures and low oxygen pressures. On the other hand, the high oxygen pressure required during the growth limits or challenges the use of conventional physical vapor deposition techniques. The deposition techniques can be classified into two groups; multi-elemental source deposition and single composite source deposition. Reactive coevaporation, molecular beam epitaxy (MBE), cosputtering, and metalorganic chemical vapor deposition (MOCVD) belong to the former category. Pulsed laser deposition and composite target sputtering belong to the latter category.

5.1.2.1. Deposition from Multi-elemental Sources

In multi-elemental source deposition processes, it is possible to control the rate and deposition times of each elemental source. Therefore, one can deposit a particular species sequentially as well as simultaneously (codeposition). In the sequential processes, one can synthesize more complicated artificial structures because the definition of a mono-layer is one atomic layer. It is also possible to control the composition and phase spreads of the films very easily.

Molecular beam epitaxy (MBE) using effusion cell sources have been very successful in making semiconductor quantum well structures. However, it is difficult to grow high-T_c superconductors because the partial pressure of oxygen required for stability is too high [27, 28]. The oxygen partial pressure in the vacuum chamber is limited because of the oxidation of the very reactive elemental sources and difficulties with the rate monitoring system. The same holds for *in situ* growth by electron beam deposition, which also utilizes multiple sources and line-of-sight deposition. Different strategies have been employed to enhance the effective oxygen pressure at the surface of the growing film with some success. In order to increase the effective oxygen pressure, several groups have used activated oxygen sources such as ozone [29], atomic oxygen [30], or the differential pumping system [31].

The significant advantage for layer-by-layer growth is in being able to monitor the formation and completion of each layer using reflection high-energy electron diffraction [32]. Techniques such as atomic absorption rate monitor have been used to accurately control the flux of individual elements [33, 34]. The atomic absorption monitor technique is highly sensitive, noninvasive, and species specific. It is also unaffected by the background pressure. Wang *et al.* [34], have used a diode laser-based monitor that offers a range of advantages including narrow spectral linewidth and tunability, which allows advanced spectroscopic techniques to be employed for increasing sensitivity and mapping atomic velocity distribution.

Kinder *et al.* have used thermal reactive evaporation from elemental sources, in conjunction with a rotating disk heater which allows intermittent deposition and oxidation in spatially separated zones, to deposit very large area ($\approx$9-in-diameter) YBCO films [35, 36]. Figure 5.3 shows a schematic of their deposition system with the large area heater. The process requires complex precision engineering, accurate control of each metal flux, and a differential pumping system. However, the potential for scaling up to large area deposition renders the reactive evaporation system cost effective.

Multi-elemental target sputtering is even more difficult to use for *in situ* growth because negative ions are generated on the target surface. Since barium is very reactive, oxide layers can be formed on the target surface, substantially reducing the sputtering yield. The plasma and the deposition rate become very unstable at high oxygen pressure.

Metalorganic chemical vapor deposition (MOCVD) is a very powerful technique for semiconductors. However, it has not yet been very successful for the synthesis of high-T_c superconductors because the decomposition temperatures of metalorganics is very high. Therefore, growth temperatures as high as 900 °C are needed, which cause severe interface reactions and limit substrate choice. Several groups reduced the processing temperature down to 650–750 °C by using nitrous oxide (N_2O) [37, 38]. The MOCVD system can be operated near 1 atm,

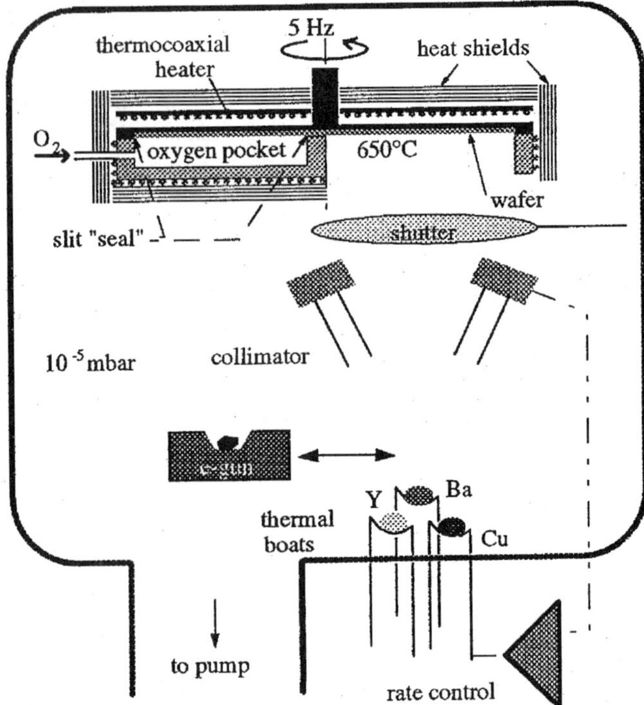

FIG. 5.3. Kinder's reactive coevaporation system with a large area substrate heater, from Reference [36].

therefore the oxygen partial pressure is not a problem. It is possible to grow large area films by this technique. However, the surface quality is not smooth and, hence, nanoscale control of interfaces, which is required for growth of heterostructures, is difficult. Furthermore, it has been observed that transporting these precursors to the substrate reproducibly is also very difficult [37]. However, to overcome this problem, MOCVD has been used with a single liquid/solution source [39]. The solution source is vaporized with a carrier gas of argon by a supersonic atomizer and is transported through a thermal vaporizer. The film deposition then proceeds as in a conventional MOCVD process.

5.1.2.2. Deposition from Composite Sources

As a single composite source deposition technique, laser ablation and composite target sputtering have been widely used. These methods utilize single target sources, which are fabricated with the desired composition by ceramic techniques. As a consequence, the growth of YBCO-thin films cannot be controlled on an atomic monolayer scale in the *c*-axis orientation (*c*-axis normal to substrate, which is the usual growth direction) because all three cations arrive simultaneously. Nevertheless, multilayered films with periods as low as the sum of the unit cells of each compound can be grown [40, 41, 42]. For *c*-axis films this translates to periods as low as 24 Å. *A*-axis multilayers have been grown with periods of 24 Å [43]. The fact that such fine-scaled multilayers can be grown is not a priori evident. It happens when the ordering energy is more important than entropy.

Basically, there is no limit of oxygen pressure in the chamber in terms of target stability because the targets are already sintered in oxygen atmosphere. However, a wide range of composition variation cannot be attained. In addition, the film composition is not necessarily the same as the composition of the source. Because the film composition depends on many variables such as geometry and pressure, optimization of processing to obtain a certain stoichiometry in the film is very important.

Pulsed Laser Deposition. Pulsed laser deposition (PLD) is one of the most widely used methods for producing superconducting thin films. The species arrive at the surface in bursts of a few tenths of a microsecond followed by a quiescent period (~0.1 s) between bursts, during which time ordering can occur. Stoichiometry in the film is easily attained with the stoichiometry of the target. Deposition rate can be varied over a fairly large range (1–250 Å/s) [44]. The laser beam cross section defines the area of the target from which the material is to be ablated. The beam size is very small (several mm^2), as the beam is focused by means of optics to increase the energy density of the beam. Therefore, the emission from the target surface and the plume seem to originate from a point source. As a result, the regions outside the influence of the plume are not consistent with the target in composition. Moreover, as the target is ablated with a high-energy laser beam, larger particles/boulders are sometimes removed from the target, instead of atoms, which then deposit on the film surface. Off-axis laser deposition with substrate rotation has been carried out and an improvement on the surface quality has been attained [45]. Scanning a fixed laser beam on a rotating target and using a shadow mask on the substrate has also been attempted to improve film surface quality [46]. Wu *et al.* have used pulsed laser deposition to grow large area YBCO films on sapphire substrates with YSZ buffer layers by a shadow mask technique [47]. They have been able to grow high-quality films on 2-in-diameter substrates.

In order to fabricate thin films and heterostructures of complex oxides with atomically smooth surfaces and sharp interfaces, well-controlled layer-by-layer growth is required. To this end, *in situ* monitoring of the film growth with reflection high energy electron diffraction (RHEED) is essential. The RHEED pattern and intensity oscillations are indispensable for digital control of growth units and are used routinely in MBE deposition of semiconductors. Tabata and Kawai [48], Kawai [49], and Koinuma *et al.* [50, 51] have used a laser-MBE system that incorporates the use of RHEED in PLD systems for the growth of perovskite oxide thin films in layer-by-layer controlled mode as in MBE. The laser-MBE systems contain many other *in situ* analysis techniques such as x-ray photoemission spectroscopy (XPS), low-energy electron diffraction (LEED), Auger electron spectroscopy (AES), ultrahigh vacuum scanning tunneling microscopy (UHV-STM), and coaxial impact collision ion scattering spectroscopy (CAICISS). A schematic of the laser-MBE system used by Koinuma *et al.* is shown in Fig. 5.4. The use of *in situ* RHEED as a diagnostic tool requires a high vacuum, which is incompatible with the relatively high oxygen partial pressures used in PLD of oxide thin films. This problem is overcome by using such activated oxidants as NO_2 [52] or O_3 [53] and, alternatively, a pulsed oxygen source [54].

Rijnders *et al.* have developed a RHEED system for monitoring *in situ* growth of oxide thin films by PLD under high pressures [55]. At the high operating pressures $\sim$400 mtorr used in PLD, the electron beam from the RHEED gun suffers increased scattering loss and oxidation of the tungsten filament. To minimize losses, the traveling path of the electrons in the high pressure region has to be kept as small as possible, while maintaining a low pressure in the electron source. The latter is achieved with a 2-stage differential pumping unit

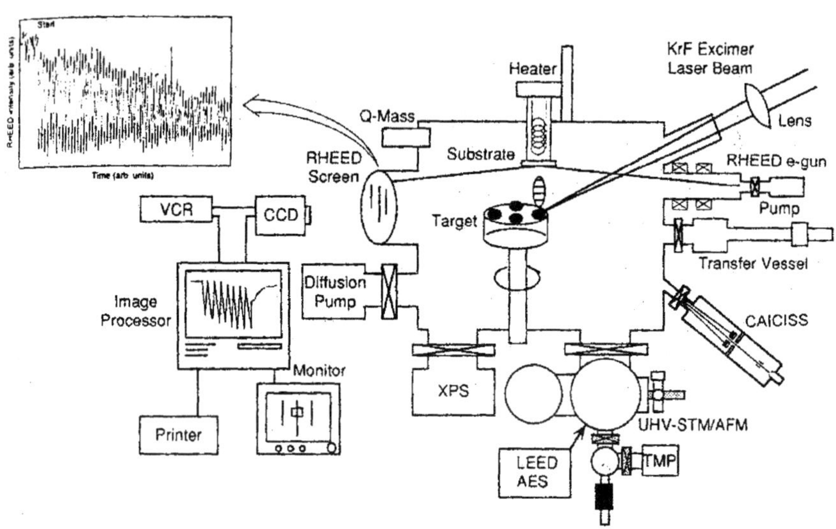

FIG. 5.4. Schematic of laser-MBE system from Reference [51].

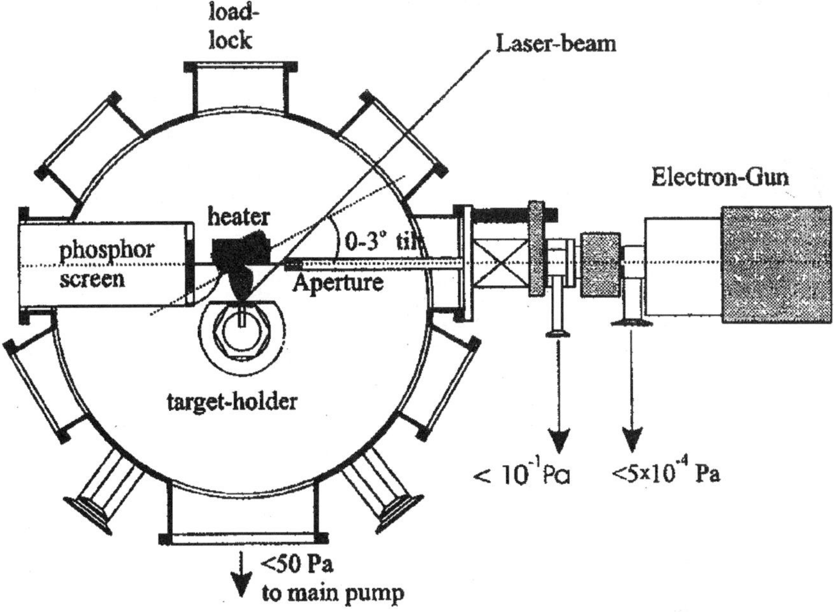

FIG. 5.5. Schematic of High pressure RHEED system from Reference [55].

that maintains a pressure of up to 400 mtorr in the deposition chamber while maintaining a low pressure ($<4 \times 10^{-6}$ torr) in the electron gun. To minimize scattering losses, a long stainless steel extension tube is used from the electron source to the deposition chamber near the substrate, as shown in Fig. 5.5. The electron beam passes through the aperture inside the differential pumping unit and the tube and reaches the substrate with minimum scattering losses. The XY deflection facility of the RHEED gun is used to direct the electron beam through the small (diameter 250 μm) aperture at the end of the tube. The tube itself is maintained at below 8 mtorr. Koster *et al.* [56] have used this technique to study the homoepitaxial growth of $SrTiO_3$ and heteroepitaxial growth of high-T_c-YBCO thin films on (100) $SrTiO_3$ substrates. Furthermore, this technique has been modified for pulsed laser interval deposition [57], which is based on a periodic sequence consisting of fast deposition of the amount of material needed to complete exactly one monolayer, followed by an interval in which no deposition takes place and the film can reorganize. Thus layer-by-layer growth is achieved by circumventing premature nucleation that leads to multilevel 2D growth.

Off-axis Sputtering. Composite target sputtering can be a very reproducible and easily controllable technique. However, under standard conditions, severe backsputtering of the substrate by high-energy particles (O^-, high energy O) causes compositional changes in the film. This problem is more severe in the sputtering of ionic compound targets (such as YBCO) rather than in a covalent bonded compound target (such as Al_2O_3). Several groups have used nonstoichiometric targets to compensate for the selective back sputtering effect [58, 59]. Alternatively, Li *et al.* have sputtered from a stoichiometric target at high pressure (600 mtorr) and high temperature (770–850 °C) [60]. The frequent collisions at high pressure result in a reduction of the kinetic energy of the particles and, consequently, a reduction of back sputtering. This approach also enables the use of high oxygen pressure for *in situ* film growth. The uniformity of the resulting film is usually poor. However, better uniformity has been obtained by using a hollow cathode sputter gun [61]. In the hollow cathode sputtering process, a cylindrical target with a magnetron plasma ring inside it is used as the deposition source. A continuous plasma sheath exists at the cathode surface. Therefore, sputtering efficiency is increased. The substrates are placed outside the cylinder on its axis. In this geometry, the negative ions do not reach the substrate but instead bombard the opposite inner wall of the target. Furthermore, by controlling the direction of gas flow, high deposition rates can be obtained. Sandstrom *et al.* have used off-axis sputtering, in which substrates are placed on the side of, and at an angle to, the sputter gun [62]. They used low pressures (6 mtorr), a rotating substrate block to obtain uniformity, and an off-stoichiometric target adjusted to improve the film composition. In spite of many such modifications, the T_cs of these *in situ* films were broad and had low onsets.

In order to overcome this problem, the 90°-off-axis sputtering technique has been widely used to grow high-quality superconducting thin films [63, 64]. In this technique, almost the exact composition of the target is obtained over large areas (2 in × 2 in) without the need for substrate block rotation. In high-pressure sputtering the species diffuse to the surface continuously so that growth is a steady-state process. The surface of the sputtered films is extremely smooth, therefore it is ideal for multilayer growth and device applications. The off-axis method employed benefits both from high pressure and off-axis geometry will be discussed in what follows.

Figure 5.6 shows the sputtering geometry and composition profile of the films across a 1.5 in × 1.5 in area, showing a comparison of off-axis (a) and on-axis geometries (b). This figure demonstrates the compositional variation of films in various positions. Composition was analyzed with an electron microprobe; results were verified in some cases by inductively coupled plasma spectroscopy and Rutherford backscattering spectrometry. When the substrate block was placed facing the sputter gun in the on-axis configuration, the composition of the film was poor in Ba and Cu due to differential backsputter etching by the high energy particles. Even when a total pressure of 400 mtorr was used the composition profile was still very inhomogeneous and the 1 : 2 : 3 composition was obtained only in a very small region (less than 1/4 in × 1/4 in).

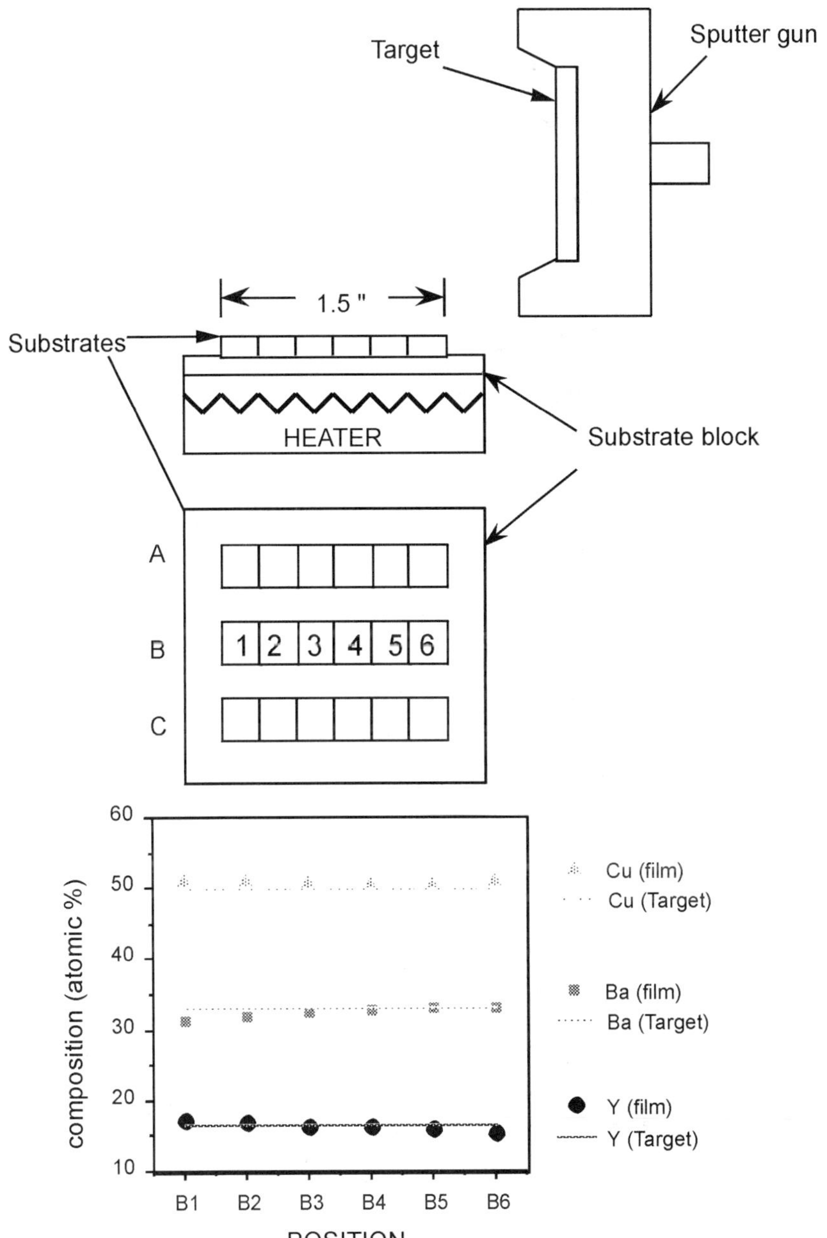

FIG. 5.6. (a) Off-axis sputtering system geometry and composition profiles [64].

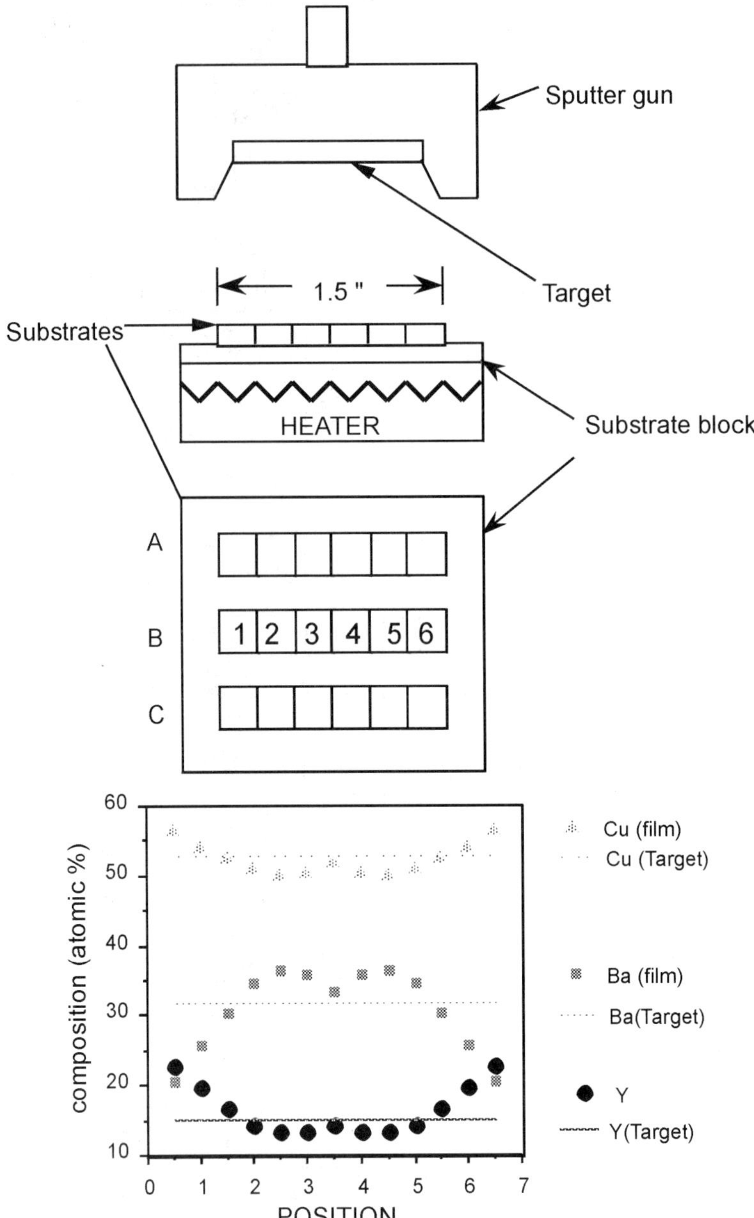

FIG. 5.6. (b) On-axis sputtering system geometry and composition profiles [64].

However, by going to the high-pressure and 90°-off-axis geometry, only a small spread in composition was found over an area of 1.5 in × 1.5 in from which high-quality films could be repeatedly obtained. In a typical run (40 mtorr Ar/10 mtorr O_2) with a target of composition $Y_{18}Ba_{33}Cu_{49}$, the measured composition of the films at the extreme positions (Fig. 5.6.(b)) varied from $Y_{19}Ba_{33}Cu_{48}$ (B1) to $Y_{17}Ba_{33}Cu_{50}$ (B6), illustrating the almost 1 : 1 correspondence between target and film composition. The slight variation in the composition profile is due to a combination of residual high energy particle bombardment and the difference in angular distribution of the different atomic species. At position B1, 1.5 in away from the sputter gun, the high energy beam still can be somewhat effective in the preferential sputtering of Ba and Cu from the film. By increasing the total pressure from 50 to 200 mtorr, increasing the Ar/O_2 ratio from 80 to 95%, and reducing the RF power from 125 W to 60 W, no compositional variation was detected. The slight enrichment of Ba concentration near the sputter gun at 50 mtorr and 125 W is due to the mass dependence of the angular distributions of the different species as reported by Wehner and Rosenberg [65]. The sputter profile of the heavy mass Ba is less directional than that of the Y and the Cu. The film composition for the off-axis geometry turns out to be rather insensitive to total pressure and Ar/O_2 ratio. In contrast to on-axis geometry sputtering the variation in thickness (as measured by a stylus on the film edges formed by patterning) is small, typically less than 10% over the 1.5 in × 1.5 in substrate block area.

Figure 5.7 is a phase diagram of the Y-Ba-Cu-O system based on the data summarized by Hammond and Borman [66]. This diagram shows the stability of YBCO equilibrium phases with respect to the molecular oxygen partial pressure and temperature. The actual phase formation sequence during *in situ* growth is complicated due to the surface growth kinetics and the presence of a plasma that activates oxygen. It is clear, however, that the presence of the oxygen plasma enables the *in situ* growth to occur at lower oxygen pressures. On the other hand the oxygenation process is affected by the growth rate, which when increased shifts the apparent phase boundary to higher oxygen pressures.

To clarify the role of the plasma in the oxygenation process, four terminal resistivity measurements on a YBCO thin film were carried out in both molecular oxygen and an oxygen plasma at reduced pressures. By noting the equality of the resistivities of the films at 400 °C under the reactive conditions it can be seen that 10 mtorr of oxygen in the presence of a plasma is equivalent in the degree of oxidation to that of the film in 1 torr of molecular oxygen without excitation. The plasmas can be excited either inductively or by the sputter gun itself. Thus, the low P_{O_2} and low temperature are adequate for the formation of the perovskite structure. Qualitatively one would expect the presence of the excited oxygen to be effective in reducing the pressure required for perovskite formation until temperatures are reached where the dissociation of O_2 on the surface is so

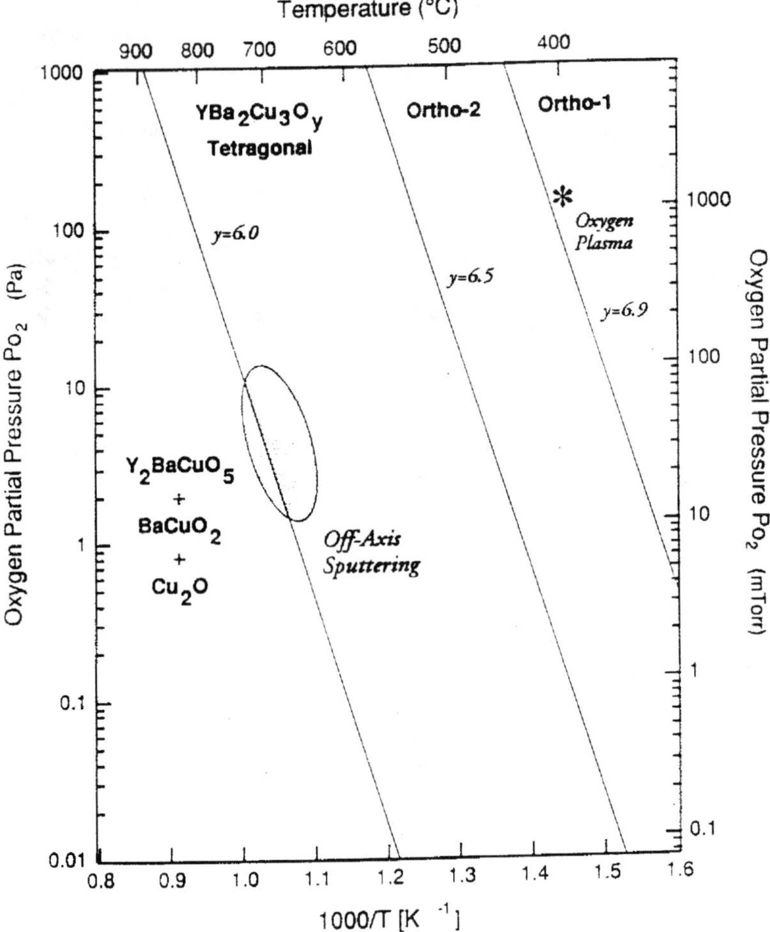

FIG. 5.7. Equilibrium phase diagram of the YBCO system from Reference [66]. The * shows that the resistivity of the YBCO thin films in 10 mtorr O_2 plasma is the same as that in 1 torr O_2.

rapid that the species of oxygen arriving at the surface no longer matters. This occurs for temperature above $\sim 750\,^{\circ}\mathrm{C}$ as was shown in a detailed study of oxygen diffusion kinetics by Yamamoto *et al* [67].

There exists an optimum growth temperature for obtaining a given property. The optimum temperature is oxygen-partial-pressure, composition, and substrate dependent. When the temperature is below a certain threshold, either unoriented or purely *a*-axis oriented films are formed. At somewhat higher temperatures, purely *c*-axis oriented films are grown. If the temperature is too high the resulting films become Y rich. There also exist possible interface reactions and phase instabilities. The optimum growth temperature for both orientations (*a*- and *c*-axis) has been found to increase with increasing P_{O_2} [68].

Rao *et al.* have demonstrated the uniform deposition of YBCO-thin films over very large areas, using a 3-in-diameter sputtering target and optimized substrate rotation in a single target 90°-off-axis sputtering technique. Two-dimensional maps of the thickness profile of YBCO films deposited on a stationary substrate from 2- and 3-in-diameter targets have been obtained by a surface profiler. These profiles were used in a computer simulation to determine the optimum center of substrate rotation during deposition that would produce the maximum area with uniform thickness. The films deposited from a 3 in target on rotating substrates displayed uniform thickness ($< \pm 5\%$ variation) and composition (<2.3 atomic % deviation from the target stoichiometry) and a consistently high transition temperature ($T_c > 88.3$ K) over an 8-in-diameter area [69].

5.1.3. MULTILAYERS

The quantum nature of superconductors permits the use of high-T_c superconductor-normal metal-superconductor (SNS) Josephson junctions in novel electronic device applications, where physical and technological limitations prevent the use of Si-based semiconductor devices. For instance, dc superconducting quantum interference devices (SQUID) using two Josephson junctions in a loop are the most sensitive magnetic flux detectors [70]. It is not physically possible for other devices to achieve sensitivity anywhere near that of a SQUID. Furthermore, rapid single flux quantum (RSFQ) logic circuits [71] that use an array of hundreds of Josephson junctions require extremely low power to perform ultrafast switching operations at speeds unattainable by Si. There are technological but no fundamental physical limitations on the switching speed of silicon. However, the speed-power ratio of Josephson logic will always be much better than that of semiconductor logic.

Artificially grown multilayers (heterostructures and artificially layered superlattices) are typically used for such superconducting electronic device applications. Sandwich-type tunnel junctions used in such devices are thin-film heterostructures consisting of a normal metal or insulator buffer layer sandwiched between two superconducting electrodes. This approach is used to solve an intrinsic material problem (extremely short superconducting coherence length along the *c*-direction) of the high T_c superconductors. A critical issue in the fabrication of multilayers and superlattices is the nano-scale control of the interfaces. Atomically sharp and clean interfaces have been obtained in multilayer films grown by PLD, sputtering, and MBE.

The HTS superlattices have been studied not only for possible device applications but also to explore many physical phenomena such as dimensionality, proximity effect, and interface pinning [32]. Triscone *et al.* [40, 41] have demonstrated the growth of *c*-axis-YBCO/Dy $Ba_2Cu_3O_7$(DyBCO) and *c*-axis YBCO/$PrBa_2Cu_3O_7$(PBCO) superlattices with a modulation wavelength, Λ, of 24 Å, where Λ is the sum of the individual layer thicknesses which forms the repeating unit of the superlattice. Other groups have also succeeded in fabricating *c*-axis-YBCO/PBCO superlattices and studied their superconducting properties [42, 72]. Eom *et al.* [43] have synthesized *a*-axis-YBCO/PBCO superlattices with a Λ of 24 Å.

5.1.4. SUBSTRATE ISSUES AND GROWTH MECHANISMS

The growth mechanism and structure of thin films is to a large extent determined by the substrate used. Furthermore, the choice of the substrate material is governed by the application involved. For instance, for YBCO thin films used in microwave applications, (001) $LaAlO_3$ or (001) MgO substrate with its lower dielectric constant is preferred over (001) $SrTiO_3$ substrates, which have a high dielectric constant. However, twinning is commonly observed in $LaAlO_3$ crystals. It has been found that (100) is an active twin plane in $LaAlO_3$ [73]. This twinning results in a rough crystal surface, strain and non-isotropic microwave properties for (001) $LaAlO_3$ wafers. To overcome the twinning problem, a new twin-free substrate, $La_{0.3}Sr_{0.7}Al_{0.65}Ta_{0.35}O_9$ (LSAT), has been developed, which is a solid solution of $LaAlO_3$ and Sr_2AlTaO_6 [74, 75]. The dielectric constant and thermal expansion coefficient of LSAT are comparable to those of $LaAlO_3$ as seen from Table 5.1.

Some of the standard oxide substrates used for the growth of HTS thin films are listed in Table 5.1. Their lattice parameters, dielectric constants, and thermal expansion coefficients are also listed. For a given substrate material, the lattice and thermal expansion mismatch with the film play an important role in determining the microstructure of the film in terms of lattice strain, misfit dislocations, and defects. The epitaxial arrangement of the film is also determined by the type of substrate used for growth. For instance, YBCO film grows with cube-on-cube in-plane epitaxy on (001) $LaAlO_3$ and (001) $SrTiO_3$ substrates, the use of MgO substrates with CeO_2 or YSZ buffer layer leads to a 45°-in-plane rotation. Therefore, by using appropriate seed and buffer layers, the in-plane epitaxy of the YBCO film can be controlled to fabricate 45°-grain boundaries. This technique is used to fabricate bi-epitaxial grain boundary Josephson junctions, where the artificially created 45°-grain boundary in the film acts as a weak link [76, 77].

TABLE 5.1
LIST OF SUBSTRATES FOR HTS-THIN-FILM GROWTH: THEIR STRUCTURE AND PHYSICAL PROPERTIES; THE PROPERTIES ALSO DEPEND ON THE CRYSTALLOGRAPHIC DIRECTION ALONG WHICH THEY ARE MEASURED

Substrate	Lattice parameter	Dielectric constant	Loss tangent	Thermal expansion coefficient (K^{-1})
$YBa_2Cu_3O_7$ (YBCO)	a = 3.82 Å b = 3.89 Å c = 11.78 Å			
Al_2O_3	Hexagonal a = 4.758 Å c = 12.99 Å	9.4–11.5	$(5.6–8.8) \times 10^{-7}$	1.0×10^{-5}
$BaTiO_3$	Tetragonal a = 3.99 Å c = 4.04 Å			
$KTaO_3$	Cubic a = 3.915 Å			
$LaAlO_3$	Cubic a = 3.792 Å	~25	6×10^{-5} at 300 K	1×10^{-5}
$La_{0.3}Sr_{0.7}Al_{0.65}Ta_{0.35}O_9$ (LSAT)	Cubic a = 3.868 Å	~22		1×10^{-5}
MgO	Cubic a = 4.216 Å	~9.8	3.3×10^{-7}	1.28×10^{-5}
$NdGaO_3$	Orthorhombic a = 5.43 Å b = 5.50 Å c = 7.71 Å	~25	$< 2.3 \times 10^{-3}$ at 300 K	1.28×10^{-5}
$SrTiO_3$	Cubic a = 3.905 Å	~300	2×10^{-2}	1.04×10^{-5}
$SrLaAlO_4$	Tetragonal a = 3.756 Å c = 12.63 Å	~16.8	8×10^{-4}	10.05×10^{-6}
$SrLaGaO_4$	Tetragonal a = 3.84 Å c = 12.68 Å	~22	5.7×10^{-5}	
YSZ (Y: ZrO_2)	Cubic a = 5.125 Å	~27	—	1.03×10^{-5}
Si	Cubic a = 5.4037 Å	~5		4.7×10^{-6}

Source: ESCETE Single Crystal Technology B. V. and MTI Corporation.

The growth mechanisms of epitaxial HTS thin films have been studied in great detail. It is known that *c*-axis-oriented YBCO films grown on lattice-matched (001) $SrTiO_3$ substrates show a high density (10^8–10^{10} cm^{-2}) of screw dislocation growth spirals [78, 79] and the density decreases with increasing miscut angle of the substrate. More recently, Haage *et al.* [80] and Mechin *et al.* [81] have also studied the growth and properties of YBCO-thin films on vicinal (001) $SrTiO_3$ substrates.

The substrate surface, terminating layer, and lattice mismatch determine the growth mode to a large extent. The stacking sequence of YBCO-thin films has been studied on different substrates using HRTEM [82, 83]. For instance, it has been shown that when one unit cell thick *c*-axis-YBCO film is deposited on TiO_2-terminated $SrTiO_3$ substrate, the film has an extra CuO_x atomic layer, which forms CuO_x precipitates in thick films. The formation of CuO_x precipitates can be avoided by depositing one monolayer of SrO on a purely TiO_2-terminated $SrTiO_3$ substrate [84]. Therefore, substrate surface preparation is a critical issue in the deposition of HTS-thin films and has been studied extensively [86–89]. A similar approach has been used [85] in controlling the growth mechanisms of other perovskite oxides such as $SrRuO_3$ as well.

5.1.4.1. *Substrate Surface Termination*

Perovskite oxides with the formula ABO_3 have two possible terminating atomic layers—AO (A-site layer) and BO_2 (B-site layer) on the {001} surfaces. Surface preparation techniques to obtain a purely single termination layer on the surface have been developed for many of the perovskite substrates listed in Table 5.1. These techniques typically include a combination of annealing and wet etching steps.

Polished $SrTiO_3$ substrates display both SrO- and TiO_2-terminated surfaces. Kawasaki *et al.* have shown that by using a NH_4F-buffered HF solution (BHF) with controlled pH it is possible to etch the more basic SrO and leave a completely TiO_2-terminated surface on the substrate [86]. Such a process results in the developments of periodic unit-cell high steps on the substrate surface. However, after etching, the steps on the substrate surface are not straight lines and meander instead. Furthermore, the pH of the etchant and the etching time need to be very carefully controlled to prevent formation of etch pits due to overetching.

Recently, Koster *et al.* [87] developed the etching method by using an intermediate water treatment step to form the Sr-hydroxide complex that can be subsequently etched relatively easily with a 30-s dip in a commercially available BHF solution. This process is claimed to be more reliable and reproducible as SrO is known to react with H_2O at room temperature to form stable $Sr(OH)_2$ while the chemically stable TiO_2 layers are not expected to react with H_2O. After etching, a final annealing step is performed in flowing oxygen to

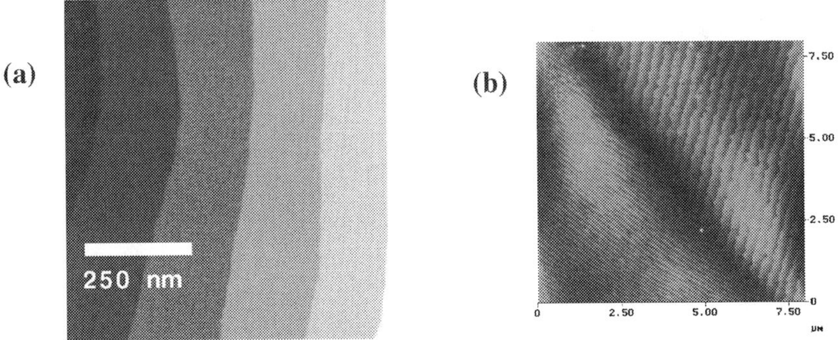

FIG. 5.8. Surface morphology of ~0.1°-miscut pretreated substrates (a) (001) $SrTiO_3$ and (b) (001) $LaAlO_3$ substrate [90].

obtain flat terraces with straight steps. The annealing temperature and time depend on the substrate miscut angle. The terrace width is determined by the substrate miscut.

Similar surface preparation techniques have been developed for other HTS substrates too. Recently, Ohnishi *et al.* have shown that by thermally annealing polished (001) $NdGaO_3$ wafers in air at 1000 °C, a purely A-site termination layer (Nd-O) can be obtained on the surface with an atomically defined step-terrace structure [88]. Similarly annealing (001) $La_{0.3}Sr_{0.7}Al_{0.65}Ta_{0.35}O_9$ (LSAT) wafers at 1300 °C in air and etching (001) $LaAlO_3$ wafers in HCl solution has been shown to produce an atomically flat surface with the B-site layer (Al-O) being the dominant termination layer at the surface [89, 90]. Figure 5.8 shows the atomic force microscope image of the surface of a pretreated (001) $SrTiO_3$ substrate and (001) $LaAlO_3$ substrate. The steps seen on the surface of these substrates are one unit-cell high.

While the A-site layer termination is desirable for the growth of smooth HTS thin films, the B-site termination layer results in CuO_x-precipitate formation [84]. The YBCO films grown on such prepared substrates with A-site layer termination show a smooth surface with numerous growth spirals and no precipitates as shown in Fig. 5.9(a). As the substrate miscut is increased, the film shows a step-terrace morphology on the film surface as shown in Fig. 5.9(b) [91].

5.2. Thin-Film Characterization Techniques

The thin films are typically characterized by structure, chemical composition, and properties. The structural characterization is carried out with a variety of

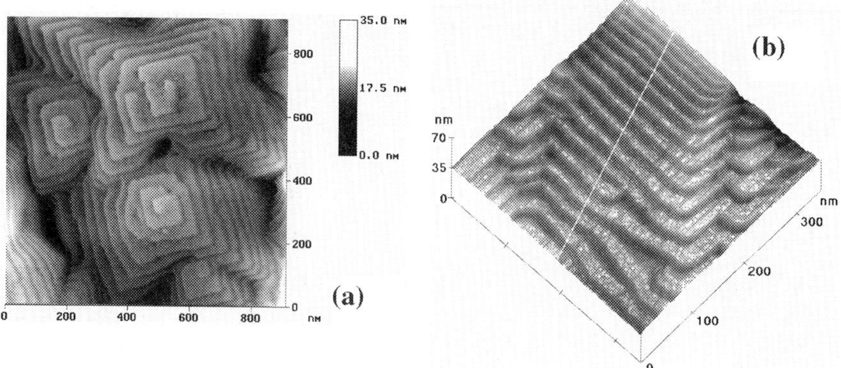

FIG. 5.9. Scanning tunneling microscopy images of surface of YBCO films showing (a) screw dislocations on (001) $SrTiO_3$ substrate and (b) step-terrace morphology on 4°-miscut (001) $SrTiO_3$ substrate (from Reference [91]).

techniques, including x-ray diffraction (XRD), transmission electron microscopy (TEM), atomic force microscopy (AFM), and scanning electron microscopy (SEM). The chemical characterization is carried out primarily with Rutherford backscattering spectroscopy (RBS) and wavelength dispersive spectroscopy (WDS). However, there are plenty of other techniques available depending upon the specific application, such as x-ray photoemission spectroscopy (XPS), Auger electron spectroscopy (AES), and secondary ion mass spectroscopy (SIMS). We will discuss some of the important chemical and structural characterization techniques used for superconductor thin films and then discuss the electrical and magnetic characterization of these films.

5.2.1. CHEMICAL COMPOSITION

Among the many chemical characterization techniques AES, XPS, and SIMS are surface sensitive, because the detected electrons and ions are emitted from surface layers $<$ approximately 15 Å deep. The RBS and electron microprobe (WDS) sample the total thickness of the film ($\sim$1 μm).

5.2.1.1. Rutherford Backscattering Spectroscopy (RBS)

This system uses a very high energy beam of low mass (typically 1.5–3.5 MeV He^{4+} ions), which gives it a high penetration depth. These low mass ions undergo classical elastic collisions with the nuclei of heavier atoms in the thin-film sample. The collisions are insensitive to the electronic configuration or chemical bonding of target atoms, but depend solely on the masses and energies involved. For incident ions of mass (atomic weight) M_0 and energy E_0, which collide with target atoms of mass M, the ion energy after collision E_1 is given by:

$$E_1 = \left[\frac{(M^2 - M_0^2 \sin^2\theta)^{1/2} + M_0 \cos\theta}{M_0 + M}\right]^2 E_0 \tag{5.1}$$

where θ is the scattering angle. The coefficient of E_0 in Eq. (5.1), called the kinematic factor, depends only on the atomic weight of the target atom, for a known incident ion species and fixed scattering angle. By measuring the number and energy of the backscattered ions, information on the nature of the elements present and their concentration and depth distribution can be obtained simultaneously, without any appreciable damage to the sample.

Figure 5.10(a) and (b) show the typical RBS spectra obtained using 1.8 MeV He^{4+} ions from a thin (~500 Å) and thick (~4000 Å) sample of YBCO deposited on silicon. For thin samples, the spectral peaks obtained from different elements in the film do not overlap. The area under a spectral peak represents the total number of atoms of a given element present within a continuous layer. From the ratio of the areas of peaks of two different elements, the ratio of their concentration in the film can be calculated according to the equation:

$$\frac{C_1}{C_2} = \frac{A_1}{A_2}\left(\frac{Z_2}{Z_1}\right)^2 \tag{5.2}$$

where C, A and Z are the concentration, area under peak, and the atomic number, respectively, of elements 1 and 2.

In thicker samples the peaks from different elements overlap each other, as seen in Fig. 5.10(b). A simulation program, RUMP [92], is available for simulating the RBS spectra of such samples. The simulation package can construct the theoretical spectrum of a sample from a knowledge of the sample structure in terms of the number of layers and thickness and composition of each layer in the sample. The simulation can be used in an interactive manner to arrive at the best set of parameters (thickness and composition of individual layers) in the sample structure that fit a given sample spectrum.

5.2.1.2. *Electron Microprobe*

The electron microprobe equipment consists of an electron gun with a column of lenses that focus the electron beam to a fine spot (~50 Å diameter). Scanning coils are used to deflect or raster the beam over any area of the specimen surface. Typical beam energies range from 10–50 keV. The electron beam excites element specific characteristic x-rays from the area of the specimen being probed. In

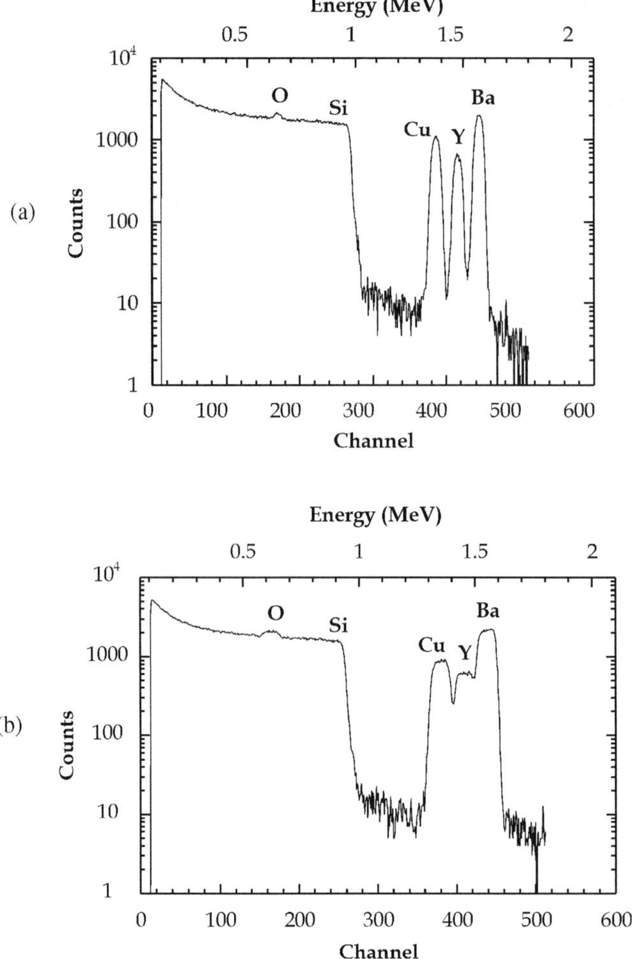

FIG. 5.10. (a) The RBS spectra of thin sample of YBCO deposited on a silicon substrate (Reference [91]). (b) The RBS spectra of thick sample of YBCO deposited on a silicon substrate from Reference [91].

wavelength dispersive analysis (WDX) the wavelength of the x-rays is dispersed and the elements in the sample are identified by their characteristic emission wavelengths. The characteristic x-rays are diffracted from single crystals with known interplanar spacings, and the diffracted beams at different angles can be precisely measured with an x-ray detector. The diffracting crystal and detector can scan the spectrum of x-rays from the sample as a function of angle. The wavelength of the diffracted beam is back calculated from the angle using Bragg's law. In energy dispersive analysis (EDS) the intensity of the peak is used to determine the composition of the film. Typically, a liquid nitrogen cooled Si or Li detector with a resolution of about 150 eV is used for energy dispersion.

5.2.2. STRUCTURAL CHARACTERIZATIONS

5.2.2.1. X-ray Diffraction

X-ray diffraction is a useful tool for the structural study of epitaxial thin films and multilayers, as it allows one to probe the crystal structure of epitaxial films non-destructively as well as to examine the layered nature of the multilayers. The thin-film samples are exposed to a monochromatic beam of x-rays from a Cu-K_α source, which is used to investigate the crystal structure. The wavelength of these x-rays is of the same order as the interatomic distance (a few angstroms), which allows atomic structure of the deposited thin film to be studied. The thin-film sample scatters the incident x-ray beam in all directions. However, due to the periodic arrangement of atoms on specific crystallographic planes in the crystalline solid thin film, the scattered x-rays mutually reinforce each other in certain directions, giving rise to a strong (high intensity) diffracted beam by constructive interference. The position (angle θ) of the diffracted beam is given by Bragg's law:

$$n\lambda = 2d \sin \theta$$

where n is the order of diffraction, λ is the wavelength of the incident x-ray beam, d is the spacing between planes that contribute to diffraction, and θ is the angle between the incident beam and the crystallographic plane. An x-ray detector such as a Geiger counter or a scintillation counter, mounted on a movable arm, detects the diffracted beam. From the intensity and position of the diffracted beam, various interplanar spacings, crystal structure, and orientation of the thin film are determined.

For thin-film- and single crystal x-ray diffraction, a 4-circle x-ray diffractometer is used. By using the four circles of the diffractometer, the sample can be oriented at different angles to the incident beam so that different crystallographic planes contribute to diffraction [93, 94]. The sample can also be rotated about its normal, and from the number of diffraction peaks obtained over a 360° scan the symmetry of the sample surface can be studied. Thus, the complete 3D crystal structure and orientation of the thin film with respect to the substrate can be determined. Using a 4-circle diffractometer, one can determine the in-plane and out-of-plane lattice parameters, as well as domain structure and crystalline quality

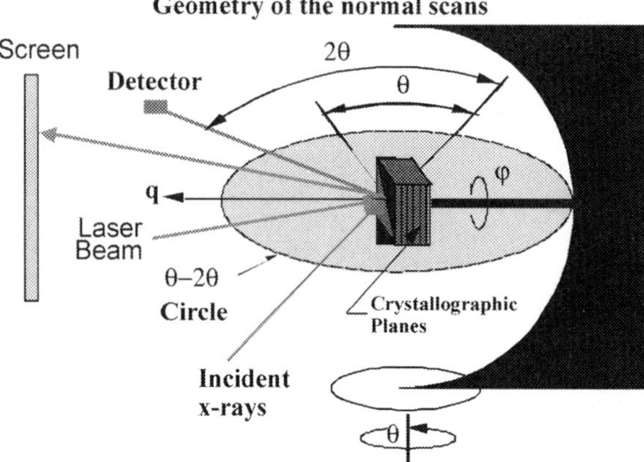

FIG. 5.11. Schematic of a normal θ–2θ scan in a 4-circle diffractometer [93, 94].

through a variety of scans such as normal θ–2θ scans, rocking curves, off-axis ϕ scans, and grazing incidence scans.

In the normal θ–2θ scan, the scattering vector **q** (the diffracting plane normal) is perpendicular to the sample surface ($\chi = 90°$), as shown in Fig. 5.11. The angle of incidence and exit is very close to the Bragg angle for the diffracting planes. From this scan, that is, lattice parameters normal to the substrate, the strain can be determined. Figure 5.12 shows a typical θ–2θ scan of a 2000 Å-thick YBCO film

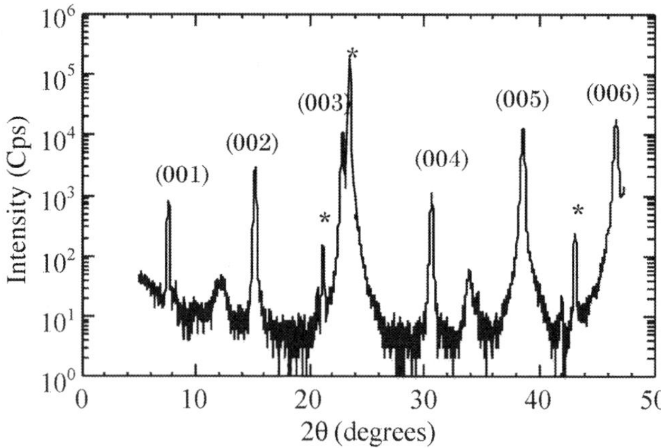

FIG. 5.12. Typical θ–2θ scan of YBCO film. The substrate ($LaAlO_3$) peaks are marked by * and the miller indices of the film peaks are indicated. The low intensity substrate peaks are the (00*l*) reflections observed from the CuK_β wavelength [91].

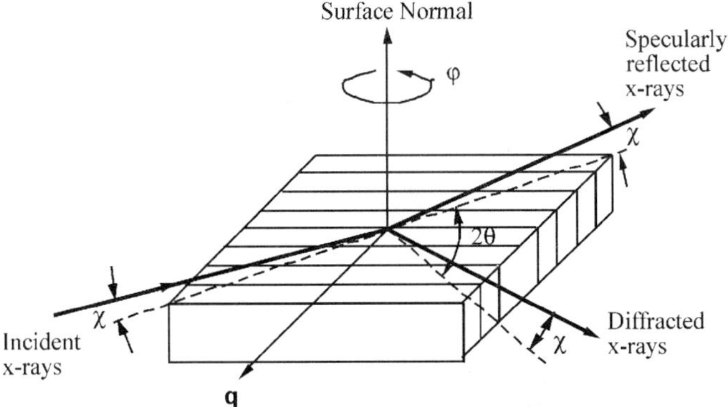

FIG. 5.13. Schematic diagram of grazing-incidence diffraction [68].

grown on the (001) $SrTiO_3$ substrate. The film is epitaxial with *c*-axis oriented out-of-plane (parallel to the [001] direction of the substrate).

Figure 5.13 shows the geometry of grazing-incidence diffraction (GID). In this case, the **q** vector is parallel to the substrate and film surface. The angle of incidence (and also of exit) χ is typically chosen between the critical angle for total reflection ($\chi_c \leq 0.5°$) and $\sim 1°$. From GID $\theta-2\theta$ scans the in-plane lattice parameter, strain, and crystallite size parallel to the substrate can be obtained. A GID Φ-scan provides information about the in-plane orientation spread. These GID scans combined with normal $\theta-2\theta$ scans are used to characterize the 3D strain states of epitaxial thin films [95, 96].

In the off-axis ϕ scans, the **q** vector is at an angle to the plane of the film. As the sample is rotated about the phi-axis over 360° (Fig. 5.14), this scan reveals information regarding the symmetry of the crystal structure of the sample. The crystallographic domain structure and epitaxy of the film can be characterized. Figure 5.15 shows the off-axis phi scans of (102) peak of a YBCO film and (101) peak of the underlying 24°-bi-crystal (001) $SrTiO_3$ substrate. The matching of the film and substrate peaks indicates that the film grows cube-on-cube with respect to the substrate, as shown in the schematic on the left [97]. Such off-axis phi scans have also been used to differentiate single domain $SrRuO_3$ thin films from multidomain films [98, 99].

5.2.2.2. *Rutherford Backscattering Spectroscopy (RBS)*

The crystalline quality of thin films and multilayers has also been studied by ion channeling investigations with Rutherford backscattering spectroscopy (RBS) using $4He^+$ ions. The ratio of the backscattered yield along $\langle 100 \rangle$ to that in a

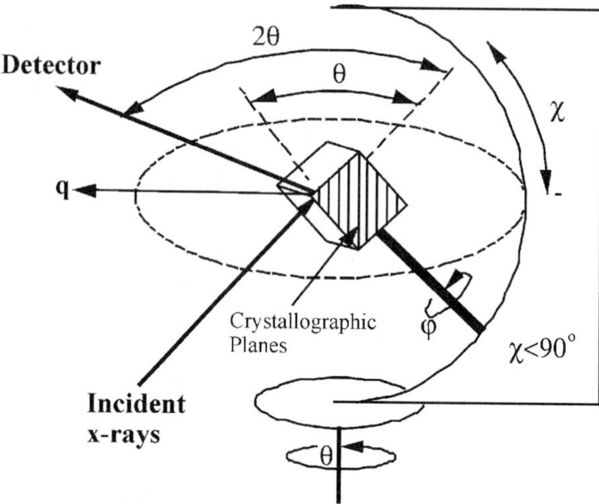

FIG. 5.14. Schematic diagram of off-axis phi scans.

random direction (χmin) gives a measure of the crystalline nature of the sample. The lower the backscattered channeling yield, the better the crystalline quality of the film [100].

5.2.2.3. Transmission Electron Microscopy (TEM)

Transmission electron microscopy (TEM) plays an important role in the structural characterization of thin films and multilayered heterostructures. High-energy electrons ($\sim$100 keV to 1 MeV) emitted from an electron gun (thermionically or by field emission) are incident upon the thin-film sample that has been prepared by a series of grinding, polishing and ion-milling steps to be thin enough to permit the transmission of electrons. During their passage through the specimen, the electron beam suffers elastic scattering from core ions and inelastic scattering at grain boundaries, defects, second phase particles etc. The former is used to form a diffraction pattern of the specimen to study the crystalline quality and epitaxial nature of the thin film. The inelastic scattering effect that leads to a spatial variation in the intensity of the primary transmitted beam is used to study the microstructure of the thin film in terms of grain boundaries, defects, second-phase particles. The primary and diffracted beams emerging from the thin-film sample pass through the objective lens to form the first image. By using the amplitude contrast of the primary beam or a single diffracted beam, bright field or dark-field images of the sample can be obtained. By using phase contrast between the primary beam and one or more of the diffracted beams, high-resolution lattice

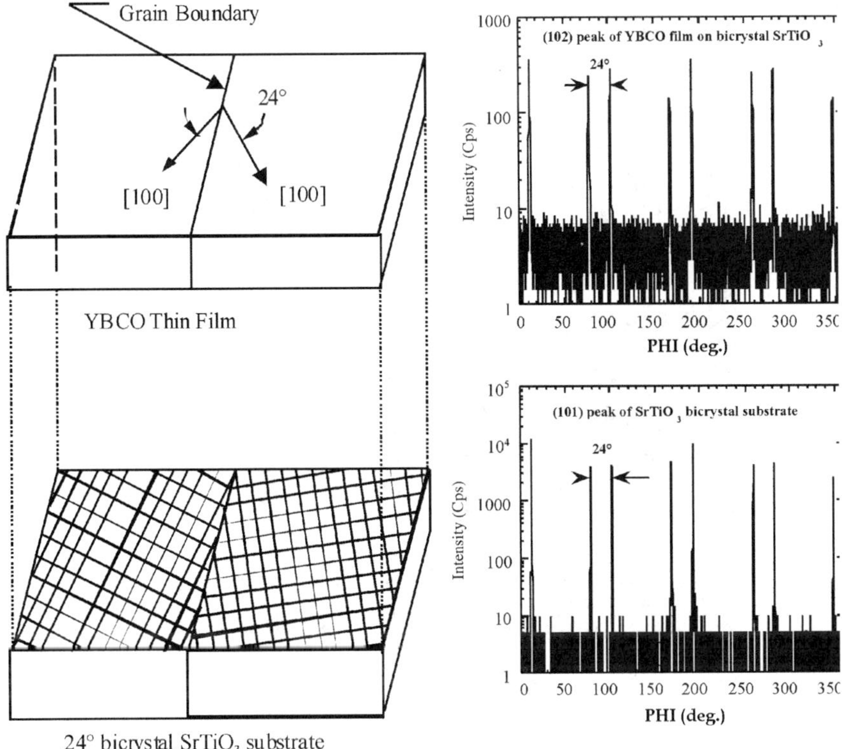

FIG. 5.15. Schematic diagram of YBCO-thin film on 24°-bi-crystal (001) $SrTiO_3$ substrate and off-axis phi scans of the (102) peak of YBCO and (101) peak of substrate showing cube-on-cube epitaxy [97].

images can be obtained. This technique is used for studying the atomic structure of interfaces in multilayer films. Such high-resolution cross-sectional TEM studies have also been used to study the stacking sequence of YBCO thin films on many substrates. Figure 5.16 shows a high-resolution cross-sectional TEM from a YBCO/$SrTiO_3$ sample [101].

5.2.3. SURFACE MORPHOLOGY

Surface morphology and roughness control are very important for making superconducting devices, multilayers, and ultrathin films. In general, the film morphology depends on lattice mismatch with substrate, texture, grain size, and film thickness. The surface morphology is typically investigated using scanning

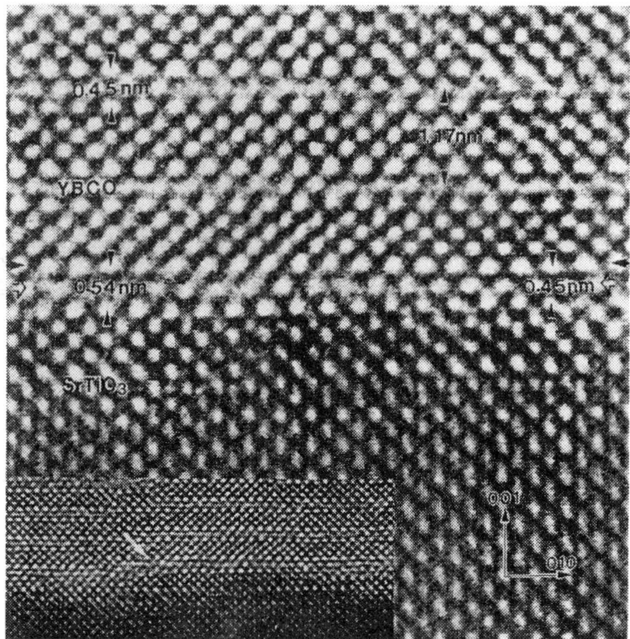

FIG. 5.16. High-resolution cross-sectional TEM image of a YBCO film on $SrTiO_3$ substrate from Reference [101].

electron microscopy (SEM), atomic force microscopy (AFM), and scanning tunneling microscopy (STM). Both AFM and STM are important for the study of growth mechanism and substrates surfaces as well.

The STM uses an atomically sharp conductive PtIr or W tip with a bias voltage applied between the tip and the sample, which is mounted on a piezoelectric scanner. When the tip is brought sufficiently close (a few nm) to the sample, electrons tunnel through from the sample to the tip or vice versa, depending upon the sign of the bias voltage. The resulting tunneling current is exponentially dependent on the tip-to-sample separation, which gives the STM its remarkable sensitivity with atomic resolution. By using a feedback loop, the tunneling current is maintained constant by adjusting the height of the scanner at each measurement point. The motion of the scanner then maps out the surface topography of the sample. Figure 5.9(a) shows the typical STM images of a 2000-Å-thick YBCO film grown on (001) $SrTiO_3$ substrate by 90°-off-axis sputtering. The spiral features seen in the image indicate a screw dislocation growth mode, which is typical of YBCO films. The terraces on the film surface are separated by unit cell high (~12 Å) steps.

5.2.4. ELECTRICAL TRANSPORT AND MAGNETIZATION MEASUREMENTS

Transition temperatures and normal state resistivities are measured by a 4-terminal dc transport method using a 4-point geometry or the Van der Pauw geometry. Figure 5.17 shows the schematic of the current and voltage contacts in the two geometries. In the 4-point geometry, the patterned bridges shown in Fig. 5.17(a) are used to determine absolute resistivities. In the Van der Pauw geometry, the resistivity of the sample can be measured without patterning and regardless of the shape of the sample [102]. Four electrical contacts are made at the circumference of the sample as shown in Fig. 5.17. The resistance R_1 is measured as the voltage output between contacts 1 and 2 per unit current through 3 and 4 [Fig. 5.17(b)]. Similarly R_2 is measured as the voltage output between contacts 2 and 3 per unit current through contacts 1 and 4 [Fig. 5.17(c)]. The resistivity of the sample is then given by the equation

$$\rho = \frac{\pi d}{\ln 2} \frac{(R_1 + R_2)}{2} f\left(\frac{R_1}{R_2}\right)$$

where d is the thickness of the film and f is a function of the ratio of R_1 to R_2 and satisfies the relation

$$\ln 2\left(\frac{R_1 - R_2}{R_1 + R_2}\right) = f \operatorname{arcosh}\left\{\frac{\exp(\ln 2/f)}{2}\right\}.$$

The Van der Pauw method assumes that the sample is homogeneous and isotropic. Figure 5.18 shows a typical resistive transition for a YBCO thin film measured in the Van der Pauw geometry. The inset shows the transition temperature to be 90.8 K. Transition temperatures are also measured by magnetic susceptibility measurements. The transition width ΔT is defined as the difference between the temperature at 90% value of the normal state resistivity near T_c and the temperature at zero resistivity.

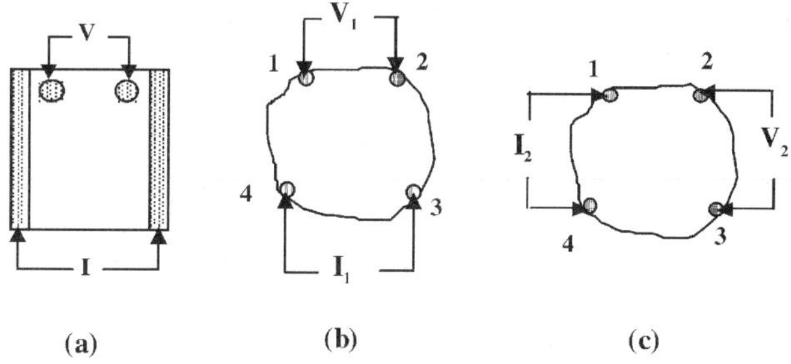

FIG. 5.17. The schematic diagram of the geometry of (a) 4-point measurement and (b) and (c) Van der Pauw measurement [68].

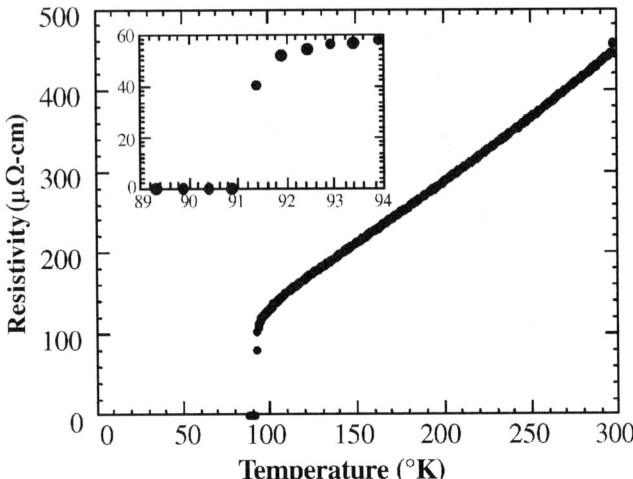

FIG. 5.18. Temperature dependence of resistivity of a YBCO thin film. The inset shows the transition to the superconducting state at 90.8 K [91].

A typical value for the critical current density of *c*-axis-oriented YBCO thin films is in the mid-$10^7\,A/cm^2$ range at 4.2 K. This makes direct transport measurements difficult. Claassen *et al.* have developed an alternative technique that makes use of a small flat coil that is pressed against the film surface and driven with an ac sinusoidal current to induce shielding currents in the film [103, 104]. When the current in the coil is sufficiently large, the screening current density reaches its critical value (J_c) and the coil impedance becomes nonlinear. By measuring the third harmonic voltage component across the coil, the nonlinear response can be monitored as the drive current is increased. The nonlinearity in the coil-film system increases abruptly when the maximum induced current equals the critical current. The relationship between coil current and screening current density can be calculated for a given coil and spacing from the film.

The J_c measurements are also made using dc magnetization methods. These measurements make use of the Bean formula, which can be readily applied for thin-film geometry. Critical currents for H parallel to the *c*-axis direction (perpendicular to the film surface) are studied as a function of temperature and applied field by measuring the induced saturation moment on a magnetometer (VSM or SQUID or extraction magnetometer). Figure 5.19(a) shows a typical dc magnetization loop for a YBCO-thin-film sample measured at 4.2 K with H parallel to the *c*-axis of the film. The dc moment is measured by ramping the field

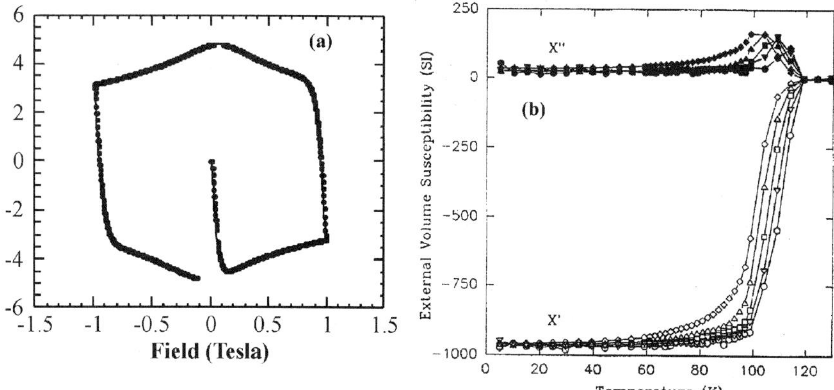

FIG. 5.19. (a) The dc magnetization hysteresis loop of a YBCO-thin-film sample from Reference 91 and (b) the real (X) and imaginary (X′) components of ac magnetic susceptibility of a thallium-based superconductor at ac field strengths of 0.01 Oe (circles), 0.02 Oe (triangles), 0.04 Oe (squares), 0.08 Oe (inverted triangles), and 0.16 Oe (diamonds) (from Reference [106]).

to a certain value, allowing the moment to decay for ~1 s, and then measuring the sample moment. The Bean formula [105]

$$J_c(H) = \frac{30M(H)}{VR}$$

is applied to relate the observed hysteretic moment $M(H)$ to the critical current $J_c(H)$ at each applied field value, where V is the film volume; R is chosen as the geometrically averaged sample radius.

The magnetometer is also used for measuring the ac susceptibility of superconductors as a function of temperature. The real component of susceptibility (χ') is used to determine the transition temperature while the imaginary component (χ'') can be used to deduce an upper limit for the lower critical field H_{c1} (T). Figure 5.19(b) shows a typical plot of the real and imaginary components of ac susceptibility of a thallium-based superconductor thin film as a function of temperature and amplitude of applied ac field [106]. These ac susceptibility measurements have also been used to identify multiple phases in the HTS materials and to investigate the inter- and intragranular coupling of bulk HTS materials. The ac measurement technique also provides information about the relaxation processes occurring in the material.

5.3. Summary

Low-temperature superconducting films have benefited greatly from advancement in vacuum science. Improvement in deposition system base pressure has enabled fabrication of reliable Nb films in institutions throughout the world. The further development of techniques such as reactive magnetron sputtering has further facilitated the development of superconducting compounds such as NbN. Both material systems have matured to the extent that they are integral processes of superconducting integrated circuit processes.

The field of HTS thin-film processing has seen rapid progress since the discovery of HTS in 1986. Several of the techniques used in HTS thin-film processing have been adapted from the low-T_c superconducting thin-film processing and/or the well-developed semiconducting IC fabrication industry. For instance, RHEED, which is traditionally used in the MBE deposition of semiconductors, has been modified to work under higher background pressures and is used in conjunction with PLD systems for monitoring the growth of HTS thin films *in situ*. However, further improvement in thin-film processing is required for commercial device application. For instance, the reproducible and reliable manufacture of Josephson junctions requires a spread of $<10\%$ in critical properties such as I_c, R_nA and I_cR_n over the entire device. The variation in these properties is due to the broad, diffuse and defective interfaces between the adjacent layers. The HTS thin-film processing needs to evolve to a stage at which high-quality films with atomically smooth surfaces and heterostructures with atomically sharp and well-controlled interfaces can be manufactured reproducibly as in the semiconductor industry.

Acknowledgment

The contribution of Rajesh A. Rao at Duke University and fruitful discussion with Dave H.A. Blank at the University of Twente are gratefully acknowledged.

References

1. Bednorz, J. G. and Muller, K. H. (1986). *Z. Phys.* B**64**: 189.
2. For a more detailed review of evaporation and MBE see Vossen, J. L. and Kern, W. (1978). In *Thin Film Processes*, New York: Academic Press.
3. Schellingerhout, J., Janocko, M. A., Klapwijk, T. M., and Mooij, J. E. (1988). Rate control for electron gun evaporation. *Rev. Sci. Instrum.* **60**: 1177.
4. Maissel, L. I. and Glang, R. (1970). *Handbook of Thin Film Technology*, New York: McGraw-Hill, pp. 4–39.

5. Huang, H-C. W., Basavaiah, S., Kircher, C. J., Harris, E. P., Murakami, M., Klepner, S. P., and Greiner, J. H. (1980). High reliability Pb-alloy Josephson junctions for integrated circuits. *IEEE Trans. Elec. Devices* **ED-27**: 1979.
6. Hollands, E. and Campbell, D. S. (1968). *J. Mater. Sci.* **3**: 544.
7. Kuroda, K. and Yuda, M. (1988). Niobium-stress influence on Nb/Al-oxide/Nb Josephson junctions. *J. Appl. Phys.* **63**: 2353–2357.
8. Lichtenberger, A. Private communication.
9. Aschermann, G., Friederich, E., Justi, E., and Kramer, J. (1941). *Phys. Zeir.* **42**: 349.
10. Gavaler, J. R., Janocko, M. A., Patterson, A., and Jones, C. K. (1971). *J. Appl. Phys.* **43**: 54.
11. Bhat, A., Meng, X., Wong, A., and Duzer, T. V. (1999). *Proc Internat. Supercond. Elec. Conf.*, p. 515.
12. Kerber, G., Abelson, L., Elmadjian, R., Hanaya, G., and Ladizinsky, E. (1997). *IEEE Trans. Appl. Superconductivity* 2638.
13. Hollands, E. and Campbell, D. S. (1968). *J. Mater. Sci.* **3**: 544.
14. Wang, Z., Kawakami, A., Uzawa, Y., and Komiyama, B. (1996). *J. Appl. Phys.* **79** (10): 7837.
15. Gray, K. E., Kampwirth, R. T., Murduck, J. M., and Capone, II, D. W. (1988). *Physica* C **152**: 445–455.
16. Moodera, J. S., Fracavilla, T. L., and Wolf, S. A. (1987). *IEEE Trans. Magnetics* **MAG-23**, (2): 1003.
17. Cukauskas, E. J., Carter, W. L., and Qadri, S. B. (1985). *J. Appl. Phys.* **57** (7): 2538.
18. Kroger, H., Smith, L. N., and Jillie, D. W. (1981). *Appl. Phys. Lett.* **39** (3): 280; Thomasson, S. L., Murduck, J. M., and Chan, H. (1989). *Extended Abstracts of ISEC '89*, 371.
19. Stern, J. A., Blumble, B., LeDuc, H., Kooi, W. J., and Zmuidzinas, J. (1997). *Proc. 9th Internat. Symp. on Space Terahertz Technology*, CIT, PC, p. 305, March.
20. Iosad, N. N., Jackson, B. D., Ferro, F., Gao, J. R., Palyakov, S. N., Dmitriev, P. N., and Klapwijk, T. M. (1999). *Proc. Internat. Supercond. Elec. Conf.*, p. 311.
21. Wu, M. K., Ashburn, J. R., Torng, C. J., Hor, P. H., Meng, R. K., Gao, L., Huang, Z. J., Wang, Y. Q., and Chu, C. W. (1987). *Phys. Rev. Lett.* **58**: 408.
22. Cava, R. J., Batlogg, B., van Dover, R. B., Murphy, D. W., Sunshine, S., Siegrist, T., Remeika, J. P., Zahurak, S., and Espinosa, G. P. (1987). *Phys. Rev. Lett.* **58**: 1676.
23. Maeda, H., Tanaka, Y., Fukutomi, M., and Asano, T. (1988). *Jpn. J. Appl. Phys.* **27**: L209.
24. Sheng, Z. Z., Herman, A. M., El Ali, A., Almason, C., Estrada, J., Datta, T., and Matson, R. J. (1988). *Phys. Rev. Lett.* **60**: 937.
25. Schilling, A., Cautoni, M., Guo, J. D., and Ott, H. R. (1993). *Nature* **363**: 56.
26. Gao, L., Xue, Y. Y., Chen, F., Xiong, Q., Meng, R. L., Ramirez, D., Chu, C. W., Eggert, J. H., and Mao, H. K. (1994). *Phys. Rev. B* **50**: 4260.
27. Hammond, R. H. and Bormann, R. (1989). *Physica C* **162–164**: 703.
28. Schlom, D. G. (1990). PhD thesis, Stanford University.
29. Berkeley, D. D., Johnson, B. R., Anand, N., Beauchamp, K. M., Conroy, L. E., Goldman, A. M., Maps, J., Mauersberger K., Mecartney, M. L., Morton, J., Tuominen, M., and Zhang, Y.-J. (1988). *Appl. Phys. Lett.* **53**: 1973
30. Missert, N., Hammond, R., Mooij, J. E., Matijasevic, V., Rosenthal, P., Garwin, E., Geballe, T. H., Kapitulnik, A., Beasley, M. R., Laderman, S. S., Lu, C., and Barton, R. (1989). *IEEE Trans. Magn.* **25**: 2418.
31. Terashima, T., Iijima, K., Yamamoto, K., Bando, Y., and Mazaki, H. (1988). *Jpn. J. Appl. Phys.* **27**: L91.
32. Eckstein, J. N., Bozozvic, I., Klausmeieir-Brown, M. E., Virshup, G., and Ralls, K. S. (1992). *MRS Bull.* **17**: 27.
33. Wang, W., Hammond, R. H., Fejer, M. M., Ahn, C. H., Beasley, M. R., Levenson, M. D., and Bortz, M. L. (1995). *Appl. Phys. Lett.* **67**: 1375.

34. Wang, W., Fejer, M. M., Hammond, R. H., Beasley, M. R., Ahn, C. H., Bortz, M. L., and Day, T. (1996). *Appl. Phys. Lett.* **68**: 729.
35. Kinder, H., Berberich, P., Utz, B., and Prusseit, W. (1995). *IEEE Trans. Appl. Supercond.* **5**: 1575.
36. Utz, B., Semerad, R., Bauer, M., Prusseit, W., Berberich, P., and Kinder, H. (1997). *IEEE Trans. Appl. Supercond.* **7**: 1272.
37. Zhang, J., Gardiner, R. A., Kirlin, P. S., Boerstler, R. W., and Steinbeck, J. (1992). *Appl. Phys. Lett.* **61**: 2884.
38. Chern, C. S., Martens, J. S., Li, Y. Q., Gallos, B. M., Lu, P., and Kear, B. H. (1993). *Supercond. Sci. Technol.* **6**: 460.
39. Matsuno, S., Uchikawa, F., Utsunomiya, S., and Nakabayashi, S. (1992). *Appl. Phys. Lett.* **60**: 2427.
40. Triscone, J.-M., Karkut, M. G., Antognazza, L., Brunner, O., and Fischer, Ø. (1989). *Phys. Rev. Lett.* **63**: 1016.
41. Triscone, J.-M., Fischer, Ø., Brunner, O., Antognazza, L., Kent, A. D., and, Karkut, M. G. (1990). *Phys. Rev. Lett.* **64**: 804.
42. Lowndes, D. H., Norton, D. P., and Budai, J. D. (1990). *Phys. Rev. Lett.* **65**: 1160.
43. Eom, C. B., Marshall, A. F., Triscone, J.-M., Wilkens, B., Laderman, S. S., and Geballe, T. H. (1991). *Science* **251**: 780.
44. Foltyn, S. R., Peterson, E. J., Coulter, J. Y., Arendt, P. N., Jia, Q. X., Dowden, P. C., Maley, M. P., Wu, X. D., and Peterson, D. E. (1997). *J. Mater. Res.* **12**: 2941.
45. Holzapfel, B., Roas, B., Schultz, L., Bauer, P., and Ischenko, G. S. (1992). *Appl. Phys. Lett.* **61**: 3178.
46. Foltyn, S. T., Muenchausen, R. E., Dye, R. C., Wu, X. D., Luo, L., Cooke, D. W., and Taber, R. C. (1991). *Appl. Phys. Lett.* **59**: 1374.
47. Wu, X. D., Muenchausen, R. E., Foltyn, S., Estler, R. C., Dye, R. C., Flamme, C., Nogar, N. S., Garcia, A. R., Martin, J., and Tesmer, J. (1990). *Appl. Phys. Lett.* **56**: 1481.
48. Tabata, H. and Kawai, T. (1993). *Thin-Solid Films* **225**: 275.
49. Kawai, T. (1993). *AIP Conference Proceedings* **288**: 191.
50. Koinuma, H. (1995). *Bull. Mater. Sci.* **18**: 435.
51. Koinuma, H., Kawasaki, M., and Yoshimoto, M. (1996). *Mat. Res. Soc. Symp. Proc.* **397**: 145.
52. Kanai, M., Kawai, T., and Kawai, S. (1991). *Appl. Phys. Lett.* **58**: 771.
53. Terashima, T., Bando, Y., Iijima, K., Yamamoto, K., Hirata, K., Hayashi, K., Kamigaki, K., and Terauchi, H. (1990). *Phys. Rev. Lett.* **65**: 2684.
54. Shaw, T. M., Gupta, A., Chern, M. Y., Batson, P. E., Laibowitz, R. B., and Scott, B. A. (1994). *J. Mater. Res.* **9**: 2566.
55. Rijnders, G. J. H. M., Koster, G., Blank, D. H. A., and Rogalla, H. (1997). *Appl. Phys. Lett.* **70**: 1888.
56. Koster, G., Kropman, B. L., Rijnders, G. J. H. M., Blank, D. H. A., and Rogalla, H. (1998). *Appl. Phys. Lett.* **73**: 2920.
57. Koster, G., Rijnders, G. J. H. M., Blank, D. H. A., and Rogalla, H. (1999). *Appl. Phys. Lett.* **74**: 3729.
58. Kawasaki, M., Nagata, S., Sato, Y., Funabshi, M., Hasegawa, T., Kishio, K., Kitazawa, K., Fueki, K., and Koinuma, H. (1987). *Jpn. J. Appl. Phys.* **26**: L736.
59. Adachi, H., Hirochi, K., Setsune, K., Kitabatake, M., and Wasa, K. (1987). *Appl. Phys. Lett.* **51**: 2263.
60. Li, H. C., Linker, G., Ratzel, F., Smithey, R., and Greek, J. (1998). *Appl. Phys. Lett.* **52**: 1098.
61. Xi, X. X., Venkatesan, T., Li, Q., Wu, X. D., Inam, A., Chang, C. C., Ramesh, R., Hwang, D. M., Ravi, T. S., Findikoglu, A., Hemmick, D., Etemad, S., Martinez, J. A., and Wilkens, E. (1991). *IEEE Trans. Magn.* **27**: 982.

62. Sandstrom, R. L., Gallagher, W. J., Dinger, T. R., and Koch, R. H. (1988). *Appl. Phys. Lett.* **53**: 444.
63. Eom, C. B., Sun, J. Z., Yamamoto, K., Marshall, A. F., Luther, K. E., Laderman, S. S., and Geballe, T. H. (1989). *Appl. Phys. Lett.* **55**: 595.
64. Eom, C. B., Sun, J. Z., Streiffer, S. K., Marshall, A. F., Yamamoto, K., Lairson, B. M., Anlage, S. M., Bravman, J. C., Geballe, T. H., Laderman, S. S., and Taber, R. C. (1990). *Physica C* **171**: 354.
65. Wehner, G. K. and Rosenberg, D. (1960). *J. Appl. Phys.* **31**: 177.
66. Hammond, R. H. and Bormann, R. (1989). *Physica C* **162–164**: 703.
67. Yamamoto, K., Lairson, B. M., Eom, C. B., Hammond, R. H., Bravman, J. C., and Geballe, T. H. (1990). *Appl. Phys. Lett.* **57**: 1936.
68. Eom, C. B. (1991). PhD Thesis, Stanford University.
69. Rao, R. A., Gan, Q., Eom, C. B., Suzuki, Y., McDaniel, A. A., and Hsu, J. W. P. (1996). *Appl. Phys. Lett.* **69**: 3911.
70. Koch, R. H., Umbach, C. P., Clark, G. J., Chaudhari, P., and Laibowitz, R. B. (1987). Quantum interference devices made from superconducting oxide thin films. *Appl. Phys. Lett.* **51**: 200.
71. Likharev, K. K. (1993). Rapid single-flux-quantum logic, in *The New Superconducting Electronics*, Weinstock, H. and Ralston, R. W. eds., Dordrecht: Kluwer Academic Publishers.
72. Li, Q., Xi, X. X., Wu, X. D., Inam, A., Vadlamannati, S., McLean, W. L., Venkatesan, T., Ramesh, R., Hwang, D. M., Martinez, J. A., and Nazar, L. (1990). *Phys. Rev. Lett.* **64**: 3086.
73. O'Bryan, H. M., Gallagher, P. K., Berkstresser, G. W., and Brandle, C. D. (1990). *J. Mater Res.* **5**: 183.
74. Tidrow, S. C., Tauber, A., Wilber, W. D., Lareau, R. T., Brandle, C. D., Berkstresser, G. W., Ven Graitis, A. J., Potrepka, D. M., Budnick, J. I., and Wu, J. Z. (1997). *IEEE Trans. Appl. Supercond.* **7**: 1766.
75. Guo, R., Ravindranathan, P., Selvaraj, U., Bhalla, A. S., Cross, L. E., and Roy, R. (1994). *Jour. Materials Science* **29**: 5054.
76. Wu, X. D., Luo, L., Muenchausen, R. E., Springer, K. N., and Foltyn, S. (1992). *Appl. Phys. Lett.* **60**: 1381.
77. Char, K., Colclough, M. S., Garrison, S. M., Newman, N., and Zaharchuk, G. (1991). *Appl. Phys. Lett.* **59**: 733.
78. Hawley, M., Raistrick, I. D., Beery, J. G., and Houlton, R. J. (1991). *Science* **251**: 1587.
79. Gerber, C., Anselmetti, D., Bednorz, J. D., Mannhart, J., and Schlom, D. G. (1991). *Nature* **350**: 279.
80. Haage, T., Zegenhaen, J., Li, J. Q., Habermeier, H.-U., Cardona, M., Jooss, Ch., Warthmann, R., Forkl, A., and Kronmuller, H. (1997). Transport properties and flux pinning by self organization in $YBa_2Cu_3O_{7-d}$ films on vicinal $SrTiO_3$ (001). *Phys. Rev. B* **56**: 8404.
81. Mechin, L., Berghuis, P., and Evetts, J. E. (1998). Properties of $YBa_2Cu_3O_{7-d}$ thin films grown on vicinal $SrTiO_3$ (001) substrates. *Physica C* **302**: 102.
82. Ramesh, R., Inam, A., Hwang, D. M., Ravi, T. S., Sands, T., Xi, X. X., Wu, X. D., Li, Q., Venkatesan, T., and Kilaas, R. (1991). *J. Mater. Res.* **6**: 2264.
83. Wen, J. G., Traeholt, C., and Zandbergen, H. W. (1993). *Physica C* **205**: 354.
84. Tsuchiya, R., Kawasaki, M., Kubota, H., Nishino, J., Sato, H., Akoh, H., Koinuma, H. (1997). *Appl. Phys. Lett.* **71**: 1570.
85. Rao, R. A., Gan, Q., and Eom, C. B. (1997). *Appl. Phys. Lett.* **71**: 1171.
86. Kawasaki, M., Takahashi, K., Maeda, T., Tsuchiya, R., Shinohara, M., Ishiyama, O., Yonezawa, T., Yoshimoto, M., and Koinuma, H. (1994). *Science* **266**: 1540.
87. Koster, G., Kropman, B. L., Rijnders, G. J. H. M., Blank, D. H. A., and Rogalla, H. (1998). Quasi-ideal strontium titanate crystal surfaces through formation of strontium hydroxide. *Appl. Phys. Lett.* **73**: 2920.

88. Ohnishi, T. *et al.* (1999). A-site layer terminated perovskite substrate: $NdGaO_3$. *Appl. Phys. Lett.* **73**: 2920.
89. Ohnishi, T. (1999). PhD thesis, Tokyo Institute of Technology.
90. Blank, D. H. A. Private communications.
91. Rao, R. A. (1999). PhD thesis, Duke University.
92. Doolittle, L. R. (1985). *Nucl. Instrum. Methods* **B 9**: 344.
93. Azaroff, L. V. (1968). *Elements of X-ray Crystallography*, New York: McGraw-Hill, pp. 360–389.
94. Schlom, D. G., Hellman, E. S., Hartford, E. H., Jr., Eom, C. B., Clark, J. C., and Mannhart, J. (1996). *J. Mater. Res.* **11**: 1336.
95. Rao, R. A., Lavric, D., Nath, T. K., Eom, C. B., Wu, L., and Tsui, F. (1998). *Appl. Phys. Lett.* **73**: 3294.
96. Rao, R. A., Gan, Q., Cava, R. J., Suzuki, Y., Gausepohl, S. C., Lee, M., and Eom, C. B. (1997). *Appl. Phys. Lett.* **70**: 3035.
97. Rao, R. A., Eom, C. B., and Hsu, J. W. P. unpublished results.
98. Eom, C. B., Cava, R. J., Fleming, R. M., Phillips, J. M., van Dover, R. B., Marshall, J. H., Hsu, J. W. P., Krajewski, J. J., and Peck, W. F., Jr., (1992). *Science* **258**: 1766.
99. Gan, Q., Rao R. A., and Eom, C. B. (1997). *Appl. Phys. Lett.* **70**: 1962.
100. See for instance, Baudenbacher, F., Hirata, K., Berberich, P., Kinder, H., and Assmann, W. (1992). Smooth YBCO films on MgO with good superconducting properties. *Proceedings of Symposium A1 on High Temperature Superconductor Thin Films of the International Conference on Advanced Materials-ICAM 1991*, Amsterdam: North-Holland.
101. Wen, J. G., Traeholt, C., and Zandbergen, H. W. (1993). *Physica C* **205**: 354.
102. Van der Pauw, L. J. (1958). *Philips Res. Repts.* **13**: 1.
103. Claassen, J. H., Reeves, M. E., and Soulen, R. J., Jr. (1991). A contactless method for measurement of the critical current density and critical temperature of superconducting films. *Rev. Sci. Instrum.* **l62**: 996.
104. Claassen, J. H. (1995). Spatially-resolved measurements of critical current density of superconducting films on 2 inch substrates. *IEEE Trans. Appl. Supercond.* **5**: 1413.
105. Bean, C. P. (1962). *Phys. Rev. Lett.* **54**: 1702.
106. LakeShore Measurement and Control Technologies. *AC Susceptometer Technical Notes.*

Fabrication of Superconducting Devices and Circuits

James M. Murduck

TRW, Space and Electronics Group, Redondo Beach, California, USA

The insertion of superconducting circuits as a technology has been only as successful as the capabilities developed to fabricate these circuits. Clearly, the greatest strides in circuit demonstration have been taken using niobium-based integrated circuit (IC) fabrication where IC technology has matured to support large-scale integration levels ($\sim$10 K gates/cm^2). To successfully fabricate large-area and dense IC, stable and reproducible processes are needed. Moreover, IC fabrication involves a number of subprocess steps. Each step has its own inherent variability, the combination of which contributes to the overall degree of circuit functionality, or yield. Fabrication groups face challenges in reducing variability or spreads in critical device parameters such as junction critical currents, values of inductance, and layer-to-layer alignment. In addition, there are design-driven needs to reduce device dimensions and parasitic inductances that put further constraints on practical circuit fabrication.

These challenges are multiplied in IC fabrication using ceramic high-temperature superconductors (HTS). Also there are inherent material issues in this family of superconductors such as maintaining film epitaxy and environmental sensitivity. As a result, the aforementioned fabrication issues are even more daunting. Current HTS circuit fabrication requires an increased understanding of the relative importance of topology and film morphology in vital circuit elements. This is the case in multilayer process elements such as interconnect crossovers,

Vol. 28
ISBN 0-12-533028-6/$35.00

ISSN 1079-4050

layer-to-layer vias and Josephson junctions, where trade-offs often exist between ease of fabrication and requirements for applications. Despite the increased challenge, if these material and processing issues can be successfully implemented, it would allow more compact, lower-power systems, making them increasingly attractive for applications.

In this chapter, the emphasis will be on the practical aspects of active-device superconductor fabrication, especially with respect to their relationship to system application. Present limitations will be noted along with the future direction of the field necessary to allow insertion of superconducting IC technology into commercial systems.

6.1 Introduction

6.1.1 A Brief History

The history of Josephson junction fabrication began in 1963 with the experimental confirmation of the Josephson effect by Anderson and Rowell [1, 2] at Bell Laboratories, one year after Josephson's theoretical predictions [3]. These rudimentary thin-film devices were the precursors of today's superconducting integrated circuit processing. The fabrication of these first junctions was accomplished by producing a thin strip of tin on a glass slide through a metal mask, heating the tin in an oxygen atmosphere to create an oxide barrier and then overlaying with a thin strip of deposited lead. A useful barrier thickness in these devices is on the order of 1 to 2 nm and must completely cover the area between the electrodes, which can range from a few to dozens of square microns. This coverage is proportionally equivalent to an area the size of a football field being perfectly covered with less than 1 cm of snowfall. With these extreme dimensional requirements it is apparent that the underlying base electrode must be smooth and the barrier growth nearly perfect to enable this device fabrication.

Early efforts to fabricate these thin-film junctions were so time-consuming and plagued with yield issues that alternate methods of creating Josephson junctions were quickly developed. One such method, point-contact, was introduced and consisted of a niobium screw contacting a superconducting plate also made of niobium [4] that could be mechanically adjusted to the desired junction parameters. These mechanical devices found their way into some of the earliest commercial products utilizing superconductivity and were integral to device fabrication understanding throughout the 1970s. Clearly, however, fabrication of larger circuits must be done in a monolithic fashion. Considerable advancement was made in thin-film fabrication of Josephson junctions in the 1970s due primarily to IBM's 100-million dollar, 14-year effort to create a computer based on Josephson effects. The junction materials of choice at the start of that project were lead and lead alloys. Researchers at IBM found that with the addition of indium and gold to the lead, smoother base electrodes could be deposited allowing more complete oxide barrier coverage. With barriers on the order of 10 atoms thick, coverage of the base electrode will be a theme repeated in whatever material system is used for fabrication. These lead alloy junctions allowed for a remarkable degree of control over junction parameters and were integrated with resistors and superconducting interconnects in monolithic integrated circuits. However, lead films stretch inelastically when cooled to liquid helium temperature and upon warming to room temperature create mounds that rupture the native oxide tunneling barrier [5]. Even with these limitations, Jaycox and Ketchen [6] successfully fabricated a monolithic dc SQUID by integrating lead alloy junctions with other IC elements such as resistors and an input coil that requires multilayer elements such as crossovers and vias.

Ultimately, the materials limitations of lead alloy junctions helped define and spur research for more robust materials. With a T_c of 9.2 K, niobium was considered an attractive candidate material. However, there were a number of difficulties that had to be resolved before it could be used as a junction electrode. As a refractory metal, niobium has a melting point above 2400 °C making it impractical to be deposited by thermal evaporation as lead alloys had been. In addition, niobium readily oxidizes and for every 1% oxygen concentration in a Nb film, the T_c is reduced by 1 K. Typical vacuum systems at the time had insufficient base pressures to prevent degradation of the critical temperature of the deposited niobium films. The resolution to both these issues is found in what is today's standard deposition equipment. Niobium is sputter-deposited, usually with a dc magnetron source, in vacuum systems capable of ultrahigh vacuum. Prior to the termination of the IBM Josephson computer program, niobium was introduced as a base electrode.

With niobium, researchers had a material that was mechanically hard, could be thermally cycled, and had T_c significantly above liquid helium. However, barrier deposition on niobium base electrodes was still problematic. Although niobium readily oxidizes, it can form various oxides and in practice was difficult to control. It was left to Gurvitch *et al.* [7] at Bell Laboratories to put the final piece in the puzzle. Rather than rely on niobium's native oxide, a thin layer of aluminum was deposited on the niobium and then oxidized. The aluminum is deposited thin enough (~5 to 10 nm) to be superconducting through the proximity effect. Aluminum has the attractive properties of both wetting (completely covering) niobium well and having a self-limiting oxide making controllable barrier fabrication possible; Nb/Al/AlO_x/Nb junctions continue to be the *de facto* standard for Nb IC processing.

In circuit fabrication in general there is often a competition between the ease of fabrication and capability of circuit operation, and superconducting circuit fabrication is no exception. One distinctive burden to insertion of superconducting electronics is the required cryogenic cooling. After the establishment of the Gurvitch process in Nb, the superconductor electronics community continued to search for materials that could be used for large-scale integration (LSI) superconducting circuits that operate at higher operating temperatures. As will be seen, materials have been developed that can reduce this burden of cooling but at the cost of significant complexity in manufacturability.

6.1.2 Nb, NbN and YBCO Material Comparison

Specifically, we will examine three processes: niobium as the most mature process; NbN with its nearly 2× improvement in T_c; and YBCO with its 10 × improvement in T_c. As these are the three material systems that are primarily being pursued for integrated-circuit fabrication, we will focus our consideration on these technologies and their issues. The goal is to examine these processes in terms of what they can currently achieve and to consider the materials-specific issues that are limiting them.

These three-materials systems provide an interesting progression of properties which in turn impacts circuit fabrication. From niobium to niobium nitride to YBCO is a progression from a metal to a semi-metal, to a ceramic. The transition temperatures also vary dramatically from 9.2 K for niobium, to 16 K for niobium nitride, and to 90 K for typical YBCO.

The coherence length, a characteristic length scale associated with the superconductor, varies in these materials from 30 nm in Nb to 5 nm in NbN. In YBCO, which is anisotropic, the coherence length is 2 nm in the *a-b* plane and 0.2 nm along the c-axis. Since junction characteristics are greatly influenced by the properties of the superconductor within a coherence length of the barrier, for junction fabrication it is critical that the film properties within a coherence length of the barrier be as optimal as possible.

Ideally, one would want a perfectly ordered superconductor within the base electrode up to the very surface of the barrier and perfectly recommencing at the opposite surface of the barrier. In reality, when films are grown, there is disorder at the interface due to imperfect lattice matching or reactive chemistry. As the film is grown there is a characteristic thickness where the bulk properties of the crystal are re-established. Materials with longer coherence lengths allow for a wider disordered region next to the barrier without degrading the junction characteristics. Considering the coherence lengths of our three superconductors and the distances from the barrier necessary for reasonably good superconducting properties to be regained in a given superconductor [8], one can gain some insight into the relative difficulty of making controllable Josephson junctions in these three material systems (Fig. 6.1). Note that for the case of niobium, greater than

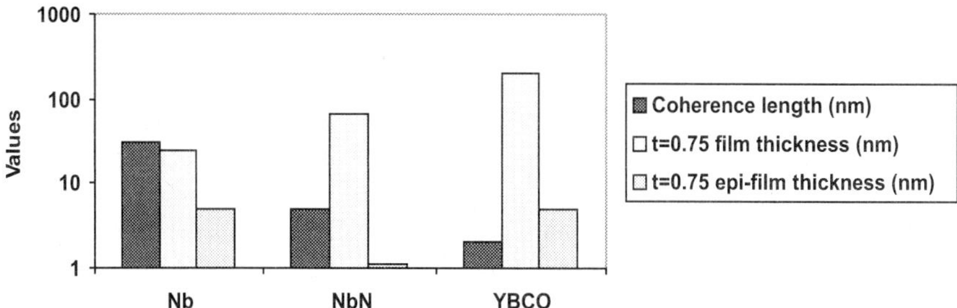

FIG. 6.1. Comparison of parameters of the three material systems used for superconducting circuit processes. t = 0.75 film thickness is the minumum thickness at which the film T_c reaches 75% of its bulk value.

75% of the film critical temperature has been regained in a distance less than the coherence length in that material (Fig. 6.1). This distance is reduced by epitaxial deposition of niobium. For NbN, the film thickness necessary to regain 75% of the bulk T_c is dramatically affected by film epitaxy as is the case for YBCO. For epitaxial film growth the distance necessary to regain the maximum critical temperature in niobium is 5 nm and for niobium nitride it is less than 1 nm; for YBCO it is 5 nm. By comparing the distance necessary to regain good superconducting properties to the distance that the junction parameters are affected by disorder, one can see the necessary strategy for fabrication. For niobium junctions, nonepitaxial niobium deposition on the aluminum oxide barrier has been as successful as one would expect from a comparison of the relatively large coherence length with respect to the thickness necessary to regain reasonable T_c (Fig. 6.1).

However, for NbN fabricating oxidized aluminum barriers produces junctions with degraded junction parameters such as reduced gap voltage. Replacing this barrier with the refractory insulator MgO dramatically improves the quality of the NbN junction due to superior lattice matching. MgO, which also has a B1 structure, has a lattice mismatch of only 4% with NbN. Talvacchio *et al.* [9] have observed using reflection high-energy electron diffraction (RHEED) that rf-sputtered MgO barriers grow initially epitaxially on NbN, although diffuse diffraction spots in low-energy electron diffraction (LEED) analysis suggested that these films were rough. This is consistent with LeDuc *et al.* [10] who proposed that the initial layer of MgO film grows in a planar epitaxial fashion. Subsequent MgO film growth proceeds by nucleation resulting in islands that account for the film roughness. This Stanski-Krastanov [11] epitaxial growth mode occurs when the initial monolayer of deposited film has a lower surface energy than the material upon which it is being deposited. This lowering of surface energy can dominate strain energy resulting from lattice mismatch. Subsequent deposition of a monolayer does not reduce the surface energy further and island growth proceeds because it minimizes strain energy. In contrast to these results, Shoji *et al.* [12] reported that MgO was observed to form amorphously on NbN. However, the difference in these results may be due to a lower substrate temperature during deposition. Transmission electron microscopy (TEM) analyses [13, 14] of another well-matched barrier material (AlN) also suggests that epitaxial growth can be continued through the dielectric barrier.

Finally, it is apparent from Fig. 6.1 that even with strict epitaxy of the YBCO junction electrodes through the barrier, junction parameters may still be degraded. This is due to the film thickness necessary to regain near-bulk values of T_c being significantly larger than the coherence length in this material system.

6.1.3 Nb, NbN and YBCO Fabrication Processes

The choice of material for superconducting IC fabrication is driven primarily by the temperature requirements of a given application. In general, superconducting circuits benefit from as low a temperature of operation as possible. Thermal noise is reduced, penetration depths decrease, and a parameter such as junction critical current becomes less sensitive to thermal variation. This must be balanced by the difficulty and inefficiency of fighting this entropy-reducing battle.

Due to the success and universal acceptance of the Gurvitch-style Nb junction, what distinguishes the various foundry and foundry-like efforts in niobium processing is typically other aspects of the circuit process. Although NbN circuit fabrication is not as common worldwide, there are ongoing efforts directed at developing aspects of NbN junction fabrication. The YBCO circuit processing represents the other extreme, where the junction process is very much in flux and there are nearly as many junction processes as there are groups researching their fabrication (Table 6.1).

6.2 Nb Circuit Process

6.2.1 Introduction

Much of the success of superconducting circuitry in recent years has occurred using niobium circuitry [41, 42] resulting in a number of material-specific advantages. Niobium has the desirable properties of a relatively long coherence length and relatively short penetration depth compared to most other type-II superconductors. This permits a control of Josephson junction properties impor-

TABLE 6.1
VARIOUS SUPERCONDUCTING CIRCUIT PROCESSES UTILIZING Nb, NbN AND YBCO

Circuit process	Institution	Ref.	Dielectric	Resistor	Ground-plane	Comments
Niobium process	ETL	[15, 16]	Nb_2O_5, SiO, MgO	Pd	bottom	
	Hitachi	[17]	SiO	MoNx	bottom	
	Hypres	[18, 19]		Mo	bottom	
	NEC	[20, 21]		Mo	bottom	4.74 μm Nb jj ($1-\sigma = 0.54\%$) 0.74 μm Nb jj ($1-\sigma = 4.8\%$)
	NIST	[22, 23]		Pd-Au	bottom	Developing Pd-Au barrier SNS junctions
	Northrop-Grumman	[24]	Sputtered SiO_2	Mo/Ti	top	
	PTB	[25, 26]		Pd	bottom	
	TRW	[27]	Bias-sputtered SiO_2	Mo, NbNx	bottom	
	UC Berkeley	[28]	ECR PECVD SiO_2	Mo	top	$1-\sigma < 0.2\%$ Nb jj I_c for chip $1-\sigma < 0.7\%$ Nb jj I_c for 4″ wafer
	Univ. Karlsruhe	[29]		Pd/Al	bottom	
Advanced niobium process	MIT Lincoln Labs	[30]	PECVD SiO_2	Ti/Pt		DPARTS, *i*-line photolithography, 150-mm wafers
	SUNY, Stony Brook	[31]	SiO	Au, Pd-Au		PARTS, electron beam lithography
	NEC	[32]	Bias-sputtered SiO_2	Mo		ECR-etching, *i*-line photolithography
Niobium nitride process	CEA-Bremable	[33]	SiO_2, Si_3N_4			MgO buffer
	Hypres	[18]				NbN/Mg/NbN (SNS)
	Kansai	[34]		Au		NbN/AlN/NbN
	TRW	[35]		Mo, NbNx	bottom	NbN/MgO/NbN
YBCO process	Conductus	[36]	SAN		bottom	IEJ, Co-YBCO edge, bi-crystal
	DERA	[37]	PBCO			CAM jj
	MagnaSensors	[38]	$SrTiO_3$		bottom	Metallic SNS jj
	Northrop-Grumman	[39]	$SrTiO_3$, SAT		top or bottom	Co-YBCO edge jj
	TRW	[40]	$SrTiO_3$, $SrTiO_3$/SAT/$SrTiO_3$	Mo/Ag	bottom	Co-YBCO, Ga-PBCO edge jj

tant to circuit design such as critical current I_c, and junction quality V_m. Even with the more forgiving properties of niobium, the critical current is still highly dependent on the insulating barrier material and the process used to fabricate it. In addition, as a refractory material Nb is quite hard and generally adheres well as a film. Moreover, Nb is typically grown as a polycrystalline film and is not particularly dependent on the crystallinity of the substrate.

In many respects, niobium-based circuit processing stands on the shoulders of previous work done in the semiconducting industry. Techniques such as reactive ion etching and photolithographic processes that had been developed for silicon-based processing have direct application in superconducting circuit processing. Even the wafers that are used for niobium technology are the same as those found in semiconductor foundries, the high quality of which is due to the tremendous demand in the semiconductor industry.

6.2.2 Process Flows

Fabrication process flows are typically a meld between application requirements and the practical reality of being able to deposit and define metals and dielectrics to realize those requirements. Fortunately, a given process flow that may have a groundplane, a junction layer, a resistor layer, and a couple layers of superconducting interconnects, is flexible enough to fulfill many of the requirements imposed by various superconductor applications. The particulars of how a process flow is accomplished has been the center of quite a bit of inventiveness by researchers throughout the world. However, there is an advantage for those designing superconducting circuitry in standardizing the fabrication process.

Sharing design modules has emerged as one of the strategies for sharing a process flow and process parameters in a given technology. This has begun for fabricating single-flux quantum (SFQ) circuits [43] but is quite process-flow specific (Fig. 6.2). However, the advantages to the designer may be so significant that careful consideration should be given to developing advanced process flows with this in mind.

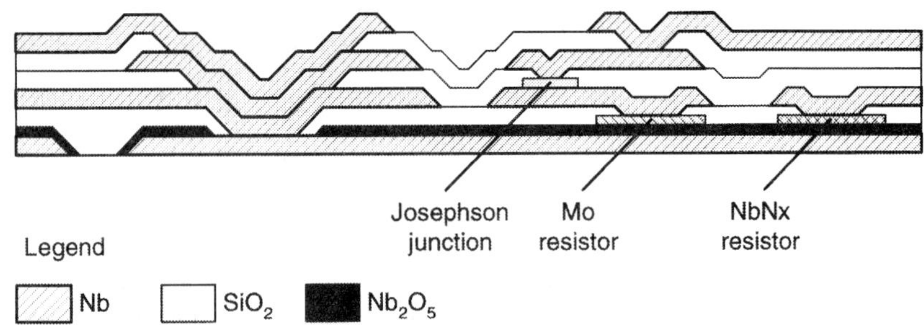

FIG. 6.2. Cross section of niobium process flow from TRW standard Nb-based junction process from Reference [44].

One process flow, SNAP (selective niobium anodization process), is an example of how a given process flow can be realized with a minimum of processing complexities while allowing as much robustness in the process as possible. The SNAP takes advantage of niobium's capability of forming an anodic oxide by simply exposing a photolithographically defined sample to an electrolytic solution of ammonia pentaborate, ethylene glycol, and de-ionized water (31 : 224 : 152) with a dc voltage applied to the niobium layer. This approach works well to form a dielectric with high dielectric integrity. However, it is difficult to extend this process to smaller junction sizes due to the isotropic nature of the wet processing along with the stress induced by the formation of the anodic oxide.

A myriad of processing techniques have been invented that address aspects of circuit processing such as ease of fabrication, planarization, reducing junction size and relieving film stress as indicated in Table 6.2. These techniques vary also in their degree of planarization. For example, chemical or mechanical polishing achieves a high degree of planarization resulting in a nearly planar surface for subsequent depositions.

The capabilities of a superconductor IC fabrication process go well beyond the performance of junctions or any other single circuit element. A given process can be subdivided into numerous processing elements such as junctions, resistors, interconnects, vias, and the necessary dielectric to isolate these structures. Each one of these elements may consist of dozens of submodule steps, all of which must be performed adequately in order to ensure successful circuit fabrication. Process details concerning dielectric integrity, contact resistance, critical current of interconnects, bridging between layers, opening of lines, insufficient clearing of dielectric, overetching of critical layers, resistor targeting, proper sizing of photoresist structures and the subsequent etching or lift-off, sufficiency of layer-to-layer registration can and does occupy much of the efforts of those maintaining a superconductor IC foundry. As any of these issues can negate the considerable time, effort, and expense that goes into a circuit fabrication, much attention is paid to increasing the yield of each step in the circuit-fabrication process.

In order to track the numerous subprocesses involved in a single wafer fabrication, techniques such as statistical process control (SPC) are utilized. A given process element can be evaluated over time to determine if a process is varying due to random variation or whether a systematic problem has arisen. A series of statistically derived rules (Western Electric rules) readily formalized this procedure. In practice, tracking metrics that are readily measurable and pertain to critical aspects of the process are most desirable. Film thicknesses and film resistivities, parameters critical to successful circuit operation, can be tracked in-process.

Film inductance is an example of a parameter vital to successful circuit operation which is difficult to measure until after a wafer has completed

TABLE 6.2
PROCESS TECHNIQUES APPLIED TO NIOBIUM-BASED CIRCUIT FABRICATION

Reference	Processing technique	Degree of planarization	Description
[45]	CLIP (Cross-Line Patterning)	Low	Provides good scaling to small junction dimensions. Cumbersome geometry for design.
[32]	MPP (mechanical-polishing planarization)	High	Neutral slurry to reduce SiO_2 etch rate for controllability. $0.24\,\mu m \times 0.24\,\mu m$ junctions demonstrated.
[31, 46]	PARTS (planarized all-refractory technology)	High	Uses chemical-mechanical polishing. $0.08\,\mu m \times 0.08\,\mu m$ junctions demonstrated.
[47]	RHEA (Resist-Hardened Etch and Anodization)	Medium	A modified SNAP that allows a greater degree of planarization.
[48]	SAWW (Self-Aligning Whole Wafer)	Low	Addresses stress-induced degradation of small junctions.
[49]	SCAN (Self-aligned Contact with Anodized Niobium)	Medium	Uses anodization to define junction and allow outside contacts in addition to dielectric other than Nb_2O_5.
[50]	SNAP (Selective Niobium Anodization Process)	Medium	Uses anodization to define junction and allow outside contacts. Considerable anodization makes small-area junction definition problematic.
[51]	SNEP (Selective Niobium Etch Process)	Low	Further developed by Yuda *et al.* [52] to include lift-off as a stress-relief mechanism.
[53]	SNIP (Self-aligned Niobium Isolation Process)	Medium	Deposits SiO as the self-aligned dielectric after the counterelectrode etch.

fabrication. As such, correlating inductance measurements with film thicknesses and resistivities allows one to respond more quickly to process drift that could result in ill-targeted inductance values.

The insensitivity of process characteristics to inevitable variations that occur in wafer fabrication often determines the success of circuit fabrication. An increasing number of groups [54, 55, 56] are employing statistical techniques of process development that have been more commonly applied in the semiconducting industry in order to develop increased robustness to variation.

6.2.3 PROCESS ELEMENTS

Critical to fabrication of a niobium circuit is the quality of the film deposition. Details of superconducting film deposition have been reviewed in Chapter 1. In this section, other elements necessary to put the superconducting films into useful circuitry will be discussed.

6.2.3.1 Junctions

As discussed earlier, the Gurvitch-style Nb junction has been key to enabling medium-scale integration (MSI) and then LSI-scale superconductor IC fabrication. These Nb junction techniques are still being developed. Vital to circuit operation are high-quality junctions with low leakage current and small junction-to-junction critical current variability. In addition, reduction of junction size and capacitance while maintaining parameter reproducibility and uniformity across ever larger wafer areas is an ongoing need in order to meet the next generation of IC applications.

As is often the case in fabrication, a deposition system-specific characterization must be made to assess the system geometry effect on the particulars of the film process. With that proviso, it is still illustrative to consider the practical aspects associated with a given Nb junction fabrication and understand how a basic process is developed.

1) A High-quality Nb film deposition process must first be developed (as outlined in Chapter 1); depositing nearly stress-free films is vital to high-quality device fabrication. This is especially true for fabrication of small-area junctions ($<2.5\,\mu m^2$) as reported by Nakagawa *et al.* [57]. Once patterned, stress relaxation can result in damage to the junction edges and in increased junction leakage current. The effect of this stress can be reduced, but not eliminated, by choice of a junction-patterning technique (Table 6.2).
2) During junction base-electrode deposition, wafers should be kept sufficiently cool during deposition so as to control grain growth and to keep the film surface reasonably smooth. Due to the excellent wetting and oxidation of the overlying aluminum layer this roughness is not as critical as it is in the case of NbN or YBCO processing. The resulting aluminum oxide barrier is typically smoother than the Nb base-electrode surface.
3) Al is deposited after the base-electrode deposition. Other materials have been explored for this purpose such as Yb [58] and Ta [59], however, Al remains the most common material of choice.
4) Al thicknesses greater than 4 to 8 nm can reduce gap voltage, but thicknesses much less than 4 to 8 nm can result in catastrophic loss of

yield. Increasing the deposition rate of the Al is one method to improve the crystallinity of the Al film and reduce junction leakage.

5) In post-Al deposition, the sample is oxidized preferably in a separate chamber that is kept under vacuum in order to avoid a partial pressure of water that can cause wafer-to-wafer reproducibility issues. For targeting purposes, it is critical that the pressure, time of oxidation, and substrate temperature be well controlled. A strength of this junction process is that by simply controlling these parameters, junction critical current densities from less than $1000\,A/cm^2$ to greater than $400{,}000\,A/cm^2$ can be targeted.
6) These barrier fabrication steps can be repeated *in situ* to obtain ultralow critical-current densities ($J_c = 5$ to $200\,A/cm^2$) [60].
7) The niobium counterelectrode is then deposited making sure that the oxygen from the thermal oxidation chamber is sufficiently evacuated. The counterelectrode thickness should be enough to allow some measure of overetch of the via through the dielectric and into the counterelectrode without endangering the few tens of nanometers near the barrier.

High-critical current densities are desirable for many applications. This is due to the characteristic frequency that can limit device operation following the relation:

$$w_j = \{(4\pi e/h)\beta_c J_c(A/C)\}^{1/2} \tag{6.1}$$

where β_c is the McCumber parameter; A is the junction area; and C is the junction capacitance. This limiting frequency can be extended by increasing the junction critical-current density. In order to obtain device-operating frequencies in the hundreds of gigahertz to THz, critical-current densities in the hundreds of kiloamps/centimeters square are needed. For the $Nb/Al/AlO_x/Nb$ system, barrier thicknesses on the order of 1 nm are needed. This equates to $\sim$3 monolayers of the AlO_x barrier, thus making it susceptible to film roughness and microshorts. These microshorts can consist of unit-cell-sized metallic defects in the oxide structure. A single monolayer reduction equates to greater than one hundred times the conductance and is thought to dominate the conductance of high-J_c junctions [61]. In practice, junction critical-current densities of greater than $400\,kA/cm^2$ have been demonstrated by simply limiting the oxygen exposure of the aluminum barrier without severe degradation of the junction I-V characteristics.

An alternate barrier material, aluminum nitride, has been explored as a candidate barrier for controllable high-J_c junctions [62]. Unlike oxidation, the nitridation of aluminum proceeds much more slowly and can be accomplished by exposure to a nitrogen plasma. Junctions fabricated in this fashion have reduced subgap leakage and hold the promise of greater control of junction critical-current density targeting.

6.2.3.2 Dielectric Depositions

Dielectric films are needed in superconducting circuitry for a number of purposes and these films impact to varying degrees capacitance, inductance, isolation, and device-to-device crosstalk. On a microscopic scale, most dielectric films used in IC processes are amorphous and are generally insensitive to considerable variation in stoichiometry. Polycrystalline films are usually avoided as they tend to be leak more [63].

Early superconducting circuitry used thermally evaporated SiO dielectric layers. De-adhesion and pinhole shorts made this dielectric problematic. Although Nb junctions are limited to processing temperatures less than 200 °C due to concerns of altering junction parameters, there have been numerous dielectrics developed by the semiconducting industry that are applicable to Nb circuit processing. A measure of how well a dielectric film serves as an insulator is the calculated dielectric defect density D as defined by the Poisson distribution:

$$Y = \exp(-DA) \tag{6.2}$$

This can be determined by measuring the fractional yield Y of nondefective planar capacitors of area A. Low-temperature (<150 °C) PECVD Si_3N_4, Nb_2O_5, ECR PECVD SiO_2 [64] and bias-sputtered SiO_2 (Fig. 6.3a) are all methods that can provide dielectric integrity of less than 1 defect/cm^2 and a degree of planarization to improve subsequent coverage of superconducting interconnects.

Single crystalline films often require too high a temperature to be readily fabricated and are critically dependent on stoichiometry, and consequently, they are not common. The exception to this is in high-temperature superconductors due to the requirement of continued epitaxial film growth through the dielectric layer in a multilayer process.

For Nb and NbN IC processing the following list of specifications should be considered in choosing a useful dielectric material:

1) complete wetting of the dielectric on the material that is to be insulated. (Note that unlike semiconductor circuitry a single "pinhole" can represent a zero-resistance superconducting short between layers.);
2) good adhesion to the adjacent films above and below;
3) low film-stress (which can affect adhesion, loss tangent);
4) stable chemistry. If an oxide, it should not have a greater affinity towards oxygen than niobium oxide or long-term diffusion will be an issue;
5) it should be insensitive to moisture in order to allow standard aqueous processing techniques;
6) amorphous; and
7) leaves a consistent and smooth surface as a template for subsequent depositions.

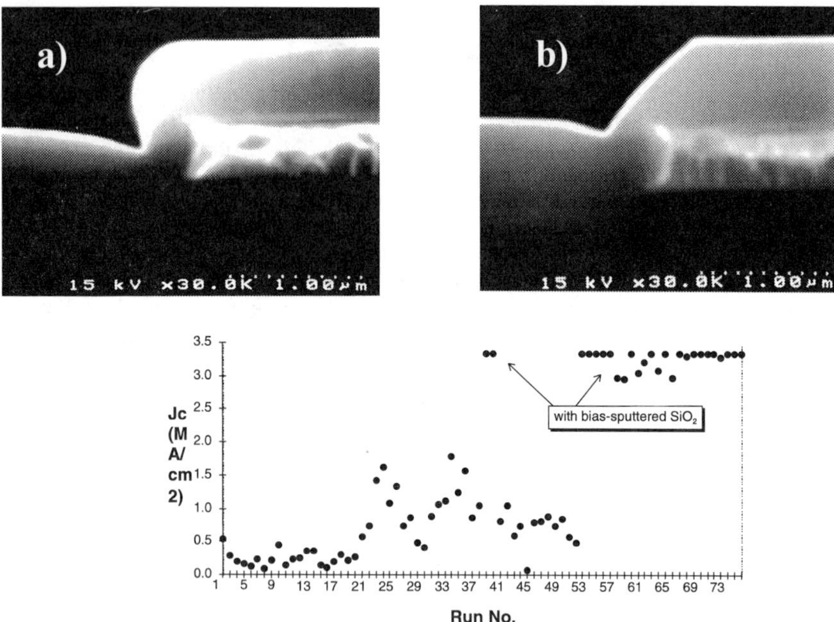

FIG. 6.3. Bias-sputtered SiO_2 demonstrating partial planarization (a) along with a trend chart indicating the increase in critical current of the interconnect crossovers (b) from Reference [35].

This last point is especially significant in maintaining consistent properties of layers that are deposited on the dielectric. For example, thin-film resistors can vary in resistivity as a function of the roughness of the underlying film, thus it is critical in developing a process to have a consistent surface from wafer to wafer.

6.2.3.3 Resistors

Thin-film resistors predate thin-film junction technology by over 40 years [65]. The maturity and relative simplicity of resistor fabrication belies its critical role in superconducting circuit fabrication. It is often the case in digital circuit fabrication that targeting and variability in resistor values is of equal significance to junction critical-current targeting and variability. Resistor materials currently in use with superconducting circuit processes include Mo, nitrogen-deficient NbN, MoN_x, Mo/Ti, Pd, Pd-Au, Pd/Al, Ti/Pt, Au, and Pd-Au.

Resistors can occupy over half the "real estate" in many SFQ circuits, and as such, directly impact circuit density and performance. This circuit element has its own parasitic inductance and capacitance associated with its geometry that must be considered. In addition, simply by the area it occupies, it lengthens the interconnects necessary to connect one gate to another, thus increasing circuit latency. Directly reducing the resistor dimensions becomes impractical at some point as vias of finite size are needed to connect it. In addition, as the resistor size decreases, the variability of the resistor values increases due to fluctuations of the critical dimension. In other words, a 0.1-μm variation in linewidth means little in a 10-μm-wide resistor (1%) whereas this variability in a 1-μm width creates an intolerable 10% resistance variation.

Resistors are generally used for two purposes in digital circuitry: as shunting resistors in parallel with junctions, and as bias resistors for the larger circuit. Shunt resistors are needed in Nb and NbN circuitry to sufficiently damp the otherwise hysteretic behavior of the junction I-V characteristics. Generally, the shunt-resistor value is on the order of the resistance of the junction in the voltage state. The dimension of the resistor should be such that its value will be well-targeted in ~2 squares of resistor. For smaller dimensions, spreading resistance at the contacts becomes troublesome; for larger dimensions, the resistor incurs excessive inductance and capacitance and occupies too great a circuit area. The other dimension, that is, the thickness of the resistor film, should be sufficiently thick as to be reasonably insensitive to small variations in surface roughness, but not so thick as to add topological concerns of step-coverage in order to contact to it. Typical resistor thicknesses range from 50 to 300 nm for these reasons.

It may not be desirable for the aforementioned reasons to form all resistors out of the same material. Often, digital circuits require resistances on the order of a couple of ohms to shunt junctions and bias resistors on the order of tens of ohms. The same resistor to satisfy both these disparate requirements would be awkward from a design perspective.

Alternate approaches currently are being explored [66] to entirely integrate the resistor with the Josephson junction. One such approach is the development of superconductor/ normal-metal/superconductor, or SNS junctions, and alternately superconductor/insulator/normal-metal/insulator/superconductor, or SINIS junctions [67]. These overdamped junctions could be directly insertable into an SFQ circuit process, and would eliminate the need for a separate resistor to shunt the otherwise hysteretic junctions. This approach is attractive in that it would simplify the overall number of process layers, reduce the circuit dimensions and reduce the inductance associated with the shunt resistor. However, in order to be useful as an RSFQ logic element, normal state resistances of greater than 1 Ω are desired in order to drive microstrip-line interconnects. In addition, the product of the critical current and the junction resistance, I_cR_m, must be sufficiently high to overcome thermally induced errors. Development of these devices is focused on in-creasing both the junction critical-current density and junction normal-state resistances to levels necessary for future applications. Theoretical treatments suggest that con-siderable progress can be made in increasing both these parameters from their current values [68].

The prospect of incorporating a separate process element (resistor) into the active device (Josephson junction) is a challenging one. This involves a fundamental change to a device whose properties (I_c uniformity, I_cR_n, V_g, V_m) are critical to circuit margin. It should be noted that in combining these two process elements, the designer loses the flexibility of independent targeting of the values of these process elements. For these reasons, it is worth considering whether the same objectives could be accomplished without modifying the active device or losing the option of individually targeting the different process elements. Further miniaturization of the resistor element would support this. However, further miniaturization of the resistor in its planar geometry is limited by the area required to form contact vias. A long-standing approach [69] to this issue that has yet to be fully explored is the use of vertical resistors. This, similar to the SNS-junction approach, would greatly reduce the inductance of the shunt resistor. Depending on the aggressive-ness of its implementation, it could conceivably reduce the circuit dimensions nearly as much as the SNS junction approach. For example, a vertical resistor consisting of a concentric ring connecting the counterelectrode of a circular junction directly to the base electrode is one approach that would be nearly indistinguishable from an SNS junction in circuit area, while maintaining the capability of independent targeting of the separate process elements.

As junction sizes decrease and critical-current densities increase, neither of these process modifications may be needed [70]. At very high critical-current densities, standard SIS junctions become overdamped exhibiting nonhysteretic I-V characteristics and require no further shunting by external or internal resistance. These devices are quantified as having a McCumber [71]-Stewart [72] damping parameter β_c less than one, where

$$\beta_c = \frac{2\pi I_c R_d^2 C}{\Phi_o^2} \tag{6.3}$$

Here, I_c is the junction critical current, R_d is the shunt damping resistance, C is the total shunt capacitance and Φ_o is the flux quantum. For SFQ applications, there is a need for overdamped, nonhysteretic junctions. For a given critical current, as junction size decreases, capacitance will decrease proportionally. In addition, subgap transport mechanisms occur that decrease the shunt resistance to the point that β_c reduces to less than one and no further resistive shunting is needed. The reason that this is not immediately and universally implemented is that this occurs at junction areas of 0.1 to 0.2 μm^2, which in turn require critical-current densities on the order of 100 kA/cm^2, both of which stringently challenge current processing capabilities. Attaining these deep submicron dimensions and critical-current densities are just beginning to be demonstrated using planarization techniques [70].

6.2.3.4 Photolithography

Photolithography is a technique common in integrated circuit fabrication that allows precise patterning of thin films into devices. It is based on the application of light-sensitive polymer film (photoresist) that is spun onto a wafer as a thick film, typically 1 to 3 μm thick. Selective areas are then exposed to light through a glass plate with patterned chrome. The exposed photoresist is then developed out similarly to common photographic processes. The remaining polymer is hardened and protects the underlying film during the subsequent etch process. Once the etch is completed the photoresist is removed by dissolution in an acetone bath or commercially available photoresist stripper. Fortunately for superconducting circuit fabrication, the basic photolithography process has been greatly developed by the semiconductor industry and many of the same techniques can be directly applied.

Aspects of this process significant to circuit fabrication are: 1) the minimum feature size resolvable (resolution); 2) the fidelity of the etched feature with respect to the initial design (called the critical dimension, or CD); and 3) the CD variation across the area of the circuit of interest.

These aspects stress circuit yield in varying ways. Minimum feature resolution helps define the minimum size of the active device. In niobium circuitry, the junction size can limit the ultimate speed of operation of a given circuit. The larger the junction, the greater the capacitance and the longer the associated RC charging time constant. Junction size is considerably larger than the minimum feature resolution in processes that do not use self-alignment or planarization. In order to contact the top electrode of the junction, a via must be made through the dielectric overlaying the junction. If any portion of the via overlaps the edge of the junction top electrode, the subsequent interconnect layer will short the top electrode to the bottom electrode. It is vital that the via be correctly aligned to the top of the electrode. Thus the minimum junction size is a sum of the minimum feature resolution of the via and the allowed layer-to-layer registration error.

The layer-to-layer alignment is a critical aspect of circuit fabrication; as seen from the foregoing, it can limit minimum junction size. It can also increase unwanted parasitic inductances and parasitic capacitances in order to ensure interconnects being connected to various process elements. Fortunately, great strides have been taken in this aspect of processing by the introduction of automated alignment equipment. These systems optically recognize designed structures, or optical alignment targets (OAT), and mechanically adjust the mask positioning until alignment is achieved. With the use of this technology, layer-to-layer registration error can be readily reduced to a 3-sigma of less than 0.4 μm. Next-generation photolithography equipment using smaller-wavelength x-rays is purported to allow less than 0.1 μm resolution and a 3-sigma of 0.0054 μm in layer-to-layer alignment [73]. These remarkable advances in photolithographic technique will only be useful if the capability exists to controllably etch and reproduce those features into the film, as will be discussed in the next section.

Due to variation in the photolithography process the critical dimension may vary. This could be due to exposure and development of the photoresist or etching of the structure in question and much effort goes into precisely controlling this aspect of the process. It is not obligatory that the critical dimension is precisely the same dimension as the device size on the mask, but that it is simply consistent from process run to process run. It is quite common that an offset, or sizing, is added to the original design in order to account for the shrinking or growth of the structure size during the processing. Clearly, this offset can limit the ultimate dimension for extremely small structures. The control of the critical dimension impacts circuit yield in a fashion similar to parameter value targeting.

In practice, the variability of the critical dimension from wafer-to-wafer and within wafer is tantamount to a direct variation in parameter values. It becomes more severe for small-dimension structures. The impacts of CD variation, especially with respect to junction critical current and resistor values, can significantly reduce the margin of circuit operation or entirely prevent operation altogether.

6.2.3.5 Etching

The most common approach to defining the structures in a given layer of niobium-based superconductor IC processes is by reactive ion etching (RIE). Reactive ion etchers accelerate reactive ions formed in a plasma in a parallel plate configuration. The reactive ions have relatively long mean free paths due to reduced operating pressures and are directionalized due to high voltages created by RF biasing of the substrate electrode. This allows RIE to provide anisotropic etching needed for small-feature definition. Thus, CF_4, or $CF_4 + O_2$, will etch anisotropically only for relatively low pressures and high RF power; SF_6 and Cl-based compounds are other reactive gases that are employed especially for fine-line resolution. The plasma created in the RIE dissociates the gas into ions and F radicals which primarily chemically, but also physically, attack the Nb. Ideally, an etch process should selectively etch only the intended film without etching any of the underlying film. Through a combination of chemistry, partial pressures of the gases, and input power used to create the etching plasma, a wide variety of relative etch rates, or selectivity, between various materials can be achieved. Due to across-wafer variation in both the film-deposition technique and the etch-rate profile, regions of the underlying layer will be etched to a greater degree than others. The higher the selectivity between the two materials, the smaller the magnitude of the undesired overetch. An alternate approach to ensure high selectivity is to intentionally insert layers that act as etch stops which are then entirely removed at a later step.

Because perfect selectivity is not attainable and limiting the overetch into the underlying film is desirable, methods of determining when a film is completely etched are employed. Numerous techniques are used that are compatible with reactive ion etching. One is a method is to monitor the spectrum of the light emitted in the plasma by the etched species. As each material emits light in the plasma of characteristic wavelengths, the completion of the etch can be determined by both the reduction of the amplitude of the signal from the film being etched and from the onset of a signal from the underlying film. This schema is successful provided there is enough area on the wafer that is being etched. This is not always the case, especially for dielectric vias. It is not uncommon to purposefully include large fields of film to be etched for no other reason than to increase the endpoint signal amplitude.

The edge profile can be controlled by the addition of oxygen to the RIE plasma. The oxygen serves to recede the photoresist pattern during the etch process, creating a progressively larger etched area and resulting in a sloped edge. This edge is desirable in allowing ordered growth of subsequent layers crossing these edges. Without planarization, the standard approach towards fabricating robust interconnect crossovers and vias is by "sculpting" the underlying layers through lithographic and etch techniques combined with conformal dielectric deposition. This approach is increasingly critical for NbN, a semimetal, compared to metallic Nb presumably due to the ductility on the microscopic level of these films. As will be discussed later, it becomes vital in ceramic YBCO, which has the added complexity of requiring film epitaxy.

In competition with this is the need to create sharp edges that allow minimum-feature resolution. Superconducting circuit processes balance these needs on a layer-to-layer basis. For example, in order to attain minimum-feature resolution, only enough oxygen to prevent polymerization of the photoresist during etch should be used. In contrast, groundplane etches that do not require small pitch will be intentionally sloped in order to facilitate subsequent crossovers of that edge.

Another aspect of etching to be considered relates to the energy of the etchant and its effect on the resulting film edge. In the RIE mode, anisotropic etching is improved by decreasing gas pressure. The ion energy of the etchant increases with decreasing gas pressure and increasing mean free path. This causes the photoresist mask to rapidly deteriorate and the elevated ion energies can have adverse effects on the film morphology of the etched edge.

One general difficulty with RIE methods is that greater anisotropy is accomplished with higher voltages, which means greater potential for both film damage and radiation damage to the photoresist. This hardened photoresist can be difficult to remove even with commercial photoresist strippers and can leave detrimental residue on the wafer surface. An alternate method of ion generation using electron cyclotron resonance (ECR) has been used to reactively etch Nb and

NbN circuitry [74]. The ECR plasma etching is performed at lower gas pressures than RIE. Since the ion energy is controlled by an external bias voltage and pressure, anisotropic etching is achieved with low ion-energy etchants. In this way film damage, especially on critical edges, can be minimized.

An alternate approach to etching is lift-off. This is accomplished by first applying photoresist and patterning it preferably with an inverted profile. Metal is then deposited through the mask in the desired locations; the metal deposition on the resist will be lifted off during the acetone dissolution of the photoresist. This approach is useful as it has the advantage of not imposing any overetch but is limited to lower temperature depositions, thinner films, and reduced resolution.

6.2.3.6 Planarization

Although not a required process element for most current applications, planarization in some form is critical to advanced processing. The motivation to planarize becomes increasingly evident from a processing perspective as aspect ratios (the height of superconducting devices divided by their planar dimension) approach one [75]. In order to obtain both minimum-feature resolution and sufficient planarity to allow interconnects to cross over the edges, advanced processes utilize planarization techniques for at least a subset of their process flow (Table 6.2). Because such planarization techniques are generally labor-intensive, other more facile methods that allow various degrees of planarization are employed for specific process steps. One such method, commonplace in semiconducting processing, is RF-bias sputtering of dielectrics. This has been extended to the use of a low-frequency (40 kHz) substrate bias eliminating the need for the conventional RF generator and substrate tuning network [76]. Under proper conditions bias-sputtered SiO_2 can cover and provide a smoother surface than underlying NbN films. This can enable the fabrication of tunnel junctions on dielectrics over a groundplane, critical to reducing circuit inductance. In addition, the use of this technique facilitates higher yield of interconnect crossovers with respect to both dielectric integrity and crossover critical currents (Fig. 6.3).

6.2.4 APPLICATION EXAMPLE: DIGITAL PROCESSING

It is didactic to consider where the future lies for fabrication. One application that superconducting electronics can uniquely fulfill is being proposed in the Hybrid Technology Multi-Threaded (HTMT) petaflop scale computer [77]. This concept as envisioned would ultimately require junctions with critical current density of $20\,\mathrm{kA/cm^2}$ and submicron junctions, but not necessarily deep submicron (0.8 μm feature size). A total junction count of 100 billion is projected, distributed over 40,000 chips interconnected among 512 MCM that would function with 100-GHz on-chip clock speed to achieve the desired computational density.

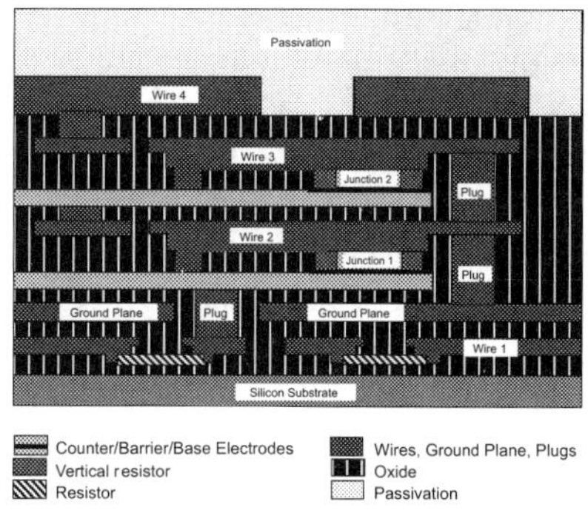

Year	1998	2001	2004	2007	SIA 1992
Minimum Feature size (μm)	1.5	1.0	0.65	0.65	0.50
Junction size (μm)	2.5	1.75	1.25	0.80	-
Junction current density	2K	4K	8K	20K	-
Gates/chip	5K	35K	100K	300K	300K
Chip size (mm^2)	100	196	400	400	250
Wafer diameter (mm)	100	150	150	200	200
Defect density (defects /cm^2)	< 2	< 0.7	< 0.2	< 0.1	< 0.1
No. of interconnects levels	3	4	5	6	3
Planarization	no	yes	yes	yes	yes
No. of I/Os	128	750	2000	2000	500
Wafer starts per month	12	60	120	240	20K
Capital Investment	17M	8M	1M	?	
Annual Operating Cost	3M	4M	5M	?	

FIG. 6.4. Ultimate process flow for HTMT along with road map from Reference [78].

The fabrication aspects of these junction sizes and current densities are well within the realm of achievability in the near term (Fig. 6.4). A further issue of latency, the time required for a signal to traverse by interconnects between operations, is critical at the proposed speeds of operation. Despite signal propagation equal to one-third the speed of light, the finite time required for the signal to propagate between a given pair of gates can dominate gate delay and be more critical at these extreme frequencies of operation. This latency can be addressed by increasing circuit density. This can be done directly by reducing the lateral design rules for linewidths, line spacing, and layer-to-layer alignment. Decreasing pitch (linewidth plus line spacing), can be advanced by creating features with sharper edges. This can be done with improved resolution photolithographic techniques coupled with anisotropic etch techniques. This strategy for increasing feature density creates difficulties of its own for either dielectric films to completely cover these sharp edges, or for superconducting interconnects to traverse these edges without significant reduction of transport properties due to disordered film growth. This further emphasizes the need for high degree-of-planarization processing techniques. In addition, superconducting circuits that are inherently low-power (I^2R heating of zero-resistance superconducting interconnects is also zero) can enable true 3D integration. A fully planarized process with vertical integration can further increase circuit densities by orders of magnitude.

6.3 NbN Circuit Process

6.3.1 INTRODUCTION

6.3.1.1 History

In 1941, Aschermann, Friederich, Justi, and Kramer reported superconductivity in the compound NbN with a superconducting transition temperature of about

16 K [79]. This work was advanced by Gavaler *et al.* [80] by reactively sputtering high-quality NbN films; NbN/native oxide/NbN tunnel junctions were inserted into a circuit process flow, but the first high-quality, high-gap NbN Josephson junctions were fabricated by Shoji, *et al.* [12]. These junctions and their subsequent refinements are quite competitive with Nb junctions in terms of parameter spreads; MgO barriers continue to be the standard for NbN junctions, although some high-quality, very high-J_c (127 kA/cm^2) junctions have been fabricated using AlN as a barrier material [34].

6.3.1.2 Why NbN?

Liquid helium has been a "friendly" cryogen in the sense that it provides stable cooling with little cavitation or boiling that can affect circuit operation. In application, cryogen must be periodically replenished although that period in the case of the best cryostats can be measured in months to years. Still, such extreme care must be taken to preserve and replenish cryogen that one must also consider, based on the application, the use of circuitry at temperatures greater than 4.2 K.

The combination of cryocoolers with NbN circuits will allow construction of systems that do not require periodic replenishment of a cryogen, enabling use in a wide variety of ground- and space-based applications. In order to illustrate this, let us consider a single LSI chip of 10,000 gates, a gate typically consisting of a dozen junctions and resistors. The nominal thermal dissipation of this circuit during performance of its electrical operation has been estimated at 10 μW. If one includes inevitable parasitic heat loads from the input/output leads along with radiation effects, the total heat load has still been estimated to be less than 100 mW [81].

More electrical input power is required by a cryocooler for a given cooling capacity in order to obtain lower operating temperatures. This involves both the ideal Carnot efficiency and the further reduced efficiency of real cryogenic refrigerators. For an operating temperature of 4.2 K used in Nb circuitry, the necessary electrical input power is estimated at 352 W for 100 mW of cooling capacity. For NbN circuits operational at 10 K, the electrical input power to the cryocooler to achieve 100 mW is reduced to 145 W. More significantly, it can be reached by a two-stage cooler. This departure from liquid cryogen cooling is a significant advancement towards developing applications that will be accepted by the user community.

One can equally consider a Nb-based circuit operating in a cryocooler. An increase in operating temperature from 4.2 to 6 K would represent a savings of 33% in cooling power simply based on a Carnot limit. Practical cryocoolers never obtain that limit and are often only a few percent of the cooling based on Carnot. In addition, further inefficiency is incurred by going from one-stage to two- or three-stage cryocoolers. Although a variety of cryocoolers can certainly obtain 4-K operation, one approach that holds particular promise for superconducting circuitry is the pulse-tube cooler due to its reduced mechanical vibrations ($<10^{-3}$ g) compared to systems with moving displacers. This aspect impacts systems that operate without shielding in the earth's magnetic field, where even miniscule displacement can equate to significant signal noise due to the extreme sensitivity of superconducting devices to magnetic field variation [82]. Commercial systems, although large compared to 10-K cryocoolers, are currently available that provide 4-K operation using only two stages [83].

Another separate motivation for use of NbN junctions is found in the fabrication of SIS mixers. Due to their higher T_c electrodes and gap voltages (5 mV compared to 2.8 mV), the corresponding gap frequency extends to 1.5 THz. These junctions must also operate with very high current densities for optimum operation. Some remarkable progress has been made by Wang *et al.* by deposition NbN/AlN/NbN tunnel junctions with J_c as high as 100 kA/cm^2 while maintaining high-quality junction parameters [34].

6.3.2 Process Flows

The differences between Nb and NbN process flows are few but significant. Due to the considerably greater ($>3\times$) penetration depth of NbN, the groundplane needs to be made considerably thicker to be able to reproduce and target inductance values on chip. This increased thickness of the first deposited layer creates an edge, once etched, that in turn requires thicker dielectric layers and thicker interconnect layers to fully cover this edge. Beveling of edges is more significant for NbN than for the more metallic Nb processing, both because the edge height is more severe and that NbN as a semimetal does not conform to contours as readily as Nb.

6.3.4 Process Elements

6.3.4.1 NbN Junction Fabrication

The NbN circuit fabrication shares the same process elements with Nb-based processing, including dielectrics, resistors, photolithography and similar dry etch recipes. A primary departure is in the area of junction fabrications as elaborated in what follows.

Similar to niobium, high-quality NbN junctions are obtainable using a number of different deposited barriers; it is just typically harder to accomplish. As one would anticipate, NbN with its relatively short coherence lengths makes nearly all aspects of NbN junction fabrication a more demanding task.

The NbN film morphology is typically textured polycrystalline in slightly tilted, columnar grains. The grains generally coalesce into larger grains with increasing film thickness. Due to the slightly tilted columnar grains, surface roughness increases with grain size and thus also with film thickness. When the base electrode film roughness is on the order of the barrier thickness, junction quality degrades. Empirically, this can occur at an RMS roughness on the order of one to two nanometers. Thus, care must be taken to grow smooth base electrodes and this is often done by keeping base-electrode thicknesses less than 200 nm. The disadvantage to this approach in practice is that using base-electrode thicknesses less than the penetration depth will result in large values of inductance of interconnections using this layer. Currently, designers can limit but not eliminate this concern through prudent circuit architecture.

In order to demonstrate the considerations that go into a NbN junction-fabrication process, the following is one example of developing a NbN junction-fabrication process:

1) NbN films are deposited as discussed in detail in Chapter 1: Fabrication Methods, Processing and Measurement for Superconducting Film Structures. The same considerations discussed in that section of optimization of film-transport properties along with film stress apply to junction deposition. A practical NbN film process includes film $T_c > 15$ K for 500-nm NbN films, resistivity <180 micro-ohm-cm, and low stress.
2) Other attributes that are junction-specific must also be considered. Film qualification must include not only film T_c but also base-electrode roughness, which should be kept to less than or on the order of the barrier thickness.
3) A reasonable starting point for junction fabrication once a NbN film process is established is depositing a series of films at various thicknesses ranging from 50 to 500 nm. By measuring the T_c of these films one should observe an increase in film T_c as a function of thickness, which is characteristic of NbN growth.
4) It is useful to measure the film roughness of the very same films. This can be readily done with any number of commercially available atomic force microscopy systems. In addition, services are available that will provide quick-turnaround analysis [84] for this purpose. Generally, film roughness increases monotonically with NbN film thickness.
5) Film roughness obtained with this film-deposition process should be less than or equal to 2-nm RMS to be useful as a base electrode.

   The increasing roughness of the base electrode must be balanced with the necessary thickness in order to obtain near-bulk values of the film parameters. One can obtain near-bulk values with thinner films by heating the substrate temperature above ambient during deposition. This can be done by both external heating or depositing at higher power. The resulting film is also rougher due to the increased substrate temperature so that this approach is not a panacea.

6) A relatively simple barrier material to begin fabricating NbN junctions is MgO, which can be deposited by RF-sputtering.
7) Even dense MgO targets formed by a hot, isostatic press are hygroscopic, which can drastically effect their use as a barrier material.
8) A pre-sputtering of the MgO target for 6 to 8 h will serve to reduce the water content in the target material and in practice, is done after every exposure of the target to the atmosphere. For this reason, load-locked systems that minimize atmospheric exposure is recommended.
9) Deposition temperatures from 100 to 300 °C help facilitate both NbN and barrier growth. Deposition temperatures much above these values degrade the junction formation as evidenced by leakiness and shorting in the resulting junction I-V characterization.

6.3.4.2 NbN Operating Temperature

Currently, much of NbN circuit fabrication is focused at 10-K operation. Indeed, there is nothing absolutely preventing 20% higher temperature of operation, further reducing the cryocooling burden by another 7%. Thermal noise will further reduce circuit margin and inductances will increase due to an increase of the penetration depth (<6%). Folded into the question of temperature of operation is the variation in parameter spreads. However, even within present NbN processing technology there is room for improving T_c. Addition of carbon by reactively sputtering in a partial pressure of methane [85] can increase T_c by one to two degrees Kelvin. Successfully developing an NbCN process operational at 14 K reduces the cooling burden input power by some 71% over that of 4.2-K operation. When one considers that 77-K operation reduces this input power by only another 25%, one needs to seriously weigh that advantage over the considerable and daunting challenges of YBCO circuit manufacturing that is outlined in the next section.

6.4 HTS Circuit Process

6.4.1 INTRODUCTION

With the advent of the discovery of high-temperature superconductors, the prospect of reducing the burden of cooling superconducting circuitry for useful

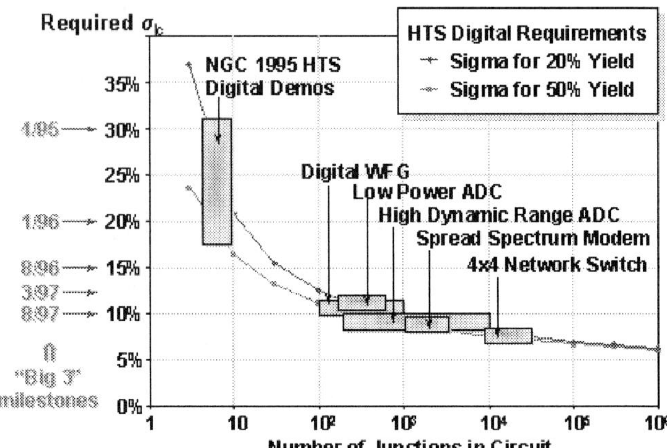

FIG. 6.5. Margin analysis of junction parameters spread and predicted circuit demonstration yield from Reference [39].

operation seemed near. Of course "high-temperature" superconductors is still an oxymoron for those accustomed to room-temperature electronics, as "high temperature" superconductor circuits typically need cooling by hundreds of degrees Celsius for 77-K operation. However, relative to Nb or NbN operation at 4 or 10 K, the cooling required for 77 K is dramatically reduced. Unlike Nb and NbN process flows, which can be likened to minor variations of a theme, HTS circuit fabrication varies dramatically from lab to lab in efforts to develop a useful circuit process. In terms of parameter spreads, it is evident that without a proven junction process with parameter spreads less than 10%, LSI or even MSI is not realizable (Fig. 6.5). It also poignantly points out where success has occurred in YBCO processing, with one- or two-junction SQUID.

As a circuit process, HTS is still in its infancy and has yet to settle on a process flow that can meet anything but the simplest of circuit requirements. A result of this is that there are disparate approaches that are being pursued in research labs throughout the world. This competition is still at a very dynamic stage and has not at all been resolved. It is generally conceded that although there are junction processes that perform better than others for given applications, the junction process that will be in use a decade from now has likely yet to be developed [86].

6.4.2 PROCESS FLOWS

The relative immaturity of HTS processing is such that present process flows are typically far simpler than corresponding LTS process flows. This is due in large part to the epitaxial requirements of the HTS films in order to avoid greatly degraded superconducting transport properties. Unlike Nb or NbN processing discussed earlier, the addition of even a groundplane beneath the junction layer has major ramifications on the HTS junction process and must be carefully developed and controlled. As an HTS film is grown, any defect that is produced is propagated into the epitaxial layers grown above that defect. The effect of these defects remains deleterious even after the typical ion beam etch removal of the area of the defect. Because the point defect often produces a larger defect structure, such as an *a*-axis grain or void, the topology of the defect replicates into the layer beneath after ion milling. The result is that film quality is increasingly difficult to maintain as the number of processing layers increases as is needed for complex circuit fabrication.

Another aspect of HTS process flows that significantly differentiates them from their LTS counterpart is in the junction structure. Although much effort has gone into producing a trilayer junction process, there have been few circuits demonstrated with this technology. Due to the anisotropic nature of HTS materials including $YBa_2Cu_3O_7$ (YBCO), the coherence length in the *a-b* plane is significantly greater than along the *c*-axis. As discussed previously, the longer coherence length eases the fabrication of junctions. For this reason, and because film growth with the *c*-axis normal to the substrate plane (denoted hereafter as "*c*-axis films") is more readily obtainable, most HTS junction processes consist of junctions with current flowing parallel to the substrate. This topology readily produces subsquare micron area junctions. One side of the junction is defined by the film thickness, which is typically 0.2 to 0.3 μ, but could be varied to less than 0.1 μ in many junction technologies. Using standard photolithography 2-μ junction widths are readily obtainable with junction areas then as low as 0.5 μ^2. This is equivalent to an LTS planar junction process with 0.8-μ diameter circular junctions.

The geometries of these HTS junctions naturally produce junctions low in capacitance and high in inductance relative to planar junctions. For practical digital circuits this puts an emphasis on lowering inductance through circuit design. The management of this parasitic inductance has strong ramifications for HTS digital-circuit fabrication.

6.4.3 PROCESS ELEMENTS

6.4.3.1 Films

High-quality, smooth film growth is critical in many aspects of circuit fabrication using YBCO. As elaborated in Chapter 1, numerous innovative deposition techniques have been developed, but no standard method has yet been established so that the development of deposition techniques remains a very active area of investigation. The properties of junctions and other process elements such as interconnect crossovers and superconducting vias have been observed to be particularly sensitive to fundamental film properties. Not unlike the early days of Nb junction development that was limited by the ability to make quality Nb films, junctions may be limited considerably by issues inherent to the reproducible deposition of HTS films.

Yet another factor that makes fabrication of HTS Josephson junctions difficult is the number of constituents that comprise the material. Niobium has but a single element, niobium nitride has two elements, and depending on its stoichiometry, it can range from a superconductor to a resistor to an insulator; YBCO, with its four different elements, has possibilities of all sorts of stoichiometries different from that of the ideal. As the number of elements increases with doping of the HTS materials, this issue of stoichiometry becomes increasingly severe. In particular, the oxygen within HTS materials is relatively mobile and its concentration can vastly change the superconducting properties of the material. One of the challenges in fabrication has been to determine ways to maintain or replenish the oxygen within these materials to obtain optimum superconducting parameters. Considering the myriad of difficulties surrounding the fabrication of HTS and Josephson junctions, there has been remarkable success in creating many different types of Josephson junctions.

6.4.3.2 Junctions

In order to form a junction, barrier thicknesses must be controlled within a fraction of a coherence length across junction areas of the order of square microns. With YBCO unit cell dimensions (1.17 nm) on the order of the coherence length, little margin of error is available. Near-atomic level smoothness must be realized on the surface of the base electrode. Perfect crystallinity is required both within the deposited film and at interfaces. Any variation from the ideal epitaxy can suppress the superconducting order parameter, thus reducing the I_cR_n product and causing considerable variation of critical junction parameters. Even without epitaxial flaws, stressed YBCO thin-film structures at interfaces can lead to localized and nonuniform oxygen loss, further degrading the superconducting transport properties [36].

It is safe to say that there is no single junction process that has been developed that satisfies all of the various requirements of superconducting applications for the HTS junctions. Rather than attempt to comprehensively categorize all of the various junction types that have been fabricated using HTS, we will examine a number of approaches that have been common to circuit fabrication. These are: bi-crystal grain boundary junctions, step-edge grain boundary junctions, SNS edge junctions, and finally, ion-beam damaged junctions (Fig. 6.6). A comment

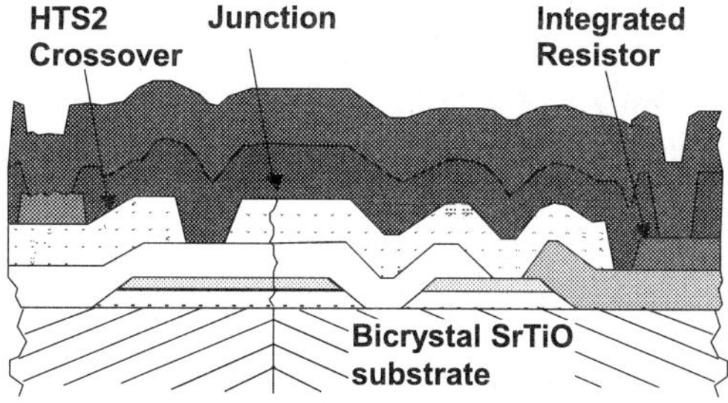

FIG. 6.6. Cross section of bi-crystal junction in multilayer process flow.

on planar junctions will be included, although in practical terms, the yield and reproducibility of this process is considerably less than these other junction processes.

The maturity of most HTS processes is such that incorporation of a junction process with other multilayer elements has only recently been developed. The fundamental building blocks of a multilayer film process for active circuit applications include: high-quality films for all superconducting and interconnect layers, high-I_c superconducting crossovers, good dielectric integrity, a stable, high-yield junction process, and an integrated resistor.

Although magnetometry is a clear example of a useful application that has met many requirements with single-layer YBCO junction processing, even this circuit could be enhanced with the availability of reliable, low-flux noise multilayer circuit processing. Low-inductance junctions are needed for the fabrication of digital circuitry. In order to reduce parasitic inductance, one is driven to tighter layer-to-layer alignments and the fabrication of junctions either above or below a superconducting groundplane. Design tolerances require controlled penetration depths and geometries and put further constraints on a junction process usable for digital circuitry. It is difficult to foresee an integrated circuit process that does not have at least the forementioned elements. To this end, present junction processes extendable to multilayer geometries will be examined in what follows in terms of their reproducibility and parameter spreads.

Bi-crystal junction. Likely the most straightforward and reliable approach to junction fabrication has been by deposition of YBCO on bi-crystal substrates. The substrate is formed by dicing a single crystal substrate into two and fusing the parts back together with an intentional misorientation of the crystal lattice. This approach was one of the first methods for fabricating HTS junctions [87] and remains one of the most reliable in terms of reproducibility of useful junction parameters. This approach to junction fabrication also is one of the simplest and requires little more than being able to fabricate a high-quality c-axis YBCO film on a commercially available bi-crystal substrate, along with subsequent definition and metallization steps. The most commonly used substrate material for this purpose is $SrTiO_3$; however, due to the high permittivity of $SrTiO_3$ ($\varepsilon_r = 1930$ at 77K) in applications where parasitic capacitance is detrimental, LaSrAlTaO (LSAT), with $\varepsilon_r \sim 25$ has been used.

Although this is the most common approach for single SQUID applications such as magnetometry, difficulties with extending this fabrication technique to complex circuits are manifold. Regardless of whether the groundplane is positioned above or below the junction it will have an ill-placed weak link directly beneath the junction that can considerably alter device inductance. Also, interconnects crossing the bi-crystal boundary (as typically needed for a multilayer coil) will include an undesirable I_c-reducing junction. Most significantly, for larger circuits, design will be greatly limited by the constrained positioning of junctions along the single bi-crystal boundary. This geometric limitation will incur greater than allowable parasitic inductance, which must be dealt with for complex digital circuitry.

Step-edge junctions. An alternate approach for inducing a grain-boundary junction in a YBCO film is by purposefully creating an edge that the YBCO film must traverse. This is the so-called step-edge junction proposed by Simon *et al.* [88] and investigated by many other groups [89]. As a *c*-axis YBCO film grows on a sloped edge of $SrTiO_3$ or $LaAlO_3$ (the most common materials for this junction type), it is energetically favorable for *a*-axis film growth as the slope approaches 45°. This "flipping" from *c*-axis to *a*-axis and back to *c*-axis creates grain boundary junctions in series [90], of which, the lower I_c junction is used in circuitry. Care must be taken in this technique to properly form the step-edge in these materials. For this reason, multilayer masking using Nb and diamond films has been utilized to ensure sharp, reproducible edge formation.

Other material systems have also been investigated that have a much different dependence on edge topology and the resulting YBCO film morphology. For steps created in MgO [91], a much shallower edge (19°) will form a grain boundary. In addition, by sculpting the bottom of the step-edge, the second inductance-increasing grain boundary can be eliminated [92].

This technique, similar to bi-crystal junctions, requires only a single deposition of YBCO to form the junction. It is well suited to magnetometry and has resulted in some of the lowest-noise SQUID fabricated in HTS [93]. In contrast to bi-crystal junctions, these junctions can be formed in any position and orientation. They can also be formed above a groundplane in a step formed in a deposited dielectric without creating grain boundaries in the underlying groundplane (Fig 6.7). The critical-current densities of these junctions are typically higher than bi-crystal junctions but have been demonstrated to decrease linearly with the deposited junction layer thickness. This method of reducing I_c, however,

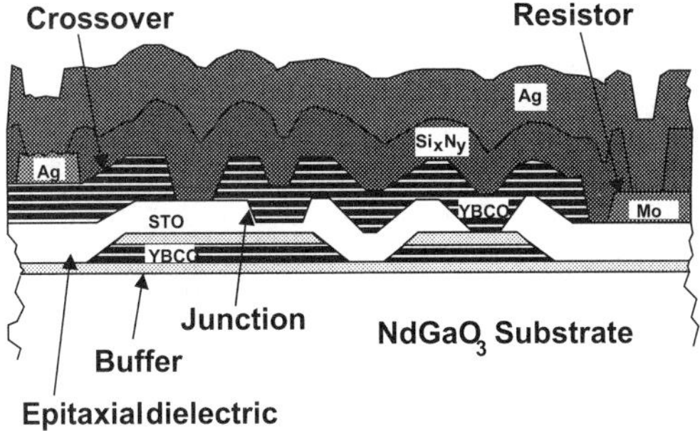

FIG. 6.7. Cross section of step-edge junction in multilayer process flow.

has the disadvantage of increasing the characteristic film inductance for circuit components. For film thicknesses necessary to target $I_c = 200\,\mu A$ for a 4-μm-wide junction, film thicknesses are on the order of one-third of the penetration depth, and as such are in a regime where characteristic film inductance increases sharply. Because this LI_c product can be the limiting consideration for digital circuitry, this becomes problematic for physically realizable inductance structures. In addition, margin must be allowed on both sides of the grain boundary junction, so that lead length for step-edge junctions is a function of twice the alignment capability dictated by the photolithographic equipment. This lead length, especially in the thin YBCO layer whose thickness is tailored to correctly target I_c, greatly increases the junction parasitic inductance. An alternate method to reduce the I_c without increasing inductance is to purposefully deoxygenate the exposed grain boundary. This can be done by annealing in an argon atmosphere at 200 to 300 °C, which will reduce junction critical current. This will also increase junction R_n, but not in proportion with the reduction I_c, thereby resulting in a smaller junction I_cR_n product.

An alternate geometry involves using steps in a substrate with a subsequent epitaxial dielectric and HTS deposition as an overlaying groundplane. However, this groundplane is no longer planar and could form weak links directly above the junctions due to the underlying topology. The possibly diverted current flow in this groundplane due to the weak links not only increases inductance but does so in a fashion that is difficult to predict.

Ramp-edge SNS. Edge junctions have been previously made using Pb and Nb and is an effective approach for making small-area junctions (Fig. 6.8). Difficulties arise in these material systems due to the necessary *ex situ* processing, that is, the base electrode edge is formed and exposed to the atmosphere and process chemicals prior to barrier and counterelectrode deposition. This same issue arises again in doing *ex situ* processing on exposed YBCO base-electrode edges with

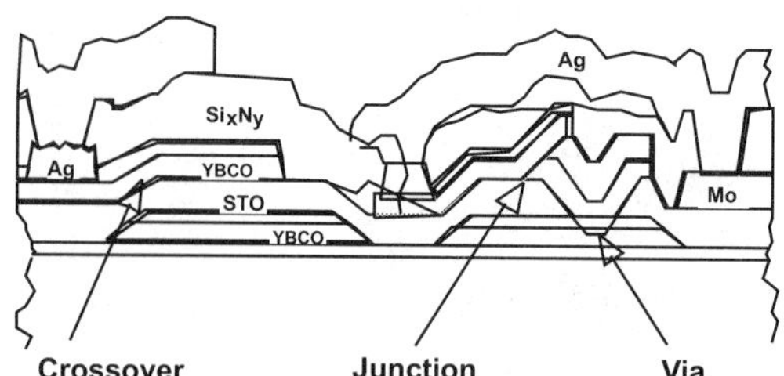

FIG. 6.8. Cross section of SNS edge junction in multilayer process flow.

the reactivity of YBCO. These edges must be carefully cleaned prior to barrier deposition to ensure good electrical contact. Early barriers consisted of noble metals such as Ag and Au that "bridged" between adjacent YBCO sections, and these type junctions still find a use in magnetometry applications [38]. At present, however, most SNS work utilizes epitaxial barriers followed by epitaxial YBCO counterelectrode deposition.

Among HTS junctions, epitaxial ramp-edge junctions have the greatest design flexibility. This junction is formed by first etching a YBCO film with dielectric overlayer to form a shallow edge (15–35°). This "ramp edge" structure is then deposited upon with a normal barrier material followed by a YBCO counter-electrode. The smoothness of the base-electrode edge [40] is critical to the fabrication of this junction. This in turn requires a smooth base-electrode film, made even more difficult to achieve when formed over a YBCO groundplane and dielectric, which can translate roughness into the base electrode film. Despite such processing challenges, some of the best I_c spread results have been achieved using this junction type.

There are several major advantages offered by this junction design. First, I_c can be controlled by varying the thickness of the barrier layer, and second, its parasitic inductance is only half as sensitive to the junction lead length compared to the previous approaches. This benefit can lead to greater circuit operating margin. Third, these junctions can be formed either above or below a superconducting groundplane without damaging the groundplane.

Interface-engineered junctions (IEJ) are a variation of this junction design. In this case the process flow creates a barrier by exposing the base-electrode edge to a sequence of plasma, chemical, and annealing treatments [94]. This approach is currently the object of intense research efforts.

Ion-beam damaged junction process. Numerous attempts at using electron and ion beams to fabricate junctions have met with varying degrees of success [95]. These junctions have been successful in demonstrating the operation of simple digital circuits. This fabrication process has the flexibility of targeting I_c

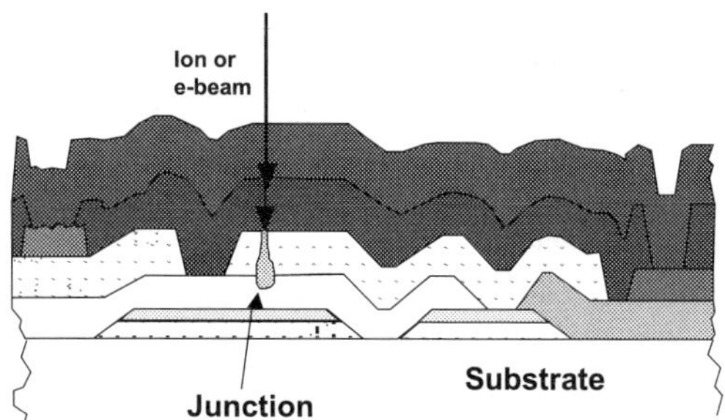

FIG. 6.9. Cross section of *e*-beam damaged junction in multilayer process flow.

by repeatedly applying the ion or *e*-beam to reduce I_c or annealing out damage to increase it (Fig. 6.9). Doubts concerning long-term stability have recently been credibly resolved. For wafer-level fabrication this technique would be beam-intensive and difficult to extend to the LSI level.

Planar junctions. All the junction types previously described (ramp-edge, bi-crystal, step-edge and ion implantation) rely on using film cross-section to achieve small junction area and coupling along the *a-b* axes (where the coherence lengths are greatest). The "price" that is paid for that geometry in terms of circuit parameters is generally higher parasitic inductance. For this reason, a fabrication process that produces junctions with a low-inductance trilayer geometry would be attractive to circuit designers. However, this is difficult to realize in practice. Topologically, this process flow is similar to the Nb-based process cross section (Fig. 6.2).

There are two primary approaches using this geometry. The first utilizes the intrinsic junctions that are formed by weak coupling between adjacent *a-b* planes. The second approach involves forming the junction electrodes from film grown with the *a*-axis normal to the substrate plane (so-called "*a*-axis films"). Both these approaches are attractive because of their low-inductance geometry but require deep-submicron lithography to reduce the junction critical current to desired levels. Furthermore, the resulting I_cR_n product is relatively small.

The first type, CAM (*c*-Axis Mesa) [96] junctions, are Likharev-type micro-bridges [97]. They require a very small, photolithographically defined junction area (~$0.5\ \mu m \times 0.5\ \mu m$). This has been achieved by a combination of *e*-beam lithography and etch-back planarization. Even with minute cross-sectional area, the resulting inductance of this geometry is much reduced relative to the previous junction types. The capacitance incurred in this junction structure is greater; however, even this has been recently reduced through further innovation [37].

The second approach uses *a*-axis films and their longer coherence length (1.5 nm). This has been explored, especially by the Japanese research community, but as yet has met with limited success. Once again, the obtaining of a sufficiently smooth base electrode is a major obstacle limiting progress. Difficulties arise in being able to consistently grow *a*-axis films (base-electrode, barrier, and counter-electrode) that are homogeneous within the junction area. The fact that *a*-axis film growth proceeds more rapidly than *c*-axis growth, any *c*-axis inclusions create large topological variations that compromise junction performance. In addition, as *a*, *b*-axis coupling results in relatively high critical current densities, the cross-sectional area must be kept very small to achieve the required critical current. Otherwise, to avoid lithographically defining submicron junction dimensions, methods of reducing the critical current density must be employed. Despite the limited success in fabricating circuits with this junction process, ultimately this may be the junction geometry necessary to extend HTS technology to the MSI or LSI level.

6.4.3.3 YBCO-compatible Dielectric Films

As discussed previously, amorphous dielectric films are commonly used in IC processing. This, however, is not an option in YBCO-based multilayer processing. Dielectric materials must be lattice matched with YBCO, have electrical properties appropriate for the given application, and must allow oxygen diffusion to underlying YBCO films. This last criterion is unique to high-temperature superconductors and is due to the high oxygen mobility within YBCO. Care must be taken in the entire process flow to reduce oxygen loss and employ methods that re-oxygenate the films. These methods include 400 to 500 °C thermal annealing in an oxygen atmosphere, plasma treatments, and exposure to ozone. The latter two methods expose the YBCO film to increased concentrations of the more reactive monoatomic oxygen. With the multiple criteria that these layers must meet, obtaining defect densities less than 100 defects/cm^2 is challenging. Dielectric integrity may improve as YBCO film quality continues to improve as evidence suggests that defects arise from outgrowths and voids generated in the YBCO underlayers [39]. Material systems that have been successfully used as insulators in multilayer structures include $SrTiO_3$, SrAlNbO (SAN) [98], SrAlTaO (ST) [39], PrBaCuO (PBCO), CeO_2 [99], and a trilayer of $SrTiO_3$/SAT/$SrTiO_3$ [100].

6.4.3.4 Vias and Interconnect Crossovers

Control of film epitaxy and stoichiometry are key to growing smooth, defect-free, superconducting YBCO films. The lattice match of the substrate to the film, deposition temperature, and deposition gas composition and pressure must be controlled and optimized. Other issues such as target composition variation from deposition-to-deposition can degrade film morphology. This would then result in costly re-optimization of the film-deposition process. In the fabrication of vias and interconnect crossovers these requirements for defect-free film growth remain the same, but accomplishing them becomes considerably more difficult. Circuit fabrication requires patterning of the initial layers for all but the most basic of applications. Epitaxial film growth must then occur over various topologies rather than on the near-perfect template of a single crystal substrate. This has numerous yield-limiting failure mechanisms. Defects occurring in the epitaxy of initial layers can propagate into the overlying structures. Crystalline damage occurring on film edges during the etching process can inhibit later epitaxial growth. Once the film is patterned, chemical reactions induced by solvents and airborne contaminants can also occur on the exposed film edges [101].

From a film-growth perspective vias and crossovers represent two distinctly different process scenarios. For vias, the YBCO film is required to grow over an etched dielectric edge. In the case of crossovers, epitaxial dielectric and subsequent YBCO is required to grow over an etched YBCO edge. Due to the greater sensitivity of YBCO to the processing environment compared to the epitaxial dielectric, successful film growth is often more difficult to achieve in the crossover-forming process. In either scenario, grain boundary formation will occur, as with step edge junctions [90], when the slope of the underlying film edge approaches 45° (Fig. 6.10a). In practice, edge angles need to be considerably less than 45° in order to avoid grain boundary formation at points where local variations cause steeper slopes. Incorporation of processing techniques can increase the critical current density that crossovers can sustain, often by several orders of magnitude, and correspondingly improve epitaxy of the film deposited along the edge as evidenced by TEM cross section (Fig. 6.10b, c).

The following guidelines should be considered in fabrication of crossovers and vias in YBCO:

1) Defects caused during epitaxial growth of the initial layers will limit critical current regardless of how careful subsequent processing is performed. Generation of outgrowths, voids, and other common defects in YBCO films (as elaborated in Chapter 1) can be more damaging in the strained growth over edges than is apparent from measurements of planar-film transport properties.
2) The most common approach for accomplishing YBCO film growth over nonplanar topologies is to simply make the structures as planar as possible. To this end, film edges are sloped to as low as 5° (with respect to the substrate plane) to allow unperturbed growth on the edge. However, this gentle sloping results in longer lead lengths required to traverse those edges. The increased lead lengths can generate increased parasitic induc-

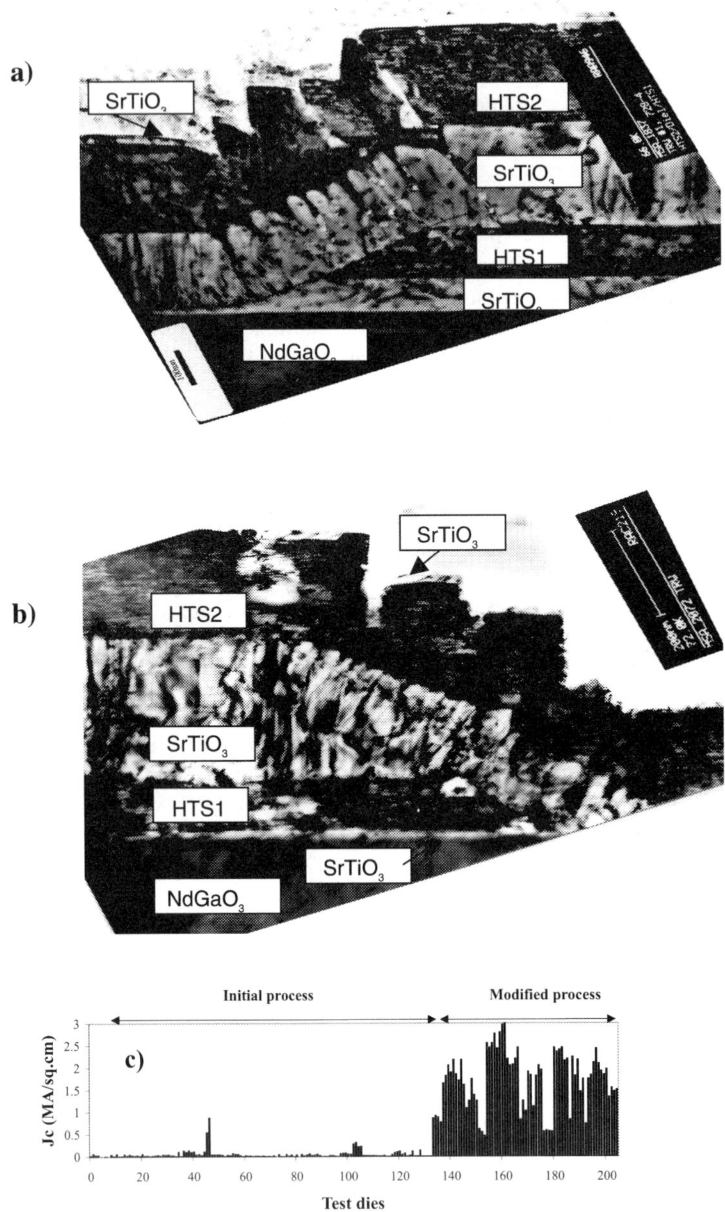

FIG. 6.10. TEM cross section of YBCO growth along SrTiO3-covered edge both before and after critical process insertions from Reference [103].

tances, which can reduce circuit operating margins. Thus, edge lengths, especially for digital circuit considerations, should be as short as possible. These conflicting requirements are in practice met with edge angles of 20 to 30°. This allows reasonably short lead lengths along with edge angles that are reasonably robust against grain boundary formation.

3) Edge damage during the etching process should be minimized. This occurs during IBE due to ion bombardment. Operating the ion source at a reduced voltage lowers ion kinetic energy, thereby reducing edge surface damage. Most ion mills using Kaufman sources cannot operate at voltages much below 500 V; RF sources are less common and more expensive, but can generate ions at considerably lower voltages. Electron cyclotron resonance (ECR) ion sources offer an even more effective method for generating a high-flux density of low-energy ions. In these systems ion formation is independent of the acceleration voltage and reasonable etch rates at low voltages can be obtained simply due to the vast number of ions generated.
4) The temperature of the wafer should be kept low during the etching process. Local heating due to the ion beam can be considerable and along with being in vacuum can de-oxygenate the YBCO film. This heat should be quickly removed from the wafer; consequently a water-cooled platen within the IBE system is common. A layer of vacuum grease or indium foil between the wafer and the platen significantly improves this heat transfer. Further improvements especially with respect to small-feature definition have been noted with substrate holders cooled to liquid nitrogen temperature during the etch process.
5) Once the sample is etched, the photoresist must be removed and the bare edges are then exposed to the chemistry involved in that process. Commercial photoresist strippers are generally not an option for this process. Acetone followed by an isopropyl alcohol rinse is most common, although using oxygen plasma strip is also possible.
6) The damage that occurs on the edges during the etching and photoresist removal should be removed prior to further deposition. This is ideally done *in situ* prior to the subsequent deposition but can be done *ex situ* if the wafer is transferred to the deposition chamber in a timely fashion. An alternate approach that does not expose the YBCO film edge to either the atmosphere or photoresist removal chemistry is using a multilayer masking and *in situ* etching prior to deposition [102]. This involves etching partially through the dielectric overlayer, stripping the photoresist and then *in situ* etching through the remaining dielectric and YBCO film immediately prior to deposition.
7) Specific to via fabrication, etches are often continued through the dielectric and well into if not through the underlying YBCO film. This is done in order to maximize the area of contact with the higher critical current density *a-b* planes in typical *c*-axis YBCO growth.

6.4.3.5 *Photolithography and Etching*

The requirements of structure definition of even simple YBCO circuits can be quite diverse. As discussed previously, the fabrication of gently sloped edges facilitates epitaxial film growth on those edges. The preferred profile of other edges that do not require subsequent epitaxial depositions is typically as sharp as possible. For magnetometry applications, this reduces sources of flux noise in the thinned and possibly damaged edge region. For digital applications, sharp-profiled edges decreases the interconnect pitch (linewidth plus necessary line-to-line spacing necessary for clearance). For the forementioned reasons controlling the YBCO edge profile from 15 to 90° is desirable and can typically be accomplished through photolithographic and etching techniques.

Photolithographic processes that are well suited for Nb circuit processing may not be necessarily optimal for YBCO. Due to the reactivity of YBCO with the trace chemicals dissolved in water, contact of the YBCO film with water is typically avoided or minimized. Passivating YBCO with $SrTiO_3$ is one method of limiting exposure of YBCO surfaces to potentially degrading chemical reactions. Unfortunately, edges that are exposed in the etching process are more difficult to passivate. These edges are then typically exposed to various chemicals involved in photolithographic processing and great care must be taken to remove this contamination before subsequent epitaxial deposition. In practice this is commonly done by ion-beam etching or wet etching in a very weak acid such as EDTA. An innovative approach towards avoiding such contamination is through multilayer masking techniques. This can be accomplished by depositing a mask material such as TiN by conventional techniques. The edge is then etched by ion-beam etching until the mask material is entirely depleted. The thickness of this mask layer is determined by the etch rate of both the TiN and the YBCO and the thickness of the YBCO that is intended to be etched [102].

Film etching is also another technique that departs from typical Nb or NbN processing. Wet etching of YBCO is readily accomplished, so much so that diligent care must be taken to avoid exposure to even slightly acidic water during processing. This technique is routinely accomplished for large-feature definition. Although nearly all acids etch YBCO quite aggressively, it is interesting to note that hydrofluoric does not. The HF quickly passivates the YBCO surface effectively preventing further etching. The etch selectivity of HF to STO and YBCO has been successfully used to etch $SrTiO_3$ out from beneath a planar film of YBCO for bolometric applications. However, due to its wicking or undercutting of the photoresist, this technique is undesirable for accurate and reproducible patterning as device sizes shrink.

Rather than standard dry etching techniques, such as RIE, which have not been effective for YBCO, argon ion-beam etching (IBE) has been the predominant method. Film selectivity by IBE is much less dramatic than is typical of RIE processing owing to the ballistic rather than chemical nature of IBE. This puts further onus on endpointing capabilities to avoid excessive overetching. Etch completion is often determined by assessing the etch rate from previous test pieces, which as a process is both time-consuming and costly. In addition, the time selected for a given etch is often overestimated to take into account wafer-to-wafer variation in the etch rate, which can lead to considerable overetching. For simple circuits, there is little penalty incurred by this. However, for multilayer processing, any overetching adds difficulty to achieving high-quality step coverage by subsequent epitaxial layers. It is important, therefore, to be able to stop etching at, or near, the desired interface. One flexible approach for endpoint detection is to employ a differentially pumped residual gas analyzer directed toward the wafer in order to sample the different species as they are etched away. This approach is essentially large-area secondary ion mass spectrometry (SIMS) so that endpoint detection of nearly any material of known composition can be effectively accomplished. Currently, commercially available units [104] are sensitive enough to endpoint film areas less than 1 cm^2.

The use of automated lithographic alignment between circuit layers can significantly broaden the range of circuit designs that can be fabricated. This is common in both semiconducting and Nb process foundries and routinely allows alignment with a precision of $3\sigma < 0.4\ \mu m$. The significance of this capability is especially important in layer-to-layer registration of digital circuitry where the lack of precise alignment forces circuit elements to have longer line lengths to accommodate larger misalignment. These longer line lengths increase circuit parasitic inductances that in turn reduces circuit margins and places tighter constraints on junction parameters such as I_c uniformity. These photolithographic tools, typically require full-size (2 in or greater diameters) wafers. In addition, these wafers must not distort during process temperature cycling and they must be flat to within 5 μm across the wafer. For these reasons and for film-growth considerations, $NdGaO_3$ and LSAT substrates reasonably satisfy these requirements.

6.4.3.6 Planarization

While YBCO film growth becomes more difficult on nonplanar surfaces, its growth on smooth and flat surfaces does not ensure defect-free growth. This is clearly evidenced by optically smooth bi-crystal substrates whose purpose is to generate grain boundaries. At the other extreme, YBCO films purposefully grown over steep step-edges can conduct supercurrent albeit with lower critical-current densities. Furthermore, film thicknesses on the order of the step-height can begin to overcome the current-limiting aspects of the grain boundaries along the step-edge. For planarization to be useful in a circuit, films must not only achieve planarity but must also retain near-perfect epitaxy in order to transport high current densities with minimal flux noise.

Another motivation for the development of HTS planarization is the reduction of pitch in the groundplane definition. As discussed in the previous sections, the present approach for fabricating interconnect crossovers requires beveling of YBCO edges in order to allow high-quality dielectric and YBCO film growth over the edges. The beveled edges have angles from 5 to 30° with respect to the substrate plane. In order to make the characteristic inductance of YBCO/insulator/YBCO structures insensitive to run-to-run thickness variations, the YBCO thicknesses should be at least equal to the penetration depth. Note that Nb and NbN film thicknesses are typically 1.5× to 3× the penetration depth. For YBCO films, penetration depths are on the order of 200 to 350 nm depending on the precise film morphology. Thus, in the worst case, for a film thickness of 350 nm and groundplane edge angle of 5°, the edge length will be 4 μm. Although this long edge length can be accommodated by today's simple HTS digital circuitry, it would be deleterious to larger-scale integration.

Planar trilayer and CAM junctions particularly need a planarization technique to incorporate them into integrated circuits. In order to reduce their critical current to useful levels for most SQUID and circuit applications (~100–200 μA), junction diameters of 2 μm or less are needed. Clearly, beveled edge lengths of 4 μm would be unacceptable for this purpose. As in the push towards smaller-dimension junctions, sharp edges are useful in controlling the junction area and thus the junction critical current and its variation. These disparate needs are met through a dielectric etchback planarization technique used by the Cambridge group [96]. This technique involves applying a planarizing material such as a spin-on photoresist or bias-sputtered dielectric. This dielectric is then etched back to the YBCO surface. A useful degree of planarization can be achieved assuming reasonably similar IBE etch rates for the dielectrics used for planarization and circuit structures. This procedure can be repeated to further improve the planarization at the cost of much processing time and labor.

6.4.3.7 Resistors and Metallic Contacts to YBCO

The HTS junctions naturally exhibit nonhysteretic I-V characteristics and as such do not need the shunting resistor that standard Nb and NbN junctions require. However, there are circuit architectures, such as resistor-coupled SFQ circuitry, which require low-inductance resistors and, hence, resistors over groundplane. The effect of the across-circuit variation of these resistors for these circuits, can add in quadrature with junction I_c variation. Although not inherently difficult to fabricate, any resistor value variation results in more stringent demands on already challenging junction requirements. Common thin-film resistor materials, such as molybdenum, can be utilized after the high-temperature depositions have been completed. This resistor can then be connected to the HTS circuits through normal metal wiring straps and annealed noble metal/HTS contacts. Care must be taken to reduce the noble metal/HTS contact resistance to values less than $10\,\Omega$-μm^2 in order not to be a major cause of variation in resistor values. Contacts made with *in situ* deposition of Au or Ag after the final YBCO deposition can readily attain these values. However, the removal of the contact Au or Ag from areas outside the contacts requires ion milling that will remove some of the top layer YBCO, thereby resulting in an undesirable thinning of that film. Alternatively, *ex situ* processing of the contacts with lift-off patterning can avoid a lossy normal metal on top of the YBCO film or having to etch off the metal and etch into the YBCO surface. This can be done by sputtering or evaporating Au or Ag after doing a brief *in situ* cleaning of the YBCO surface and annealing the sample in an oxygen furnace at temperatures near 550 °C [105]. This anneal can also serve to provide re-oxygenation of the lower HTS layers that may have been depleted during processing.

6.4.4 Application Examples

6.4.4.1 Digital Considerations

In fabrication of digital circuitry, low-inductance junctions are needed. In order to reduce the inductance, and especially the parasitic inductance, one is driven to tighter layer-to-layer alignments and the fabrication of junctions either above or below a superconducting groundplane. Design tolerances require controlled penetration depths and geometries and put further constraints on a usable junction process for digital circuitry.

The primary unity in the majority of logic schemes for superconductors involves a loop of inductance L and its interaction with either an applied or an induced flux. A common condition arising in SFQ and other logic families is that the inductance-critical current product should be on the order of one flux quantum or less. In order to satisfy $L_pI_c < \Phi_o$, both junction critical current and total inductance need to be relatively small; reasonable values that satisfy this relation are $200\,\mu A$ and 10 pH, respectively. This pair of numbers arises because of the conditions on physically producing the inductive loop and in potential mutual coupling concerns. The parasitic inductance is taken as the inductance generated primarily by the junction leads that is not useful for signal processing. At $I_c = 200\,\mu A$, the allowable parasitic L_p must then be less than 1.2 pH. This parasitic inductance is a consequence of the penetration depth, finite junction lead length, junction critical-current density, layer-to-layer alignment, and the film thicknesses of groundplane, junction layer and interceding dielectric. In HTS processing, even monitoring thickness of these film layers during deposition provides difficulty due to their epitaxial growth and elevated processing temperatures. One factor, the penetration depth, is an inherent function of film morphology and is not readily targetable. However, in order to optimally design digital circuits, the value and variation of the penetration depth needs to be accurately and empirically determined for a given process flow (see Chapter 3: Superconductor Electronics For Digital Devices, Circuits & Systems).

Digital circuitry in YBCO has enjoyed limited success with current levels of I_c variation for junction processes with useful geometries, consistent with Fig. 6.5. This application continues to be the most daunting of challenges for this material system.

6.4.4.2 Magnetometry

In contrast, the development of HTS films and junctions operational at the boiling point of liquid nitrogen had immediate and practical ramifications to this field of magnetometry. For example, SQUID magnetometry has been applied to such diverse applications as aircraft-based geophysical exploration, characterization and exploration of brain activity, failure analysis of critical aircraft parts and failure analysis of integrated microcircuits. Niobium-based SQUID have been well established in meeting many of the requirements that could operationally use liquid He temperatures. The motivation for HTS device insertion is due to cost of operation and convenience, and for numerous applications were quickly realized due to the higher temperature of operation. In addition, reduced thermal insulation would be required, which brings the SQUID sensor closer to the room-temperature object being measured, resulting in the potential for greater sensitivity. Relative to digital circuit fabrication, the fabrication requirements are less severe and present a credible opportunity for insertion of the less-mature HTS processing.

SQUID measurements in the ideal case approach the quantum limit in their sensitivity. However, real-world processing issues can prevent that measure of sensitivity from being obtained. Fortunately, there is correlation between film quality and reducing $1/f$ noise in magnetometers. The ideal that is sought for in film deposition is a purely *c*-axis oriented film of high critical-current density. Even small inclusions of *a*-axis grains can significantly increase $1/f$ flux noise. Large *a*-axis inclusions can be readily observed by SEM and methods of eliminating these are dealt with in Chapter 1. A more sensitive technique of identifying *a*-axis inclusions that are not necessarily evident by SEM is by performing a ϕ-scan in an *x*-ray diffraction unit.

Once the film morphology is under control in the deposition process, a common follow-up is to optimize films with respect to critical-current density. The motivation behind striving towards high critical-current density in these films is not due to a need for large supercurrents in these devices. The actual supercurrent used in the operation of these devices is several orders of magnitudes less than the films can typically sustain. A mechanism that accounts for higher critical current densities is "pinning" of flux vortices, making it energetically unfavorable for them to move. This same pinning allows "magnetically quiet" devices to be fabricated as it is the movement of these flux vortices during device operation that is a source of $1/f$ noise.

Once it is understood that $1/f$ noise arises from movement of flux vortices, there are further implications to device processing, especially with respect to edge formation. As the strength of the pinning of the flux vortices decreases with decreasing film thickness, gently tapered edges should be avoided. Entry of flux vortices in the thinned edge area is a noise source; this area can be minimized on a given device by fabricating edges as sharply as possible. These near-vertical edges provide large edge-pinning forces. Recent magnetometer designs that include a grid-work of narrow lines put further emphasis on the importance of making high-quality edges.

In order to create vertical device edges, anisotropic etching by argon ion-beam etching is the most common method. Factors such as angle of ion-beam etching and argon pressure during ion-beam etching do affect the resulting edge angle after ion-beam etch; however, steepness of the photolithographic mask has the most direct bearing on the resulting edge topology. High-contrast photoresist with minimal or no post-baking provides a sharp-walled photolithographic mask with angles greater than 88° with respect to the film surface. In contrast, wet etching in slightly acidic solutions is an isotropic etch process and typically undercuts the photolithographic mask creating a shallow and undesirable edge for this purpose.

Finally, another fabrication-related aspect of YBCO-based magnetometry is choice of junction. The single most significant parameter for this choice is I_cR_n at the temperature of operation. Due to the low-level integration the choice of junction is not constrained by requirements of junction uniformity, nor ultralow inductance as is the case in digital fabrication. Grain boundary junctions, especially bi-crystal junctions, perform admirably for this application due to both their simplicity of fabrication and their resulting electrical characteristics (see Chapter 6: SQUID Magnetometers).

6.5 Summary

Superconducting circuit fabrication has benefited from the rapid development of techniques for the semiconducting IC industry. It has been material issues that have often paced the progress of superconducting circuit fabrication. One theme that has repeated itself through the short history of Josephson junction fabrication is the need for junction electrode material, especially base-electrode films, that are both sufficiently smooth to allow complete barrier coverage and robust to the operational environment. This issue has been successfully addressed in the Nb/Al/AlOx/Nb system, reasonably met in the NbN/MgO/NbN system, and found wanting in YBCO. That YBCO circuits have yet to reach a high level of maturity should not be a surprising realization considering the challenges inherent to the material and the limited amount of time since its discovery. Historically, it has been a materials development that has improved junction processes to the point of greater utility. This was accomplished in lead by alloying, in niobium by overlaying with Al, and in niobium nitride by finding a compatible barrier material in MgO. It will likely again be needed before a circuit process useable for large-scale integration will be developed in YBCO.

Acknowledgements

I gratefully acknowledge my colleagues J. Luine, R. Hu, L. Eaton, M. Johnson, G. Dantsker, E. Ladizinsky, G. Kerber along with the volume Editor for their critical reading and most useful comments.

References

1. Anderson, P. W. and Rowell, J. M. (1963). *Phys. Rev. Letters* **10**: 230.
2. Rowell, J. M. (1963). *Phys. Rev. Letters* **11**: 200.
3. Josephson, B. D. (1962). *Phys. Letters* **1**: 251.
4. Zimmerman, J. E. and Silver, A. H. (1966). *Phys. Rev.* **141**: 367.
5. Anacher, W. (1980). *IBM J. Research and Development* **24**: 107.
6. Jaycox, J. M. and Ketchen, M. B. (1981). *IEEE Trans. Magnetics* **17**: 400.
7. Gurvitch, M., Washington, M. A., and Huggins, H. A. (1983). *Appl. Phys. Lett.* **42**: 472–474.
8. Ruggiero, S. T. and Rudman, D. (1990). San Diego, CA: Academic Press, Inc.
9. Talvacchio, J., Braginski, A. I., Janocko, M. A., and Gavaler, J. R. (1986). *Bulletin of the American Physical Society* **31** (3): 438.
10. LeDuc, H. G., Stern, J. A., Thakoor, S., and Khanna, S. K. (1987). *IEEE Trans. Magnetics* **23**: 863.
11. Bauer, E. and Poppa, H. (1972). *Thin Solid Films* **12**: 167.
12. Shoji, A., Aoyagi, M., Kosaka, S., Shionoki, F., and Hayakawa, H. (1985). *Apply. Phys. Lett.* **46** (11): 1098.
13. Miller, D. J., Gray, K. E., Kampwirth, R. T., and Murduck, J. M. (1992). *Europhysics Letters* **190**: 27–32.
14. Murduck, J. M., Vicent, J., Schuller, I. K., and Ketterson, J. B. (1987). *J. Appl. Phys.* **62** (10): 4216.
15. Kurosawa, I., Nakagawa, H., Kosaka, S., Aoyagu, M., and Takada, S. (1989). *IEEE J. Solid-State Circuits* **24**: 1034.
16. Kurosawa, I., Nakagawa, H., Kosaka, S., Aoyagu, M., and Takada, S. (1991). *IEEE J. Solid-State Circuits* **26**: 572.

17. Hioe, W., Hosoya, M., Kominami, S., Yamada, H., Mita, R., Takagi, K. (1995). *IEEE Trans. Appl. Superconductivity* **5**: 2992.
18. Radparvar, M. (1995). *Cryogenics* **35**: 535.
19. Hypres design rules, available from Hypres, Inc., Elmsford, New York.
20. Ishida, I., Tahara, S., Hidaka, M., Nagasawa, S., Tsuchida, S., and Wada, Y. (1991). *IEEE Trans. Magn.* **27**: 3113.
21. Tahara, S., Nagasawa, S., Numata, H., Hashimoto, Y., Yorozu, S., and Matsuoka, H. (1995). *Ext. Abstr. 5th Int'l Superconductive Elec. Conf.*, **23**.
22. Savageau, J. E., Burroughs, C. J., Booi, P. A. A., Cromar, M. W., Benz, S. P., and Koch, J. A. (1995). *IEEE Trans. Appl. Superconductivity* **5**: 2303.
23. Benz, S. P. and Burroughs, C. J. (1997). *IEEE Trans. Appl. Superconductivity* **7**: 2434.
24. Kang, J., Miller, D., Przybysz, J. X., and Janocko, J. (1991). *IEEE Trans. Magn.* **27**: 3117.
25. Khabipov, M., Dolata, R., Buchholz, F-Im, Kessel, W., and Niemeyer, J. (1995). *Ext. Abstr. 5th Int'l Superconductive Elec. Conf.*, **31**.
26. Dolata, R., Khabipov, M., Buchholz, F-Im, Kessel, W., and Niemeyer, J. (1995). *Inst. Phys. Conf. Ser.* No. 148, p. 1709.
27. Abelson, L., Elmadjian, R., Ladizinsky, E., Lee, A., and Chan, H. W. (1993). *Ext. Abstr. 4th Int'l. Superconductive Elec. Conf.*, 270.
28. Meng, X., Bhat, A., Zheng, L., Wong, A., and Van Duzer, T. (1999). *Proc. of the 7th Int. Superconductive Elec. Conf.* 235–7.
29. Lochschmied, R., Herwig, R., Neuhaus, M., and Jutzi, W. (1997). *IEEE Trans. Appl. Superconductivity* **7**: 2983.
30. Ketchen, M. B., Pearson, D., Kleinsasser, A. W., Hu, C.-K., Smyth, M., Logan, J., Stawiasz, K., Baran, E., Jaso, M., Ross, T., Petrillo, K., Manny, M., Basavaiah, S., Brodsky, S., Kaplan, S. B., Gallagher, W. J., Bhushan, M. (1991). *Appl. Phys. Lett.* **59**: 2609–261.
31. Chen, W., Rylyakov, A. V., Patel, V., Lukens, J. E., and Likharev, K. K. (1998). *Appl. Phys. Lett.* **73**: 2817–2819.
32. Numata, H., Nagasawa, S., Tanaka, M., and Tahara, S. (1999). *IEEE Trans. Appl. Supercond.*, in press July 1999.
33. Caputo, J. G., Febrvre, P., Karpov, A., Larrey, V., Miletto-Granozio, V., Salez, M., and Villegier, J. C. (1999). *IEEE Trans. Magnetics* **9**: 3216.
34. Wang, Z., Terai, H., Kawakami, A., and Uzawa, Y. (1999). *Proc. of the 7th Superconductive Elec. Conf.*, 244–246.
35. Kerber, G. L., Abelson, L. A., Elmadjian, R., Hanaya, G., and Ladizinsky, E. (1997). *IEEE Trans. Appl. Superconductivity* **7**: 2638.
36. Moeckly, B. H. and Char, K. (1997). *Appl. Phys. Lett.* **71** (17).
37. Hirst, P. J., Satchell, J. S., Atkin, I. L., Wooliscroft, M. J., and Humphreys, R. G. (1999). *Proc. of the 7th Int. Superconductive Elec. Conf.*, pp. 285–7.
38. DiIorio, M. S., Yang, K-Y., and Yoshizumi, S. (1995). *Appl. Phys. Lett* **67**: 1926.
39. Talvacchio, J., Forrester, M. G., Hunt, B. D., McCambridge, J. D., and Young, R. M. (1997). *IEEE Trans. Appl. Supercond.*, **7**: 2051–2056.
40. Murduck, J., Pettiette-Hall, C. L., Hu, R., Salazar, O., McGerr, M., Daly, K., and Chan, H. (1999). *IEEE Trans. Appl. Supercond.* **9** (2): 3354.
41. Kurosawa, I., Nakagawa, H., Aoyagi, M., Kosaka, S., and Takada, S. (1991). *RAM IEEE J. Solid-State Circuits* **26**: 572.
42. Lochschmied, R., Herwig, R., Neuhaus, M., and Jutzi, W. (1997). *IEEE Trans. Appl. Superconductivity* **7**: 2983.
43. http://gamayun.physics.sunysb.edu/RSFQ/Lib/
44. Abelson, L., Elmadjian, R., and Kerber, G. (1999). *IEEE Trans. Appl. Supercond.* **9** (2): 3228.

45. Aoyagi, M., Shoji, A., Kosaka, S., Shinoki, F., and Hayadiawa, H. (1986). *Advances in Cryogenic Engineering Materıals* **32**: 557–563.
46. Ketchen, M. B., Pearson, D., Kleinsasser, A. W., Hu, C.-K., Smyth, M., Logan, J., Stawiasz, K., Baran, E., Jsao, M., Ross, T., Petrillo, K., Manny, M., Basavaiah, S., Brodsky, S., Kaplan, S. B., and Gallagher, W. J. (1991). *Appl. Phys. Lett.* **59** (20): 2609–2611.
47. Lee, L. P. S., Arambula, E. R., Hanaya, G., Dang, C., Sandell, R., and Chan, H. (1991). *IEEE Trans. Magnetics* **27**: 3133–3136.
48. Blamire, M. G., Evetts, J. E., and Hasko, D. G. (1989). *IEEE Trans. on Magnetics* **25**: 1123.
49. Imamura, T. and Hasuo, S. (1988). *J. Appl. Phys.* **64** (3): 1568–1588.
50. Kroger, H., Smith, L. N., and Jiltie, D. W. (1981). *Appl. Phys. Lett.* **39**: 280–282.
51. Huggins, H. A. and Gurvitch, M. (1985). *J. Appl. Phys.* **57**: 2103.
52. Yuda, M., Kuroda, K., and Nakano, J. (1987). *J. Appl. Phys.* **26** (3): L166.
53. Shoji, A., Shinoki, F., Kosaka, S., Aoyagi, M., and Hayakawa, H. (1982). *Appl. Phys. Lett.* **41** (11): 1097–1099.
54. Heinsohn, J.-K., Hadfield, R., and Dittmann, R. (1999). *Proc. of the 7th Int. Superconductive Elec. Conf.,* 305–307.
55. Eulenburg, A., Romans, E. J., and Pegrum, C. M. (1999). *IEEE Trans. Magnetics* **9** (2): 2402.
56. Murduck, J., Pettiette-Hall, C., Cordromp, J., Sergant, M., Burch, J., LaGraff, J., Salazar, O., Aquilino, H., Hu, R., Elmadjian, R., and Chan, H. (1997). *Proc. of the 10th Int. Symp on Superconductivity* **2**: 1185–1190.
57. Nakagawa, H., Nakaya, K., Kurosawa, I., Takada, S., and Hayakawa, H. (1986). *Jpn. J. Appl. Phys.* **25**: L70.
58. Morohashi, S. and Hasuo, S. (1986). *J. Appl. Phys.* **60**: 3774.
59. Morohashi, S. and Hasuo, S. (1987). *Ext. Abst. 1987 Int. Superconct. Electon. Conf.,* Tokyo, 305.
60. Murduck, J. M., Porter, J., Dozier, W., Sandell, R., Burch, J., Bulman, J., Dang, C., Lee, L., Chan, H., Simon, R. W., and Silver, A. H. (1989). *IEEE Trans. MAgnetics* **25**: 1139–1142.
61. Kleinsasser, A. W., Miller, R. E., Mallison, W. H., and Arnold, G. B. (1994). *Phys. Rev. Lett.* **72**: 1738–1741.
62. Kleinsasser, A. W., Mallison, W. H., and Miller, R. E. (1995). *IEEE Trans. Appl. Supercond.* **5**: 2318.
63. Harrop, P. J. and Campbell, D. S. (1970). *Handbook of Thin Film Technology,* Chap. 16, L. I. Maissel and R. Glang, eds., New York: McGraw-Hill.
64. Meng, X., Bhat, A., and Duzer, T.V. (1999). *IEEE Trans. Magnetics*, **9** (2): 3208.
65. Kruger, F. (1919). British Patent 157, 909.
66. Wang, Q., Kikuchi, T., Kohjiro, S., and Shoji, A. (1997). *IEEE Trans. Appl. Supercond* **7**: 2801–2804.
67. Balashov, D., Khabipov, M.I., Buchholz, F.-Im., Kessel, W., and Niemeyer, J. (1999). *Proceedings of the 7th Int. Supercond. Elec. Conf.,* 238.
68. Brinkman, A., Golubov, A. A., Rogalla, H., and Kupriyanov, M. Yu. (1999). *Proceedings of the 7th Int. Supercond. Elec. Conf.*, 269.
69. Wolf, P. (1976). *IBM Technical Disclosure Bulletin* **18** (8).
70. Patel, V., Tolpygo, S. K., Chen, W., and Lukens, J. E. (1999). *Proceedings of the 7th Int. Supercond. Elec. Conf.*, 229 .
71. McCumber, D. E. (1968). *J. Appl. Phys.* **39**: 3113.
72. Stewart, W. C. (1968). *Appl. Phys. Lett.* **12**: 277.
73. SAL NanoLithography, http://www.XRAYLITHO.com.
74. Aoyagi, M., Maezawa, M., Nakagawa, H., and Kurosawa, I. (1997). *IEEE Trans. Appl. Supercond.* **7**: 2644.
75. Ketchen, M. B. (1991). *IEEE Trans. Magentics* **27**: 2916–2919.

76. Kerber, G.L., Abelson, L.A., Elmadjian, R. N., Hanaya, G., and Ladizinsky, E.G. (1997). *IEEE Trans. Appl. Supercond.* **7**: 2638–2643.
77. http://htmt.cacr.caltech.edu/Overview.html
78. Abelson, L., Herr, Q., Kerber, G., Leung, M., and Tighe, T. (1999). *Proc. of the 7th Int. Superconductive Elec. Conf.*, p. 278.
79. Aschermann, G., Friederich, E., Justi, E., and Kramer, J. (1941). *Phys. Zeir.* **42**: 349.
80. Gavaler, J. R., Janocko, M. A., Patterson, A., and Jones, C. K. (1971). *J. Appl. Phys.* **43**: 54.
81. (1997). *New Research Opportunities in Superconductivity IV*, February 10–12, Monterey, CA, 61.
82. Hohmann, R., Lienerth, C., Zhang, Y., Bousack, H., Thummes, G., and Heiden, C. (1999). *IEEE Trans. Magnetics* **9** (2): 3688.
83. www.cryomech.com.pulsetube.html
84. For example, Advanced Surface Microscopy, Inc., see http://www.a1.com.asm
85. Cukauskas, E. J., Carter, W. L., and Qadri, S. B. (1985). *J. Appl. Phys.* **57** (7): 2538.
86. (1997). *New Research Opportunities in Superconductivity IV*, February 10–12, Monterey, CA, 38.
87. Chaudhari, P., Mannhart, J., Dimos, D., Tsuei, C. C., Oprysko, M. M., and Scheuermann, M. (1988). *Phys. Rev. Lett.* **60**: 1653–1656.
88. Simon, R., Bulman, J. B., Burch, J. F., Coons, S. B., Daly, K. P., Dozier, W. D., Hu, R., Lee, A. E., Luine, J. A., Platt, C. E., and Zani, M. J. (1991). *Appl. Phys. Lett.* **58**: 543–545.
89. Herrmann, K., Zhang, Y., Muck, H. M., Schubert, J., Zander, W., and Braginski, A. I. (1995). *J. Appl. Phys.* **78**: 1131–1136.
90. Jia, C. L., Kabius, B., Urban, K., Herrmann, K., Cui, G. J., Schubert, J., Zander, W., Braginski, A. I., and Heiden, C. (1991). *Physica C* **175**: 545.
91. Foley, C. P., Lan, S., Sankrithyan, B., and Wilson, Y. (1997). *IEEE Trans. Appl. Supercond.* **7**: 2051–2056.
92. Mitzuka, T., Yamaguchi, K., Yoshikama, S., Hayashi, K., and Enomoto, Y. (1993). *Physica C* **218**: 229–39.
93. Dillman, F., Glyantsev, V. M., and Siegel, M. (1996). *Appl. Phys. Lett.* **69**: 1948–1950.
94. Moeckly, B. H. and Char, K. (1997). *Appl. Phys. Lett.* **71** (17).
95. Pauza, A. J., Booij, W. E., Herrmann, K., Moore, D. F., Blamire, M. G., Rudman, D. A., and Vale, L. R. (1997). *J. Appl. Phys.* 5612.
96. Goodyear, S. W., Chew, N. G., Humphreys, R. G., Satchell, J. S., and Lander, K. (1995). *IEEE Trans. Appl. Supercond.* **5**: 2923.
97. Likharev, K. K. (1972). *Sov. Phys. JETP* **34**: 906–912.
98. Li, H. Q., Ono, R. H., Vale, L. R., Rudman, D. A., Liou, S. H., and Mallison, W. H. (1997). *IEEE Trans. Appl. Supercond.* **7**: 2169–2172.
99. Sato, H., Akoh, H., Takada, S., and Miyagawa, R. (1997). *IEEE Trans. Appl. Supercond.* **7**: 2165–2168.
100. Pettiette-Hall, C. L., Murduck, J., Burch, J. F., Sergant, M., Hu, R., Cordromp, J., Luong, M., and Ellis, R. (1997). *IEEE Trans. Appl. Supercond* **7**: 2057–2062.
101. Rosamilia, J. M., Miller, B., Schnemeyer, L. F., Waszczak, J. V., and O'Bryan, H. M. (1987). *J. Electrochem. Soc* **134**: 1863–1864.
102. Blank, D. H. A., Rijnders, G. J. H. M., Bergs, R. M. H., Verhoeven, M. A. J., and Rogalla, H. (1997). *IEEE Trans. Magnetics* **7** (2): 2940.
103. Pettiette-Hall, C. L., Murduck, J., Burch, J. F., Sergant, M., Hu, R., Cordromp, J., and Aquilino, H. (1999). *IEEE Trans. Appl. Supercond* **9** (2): 1998.
104. http://www.hiden.co.uk/
105. Ekin, J. W., Clickner, C. C., Russek, S. E., and Sanders, S. C. (1995). *IEEE Trans. Magnetics* **57** (2): 2400.

Microwave Magnetic Film Devices

DOUGLAS B. CHRISEY

Plasma Processing Section, Naval Research Laboratory, Washington, DC, USA

PAUL C. DORSEY

Komag, Inc., Milpitas, California, USA

J. DOUGLAS ADAM

Northrop Grumman STC, Baltimore, Maryland, USA

HARRY BUHAY

Northrop Grumman STC, Pittsburgh, Pennsylvania, USA

7.1. Introduction to Applications of Magnetic Films in Microwave Devices

7.1.1. INTRODUCTION TO THE PROPERTIES OF FERRITES

Magnetic film microwave devices are based on a broad class of ferrimagnetic materials commonly referred to as ferrites. Both ferro- and ferrimagnetic materials exhibit the hysteresis effect (see Fig. 7.1), which is a nonlinear relationship between the applied magnetic field (H) and the magnetic induction (B) of the materials. In ferromagnetic materials the magnetic moments of all the atoms are oriented in the same direction whereas ferrimagnetism arises from the antiparallel alignment of the magnetic moments of ions on different sublattices in the crystal. The oppositely directed magnetic moments do not exactly cancel, so

ISSN 1079-4050
Vol. 28
ISBN 0-12-533028-6/$35.00

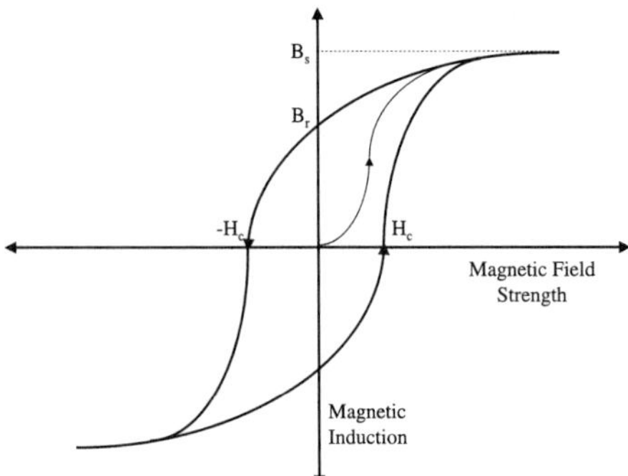

FIG. 7.1. Schematic diagram of a typical magnetic hysteresis loop of an unpoled ferrite. As the external magnetic field is increased, the magnetization increases to saturation (B_s). When the external field is removed the remanent magnetization (B_r) remains. If the external field is reversed, the magnetization goes to zero at coercive field (H_c).

that a net magnetic moment results. Néel was the first to describe this type of magnetism that ferrites exhibit, that is, ferrimagnetism [1]. The magnetic hysteresis loop consequently governs many of the properties associated with a particular ferrite. Besides the properties indicated in Fig. 7.1, other relevant properties include the permeability (μ), dielectric loss tangent (tan δ), anisotropic field (H_a), ferrimagnetic resonance frequency (f_o) and linewidth (ΔH), etc. Features common to all ferrites include the following properties:

1. They are all oxides.
2. They are all based on Fe_2O_3 as the major compositional component.
3. They exhibit a spontaneous magnetic induction in the absence of an external magnetic field.

The important properties that make ferrites useful for microwave device applications are their strong magnetic coupling, high resistivity, and low loss characteristics. Other materials, which could fit under the title of magnetic films for microwave devices, could include superconductors, colossal magnetoresistance materials, and spin-polarized ferromagnetic magnetic metals, but for the sake of space and completeness these topics will be omitted from this chapter.

Ferrites have a large variety of crystal structures that result in a wide range of properties and applications. The three dominant crystal classes are the garnets, spinels, and hexagonals. In Table 7.1, a summary of the crystal structures, compositions, and magnetic properties of these classes of ferrite materials are given. The magnetic properties of most ferrites can be tailored to the specific application of interest because of the substitutional solubility of the transition metal ions. In general, garnets have good magnetic properties and an extremely narrow ferrimagnetic resonance linewidth, spinel ferrites offer higher saturation

TABLE 7.1
SUMMARY OF FERRITE CRYSTAL STRUCTURE TYPES

Type (structure)	Ferrite composition	MeO	Common ferrite materials	Typical magnetic properties $4\pi M_s(G)$, H_a(Oe), ΔH(Oe)
Garnet (cubic)	5 Fe_2O_3 : 3 Me_2O_3	Rare earth metal oxide	$Y_3Fe_5O_{12}$	1750, 82, 0.5 (X-band)
Spinel (cubic)	1 Fe_2O_3 : 1 MeO	Transition metal oxide	(Ni,Zn)Fe_2O_4, (Mn,Zn)Fe_2O_4	3800, 300, 77 (X-band)
Magnetoplumbite (hexagonal)	6 Fe_2O_3 : 1 MeO	Divalent metal oxide from group II A–BaO, CaO, SrO	$BaFe_{12}O_{19}$, $SrFe_{12}O_{19}$	4700, 16500, 26 (55 GHz)

magnetization and higher Curie temperatures, and last, hexagonal ferrites have a very large magnetocrystalline anisotropy, which reduces the field needed for resonance. The coercive field (H_c) is the most important property of magnetic materials and is the first criterion used in selecting particular ferrites for different applications. Materials with low H_c are called magnetically soft while those with higher values are called magnetically hard materials.

A ferrite can, in principle, be magnetized in any direction, but in certain directions the magnetization process is easier. Optimized ferrite magnetic properties will also depend on the crystal anisotropy and magnetostriction. The magnetocrystalline anisotropy is that part of the crystal energy that is dependent on the direction of the magnetization in the lattice whereas the magnetostriction is crystal energy that is dependent on the stress applied.

7.1.2. MAGNETIC MICROWAVE DEVICES

Ferrites are very useful for magnetic microwave devices because of their strong coupling to magnetic fields and the low loss for electromagnetic wave transmission. In common homogeneous, linear and nonmagnetic media, the transmission of electromagnetic waves is isotropic. Under a magnetic field, where some or all of the moments are aligned, ferrites become anisotropic to the transmission of electromagnetic waves, which results in a nonreciprocal rotation of the plane of polarization, phase shift, and displacement of the microwave field pattern. These nonreciprocal properties are extremely useful in fabricating important microwave devices such as isolators, circulators, power limiters, filters, and phase shifters. In fact, ferrite nonreciprocal devices provide unique circuit functions that cannot be reproduced with any other materials.

Nonreciprocal magnetic ferrite microwave device functions are extremely simple to visualize schematically, but almost impossible to describe analytically due to the complexity of modeling a vector magnetic potential (**A**) as compared to modeling of a scalar potential (ϕ) for simple passive microwave devices. Schematic diagrams of the circuit function for an isolator, circulator and phase shifter are shown in Fig. 7.2. An isolator and a circulator are very similar in function. An isolator allows power flow in one direction while limiting the power flow in the opposite direction. Isolators are used in microwave systems to isolate sources from the undesirable effects of reflection from mismatched loads. In a multiport circulator, the power flows in a cyclic order, that is, from port 1 to port 2 only and from port 2 to port 3, only, and so on. Circulators are used in transmit/receive (TR) modules to allow a single antenna to be used for both functions, that is, isolate the large transmitted signal from the small received one. They can also be used with reflection-type negative-resistance amplifiers to separate the amplified signal from the input signal. Nonreciprocal phase shifters introduce a differential phase shift in the path of a transmitted wave compared to that transmitted in free space. Microwave ferrite phase shifters are used extensively for switching and control of microwave signals in phased array antennas for electronic scanning. In all cases, the performance can be modulated with an applied magnetic field.

For the conventional microwave components already listed here, bulk ceramic or single crystals of spinel and garnet magnetic oxides are typically used, but these are not conducive to low-cost, high-volume manufacturing. There is a strong need, and thus a market, for high-quality thin-film ferrite devices. A major reduction in size, weight, and cost of ferrite devices can be achieved through developing techniques to eliminate the bulky external magnets present in all current devices. One approach to address the external magnet involves utilizing ferrites with high internal anisotropy fields, which can effectively self-bias the ferrite device, for example, hexagonal ferrites. The ferrite must also have a reasonably narrow linewidth and that can only be obtained by a single crystal or film with a high degree of crystallographic orientation.

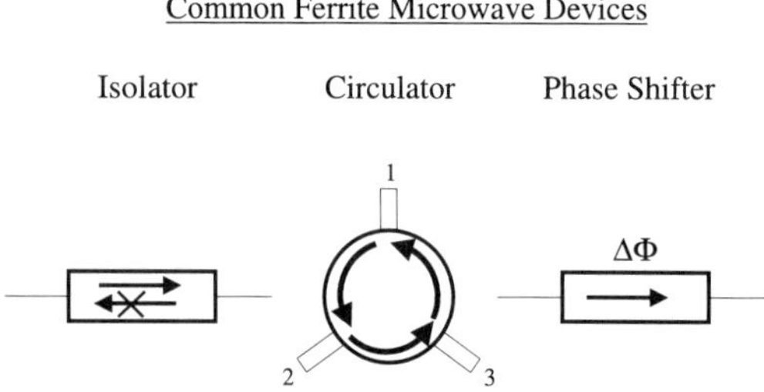

FIG. 7.2. Schematic examples of three common nonreciprocal ferrite devices. Nonreciprocal devices provide unique circuit functions that cannot be reproduced with any other materials.

7.1.3. FERRITE FILM DEVICES

The benefits and potential applications of ferrites for microwave devices have been known for decades, but their application in thin-film form has been hampered due to complicated problems in processing them [2]. In particular, ferrite film microwave devices require very thick films of multicomponent oxide ceramics. The problems associated with the latter have been troubling researchers for decades, as well, because the general category of thin-film oxide ceramics have a wide range of important properties and applications; these include superconductors, magnetics, optoelectronics, piezoelectrics, tribological materials, photonics, etc. The problems with processing oxide ceramics are unlike most any of those that exist in conventional electronic and semiconductor thin-film processing, including accurately and precisely controlling the stoichiometry of 3 or more components during deposition, high processing temperatures (typically ~30–60% of the ceramic melting point), high pressures of reactive gases (O_2), complicated crystal structures and phase diagrams (large unit cells), and anisotropic properties requiring highly oriented growth. The problems associated with the film-substrate's lattice match, temperature coefficient of resistance, and chemical inertness issues are, by themselves, almost equally as large. Furthermore, as the nonreciprocal effect is a volume effect, ferrite films used for microwave devices must be very thick (depending on their operational frequency) in order to minimize insertion losses (e.g., ~100-μm thick at 9 GHz and ~50-μm thick at 35 GHz), which means that the tolerances for all of the forementioned processing issues are even tighter because the film deposition must either be at a high rate or over long periods of time. Defects in ferrite films, such as dislocations, cracks, voids, compositional inhomogeneity, impurities, strain, and vacancies will increase losses and degrade performance. These seemingly insurmountable difficulties existed in 1987 and since that time a significant amount of progress has been made, curiously, on the coat tails of the high T_c superconductor (HTS) revolution. The progress falls into two categories: improving existing techniques for ceramic thin-film processing and developing novel approaches. An argument could be made that it was the potential applications of HTS materials and the excitement of working on something with absolutely no dc electrical resistance that catalyzed researchers to develop new methods for HTS film deposition; that is, methods that could be applied with equal success to the other oxide ceramic materials listed here. It is interesting to note that in June of 1987, about the time of the discovery of HTS, Glass wrote in the *Microwave Journal* that "ferrite films for microwave devices have been waiting in the wings for perhaps 20 years" [3]. What is more, there have been order of magnitude reductions in the cost and size of conventional microwave circuits and components through improved design, fabrication and production techniques. In 1990 it was stated that magnetic materials like ferrites have awaited the appropriate material processing techniques to deposit them in thin-film form in order for their properties to be exploited on a scale similar to semiconductors [4]. The performance of many common microwave circuits is limited by components that are poorly implemented or cannot be implemented using conventional technology. This imbalance in maturity exists between ferrite components and common monolithic components (e.g., amplifiers and switches).

To approach the issue of fabricating ferrite films for microwave devices it is useful to consider schematically the potential issues of processing them as listed here and integrating them in monolithic microwave integrated circuit (MMIC) technology, that is, with ferrite films and active devices residing on a common semiconducting substrate. The benefits of integrating ferrites with MMIC technology include, smaller size, higher reliability, potentially lower cost at large production numbers, and lower parasitics and better reproducibility, which contributes to better performance especially at high frequencies. Thus, there is a strong market incentive for fabrication of planar thin-film nonreciprocal devices on semiconducting substrates. Unfortunately, semiconductor integration is the most difficult possible implementation of ferrite film microwave devices because of the low processing temperatures allowed and the poor film-substrate match. Despite the technological hurdles involved, significant progress has been made through processing breakthroughs as described later in the chapter. The problems associated with the processing of ferrite films are associated with the way in which ceramic films nucleate and grow from the vapor flux. For almost all applications, single-phase materials are required and in some cases oriented or epitaxial crystal orientation is preferred. This is because single crystal thin films lead to improved performance at higher frequencies [5]. For magnetostatic wave devices, epitaxial films are required because polycrystalline films have poor transmission [6]. Figure 7.3a defines the different possible crystal orientations and Fig. 7.3b shows the evolution of potential defects commonly found with thicker film growth using the HTS material YBCO as an example. In Fig. 7.3a, the crystallographic differences between random, oriented and epitaxial film growth are schematically illustrated. Epitaxial films provide the best approximation to single crystal ferrite properties with the exception of liquid phase epitaxial yttrium iron garnet (YIG) on $Gd_3Ga_5O_{12}$ (GGG). In Figure 7.3b, the problems typically associated with growing thick ceramic films are illustrated. In this case, the example demonstrates what is observed with YBCO as a function of thickness, but the effects are qualitatively similar to what is observed for ferrites and other oxide ceramics. Simply stated, as the film grows thicker, the positive influence of the substrate for crystal growth and registration is lost and instead the cumulative effects of various defects can become dominant in determining subsequent film growth. On the other hand, epitaxial ferrite films for very low-loss hybrid-MIC applications (i.e., connecting off-chip to discrete passive microwave devices) do not necessarily require low temperatures because substrates for epitaxial ferrite growth are typically capable of withstanding

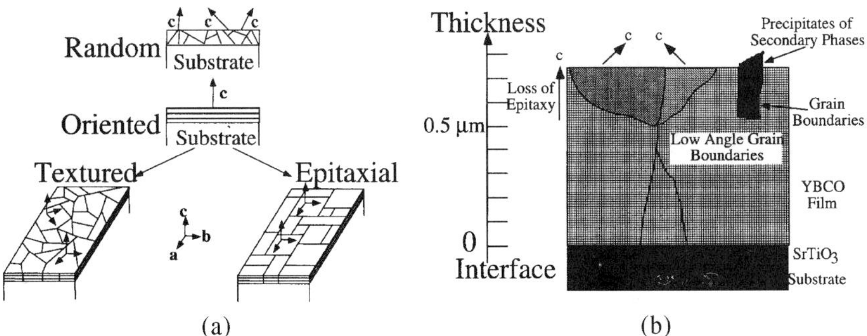

FIG. 7.3. Schematic diagram of: (a) the possible crystal orientations of a single phase anisotropic material, and (b) the evolution of common ceramic defects with thickness for a YBCO film.

relatively high substrate temperatures. However, controlling stoichiometry, microstructure, and stress is still an issue if microstructural defects and high losses are to be avoided. In either case, combining the appropriate deposition technique with a defined deposition process and substrate material is a difficult task that needs to be optimized for each specific application.

7.2. Current Approaches to Fabricate Ferrite Films

7.2.1. INTRODUCTION

The ideal approach to fabricate ferrite films for microwave devices would have to have a high deposition rate ($\sim$1–10 μm/min), a large deposition area ($>1\ cm^2$), result in films with narrow line widths ($\sim$200 Oe for circulator/isolator applications and $\sim$1–10 Oe for filter applications), a well-controlled anisotropy field, low dielectric loss tangents ($\tan\delta < 0.001$), a high degree of crystallographic orientation, uniform grain structure and stoichiometric control throughout the entire thickness, near bulk magnetization, a low deposition temperature ($<1000\,^{\circ}C$ for Si and $<600\,^{\circ}C$ for GaAs), and the ability to produce thick films ($\geq$100 μm that are smooth). While great strides have been made to prove the viability of thin-film ferrite microwave devices, no one technique possesses all the required attributes for commercial fabrication line insertion. Figure 7.4 shows a schematic diagram of four popular and successful techniques for ferrite film growth—spin spray, jet vapor deposition, pulsed laser deposition, and sputtering. These four techniques vary significantly in their approach, especially in the area of deposition pressure, temperature, and rate as well as cost. The spin spray technique also could be

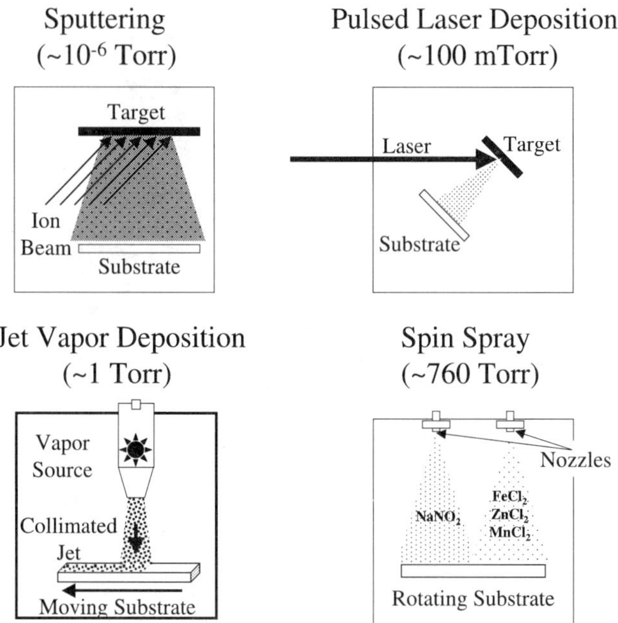

FIG. 7.4. Four popular and successful ferrite film deposition techniques. Sputtering, pulsed laser deposition, jet vapor deposition, and spin spray vary considerably in their approach, required equipment, cost, and vacuum and temperature requirements.

characterized as chemical precipitation whereas the others are physical vapor deposition techniques. In what follows we briefly summarize the advantages and disadvantages of the more popular and successful approaches used for ferrite film deposition and list some of the others that have been employed to a lesser extent.

7.2.2. Spin Spray

Researchers at the Tokyo Institute of Technology demonstrated in the late 1980s a novel electroless plating technique called spin spray for preparing spinel ferrite thin films at atmospheric pressure and temperatures $<100\,^{\circ}\mathrm{C}$ [7–11]. Here chloride solutions of the spinel transition metal constituents (Fe, Ni, Zn, etc.) and a separate solution of an oxidizing solution (e.g., sodium nitrite and ammonium acetate) are sequentially sprayed on a spinning substrate that is heated to $100\,^{\circ}\mathrm{C}$. The film is thus built up layer-by-layer by successive deposition and oxidation steps. By this approach polycrystalline films of mixed Ni, Zn, Co, and Mn spinels were produced. These films could be used for nonreciprocal devices, but not for magnetostatic wave devices because the propagation losses were too high (>20 dB). The deposition rate can be improved to several microns per minute by using a laser to produce localized superheating during deposition. Improvements in the dielectric loss tangent are needed before this technique can be commercialized; in particular, the presence of Fe^{2+} ions in the films yields an unacceptably high conductivity. However its low processing temperature and equipment costs separate it significantly from all other techniques.

7.2.3. JET VAPOR DEPOSITION (JVD)

Jet-vapor deposition takes place in a low pressure deposition chamber (~1 torr). The two main components are the jet source and the substrate motion mechanism [12–15]. The jet source comprises a source of volatilized material (atomic or molecular constituents of the intended films) that is combined with an inert carrier gas and expelled through a nozzle in the form of a sonic jet. This collimated jet is aimed at a moving substrate to deposit the volatilized material. The material is transported to the substrate in a line-of-sight from the orifice by convection in the inert gas. The expanding gas travels at sonic speeds. The deposition is axially symmetric, but nonuniform in thickness, thus requiring the substrate to be rastered to provide a uniform coating. The strengths of JVD include, economically high deposition rates and collimated flux, simple and inexpensive apparatus, usefullness with virtually any material, uniform films are produced over large areas, versatility for synthesis of multicomponent materials, negligible transport limitations, and environmental friendliness. The disadvantages include poor control over film deposition, poor surface morphology, and difficulties in depositing on heated substrates. The JVD has many of the benefits of plasma-arc spray, but with smaller atomic or molecular components as compared to molten powders [16, 17].

7.2.4. PULSED LASER DEPOSITION (PLD)

Pulsed laser deposition is an especially well-suited physical vapor deposition technique for ferrite film growth. This is because ferrites are multicomponent oxides with a complex and anisotropic crystal structure and because PLD has the unique ability to reproduce in thin film form the stoichiometry of the starting material. Thus, many different ferrite film compositions and substitutions can be rapidly investigated for solid solution ferrites. Unlike spin spray and JVD, PLD can readily produce oriented or epitaxial thin films on lattice-matched substrates or buffer layers, which is very important for low losses and to make use of the large uniaxial anisotropy of the hexagonal ferrites.

Nathaway *et al.* and Ogale *et al.* were the first researchers to apply PLD to the growth of spinel ferrite films. Thin films of Fe_3O_4, $Ni_xFe_{3-x}O_4$, and $Zn_xFe_{3-x}O_4$ were deposited with a ruby laser [18–21]. Although no magnetic properties were presented in their published research, they demonstrated that crystalline and textured ferrite films could be deposited by PLD. Shortly thereafter, PLD was successfully applied to produce high-quality epitaxial thin films of Mn-Zn spinel, YIG, and the hexagonal $BaFe_{12}O_{19}$, that is, all three classes listed in Table 7.1 above [22–25]. In particular, the beneficial effects of depositing ferrites on lattice-matched substrates and elevated temperatures ($>850\,^{\circ}C$) for achieving bulk-like magnetic properties and low losses as indicated by FMR linewidths approach single crystal values. Even more recent articles on the PLD of ferrites extend this work to other materials, multilayers, substitutions, and effects of strain and buffer layers [26–35]. Most of the benefits of PLD for high-quality epitaxial ferrite film deposition have already been listed. The disadvantages of PLD include the presence of macroscopic particulates, initial industrial set-up costs, and low deposition rates. These disadvantages have been addressed to a large extent, but there is still hesitation to insert the first laser-based deposition technique into an industrial fabrication line. Thus, PLD has remained an outstanding research tool for ceramic thin film deposition including ferrites.

7.2.5. SPUTTERING

Sputtering is an accepted industrial electronic thin film deposition technique, but in applications to ferrite films for microwave devices, it falls short because of the slow deposition rate and difficulties with multicomponent oxides. Still, there are applications in magnetostatic wave devices and magnetic recording where sputtering could prove useful and for that reason there exists a substantial database [36–42]. In Fig. 7.3, ion beam sputtering is illustrated, but it should be noted that there are a multitude of different approaches to sputtering including rf diode sputtering, facing-targets sputtering (FTS), off-axis sputtering, and rf magnetron sputtering. Because of the energetics of the process, sputtering can produce unique microstructures and magnetic properties. Other disadvantages of sputtering for ferrite film growth include the high processing temperatures required, low deposition rates, and difficulty in processing a multicomponent oxide.

7.2.6. LIQUID PHASE EPITAXY

The existence of liquid phase epitaxy (LPE) for ferrite film growth is useful in many ways, but especially because it demonstrates the potential for ferrite film microwave devices. Unfortunately, the high processing temperature and strict requirement for lattice match preclude its widespread application to other classes of ferrites [43–45]. The use of the nonmagnetic GGG is ideal as a substrate for the epitaxial growth of substituted YIG films because of its close lattice match, chemical inertness, and high melting point. The LPE YIG films have nearly intrinsic (single crystal) properties. Thus, the LPE of YIG on GGG is a relatively unique area for ferrite film growth.

7.2.7. OTHER APPROACHES

There are many other methods to fabricate ferrite films for microwave devices. These include direct chemical precipitation, sol-gel deposition, arc plasma spray, molecular beam epitaxy (MBE), chemical vapor deposition, spray pyrolysis, and tape casting [46–52]. Except for the last one, these techniques each have drawbacks that limit their application to research environments until further process development is achieved. Tape casting of ceramic films has been a common and economical fabrication technique for many years, although most efforts have been focused on small-grained, electrically isotropic materials. The approach involves forming a slurry of the desired material with an organic binder and solvent and spreading the material to a uniform thickness for drying to produce green sheets. The sheets are then heated to evaporate the binder and sintered at high temperatures. Tape casting is an especially interesting approach for ferrite films for microwave devices because it approaches the issue of ferrite film processing by applying bulk-like techniques, that is, start with what works well for bulk and make it smaller (thinner).

7.3. Ferrite Film Progress

7.3.1. INTRODUCTION

The magnetic and dielectric properties of a ferrite affect the applicability of the material to any particular microwave device and the performance of the device once fabricated. For example, properties such as the ferrimagnetic resonance (FMR) linewidth (ΔH), saturation magnetization ($4\pi M_s$), uniaxial anisotropy field (H_a) and Curie temperature (T_c) affect the losses, bandwidth, operating frequency, and temperature stability of microwave ferrite devices, respectively. Ultimately, the magnetic and dielectric properties are dictated by the composition and microstructure of the ferrite. Resonance microwave devices (e.g., filters) require low losses so single crystal films are necessary. On the other hand, off-resonance devices such as control components (e.g., isolators and circulators) can tolerate varying degrees of magnetic loss in the ferrite so both polycrystalline or single crystal films can be used.

The practical application of ferrite thin films to a wide range of microwave devices, which would typically be designed around bulk ferrites, has been extensively demonstrated in the literature [6,53–57]. These devices include resonators, filters, phase shifters, and circulators. Almost all of these devices have incorporated YIG, as it has been the only readily available high-quality ferrite thick film. The LPE YIG films have been grown up to 100 μm and exhibit a near intrinsic FMR linewidth of about 0.5 Oe or less. Some of the LPE YIG device configurations employ the same or similar design techniques used for bulk ferrites, while other designs take advantage of magnetostatic wave (MSW) technology [57–60]. These LPE YIG films are ideal for MSW applications because propagation losses are related to the FMR linewidth (i.e., a larger linewidth in general means larger propagation losses) and high-quality single-crystal films provide a more uniform effective internal field that affects propagation characteristics. Although YIG is an ideal ferrite in many respects, it is not suitable for many microwave device applications. It has a low T_c, a relatively small $4\pi M_s$ and small magnetic anisotropy terms, which put limits on the temperature stability, bandwidth, and millimeter-wave applications of devices incorporating YIG. In addition, LPE YIG films require single crystal substrates and high processing temperatures, which makes the technique incompatible with many applications such as MMIC. In order to apply planar ferrite technology to all of the applications currently served by bulk ferrite microwave devices and to make ferrite films competitive with both bulk ferrite and semiconductor technology, a variety of ferrite films on single-crystal, amorphous and semiconducting substrates must be made available. The following sections review the past and current status of ferrite film technology relating to garnets, spinels, and hexagonals and the approaches that are being used to remove these limitations.

7.3.2. GARNETS

There have been a wide range of techniques applied to the processing of garnet films including rf magnetron sputtering, ion beam sputtering, rf diode sputtering, chemical vapor deposition, reactive ion-beam sputtering and sol-gel, but the most successful application of thin film techniques to garnets for microwave applications has been LPE [3, 61–71]. Epitaxial growth techniques (i.e., growing single-crystal films on single-crystal substrates) require suitable substrates in order to produce high-quality films. The substrate should be of high quality, able to withstand the deposition environment, and have a lattice parameter and coefficient of thermal expansion (CTE) closely matched to the film with a crystal structure from the same crystallographic space group. In the case of YIG, there was active research on developing garnet crystals for use as laser hosts at about the same time YIG film growth was being investigated. The GGG is a paramagnetic oxide crystal that has a close lattice match with YIG. The cubic lattice parameter for YIG is 12.376 Å with a CTE of 10.4×10^{-6}/C, while the lattice parameter for GGG is 12.383 Å with a CTE of 9.18×10^{-6}/C [72]. Although the mismatches are relatively small, they can produce stress-induced magnetic anisotropy in the film, or in thicker films, stress relief can lead to film fractures or delamination. Lattice mismatch can be minimized by incorporating some Pb from the melt into the film lattice or by substitutions such as La for Y [72–75]. The highest quality YIG films have been achieved using the dipping method of LPE with an isothermal growth process and substrate rotation. The isothermal growth provides for homogeneous films, while rotation assists in the deposition of films with a uniform and controllable thickness. High-quality films using this technique have been available since the 1970s and further improvements in the processing by Glass led to the development of high-quality thick films up to a thickness of about 100 μm at a rate of 0.6 μm/min [76]. These films exhibit a near intrinsic Kasuya-LeCraw FMR ΔH at X-band (~0.2 Oe), indicating that the defect density in the LPE YIG films is very low. Specifically, the narrow linewidth means that the microwave loss contributions due to fast or slow relaxation processes from the valence exchange mechanism or impurity ions, or two magnon scattering induced by magnetic inhomogeneities, such as pits, voids, vacancies or surface imperfections, have been minimized. Although LPE has met many of the requirements for a practical means of producing high-quality single-crystal garnet films for microwave applications, the use of LPE and the need for single-crystal GGG substrates limit the applicability of these garnet films to discrete planar microwave devices as opposed to being fully compatible with MMIC technology. One novel approach to overcome this limitation involves bonding thick single-crystal YIG films to semiconductor substrates [77]. The YIG film is grown using LPE on a GGG substrate to achieve a high-quality film with optimum magnetic properties and then the film is transferred and bonded to a metallized silicon substrate using processing temperatures no greater than 200 °C. The exposed GGG substrate is removed by grinding and then treated to chemical-mechanical polishing to improve the surface finish. The insertion loss and isolation at X-band for a Y-junction circulator, which was fabricated using this procedure, were 1 dB and 20 dB over a 1 GHz bandwidth, respectively. A similar approach involves a lift-off technique that is performed by forming a sacrificial layer at the ferrite/substrate interface using energetic He ion implantation [78]. The substrate and film are then separated, preferentially by acid etching the sacrificial layer. Results have been demonstrated for a Bi-substituted YIG film for optical applications with only minor changes occurring in its magnetic properties due to the lift-off procedure. These results demonstrate some of the potential methods of overcoming the inherently difficult task of monolithically fabricating passive microwave components that incorporate high-quality single-crystal garnets with microwave integrated circuits.

One of the more recent promising approaches to garnet film growth has been pulsed laser deposition (PLD). Initial research was carried out on both Bi-substituted YIG and pure YIG films on (111) GGG single-crystal substrates [24, 25]. The Bi-substituted YIG films were grown using an ArF excimer laser in oxygen at a substrate temperature of about 500 °C, which was required to grow single-phase garnet films depending on the background pressure. Increasing background pressure from 27 to 133 Pa appeared to decrease the temperature required for crystallization slightly and produced denser, more uniform films. At substrate temperatures lower than about 500 °C, the films were amorphous. The single-phase Bi-substituted YIG films had a $4\pi M_s$ value of 1500 G and a uniaxial anisotropy field of 985 G, in the film plane, based on FMR measurements at 9.47 GHz. FMR linewidth values, which would have given some indication of crystalline quality for microwave applications, were not reported because these films were intended for magneto-optic applications. The FMR ΔH values have been reported for pure YIG films grown on (111) GGG using a KrF excimer laser in oxygen pressures ranging from 50 to 1000 mtorr and substrate temperatures of 700 to 850 °C. Over this range of deposition conditions the stoichiometry of the film matched the target, but film epitaxy improved at lower pressures and higher temperatures (i.e., 50 mtorr and 850 °C). Under these conditions, FMR ΔH values of about 1 Oe were measured and they approach the best values achieved for high-quality LPE YIG films. Not coincidentally, low oxygen pressures during deposition (~50 mtorr) produced denser films with a smoother surface morphology as compared to high oxygen pressures. In both of these cases the deposition rates were only 0.7 to 5.5 Å/s for 0.5- to 1.0 μm thick films, which are impractical for microwave device applications. High deposition rates (25 μm/hr) and thick (53-μm) epitaxial YIG films on (111) GGG have been demonstrated by researchers using PLD at Westinghouse [79]. High rates were achieved by using a high laser repetition rate of 150 Hz with a KrF excimer laser and a target to substrate distance of 5 cm. The best epitaxial films were achieved at a substrate temperature of 850 °C and a background oxygen pressure of 50 mtorr. Double-crystal x-ray rocking curves showed a slightly larger lattice parameter and full-width at half-maximum (FWHM) for the PLD YIG film as compared to a thick LPE YIG film (see Fig. 7.5). These differences were explained as possibly due to a nonuniform distribution of excess Fe in the film. The excess Fe might also explain the slightly larger linewidth of 5.7 Oe at 9.0 GHz as compared to LPE YIG films. Another possibility that might explain these results, is the presence of a nonuniform strain through the thickness of the film. The thick PLD YIG film had $4\pi M_s$ and dielectric loss tangent (tan Δ) values of 1800 G and 0.0002, respectively. These values, as well as the narrow FMR ΔH demonstrate

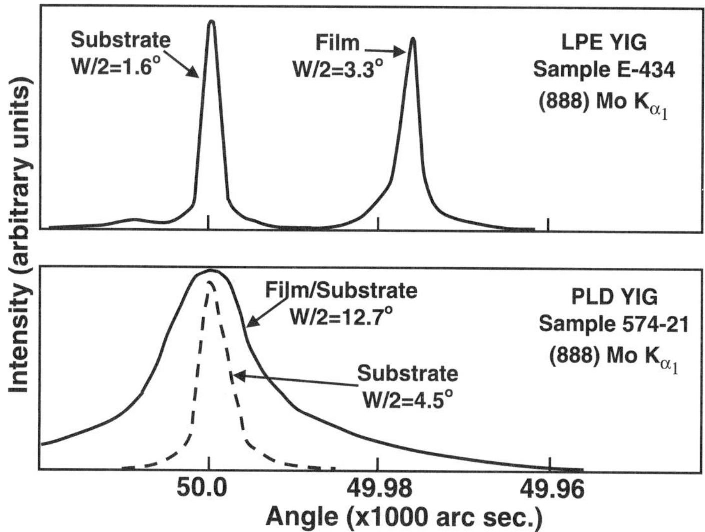

FIG. 7.5. Double-crystal x-ray rocking curves. These show a slightly larger lattice parameter and full-width at half-maximum (FWHM) for the PLD YIG film as compared to a thick LPE YIG film.

the ability to achieve fairly high-quality epitaxial garnet films at high rates using PLD. However, due to the success of LPE YIG films, most of the recent microwave garnet film development has focused on polycrystalline YIG films deposited on amorphous or semiconductor substrates.

Polycrystalline YIG films on amorphous or semiconductor substrates are a much more attractive application of PLD because LPE already meets the requirements as a practical means of producing high-quality single crystal garnet films for microwave applications. In addition, achieving garnet films on amorphous or semiconductor substrates provides larger cost and size benefits in the form of MMIC compatible microwave components. Initial studies undertaken using (100) single crystal silicon, (100) and (111) magnesium oxide (MgO), fused quartz, and amorphous glass showed promising results for achieving this goal [80]. On glass, the films were weakly magnetic when grown at temperatures below 720 °C. Post annealing of the samples below 700 °C had little effect, but higher temperatures fully magnetized the films, which were several microns thick. A rapid thermal anneal (RTA) at 700 °C for 4 to 6 min had a similar effect with slightly higher magnetization values than the furnace-annealed samples. The FMR ΔH values for these films were 55–160 Oe, which are comparable to bulk polycrystalline YIG. These YIG films on (100) silicon wafers were amorphous and nonmagnetic at a substrate temperature below 800 °C but above 800 °C the films were polycrystalline with effective saturation magnetization values of

1500 G. The FMR ΔH values at X-band ranged from 180–190 Oe. A similar trend was seen for YIG films grown on fused quartz. A substrate temperature of 850 °C was required to achieve polycrystalline magnetic films with slightly larger FMR ΔH values of 168–280 Oe depending on the applied magnetic field orientation. Polycrystalline YIG films were also obtained on (100) and (111) MgO substrates with FMR ΔH values ranging from 212–270 Oe. Although these studies did demonstrate the feasibility of using PLD to deposit polycrystalline garnet films on a variety of semiconductor, crystalline and amorphous substrates, the rates, thickness, and areas of the films were not sufficient for practical applications. Research aimed at overcoming these obstacles has been conducted on polycrystalline YIG films which were deposited onto 3 in. diameter Si and epi-GaAs on Si substrates using PLD [79, 81–82]. An off-axis orientation of the target with respect to the substrate and a 7 cm target to substrate distance were used in order to uniformly coat the large wafers. This limited the rate to 4 μm/hr, but films as thick as 100 μm were still achieved. Films were deposited in an amorphous state at low temperatures (i.e., 400–500 °C) onto metallized substrates and then subjected to a RTA at 850 °C for 20 s to achieve crystallization. The X-ray diffraction (XRD) pattern for a PLD polycrystalline YIG film on an Au-coated Si wafer in Fig. 7.6 shows good agreement with the standard YIG powder diffraction pattern. The $4\pi M_s$ value and FMR ΔH for a 100 μm PLD YIG film were measured to be 1680 G and 84 Oe, respectively, which agree well with the reported bulk values of $4\pi M_\sigma = 1730$ G and $\Delta H = 40$–100 Oe [83]. The 100 μm thick polycrystalline PLD YIG films were deposited on capped epi-GaAs on Si in a similar manner. These films exhibited a narrower FMR ΔH of about 35 Oe at X-

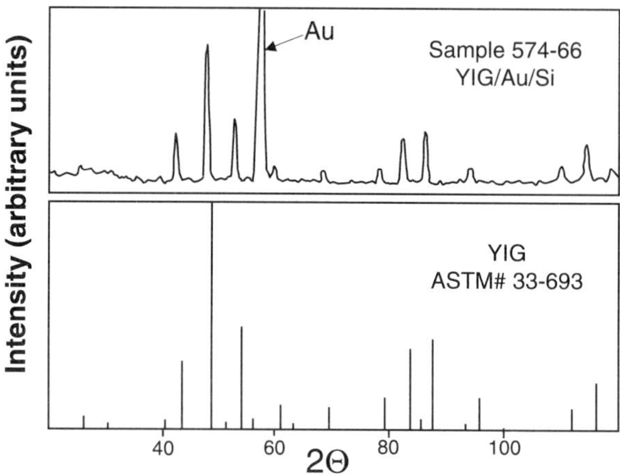

FIG. 7.6. The x-ray diffraction pattern for a PLD polycrystalline YIG film on an Au-coated Si wafer. This shows good agreement with the standard YIG powder diffraction pattern.

band. The microwave performance for circulators designed with these films, which will be discussed in a later section, demonstrate the feasibility of directly integrating passive microwave components with GaAs MMIC using polycrystalline garnet films.

7.3.3. SPINELS

Some of the earliest work on developing single-crystal spinel ferrite films involved CVD. A review by Mee *et al.* in 1969 summarized the state-of-the-art at that time in CVD growth of spinel ferrites which included: 1–150 μm of $MnFe_2O_4$, $NiFe_2O_4$ and $Mn_xNi_{1-x}Fe_2O_4$ on MgO, 5–50 μm of $CoFe_2O_4$ on MgO, and 1–10 μm of $MgFe_2O_4$ and $LiFe_5O_8$ on MgO [2]. The CVD was also used to grow 1–40 μm of YIG on GGG at a rate of ~6 μm/hr with FMR ΔH values, in the range of 0.6–1.6 Oe for films with a maximum thickness of 3.5 μm, indicating crystalline quality that is only slightly inferior to LPE YIG films. The LPE has several advantages over CVD, however, including lower cost, simpler operation, faster growth rates, better control over film composition and growth at lower temperatures [3]. Consequently, attempts were made to apply the same isothermal dipping LPE technique, which was so successful in the case of YIG, to the development of low-loss single-crystal lithium ferrite ($Li_{0.5}Fe_{2.5}O_4$) films [84]. Lithium ferrite, a spinel ferrite, only contains Fe in a 3+ valence state so microwave losses should theoretically be low as with YIG. This makes lithium ferrite an attractive alternative to YIG because lithium ferrite has a higher Curie temperature and saturation magnetization value, but this potential was never realized in bulk lithium ferrite single crystals. In the LPE process, nonmagnetic spinels with lattice parameters ranging from 8.083–8.381 Å were used as substrates to achieve a lattice match for lithium ferrite, which has a cubic lattice parameter of 8.33 Å. The resulting films were about 5 μm thick and cracked due to differences in CTE between film and substrate, but the best results were achieved on $MgGa_2O_4$ substrates at a growth temperature of 935 °C. The XRD indicated the films to be epitaxial spinel with a lattice parameter of 8.335 Å. Magnetic resonance values from FMR data at 9.07 GHz were consistent with bulk literature values and ΔH values as low as 7.8 Oe were achieved. These results were very encouraging, but further research was required to overcome the instability of the LPE melt used for the spinel growth and to utilize substrates with a more closely matched CTE.

Further research on epitaxial spinel ferrite films for microwave applications gained renewed interest recently with the development of PLD. The ability of PLD to deposit multicomponent oxides over a broad range of temperatures, gases, and pressures appeared to be ideal for investigating ferrite film growth. Other new techniques such as spray pyrolysis have also been used but the majority of recent epitaxial spinel ferrite work has focused on PLD [51]. Some of the first work related to PLD ferrite films was published by Nathaway *et al.* in 1989 [18]. In this work, zinc ferrite films were deposited on single-crystal sapphire substrates as a function of substrate temperature and oxygen pressure. The conditions necessary to obtain good-quality, stoichiometric zinc ferrite films as dictated by morphological, compositional and microstructural studies were 450 °C and 50 mtorr; however, no magnetic measurements were performed on these films. Shortly thereafter, numerous studies focusing on the magnetic and microwave properties of epitaxial PLD ferrite films were undertaken for a number of different spinel ferrite compositions including nickel ferrite, nickel-zinc ferrite, manganese-zinc ferrite, and lithium ferrite [22, 31, 85–88]. These studies demonstrated the effect of microstructure and composition on the magnetic and dielectric properties of spinel ferrite films and the ability to control these properties using PLD. Typically, these studies involved relatively thin films (0.5–2 μm) deposited at rates of 1–2 μm/hr on single-crystal substrates such as MgO over a range of substrate temperatures from 400–800 °C and oxygen pressures from 10–1000 mtorr. Although significant changes occur in film microstructure due to both temperature and pressure, most PLD spinel films grown in this range of conditions showed varying degrees of epitaxial quality. One important study by Williams *et al.* demonstrated the effect of microstructure on FMR ΔH, coercive field (H_c) and stress and the effect of composition and cation distribution on magnetization and magnetic anisotropy (K_1) [22]. Microstructurally, spinel ferrite films grown using PLD at low temperatures exhibit a columnar microstructure, which develops through the vertical growth of small crystallites, separated by low angle grain boundaries. As substrate temperature is increased to 800 °C, the film microstructure becomes denser with very few low angle grain boundaries. The XRD similarly shows an increase in the intensity and sharpness of diffraction peaks at higher substrate temperatures, indicating a decrease in both dislocation density and misorientation between film and substrate as well as a reduction in internal strains. These reductions in defects and strain with increasing substrate temperature caused a corresponding decrease or improvement in H_c, K_u and ΔH of the spinel films. The other important factor affected by substrate temperature in the spinel films was the Fe^{2+} concentration that caused changes in the net magnetization and the anisotropy of the spinel B (octahedral sites) sublattice. Consequently, the overall anisotropy of the spinel, which is the net anisotropy of both the A (tetrahedral sites) and B sublattices, is then affected. In general, the static magnetic properties (i.e., $4\pi M_s$ and K_1) of PLD spinel films in this study and others match bulk spinel properties, but the magnetic losses in epitaxial PLD films can be greater than bulk single-crystal spinels. Research by Srivastava *et al.* has shown that the magnetic losses in some PLD films have not achieved the intrinsic limits of the materials and that the increased linewidth qualitatively agrees with a two magnon scattering model using inhomogeneities in the submicron range and volume fractions below 1% [89]. Similar results were also found for PLD YIG films, but to a lesser extent, indicating that further research is required to improve the microstructure of epitaxial PLD spinel films in order to achieve the quality necessary for low loss microwave devices.

The more practical issues of rate and thickness have been investigated in thick epitaxial (12-μm) NiZn-ferrite ($Ni_{0.6}Zn_{0.4}Fe_2O_4$) films that were pulsed laser deposited at relatively high rate (12 μm/hr) [34]. The films were deposited on (100) MgO substrates and near bulk magnetization ($4\pi M_s = 5000$ G) was achieved at a substrate temperature of 700 °C. Although the films were (100) oriented, film quality was limited by a large compressive stress on the order of 10^{10} dynes/cm^2, which was measured using FMR and XRD techniques. The compressive stress produced delaminating regions in the film and caused a loss of in-plane orientation (i.e., the film exhibited a fiber texture). The magnitude of the compressive stress could be reduced nearly an order of magnitude by increasing the oxygen pressure during deposition from 50–200 mtorr, but this was accompanied by a degradation of film surface morphology (i.e., formation of a nodular surface (see Fig. 7.23) and a more polycrystalline film structure in XRD scans. Further research was then done on deposition of polycrystalline NiZn-ferrite films and MnZn-ferrite ($Mn_{0.4}Zn_{0.6}Fe_2O_4$) films directly onto Si and GaAs substrates for MMIC applications. The range of substrate temperatures used for deposition was 400–650 °C in a background pressure of 10–50 mtorr, which was necessary to maintain a relatively smooth surface morphology. In the case of NiZn-ferrite, pure oxygen was used as the deposition gas while air was used for the deposition of MnZn-ferrite. Table 7.2 shows a summary of the FMR and vibrating sample magnetometer (VSM) data for MnZn-ferrite films grown on silicon substrates in 10 mtorr of air at a substrate temperature of 650 °C, which was necessary to fully crystallize the ferrite film. The magnetic properties for these thick films at high rates were acceptable for microwave device applications and agree with the bulk values for MnZn-ferrite, which are $4\pi M_s = 4000$ G and

TABLE 7.2
SUMMARY OF AS-DEPOSITED PROPERTIES FOR PLD MnZn-FERRITE FILMS GROWN ON Si SUBSTRATES AT 650 °C IN 10 MTORR OF AIR. THE FMR ΔH VALUES WERE MEASURED WITH THE APPLIED MAGNETIC FIELD NORMAL (⊥) AND PARALLEL (∥) TO THE FILM PLANE.

Film thickness (μm)	ΔH (Oe) @ 9.8 GHz (⊥/∥)	$4\pi M_s$ (Gauss)	Deposition rate (μm/hr cm^2)
2	275/175	4100	–
33	284/518	4240	7.1
45	224/365	3920	8.1
72	370/460	4000	17.3

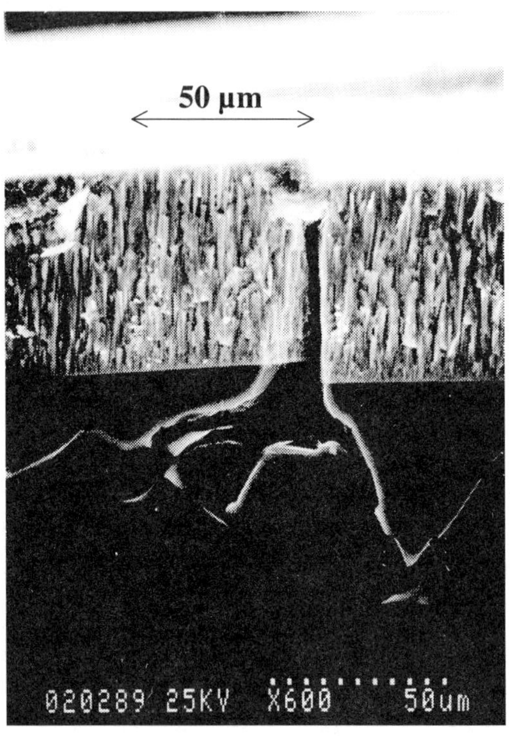

FIG. 7.7. Cross-section SEM micrograph of a 48 μm thick PLD MnZn-ferrite film, deposited on a Si substrate at 650 °C in 10 mtorr of air, showing cracking due to compressive strain.

$\Delta H = 200$ Oe at 9.4 GHz [83]. However, the films were not usable in actual devices because the compressive stress in these films caused fracturing of the Si substrates for thick films as shown in Fig. 7.7. The NiZn-ferrite films, which were deposited in 50 mtorr of oxygen, also exhibited a high degree of compressive stress on Si substrates, causing a bowing, but not a catastrophic fracturing of the Si substrate. This was due, in part, to the ability to crystallize the NiZn-ferrite films at lower temperatures than the MnZn-ferrite films, but differences in CTE between the two ferrites may also have been important. The results for some thin NiZn-ferrite films on Si substrates and thick NiZn-ferrite films on low dielectric loss alumina substrates are shown in Table 7.3. The tan δ values are comparable to bulk polycrystalline NiZn-ferrite, but the FMR ΔH values and $4\pi M_s$ value for the film deposited at 550 °C are about four times greater and 20% lower, respectively, than the values reported in the literature for bulk NiZn-ferrite (bulk NiZn-ferrite: ΔH ~100 Oe and $4\pi M_s = 5000$ G) [35]. The XRD patterns for thick PLD NiZn-ferrite films on alumina substrates are shown in Figs. 7.8 and 7.9 along with the standard powder x-ray diffraction pattern for Ni-ferrite,

TABLE 7.3.
SUMMARY OF AS-DEPOSITED PROPERTIES FOR PLD NiZn-FERRITE FILMS GROWN IN 50 MTORR OF OXYGEN. THE FMR ΔH VALUES WERE MEASURED WITH THE APPLIED MAGNETIC FIELD NORMAL ($\perp$) AND PARALLEL ($\parallel$) TO THE FILM PLANE.

Substrate	Substrate temperature (°C)	Film thickness (μm)	$4\pi M_s$ (gauss)	ΔH (Oe) @ 9.8 GHz ($\perp$/$\parallel$)	Tan Δ @ 9.3 GHz	Deposition rate (μm/hr)
Si	650	3	3699	610/460	–	4.9
Si	550	3	3620	590/136	–	4.9
Si	450	3	3348	920/1011	–	4.9
Si	400	3	Non-magnetic	–	–	4.9
Alumina	450	27	3700	1200/1000	0.0011	14.6
Alumina	550	28.5	4050	750/400	0.00095	15.4

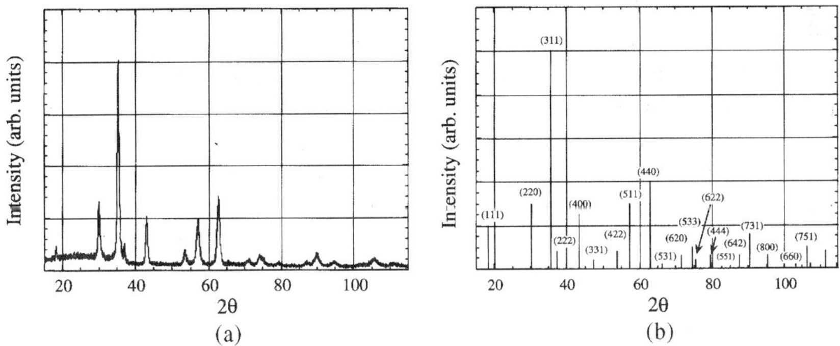

FIG. 7.8. (a) XRD pattern for an as-deposited 27 μm thick PLD NiZn-ferrite film deposited at 450 °C in 50 mtorr of oxygen on an alumina substrate; (b) Standard powder diffraction pattern for Ni-ferrite ($NiFe_2O_4$).

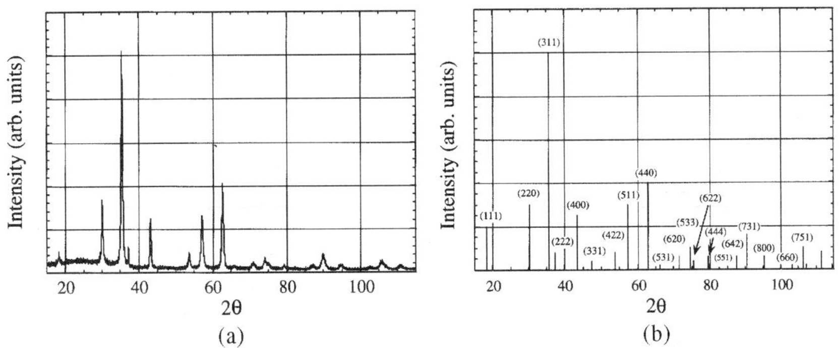

FIG. 7.9. (a) XRD pattern for an as-deposited 28.5 μm thick PLD NiZn-ferrite film deposited at 550 °C in 50 mtorr of oxygen on an alumina substrate; (b) Standard powder diffraction pattern for Ni-ferrite ($NiFe_2O_4$).

indicating that the films are single-phase polycrystalline NiZn-ferrite in the as-deposited state. The lower than expected $4\pi M_s$ and large FMR ΔH values at X-band may be due to the presence of voids in the films, which would increase magnetic losses and result in an overestimation of film volume.

In order to demonstrate the feasibility of integrating polycrystalline NiZn-ferrite films directly with microwave integrated circuits, NiZn-ferrite films were PLD-deposited at high rates onto metallized epi-GaAs on Si substrates followed by an RTA to improve film quality. However, the negligible improvements in magnetization seen after the RTA would be consistent with the hypothesis that the smaller than expected $4\pi M_s$ values and higher magnetic losses in the PLD films are due to voids, not incomplete crystallization of the NiZn-ferrite. The as-deposited NiZn-ferrite films on metallized GaAs are polycrystalline with essentially no preferred texture (see Fig. 7.10), as was seen for the NiZn-ferrite films on alumina substrates. A summary of processing conditions and film properties are shown in Table 7.4 and the in-plane VSM hysteresis loop for a 68-μm-thick NiZn-ferrite film is shown in Fig. 7.11. Although FMR ΔH values are somewhat large in these films, >20 dB isolation, a 2-dB insertion loss, and ~3-GHz bandwidth were still achieved for a 35 GHz circulator on GaAs incorporating a PLD NiZn-ferrite film grown under these conditions. The circulator was fabricated using a NiZn-ferrite film that was deposited through a shadow mask onto a ground-plane on GaAs. The resulting thick puck-like film is shown in Fig. 7.12. The reasonably good performance of the circulator fabricated from this film further supports the argument that the origins of the high magnetic loss in the PLD NiZn-ferrite films are caused by microstructural factors such as voids since FMR ΔH is affected by two magnon processes correlated to microstructure while circulator devices operate at off-resonance.

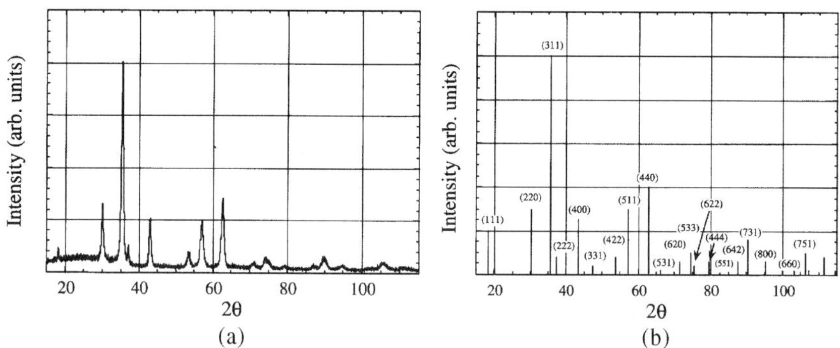

FIG. 7.10. (a) XRD pattern for an as-deposited 200 μm thick PLD NiZn-ferrite film deposited at 550 °C in 50 mtorr of oxygen on a metallized epi-GaAs on Si substrate; (b) standard powder diffraction pattern for Ni-ferrite ($NiFe_2O_4$).

TABLE 7.4

SUMMARY OF PROPERTIES FOR PLD NiZn-FERRITE FILMS GROWN ON METALLIZED EPI-GaAs ON Si SUBSTRATES IN 50 MTORR OF OXYGEN; THE FMR ΔH VALUES WERE MEASURED WITH THE APPLIED MAGNETIC FIELD NORMAL ($\perp$) TO THE FILM PLANE.

Substrate temperature (°C)	20-s RTA @ (°C)	Film thickness (μm)	$4\pi M_s$ (gauss)	ΔH (Oe) @ 9.8 GHz ($\perp$)	Deposition rate (μm/hr)
400	650	40	3000	550	6.6
450	650	20	3000	550	3.5
450	850	100	4000	1200	20
550	850	200	3600	1900	40
550	850	63	4560	786	–

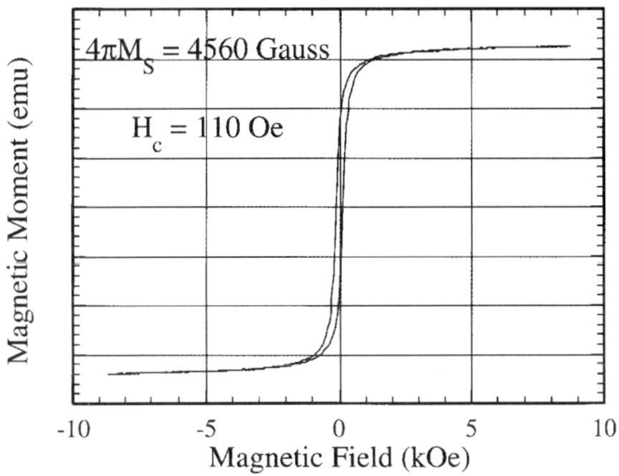

FIG. 7.11. VSM hysteresis curve for a 63 μm thick PLD NiZn-ferrite film deposited at 550 °C in 50 mtorr of oxygen followed by an RTA at 850 °C for 20 s. The applied magnetic field is in the film plane.

One of the more promising approaches for direct integration of passive microwave components with MMICs has been the spin-spray technique, which is capable of producing polycrystalline spinel ferrite films at temperatures <100 °C and allows for deposition on even heat sensitive substrates such as GaAs without a protective capping layer [90,93]. An early attempt at fabricating a junction circulator using a 0.8 μm thick spin-spray NiZn-ferrite ($Ni_{0.14}Zn_{0.14}$-$Fe_{2.70}O_4$) film deposited at 300 Å/min on a glass substrate showed very good potential with an in-plane coercivity of 37 Oe and a saturation magnetization of 478 emu/cc [92]. However, a complete analysis of the success of the device was not possible because the film was so thin. Attempts at thicker films and higher rates resulted in the delamination of the film from the substrate at a thickness

FIG. 7.12. A 40 μm thick PLD NiZn-ferrite film puck on a ground plane on GaAs as part of a 35 GHz circulator. The film was deposited at 550 °C in 50 mtorr of oxygen through a shadow mask.

~1.5 μm. In order to overcome this thickness limitation, laminated structures were fabricated using spin-spray layers of NiZn-ferrite, Fe_3O_4 (magnetite), and dextran (~100 Å) [93]. The thin (~300 Å) magnetite layers were deposited between the substrate and all other subsequent layers as a means of improving adhesion between the layers. The dextran layers did appear to relieve the stress, but the thickest film achieved was still no greater than 5.2 μm. Later work, which focused on depositing the NiZn-ferrite films on glass, 2 in-diameter (100) Si wafers and 3-in-diameter (100) GaAs wafers, both with evaporated gold layers, was more successful. Films deposited on all of these substrates exhibited (111), (110) and (211) texturing of the NiZn-ferrite with (111) and (110) orientations being the most common. The best results for microwave applications were achieved for a 4 μm thick NiZn ferrite film deposited at 2 μm/hr on an Au-coated GaAs wafer [94]. The $4\pi M_s$ of the film was measured to be 5620 G, which is consistent with bulk values and the in-plane and out-of-plane ΔH values were measured to be 260 and 83 Oe, respectively, at 9 GHz. Thicker films of up to 25 μm have been demonstrated as well using multiple deposition steps in order to maintain film adhesion [44]. Although the magnetic properties of these and other spin-spray films have exhibited excellent magnetic properties, further research is needed to overcome the poor performance of spin-spray films in terms of their dielectric loss [95]. Typical bulk microwave devices utilize spinel ferrites with a tan δ value of 0.001 or less and a resistivity of $10^6\,\Omega$ cm. In comparison, the tan δ values at 9.5 GHz for 0.5–2 μm thick spin-spray NiZn-ferrite films were measured to be about 1.0 or greater. This high dielectric loss was attributed to the high ferrous content of the films. In the spin-spray process, control of the cation ratios in the films is difficult and high dielectric losses are due to the mixture of valence states in the same lattice site. The dielectric losses could be reduced to a certain extent by incorporating the maximum amount of Ni and Zn in the films followed by a low temperature anneal, but annealing also caused an increase in the FMR ΔH and an approximate one-third reduction in magnetization. The behavior of the annealed film was indicative of a conversion from the pure spinel phase in the as-deposited film to a spinel and maghematite (γ-Fe_2O_3) mixture. These results demonstrate that the spin-spray technique has very high potential as a technique for depositing films at high rates over large areas with bulk magnetic properties; however, limitations in the ability to control composition and ferrous ion content of the spinel films currently limits its practical application for microwave devices.

Several other approaches have been applied to the deposition of polycrystalline spinel ferrite films such as rf diode sputtering, microwave plasma spray, jet vapor deposition (JVD), facing-targets sputtering (FTS), ion beam sputtering, rf magnetron sputtering, and plasma enhanced metalorganic chemical vapor deposition (MOCVD), but in most cases the films are thin, deposition rates are low and microwave properties have not been reported because the films are intended for nonmicrowave applications [3, 37, 96–103]. In those cases where magnetic films for microwave applications were the intended purpose, some promising results have been reported. Rates as high as 0.1 μm/min were achieved for Ni-ferrite and YIG films deposited at 200 °C on alumina substrates using an FTS setup [103]. The as-deposited ferrite films were amorphous, but the films crystallized into the appropriate single-phase spinel or garnet structure after annealing for several hours at 800 °C. Most important, the composition, structure, and magnetic properties reportedly matched the bulk material for these annealed films. In another study using rf sputtering, the deposition rates were lower (~60 Å/min), but microwave properties were measured [37]. Amorphous films of YIG, lithium ferrite, and strontium hexaferrite were sputtered onto alumina and silicon substrates and then crystallized by annealing at 800, 750, and 900 °C, respectively, with the resulting $4\pi M_s$ values only slightly lower than the bulk materials. FMR ΔH values of about 60 Oe were measured between 6 and 30 GHz for YIG films on bare alumina substrates and ΔH values for YIG films on metallized alumina substrates were about 100 Oe. In either case, the FMR ΔH values for these polycrystalline YIG films were not much different than the values for bulk polycrystalline YIG. It was not possible to measure ΔH values for the strontium hexaferrite films due to the high field requirements, but the FMR ΔH values for lithium ferrite films were measured to be about 500 Oe, indicating some microstructure contribution to magnetic losses. Another approach that has been researched for microwave applications, largely due to the high rates (~100 μm/hr) that can be achieved with this process, is JVD [12–15]. Initial studies were conducted on Ni-ferrite ($NiFe_2O_4$) films (>25 μm) deposited on alumina substrates at 600 °C. The as-deposited films are amorphous and essentially nonmagnetic with a very large compressive stress on the order of 10^9 dynes/cm^2. After several annealing cycles in air at temperatures of up to 1100 °C, the films become crystalline and magnetic with $4\pi M_s$ values of 2500 G ($4\pi M_s$ of bulk Ni-ferrite ~3200 G). The annealing also relieved the compressive stress in the films but was accompanied by a cracking and segmentation of the film structure. Improvements in the process led to the deposition of a 14 μm thick film with a $4\pi M_s$ value of 2900 G in the as-deposited state and a much reduced compressive stress. Subsequent annealing increased $4\pi M_s$ to 3000 G and again relieved the compressive stress in the film. Further work is needed to measure the dielectric and magnetic losses in these films and to determine if the approach is compatible with MMIC technology, but the improved results did show strong potential.

7.3.4. HEXAGONALS

Like the spinel and garnets, early attempts at single-crystal hexaferrite films involved the use of LPE [104–108]. These attempts, however, met with only limited success largely due to the unavailability of suitable substrates on which the success of the LPE technique crucially depends. Stearns and Glass attempted to solve the lattice mismatch problem by growing hexagonal ferrite films on other hexagonal ferrite substrates using the isothermal dipping method of LPE [104]. The XRD and FMR experiments on the resulting films confirmed that LPE could be used in this manner to grow epitaxial hexagonal ferrite films. In a further refinement of these results, Stearns and Glass then used the same technique to grow hexagonal films on nonmagnetic spinel substrates of $MgAl_2O_4$, $MgGa_2O_4$, and $Mg(In,Ga)_2O_4$ [105]. Characterization of the films showed that there were actually two phases present, a spinel phase and a hexagonal phase. The intent was to grow the *Y*-type hexagonal ferrite, $Ba_2Zn_2Fe_{12}O_{22}$, but the films that actually crystallized were *M*-type hexagonal and the accompanying spinel phase was attributed to either zinc ferrite or zinc-substituted magnetite. It is likely that the unexpected formation of the *M*-type hexagonal was due to the depletion of zinc from the melt by the formation of the spinel phase. Further research yielded conditions for which *W*-type hexagonal ferrite could be deposited on $Mg(In, Ga)_2O_4$ spinel substrates by using intermediate quantities of ZnO in the melt. The epitaxial relation for these films with respect to the substrates was found to be: hexagonal basal plane parallel to spinel substrate (111) and hexagonal *a*-axis parallel to spinel ⟨101⟩. On $ZnGa_2O_4$ spinel substrates, ZnO was not even required in the melt to deposit *M*-type hexagonal ferrites. The FMR ΔH values at 35 GHz for the *W*- and *M*-type hexaferrite films obtained in this manner were 42 and 26 Oe, respectively. Dötsch *et al.* attempted to grow epitaxial barium hexaferrite ($BaFe_{12}O_{19}$) films using LPE as well [108]. The films were grown on $Sr_{1.03}Ga_{10.81}Mg_{0.58}O_{19}$ substrates, which have these hexagonal lattice parameters, $a = 0.5822$ nm and $c = 2.3061$ nm. Typically, 2 μm thick films were achieved at a temperature of 960 °C and a growth rate of 0.7 μ/min with films up to 8 μm thick reported. The $4\pi M_s$ and H_a values for the films were measured in a VSM to be 4600 G and 16000 Oe, respectively, which are only slightly lower than bulk values of $4\pi M_s = 4500$ G and $H_A = 17.3$ kOe [83]. The FMR ΔH at about 53 GHz was 62 Oe in these films and the resonance values as a function of frequency and applied magnetic field were in good agreement with the VSM values. These results clearly demonstrated the use of LPE for hexagonal ferrite growth. Unfortunately, the LPE process, the melts and the substrates were much more complicated due to the complex hexagonal crystal structure and phase diagram making LPE not very practical for device applications incorporating hexagonal ferrite films.

A more practical approach to epitaxial hexaferrite film growth recently involved using commercially available single-crystal sapphire substrates. One approach to film growth has been rf-sputtering of barium hexaferrite films which has been demonstrated on *c*-, *r* and *a*-plane sapphire substrates [109–110]. Typically, the films have been deposited in 3–8 mtorr total pressure of an argon-oxygen gas mixture at temperatures from ambient to 200 °C. In the as-deposited state, the films are amorphous and nonmagnetic, but subsequent annealing in an oxygen atmosphere at temperatures between 850 and 900 °C produce crystalline magnetic films. The $4\pi M_s$ and H_a values for the annealed films were reported to be close to the bulk or somewhat lower indicating some portion of the film to be misoriented and nonmagnetic. Thicker films for practical microwave applications have been obtained by using a modified version of this technique [111]. A barium hexaferrite target was rf-sputtered onto a *c*-plane sapphire substrate at a rates of about 0.45 μm/hr up to a thickness of 0.35 μm and then annealed at 800 °C for 2.5 hrs. This substrate/film structure was then used as the substrate for the LPE growth of a 15 μm thick barium hexaferrite film. The XRD of the resulting film indicated a high degree of *c*-axis orientation with *c*-axis of the film parallel to the *c*-axis of the substrate and *a*-axis of the film rotated by 30° in the plane relative to the substrate. The FMR measurements at 60 GHz on this film revealed a linewidth of 41 Oe and resonant field values consistent with bulk barium hexaferrite.

Epitaxial barium hexaferrite films have also been grown on *c*-plane sapphire using PLD [23]. The best quality films were achieved at high substrate temperatures of 900–950 °C and high oxygen pressures of several hundred mtorr. The XRD scan for a PLD barium hexaferrite film on (0001) sapphire shown in Fig. 7.13 exhibits only the basal plane peaks of barium hexaferrite, indicating the high degree of *c*-axis orientation for the film. The $4\pi M_s$ and H_a values for a 0.5-μm-thick film were measured to be 4400 G and 16.5 kOe, respectively, which are consistent with bulk barium hexaferrite. Annealing of

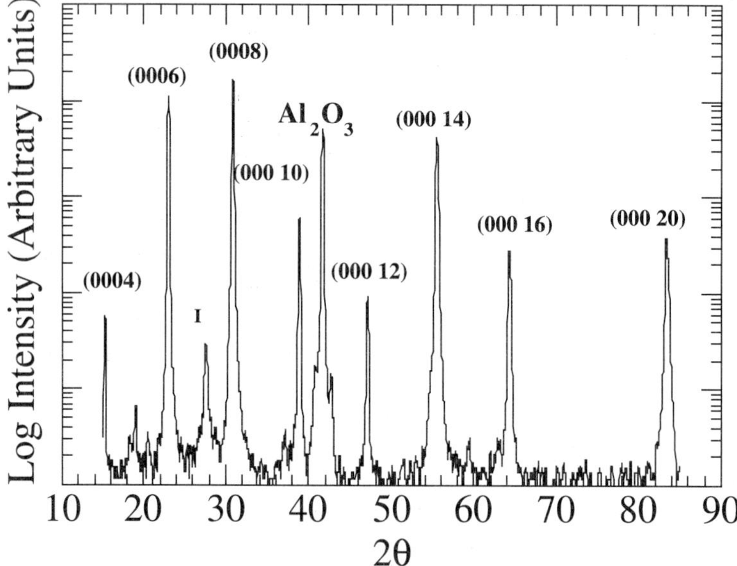

FIG. 7.13. XRD pattern for a 1 μm thick PLD barium hexaferrite film grown on a (0001) single-crystal sapphire substrate at 920 °C.

the films for 2 hr at 900–1000 °C caused some improvements in film quality. The coercive field was reduced from 700–100 Oe; FMR ΔH values at 50 and 58 GHz were reduced from 280 and 220 Oe to 88 and 66 Oe, respectively, which indicated that annealing improved crystalline quality. Thicker barium hexaferrite films have been grown on *c*-plane (0001) sapphire substrates using the same process [112]. The films were grown at 920 °C using a range of oxygen pressures from 50–1000 mtorr. Films grown using oxygen pressures of 75 mtorr or greater developed a "soot-like" appearance as the thickness of the film increased. The XRD indicated that this loosely adherent material, which may have been the result of gas phase reactions in the laser generated plume or increased particulates at high pressures, was randomly oriented polycrystalline barium hexaferrite. At a pressure of 50 mtorr, however, *c*-axis orientation and a relatively smooth surface morphology were maintained to a thickness of 15 μm. The magnetic properties of a 15 μm thick barium hexaferrite film were measured using a VSM to be $4\pi M_s = 4200$ G and $H_a = 15900$ Oe, which were consistent with experimental resonant field values measured from 85–90 GHz. The magnetic and dielectric losses of the film were also measured using FMR and cavity perturbation techniques, respectively. The FMR ΔH at 85 GHz was measured to be 400 Oe, but after annealing the film for 2 hr at 950 °C, ΔH was reduced to about 200 Oe. Tan δ for a 12 μm thick film was 0.004. Films grown at 50 mtorr and greater than

15 μm thickness began to crack and delaminate during deposition and after deposition when the film is cooled. The cracking and delamination of the film that occurred when cooling clearly occurred due to strains caused by CTE (α) mismatch between film and substrate (sapphire: $\alpha_a = 4.48 \times 10^{-6}/^\circ C$, $\alpha_c = 5.58 \times 10^{-6}/^\circ C$ and barium hexaferrite: $\alpha_a = 9.99 \times 10^{-6}/^\circ C$, $\alpha_c = 12.2 \times 10^{-6}/^\circ C$), but the cause of the defects that occurred during growth are less clear [113–114]. Possibilities include stress during growth caused by factors such as lattice mismatch, formation of a secondary phase in the film, or possibly vacancies in the crystal lattice. Evidence of a secondary phase (α-$BaFe_2O_4$) has been observed in the XRD and Rutherford backscattering spectroscopy (RBS) data and complete oxidation of the film may not occur at low background pressures. One approach that is being used to overcome this limitation in thickness due to strains in the barium hexaferrite film is to initially deposit a thin film at high temperatures as was described here so as to obtain an epitaxial barium hexaferrite film and then deposit the bulk of the film at low temperatures to minimize stress [115]. A lift-off procedure for thick ferrite films (40–50 μm) is then applied to separate the film and substrate. This procedure uses the large shear and normal stresses present at the interface between the ferrite film and the growth substrate to provide a plane-of-weakness for film peeling. The magnitude of the stresses depends upon the thickness of the film, the differences in CTE between the film and substrate, and the change in temperature experienced by the system. An appropriate choice of processing temperatures leaves the film highly stressed at the interface, such that the application of an external mechanical shock or shearing force detaches the film from the substrate. The resulting free-standing film can then be post-annealed to crystallize the bulk of the film before being transferred to another surface, such as metallized Si or GaAs, for additional processing or device fabrication.

Although epitaxial barium hexaferrite films are ideal for many microwave applications, a c-axis-oriented polycrystalline structure would be preferred for magnetless microwave applications. Ideally, barium hexaferrite films for magnetless applications are composed of c-axis-oriented polycrystalline grains in order to achieve a high remanence ratio (Br/Bs), which is necessary to minimize or eliminate the applied magnetic field required to saturate the barium hexaferrite film. The high remanence ratio is achieved by depositing films with an approximate 5000 Å grain size, which is needed to maintain single-domain particles while still minimizing magnetic losses [116]. These characteristics are similar to the requirements for recording media except that low-noise media require grains of only a few hundred angstroms in size. Typically, those grain sizes achieved in barium hexaferrite recording media have been closer to 1000 Å. A large body of work has been done on growing oriented barium hexaferrite films directly on amorphous or semiconductor substrates for magnetic recording applications using many techniques such as rf diode sputtering, rf magnetron sputtering, dc magnetron sputtering, MOCVD, facing-targets sputtering, sol-gel pyrolysis and arc discharge evaporation [38, 49, 117–138]. Although these films, which are thin (<1 μm) and deposited at low rates (<1 μm/h), are not suitable for microwave applications, there is information to be learned from the processing used to produce these films because many of the film characteristics needed are the same for both applications. Barium hexaferrite films for magnetic recording have been processed following two basic approaches: (1) sputter deposition at elevated temperatures; or (2) sputter deposition at low temperatures followed by post-deposition annealing. In the case of (1), deposition temperatures on thermally oxidized silicon substrates are typically 600 °C or greater for producing films with reasonably good crystalline and magnetic properties. Further efforts to reduce the deposition temperature have seen some success by using Pb-substituted barium hexaferrite (i.e., a lower melting temperature) and seed layers such as ZnO or a spinel phase to promote a lower crystallization temperature. In the case of (2), *c*-axis-oriented barium hexaferrite films have been produced by depositing an amorphous barium iron oxide film followed by a post-deposition anneal at 700 °C or higher. In both approaches, the temperatures required for crystallization are not directly compatible with MMIC technology, but these two approaches do make it conceivable that a substrate with suitable properties such as CTE could be used with one of these approaches to deposit thick barium hexaferrite films. These films could then be transferred to semiconductor substrates for MMIC applications using a ferrite film lift-off procedure as previously described.

7.4. Monolithic Integration of Ferrite Film Devices with Semiconductors

7.4.1. Introduction

Ferrite devices play a key role in most microwave and millimeter wave systems where they provide duplexing, isolation, switching, phase-shifting, tunable filtering, and power-limiting functions. While much effort has been directed towards the size reduction and integration of active semiconductor devices, relatively little work has been directed towards achieving comparable size and cost reductions for ferrite and other passive components. Integration of multiple functions on the same chip is only economically viable in large cost-driven applications. Major markets are projected for active aperture electrically scanned antennas (ESA) for ground and airborne radar in the 1–20 GHz range, and cellular/PCS systems in the 800–2800-MHz range. Future scenarios for communications systems include satellite-based, low earth orbit, high data, rate communications systems where electronically steered phased array antennas would be required as a consumer product. Modeling results indicate that losses in a ferrite film device increase with decreasing film thickness and decrease with increasing frequency. Most devices operating directly at microwave or millimeter wave frequencies require very low loss, typically less than one dB, thus major emphasis on device application for ferrite films has been directed towards operation at frequencies >10 GHz. Device activity has been focused on circulators, as this function cannot be performed effectively by other technologies [139].

The feasibility of complete monolithic integration of circulators and other ferrite devices with active circuits on Si and GaAs wafers has been demonstrated. However, the goal of monolithic integration remains controversial. Concerns exist because of the high cost of semiconductor chip real estate and yield degradation due to increased processing complexity. An example of an X-band (8–12 GHz) transmit/receive module for an electronically steered active aperture antenna is shown in Fig. 7.14. The circulator occupies a significant volume and because it is a hybrid component that must be inserted and bonded into the module, assembly costs are higher. In addition, the monolithic integration approaches discussed here are expected to be applicable to microwave wafer scale integration (WSI), and other future, planar transmit/receive, module approaches.

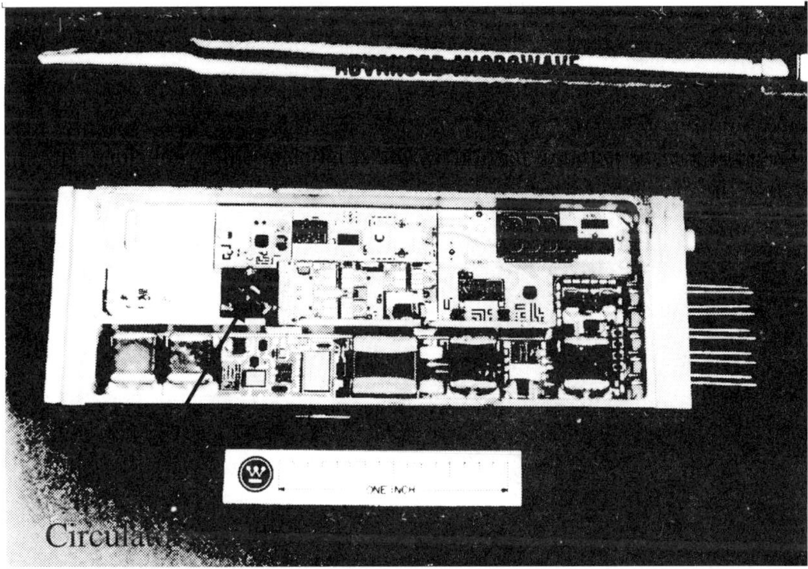

FIG. 7.14. Early X-band (8–12 GHz) transmit/receive module for an electronically steered active aperture antenna showing a circulator that occupies a significant volume and adds to module assembly cost.

The work described here builds upon an earlier demonstration of the compatibility of ferrite films deposited by spin spray, with GaAs devices. Although the low deposition temperature of spin-spray makes it attractive for monolithic integration with microwave/mm wave semiconductor circuits, attempts to achieve high resistivity and low tan have not been successful. Because of this, the discussion will focus on ferrite films grown by pulsed laser deposition [24, 79].

The techniques described here for YIG and NiZn-ferrite film devices are also applicable to uniaxial hexagonal ferrites and will result in very compact magnetless circulators for millimeter-wave applications. It appears that a fully integrated millimeter-wave T/R module, shown in Fig. 7.15, is now technically feasible through application of thin-film technology and photolithography. This approach is increasingly attractive at W-band (94 GHz) where the spacing between antenna elements ($\lambda/2$) is only 1.6 mm, and 3000-planar WSI elements would be obtained on a 4-in substrate. This development could make affordable active ESA available for radar and missile seeker applications [140].

Two critical material issues to be resolved for successful monolithically integrated ferrite device fabrication are:

(a) development of a compatible (low thermal budget) deposition process for thick films of YIG (with acceptable $4\pi M_s$ for X-band devices) or a spinel having high $4\pi M_s$ (required for frequencies above X-band) on metallized (ground plane) semiconductor substrates; and

(b) reduction of catastrophic stress expected between the semiconductor substrate and a thick ferrite film due to thermal expansion coefficient mismatch (see Table 7.5).

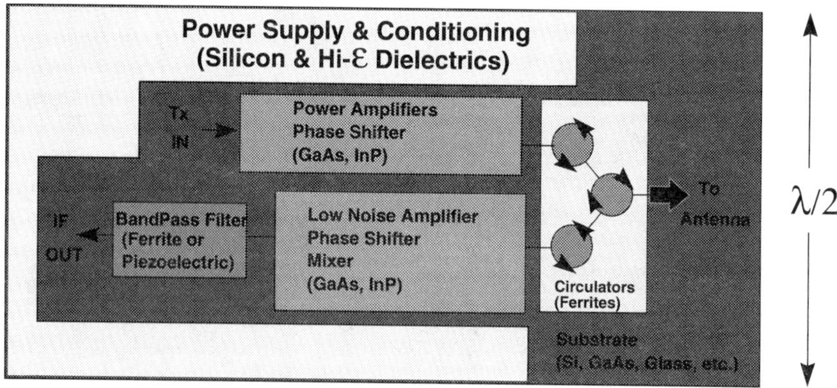

FIG. 7.15. A monolithic millimeter-wave front end "A T/R module on-a-chip" is technically feasible due to the thin film ferrite and photolithographic fabrication techniques described here.

TABLE 7.5
COEFFICIENTS OF THERMAL EXPANSION (CTE), STRUCTURE, AND LATTICE PARAMETERS FOR CANDIDATE SEMICONDUCTOR SUBSTRATES AND FERRITE FILM MATERIALS

Material	CTE (10^{-6}/°C)	Structure	a (Å)	c (Å)
Silicon	3.8	Cubic	5.4309	–
GaAs	6.8	Cubic	5.6533	–
$Y_3Fe_5O_{12}$	10.4	Cubic	12.376	–
Ni-ferrite	7.5–8	Cubic	8.339	–
$BaFe_{12}O_{19}$	7	Magneto-plumbite	5.892	23.198

Figure 7.16 displays some important temperature limitations related to GaAs integration. The common practice in ceramics manufacture of bulk ferrites is to prepare a low-density precast tape or block of ferrite followed by a lengthy firing at temperatures exceeding 1200 °C, and is unacceptable for GaAs integration [141]. Liquid phase epitaxy (LPE), commonly used to deposit YIG films, also requires a corrosive melt at a high temperature (~900 °C) [44]. However, ferrite films with properties near that of bulk have been prepared by vacuum deposition at considerably lower temperatures (e.g., for YIG, 700–850 °C), but these temperatures were still too high for long-term heating of GaAs [142]. Unprotected or protected (phosphorus-doped glass) GaAs decomposes or degrades at a temperature below 700 °C. This leads to arsenic contamination and reactions, which can cause degradation of the ferrite properties and physical damage to the

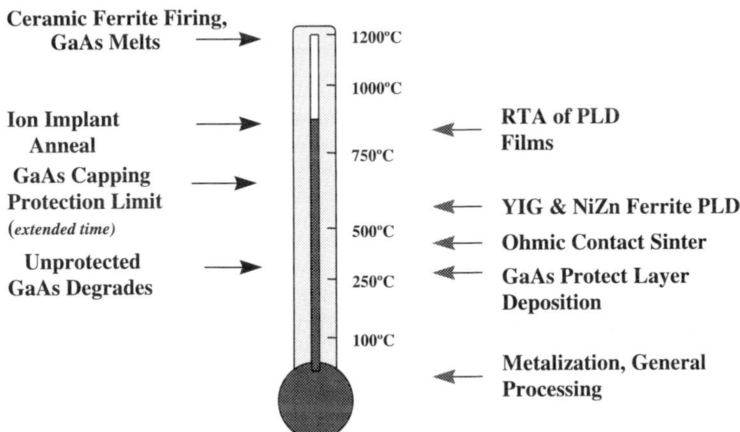

FIG. 7.16. Critical temperatures for both ferrite films and semiconductor device fabrication. Low thermal budget ferrite film deposition is essential for integration compatibility with semiconductor materials.

film. An additional complication with GaAs is its fragility, which results in wafer breakage during deposition and photolithographic processing. These issues are less severe for Si integration. Initial processing development on silicon wafers avoided these problems and provided a high device yield for evaluations of early demonstrations of integrated circulators.

Since the magnetization of ferrite materials arises from an ordered crystalline structure, a high-temperature deposition or post-deposition anneal is required. Polycrystalline thin films of other compound oxide materials have successfully been achieved using a rapid thermal anneal (RTA) at a considerably higher temperature than deposited [79]. This led us to consider a 2-step deposition process involving deposition at a low temperature followed by an RTA at the highest allowable temperature. A high-temperature compatible metallization was also developed with conductivity that does not degrade, important for good circulator microwave performance, with extended heating up to 700 °C and for RTA of up to 850 °C. A new triple-layer coating (capping layer) for GaAs was also developed that prevented contamination and reaction of the ferrite with GaAs, and Si contamination of an epitaxial GaAs layer on Si, during RTA at 850 °C [143]. Using these new processes we were able to successfully demonstrate a 2-step process for YIG device integration on metallized Si and GaAs involving deposition at 550 °C and RTA at 850 °C for 20 s.

7.4.2. FERRITE CIRCULATORS

Circulators are nonreciprocal devices that are most commonly used to allow a radar, or communications, transmitter and receiver to use a single antenna, as shown in Fig. 7.17 [139]. In addition to directing the transmitted and received energy, the circulator also provides isolation, thus ensuring that impedance mismatches do not degrade system performance. An example of a microstrip circulator integrated on a semiconductor substrate is shown in Fig. 7.18. Modeling of ferrite circulators is now well developed and can be used to design and optimize devices for the required device bandwidth, insertion loss and isolation, utilizing available ferrite materials [144].

Modeling of the loss contributions in a thin-film YIG circulator at 10 GHz, Fig. 7.19, showed that conduction losses dominate and that film thickness of around 100 μm is necessary for total losses <1 dB. Here bulk values of gold resistivity, dielectric $\tan\delta_d = 0.001$ and magnetic $\tan\delta_m = 0.001$ were assumed. Microstrip circulators are typically 0.5–1 mm thick and achieve insertion losses <0.5 dB in X-band (8–12 GHz). Note that circulators operate far from ferromagnetic resonance (FMR) and thus ΔH, the FMR linewidth, does not represent the magnetic loss. Use of copper or silver in place of gold would result in only a small reduction in conduction loss, and would not justify the increased processing

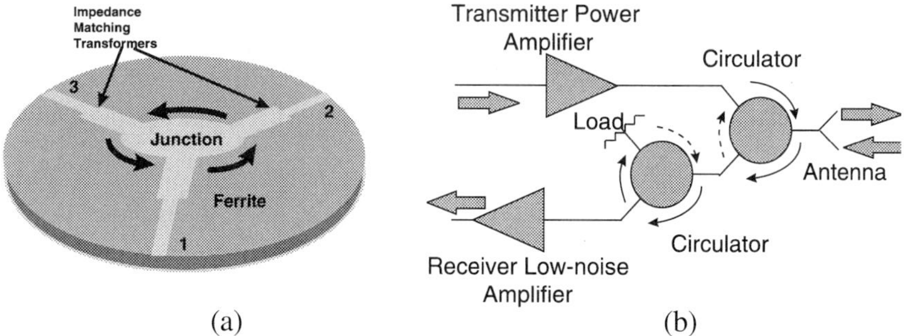

FIG. 7.17. Ferrite circulators are unique nonreciprocal devices that behave like a turnstile for microwave energy. (a) Microstrip circulator, typically fabricated using gold metallization for low resistivity on an approximately 0.5 mm thick ceramic ferrite. The ferrite is biased above magnetic saturation by a magnetic bias field applied normal to its surface. An input signal at port 1 exits at port 2 with less than 1-dB loss while port 3 is isolated with an attenuation of more than 20 dB relative to port 1; (b) Circulators allow a transmitter and receiver to use a common antenna. They also provide isolation for reflections from components so that each component appears to operate into a matched impedance. Multiple circulators are used to provide the desired amount of isolation.

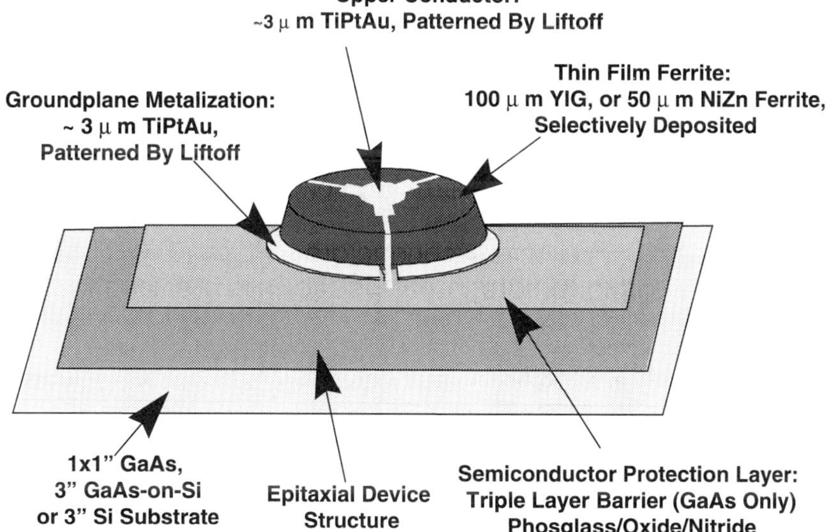

FIG. 7.18. Key elements for integrated circulator fabrication. The processing sequence is: protect semiconductor, fabricate groundplane, deposit and anneal the ferrite and fabricate upper conductor.

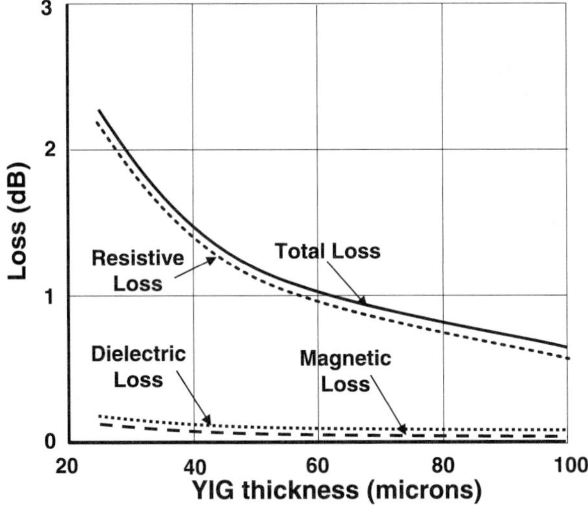

FIG. 7.19. Modeling of the loss contributions in a thin-film YIG circulator at 10 GHz. This showed that conduction losses dominate and that film thickness of around 100 μm is necessary for total losses <1 dB. Calculated magnetic, dielectric, conduction, and total insertion loss vs YIG film thickness for an X-band microstrip circulator. Gold metallization and electric and magnetic tan $\delta = 0.001$ were assumed. The insertion loss is dominated by conduction losses in the junction conductor. Only loss due to the circulator junction is shown here; a practical device would require at least one matching transformer section on each port that could more than double the loss shown in the figure.

complexity involved. Superconducting materials could significantly reduce conductor losses, but would significantly increase the size, weight, and power required [145].

For a given film thickness, the conduction losses decrease with increasing frequency. The diameter of the circulator (and hence the substrate area required) also decreases with increasing frequency as shown in Fig. 7.20. Thus film devices are more attractive with increasing frequency, through the mm-wave range. The port impedance is also shown to increase linearly with frequency in Fig. 7.20. At 10 GHz, the port impedance is approximately 7 Ω for 100 μm thick YIG film, requiring the use of multi-section transformers to achieve the required bandwidth at minimum loss. As each transformer section is $\lambda/4$ long, impedance matching can significantly increase the substrate area required by the circulator. A port impedance of 50 Ω eliminates the need for matching transformers and thus allows a more compact circulator to be designed. Figure 7.21 shows the variation of YIG film thickness with frequency required to yield a 50-Ω port impedance, thereby reinforcing the desirability of film devices for mm-wave operation.

The work described here initially focused on 10 GHz operation, but evolved to higher frequencies (20 GHz and 35 GHz) as modeling and experimental results

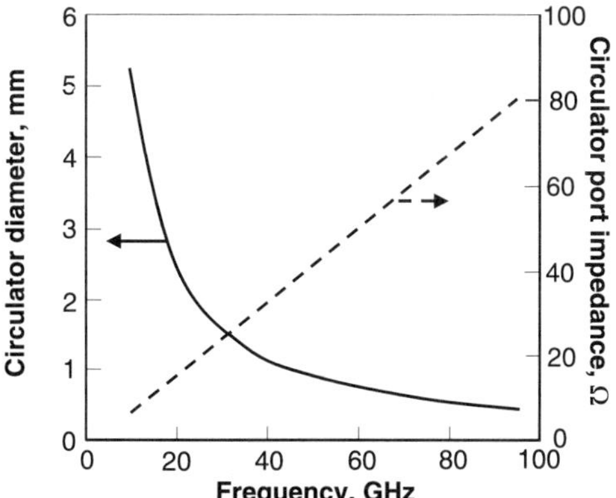

FIG. 7.20. Calculations on a 100 μm thick circulator show that the area occupied by the junction decreases with increasing frequency and, because the port impedance increases with frequency, it is also possible to minimize the area occupied by matching transformers.

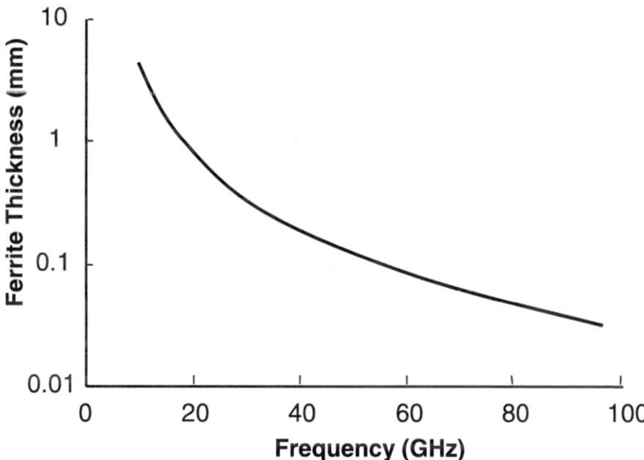

FIG. 7.21. Calculated YIG circulator thickness vs frequency for a 50-Ω port impedance.

became available. Although YIG was a good choice for initial development, high $4\pi M_s$ and controllable anisotropic ferrite such as lithium ferrite and hexagonal ferrites are better suited to mm-wave circulator applications.

7.4.3. PROCESSING/INTEGRATION

Process integration requires the development of techniques, processes, and designs suitable for integrating ferrite components and active semiconductor devices on a single monolithic semiconductor chip. A general process for ferrite film integration involves the following elements.

(a) Substrate preparation. Semiconductor substrates suitable for microwave applications include: (1) several bulk III-V compounds, of which only GaAs is considered here; (2) high-resistivity silicon, as has been developed and proven in the MICROX technology; and (3) epitaxial GaAs on high-resistivity silicon substrates [143, 146].
(b) The ferrite deposition process. Spin spray and pulsed laser deposition (PLD) of ferrite materials and many alternative methods and materials have been investigated. The only method that was proven to provide ferrite films of the required magnetic and dielectric properties, along with the thickness required for low loss operation, is pulsed laser deposition of thick YIG films.
(c) The ferrite activation/anneal process. Most methods of ferrite deposition compatible with semiconducting processing require low temperature (450–550 °C) deposition followed by an annealing or activation step to convert the deposited material into a crystalline, magnetic ferrite. In the case of pulsed laser deposition of YIG films, a low thermal budget semiconductor compatible process involves deposition at 550 °C followed by an additional high-temperature rapid thermal anneal (e.g., 850 °C for 20 s).
(d) Patterning of the ferrite. This step of the process involves shaping the ferrite to the desired form, leaving the remainder of the substrate free for the formation of active semiconductor devices and MMIC.
(e) Provision of a metal ground plane on one side of the ferrite. When the ferrite is to be integrated with semiconductor devices and circuits, the ground plane must be patterned to be compatible with those devices. This thick metal layer was found to be critical in accommodating the CTE mismatch between the YIG film and the silicon substrate, Table 7.5.
(f) Patterning of the ferrite electrode and interconnects. In this step, the functionality of the ferrite device is defined by the metal pattern. Additionally, this metal pattern interacts with the ground plane metal to provide test pads and interconnects with the remainder of the monolithic chip.
(g) Fabrication of active microwave devices on the semiconductor substrate. For microwave applications, this fabrication involves electron beam lithography. Other commonly used processes include patterned etching of the substrate, ion implantation of selected areas, deposition of capacitor dielectrics, formation of submicron metal patterns, and definition of thick metal interconnects for low microwave loss. The key elements in the fabrication of ferrite circulators for integration are illustrated in Fig. 7.18.

7.4.3.1. Substrates

Practical substrates for MMIC include III-V semiconductors, notably GaAs, high-resistivity silicon, and epitaxial GaAs on high resistivity silicon. Of these, GaAs can be expected to provide superior microwave performance because of its inherent high electron mobility and low dielectric loss (semi-insulating properties). In the ferrite integration process, however, GaAs is a difficult material because of its fragility and limited stability at high temperatures. High-resistivity silicon possesses good strength and temperature stability, but applicability to 20 and 35 GHz circuits is marginal. A good compromise is epitaxial GaAs on high-resistivity silicon, providing substrate strength with high electron mobility, with a penalty of higher substrate cost. Although YIG on silicon structures survive processing, they are under significant strain due to the difference in thermal expansion coefficients as given in Table 7.5. As a result of the experience reported here, it should be possible to achieve high-yield processing of Ni or Li spinel ferrites, or barium hexaferrites, on GaAs substrates for mm-wave applications.

7.4.3.2. Ground Plane Metal

Requirements for the ground plane metal include the following items:

(a) at least 3 μ thick and highly conductive (less than $5 \times 10^{-6}\,\Omega$-cm) in order to reduce microwave conductive losses;
(b) nonreactive to the ferrite material and to the substrate (or isolated from these materials by a suitable barrier film);
(c) stable for a long period of time (about 24 h) during the PLD process at 500–600 °C and for the shorter time period (20 s) of the ferrite anneal processes at 850 °C;
(d) strongly adherent to the substrate (or barrier film) and to the ferrite (or barrier film) during temperature cycling; and
(e) that it not cause excessive mechanical stress in either the ferrite or the substrate by virtue of its coefficient of temperature expansion and its elastic modulus.

Many ground plane metal structures were tested for stability under the processing conditions of ferrite deposition (tens of hours at a temperature of about 550 °C) and of ferrite anneal/activation (tens of seconds at 850 °C in an oxygen atmosphere). As a result of this investigation, the ground plane structure consisting of Ti/Pt/Au was chosen as best suited to the process. In this structure, the thin titanium film provides adhesion to the substrate and the thick gold film provides low electrical resistance and chemical stability (and helps to relieve stress in the substrate and ferrite film). The platinum layer prevents interaction between the titanium and gold under processing conditions.

A major source of difficulty in the integration of ferrites with semiconductors has been the mechanical stresses generated during ferrite deposition and processing, see Table 7.5. These stresses always result in some cracking of the YIG film on a silicon substrate, and they often result in GaAs substrate breakage as well. Silicon and YIG are relatively high-strength materials and can tolerate the stress. The possibility that the ground plane metal (or some other material) might be used to reduce these stresses was investigated and it was shown that a thick gold ground plane does assist in relieving stress and is essential to achieving monolithic integration of ferrites and semiconductors.

7.4.3.3. Metal Barrier/Capping Films

Barrier films are used to prevent chemical interactions between the substrate and the atmosphere, the substrate and the ground plane metal, or the ferrite and either ground plane metal or top electrode metal.

Requirements of the barrier film include the following items:

(a) that it be chemically inert under the processing conditions;
(b) it should prevent interactions between the other materials;
(c) it should not contribute to microwave losses; and
(d) the thin film should be compatible with conventional semiconductor manufacturing techniques.

For silicon substrates, a thermally grown oxide film, 0.5–1 μ thick, serves as an effective barrier film. Effective barrier films are more difficult to produce on GaAs and GaAs/Si substrates. Because a bare GaAs surface will decompose at temperatures above about 450 °C, the barrier film must be deposited at low temperature, but must be able to function for a long deposition time (typically 24 h for PLD) at 550 °C and for a short time (20 s) at temperatures to 850 °C, which is the YIG or NiZn rapid thermal annealing temperature. In many cases, a chemical-vapor-deposited film of low-temperature oxide, heavily doped with phosphorus, and 2500 Å thick, was sufficient. In other cases, this film failed and a more reliable film was sought. Eventually, it was determined that the low-temperature, phosphorus-doped oxide covered by a sputtered film of silicon dioxide and silicon nitride, 1-μ thick, provided a reliable barrier to GaAs decomposition and prevented interaction of the GaAs with the ground plane metal.

7.4.3.4. Wafer-Scale PLD

The pulsed laser deposition (PLD) technique has received widespread attention in the thin-film research community because of its simplicity and demonstrated versatility in depositing advanced compound materials (superconductors, ferroelectrics, ferrites, phosphors, etc.) that are difficult to prepare by conventional techniques such as sputtering and evaporation [24]. One particularly important feature is its ability to readily deposit doped or atomically substituted coatings often used to tailor specific properties (e.g., magnetization for this case), for specific device requirements.

Pulsed laser deposition capabilities have evolved to better meet the requirements of fabricating thick film ferrite devices. Unique equipment features, shown in Fig. 7.22, have been developed including: (a) rotating radiant-heated substrate holder compatible with coating 3-in-diameter silicon wafers; (b) automated scanning of 2-in-rotating target materials to allow long uninterrupted ablation times needed to deposit films 100 μm thick; and (c) an "intelligent" window to

FIG. 7.22. PLD chamber for large area, thick ferrite film deposition. Key features include: radiantly heated substrate holder for a 3-in diameter semiconductor wafer, scanned 2-in diameter PLD target for >100 μm thick film deposition, "intelligent" window for smooth films at low pressure.

allow thick-film deposition at low pressure, which achieves smoother thick-film surfaces more suitable for use with lithographic patterning of device structures. The net result has been establishment of a semiproduction capability to deposit thick ferrite films on a wafer-scale, with reasonable throughput. Focus of film deposition by PLD has been on achieving high-quality, 100 μm thick YIG films useful for demonstrations of thick-film circulator devices at 20 or 35 GHz.

7.4.3.5. Low Thermal Budget Process/YIG Deposition on Metallized Semiconductor Wafers

Earlier work reported low magnetization for YIG films prepared at temperatures $<850\,^\circ$C and loss of orientation and film mechanical quality for Ba-M films exceeding 15 μ thickness [112, 142]. Reported PLD results also have been for films deposited at modest rates (~4 μm/hr) on small area (1×1 cm^2) insulating substrates. Little work has been reported on deposition of thick YIG or spinel films. Here we address these issues and present results on the development of PLD as a technique to economically prepare thick magnetic films of sufficient quality, on 3-in-diameter semiconductor substrates, for use in future low-cost microwave and mm-wave device production.

Small (1/4 in $\times$ 3/4 in) gadolinium gallium garnet, GGG (111), substrates with a close-up PLD geometry were used to establish a baseline for the quality of thick YIG films deposited by high rate PLD. Typical deposition conditions used were 200 mJ (KrF) energy, 150 Hz repetition rate, 5 cm substrate-to-source distance, 50 mtorr oxygen pressure, and 850 °C temperature. A deposition rate of 25 μm/h was attained and used to prepare thick films. At 850 °C on GGG (111), which has a close crystallographic lattice match, a 53 μ thick YIG film was deposited that was highly epitaxial (x-ray rocking curve FWHM = 12 arc sec) and had a narrow (5.7 Oe) FMR linewidth measured at 9 GHz. The magnetization and dielectric loss (tan δ) were determined to be 1800 G and 0.0002, respectively. The quality of these thick films is close to that achieved for YIG prepared by liquid phase epitaxy (LPE) at a competitive deposition rate. With modest improvement in the linewidth, demonstrated to be sensitive to oxygen pressure, PLD films could be employed in narrow bandwidth microwave filters [24].

The next step was to develop the Si-compatible, wafer-scale, PLD deposition process that would allow integration of quality thick YIG films. Our initial interest was to develop integrated circulator devices, which require thick films with bulk magnetization and dielectric loss properties but not exceptionally narrow linewidth. Thick polycrystalline YIG films meet these requirements. To demonstrate the suitability of PLD for this application, we developed a low thermal budget deposition process, not achievable by LPE, involving two steps: (1) low-temperature deposition (550 °C) followed by (2) rapid thermal anneal (RTA) at 850 °C for 20 s. The X-ray diffraction data for polycrystalline YIG deposited on Au-metallized Si(100) using this type of 2-step integration compatible process yielded excellent agreement with the intensities of the standard powder diffraction pattern (ASTM #33-393) for YIG, suggesting a "perfect" polycrystalline structure (see Fig. 7.6).

7.4.3.6. Low-Pressure YIG PLD Process Development

The last PLD development involved defining a low-pressure process that attains a practical deposition rate for coating a 3-in-semiconductor wafer using an "Intelligent Window" to allow thick film deposition on wafers at low oxygen pressure. Basically, this equipment keeps the laser input window of the vacuum chamber from becoming coated. Preventing severe incident laser power attenuation and, consequently, low deposition rate over long durations (typically 20–30 h) required deposition of thick films for device fabrication.

Although film smoothness was best at 10 mtorr, the lowest pressure attempted, the rate was reduced by approximately a factor of two. This would require about 50 h to deposit 100 μm on a wafer, which is considered too long with respect to maintaining a low thermal budget for processing devices involving GaAs. The rate was doubled by depositing at an intermediate oxygen pressure of 20 mtorr, at which smoothness of the film was only slightly worse than 10 mtorr, and raising the laser pulse rate (approximately 25% higher average power). Figure 7.23 illustrates the smoothing effect of lower PLD process pressure achieved by suppressing growth of "nodules" in the YIG film (here only 25 μm thick), which form at high pressure and result in a rough surface. With this low pressure PLD process 100 μm thick, YIG films can be achieved at a respectable deposition rate of 3.5 μm/hr over a 3-in-wafer area.

To test the optimized thick film PLD process, a thick (100 μm) YIG film was selectively deposited on a silicon wafer with a patterned ground plane. The final thickness of the film for a total deposition time of 31 h was approximately 102 μm. The surface of the as-deposited film was specularly reflecting, indicating the desired improvement over previous films deposited at 50 mtorr, which were diffuse reflecting. Under microscopic examination, a low density of YIG nodules was visible. Cracking in the film appeared about the same as for films made at 50 mtorr. Post-annealing the coated wafer using the standard RTA for YIG caused little change in its physical appearance.

7.4.3.7. Ferrite Film Patterning

Crucial to the fabrication of discrete circulators on a single substrate carrier is the ability to selectively (pattern) deposit the thick films. This required the use of a shadow (contact) mask, which is compatible with the high temperatures required to deposit crystalline ferrite films. Figure 7.24 illustrates the selective deposition

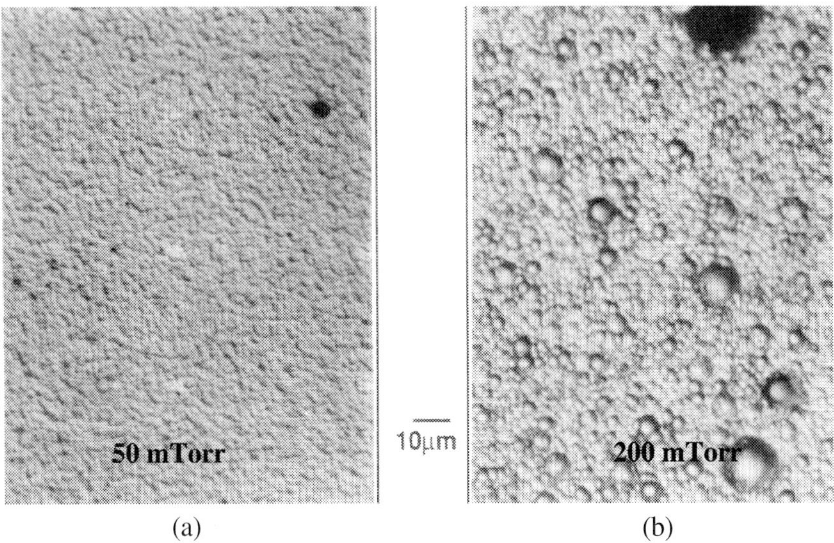

FIG. 7.23. Smoothing effect of reduced PLD process pressure (50 mtorr) (a) suppressed the growth of "nodules" in the YIG film that form at high pressure (200 mtorr) and result in a rough surface; (b) operation at 10 mtorr resulted in completely smooth films but the deposition rate was reduced by 40%.

process developed and used routinely to deposit 100 μm thick YIG films. At the left (Figure 7.24a) is a photograph taken during selective PLD of a thick ferrite film onto a 3-in wafer, heated to 550 °C, though a silicon shadow mask shown in the right photograph (Figure 7.24b). "Cut-outs" in the silicon shadow masks are created using a commercial programmable laser cutting tool.

Figure 7.25 shows an epitaxial GaAs-on-Si wafer with an array of 100 μm thick YIG (dark circles and rectangles) patterns defined by the shadow mask during selective PLD. Visible in this figure are the thick gold ground planes, protruding from under the dark ferrite patterns. The different diameter dark circles represent YIG areas from which 20 and 35 GHz circulators are fabricated by photolithographic patterning electrodes and matching networks on top of the thick YIG film patterns.

7.4.3.8. Ferrite Film Activation

The conversion of amorphous PLD YIG to a crystalline, magnetic film or reduction of the linewidth of PLD NiZn-ferrite is accomplished by rapid thermal (RTA) in an oxygen atmosphere, at 850 °C for 20 s. This temperature excursion is marginally high for GaAs and the surface must be sealed to prevent loss of

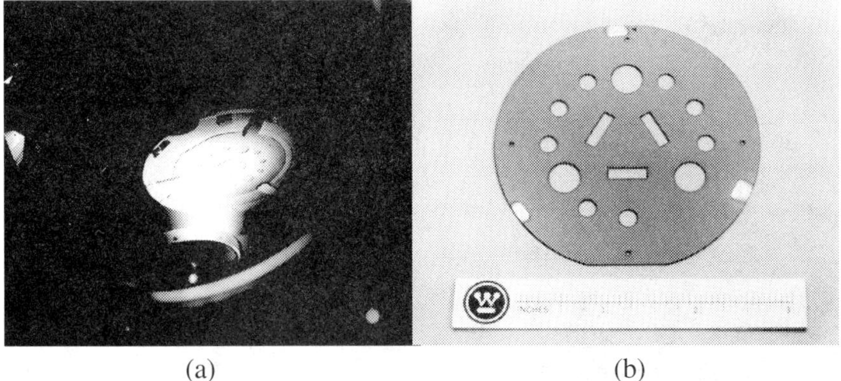

(a) (b)

FIG. 7.24. Selective PLD for patterned circulator deposition used a silicon shadow-mask with laser machined openings. (a) Selective PLD deposition in-process showing shadow mask and substrate illuminated by the plasma; (b) detail of silicon shadow mask. Gold coating of the shadow mask was necessary to prevent cracking of the silicon.

FIG. 7.25. A 3 in epitaxial GaAs-on-Si wafer with an array of 100 μm thick YIG (dark circles and rectangles that are 20 and 35 GHz circulators, and microstrip test structures) patterns defined by the shadow-mask during selective PLD.

arsenic through volatilization and reaction with oxygen. During this anneal, appreciable shrinkage of the ferrite films occurs (the thickness decreases by about 10%) and stresses are created in the ferrite film and in the substrate. These stresses may be relieved in several ways. Typically, the ferrite film develops cracks, (see Fig. 7.26) and the substrate is warped. Both of these phenomena

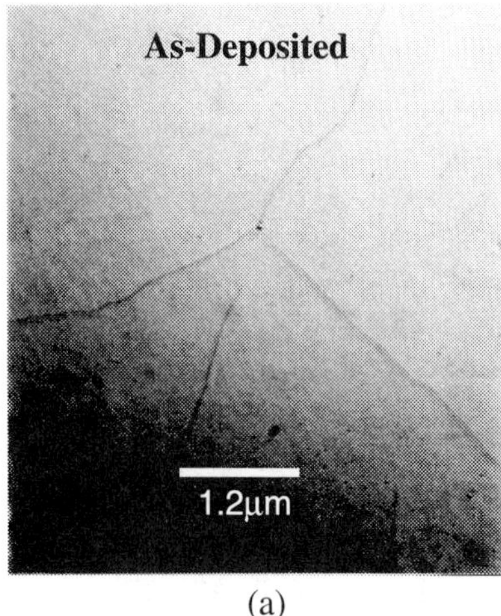

(a)

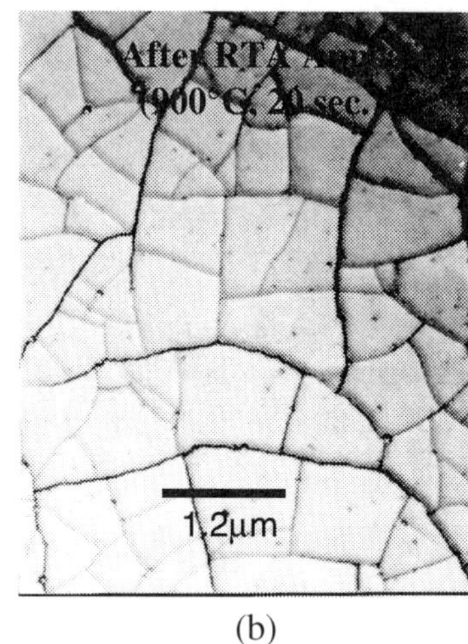

(b)

FIG. 7.26. Minor cracking of PLD YIG films onto silicon occurs during deposition (a). More severe cracking occurs when the film is annealed by RTA at 900 °C for 20 s to develop its magnetic properties (b).

contribute to difficulties in the ensuing processes. If there are discontinuities in the barrier film that seals a GaAs surface, arsenic is lost from the substrate and can be seen as a deposit of arsenic oxide on the wafer and on the annealing oven walls. In many cases, the stresses developed in GaAs wafers were catastrophic, causing wafer breakage, delamination of the ferrite film, cracking of the barrier films, and GaAs decomposition.

7.4.3.9. Recoat Deposition Process

As mentioned earlier, "hairline" stress cracks due to thermal expansion mismatch between the film and substrates are observed in the YIG films after cooling from a deposition temperature of 550 °C. Under microscopic examination, it was clear that the cracks expose the underlying ground plane metal, which led to electrical shorts with an electrode film processed on top of the YIG. These stress cracks are widened (~5 μm) by the application of the post-deposition rapid thermal anneal (RTA) at 850 °C, but are not too wide to prevent devices from being fabricated. This was possible with a recoat procedure that deposited nonconducting material in the cracks and partially sealed them. After the RTA, wafers were remounted in the PLD system and an amorphous YIG film was deposited for approximately 1 hr (approximately 5 μm thick) under conditions similar to those used to make the initial thick YIG film, except unheated. A cleaved section through a recoated and metallized film is shown in Fig. 7.27. After recoating, shorting between the

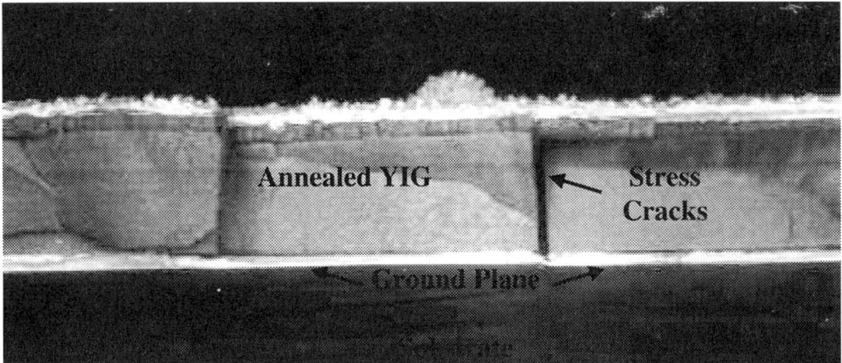

FIG. 7.27. Cross section through a YIG film circulator showing gold groundplane, 100 μm YIG film with cracks, coating layer to partially fill the cracks and prevent shorting of the top gold layer.

top and bottom gold films was significantly reduced, giving a high functional yield of finished devices.

7.4.3.10. Improved Resists for Morphology

Photolithographic patterning of the top electrode is one of the more difficult processes in the formation of thin film ferrite devices. The metal that is to form the top electrode is thick, approximately 3 μ, in order to minimize microwave conduction losses. Since the metal is defined by lift-off (chemical or physical etching of thick gold is quite difficult) the photoresist must also be thick (at least 5 μ). Thick resist is required also because ferrite films of 100 μ thickness are very rough (about 10 μ asperities) and contain many cracks that are several microns wide. Resists thicker than 5 μm are available, but they are not capable of resolving the finer details of the top electrode.

Photoresist is almost universally applied by spinning, that is, the resist resin, contained in a solvent, is puddled on the wafer surface and the wafer is spun at several thousand rpm to cause the resist to flow into a thin film. Even small nonplanarities on the wafer surface can result in uneven resist coating during spinning, and the 100 μm high ferrite shapes have a profound effect on coverage. Fortunately, the shadow mask method of defining the ferrite shapes results in sloped edges on the ferrite structures and the resist film can cover these edges fairly well if the operator is careful and experienced. However, the resist is still thinner at the top edges of the ferrite structure, asperities may not be completely covered, and subsequent metal patterning by lift-off can be an arduous process. On the other hand, the resist is always thicker at the bottom edges of the ferrite shapes and here it is more difficult to expose and develop.

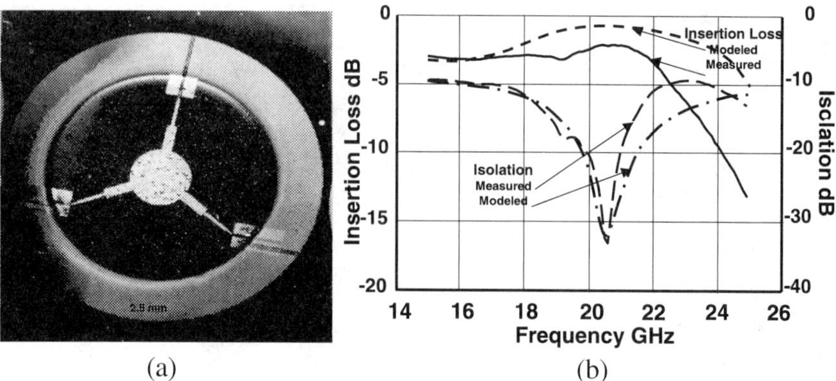

FIG. 7.28. A 20 GHz YIG film circulator on a GaAs wafer; (a) device fabricated using 'high-pressure' PLD that resulted in a rough YIG surface; (b) measured and calculated insertion loss and isolation.

Several methods were investigated for alleviating these difficulties. One of these methods involved use of two different types of resist, patterning a conventional resist first, then overcoating this resist with an "image reversal resist," and separately exposing and developing the second resist. This method has shown promise because the image reversal resist can be made to develop with a re-entrant edge profile that greatly facilitates metal lift-off. A second experimental method has been to apply the photoresist by spraying rather than spinning. This method alleviates some of the problems of uneven topography, but is not sufficiently automated to ensure consistent results.

Significant progress has been made in reducing the asperities in the YIG surface and smoother films have been achieved through deposition at lower pressure; compare a YIG circulator pattern made at "high pressure" in Fig. 7.28a with the "low-pressure" YIG in Fig. 7.29a.

7.4.4. Circulator Results at X-Band, 20 and 35 GHz

A thin-film ferrite X-band circulator fabricated on a semiconductor substrate is shown in Fig. 7.30. The rf performance of this early circulator was somewhat disappointing, with the insertion loss of 3 dB at 8 GHz higher than was expected. Analysis of the material properties and modeling of the known sources of loss have led us to believe that high insertion loss was due to defects and cracks in the YIG film.

In subsequent refinements, the concentration was on 20 and 35 GHz devices because higher frequency devices are more readily integrable on semiconductor

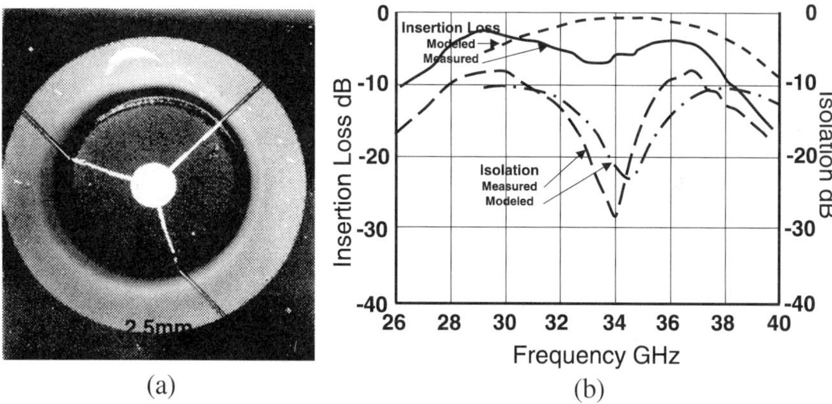

FIG. 7.29. A 35 GHz YIG film circulator on a GaAs-on-silicon wafer; (a) device fabricated using 'low-pressure' PLD that resulted in a relatively smooth YIG surface; (b) measured and calculated insertion loss and isolation.

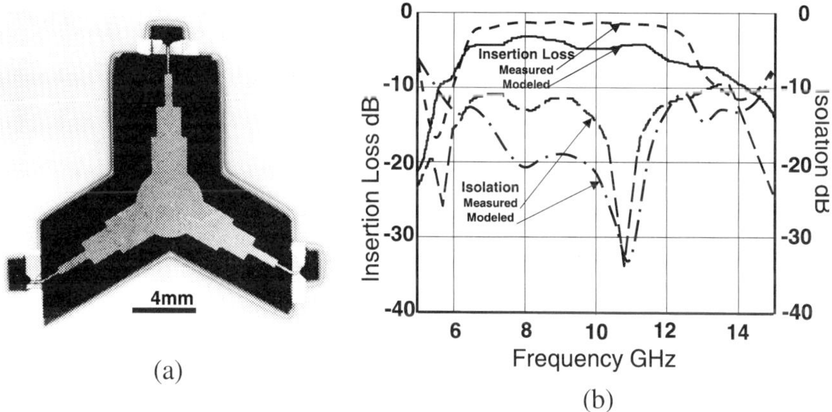

FIG. 7.30. X-band (8 to 12 GHz) YIG film circulator on a silicon substrate; (a) fabricated device; (b) measured and modeled performance. The "Y" YIG film pattern was used to reduce the YIG film area and hence minimize stress deformation. Notice that three matching transformer sections were necessary to transition between 50 Ω and the 7-Ω junction impedance.

chips due to their smaller size as well as acceptable insertion losses with 100 μm or thinner ferrite films.

A completed wafer of 20 and 35 GHz devices is shown in Fig. 7.25. In addition to circulators, the wafer contains impedance-matching test structures. The circulators are of two types—some are provided with probe pads on top of the ferrite film, others have the probe pads on the substrate wafer with the

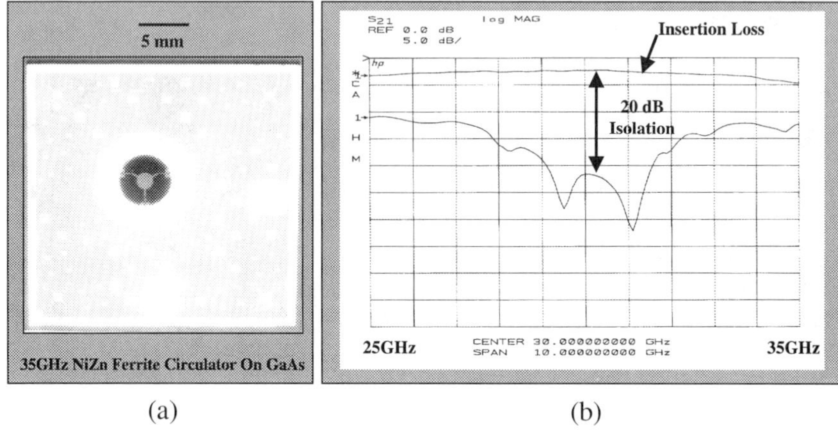

(a) (b)

FIG. 7.31. A 35 GHz Ni ferrite film circulator on a GaAs wafer; (a) completed circulator with FET fabrication in-process; (b) measured insertion loss and isolation.

coplanar waveguide interconnect traversing the side of the ferrite structure down to the substrate surface and ground plane. A 20 GHz YIG film circulator and its measured performance are shown in Fig. 7.28a,b, respectively. This device was fabricated using high-pressure PLD, resulting in a rough surface (see Fig. 7.23b). The surface roughness of the ferrite appeared to have a significant effect on processing yield but no effect on microwave performance. The measured isolation is close to that modeled but the insertion loss is approximately 2 dB, which is significantly higher than the predicted <1 dB. Similar results for a 35 GHz YIG film circulator are shown in Fig. 7.29. Again the measured and modeled isolation are in reasonable agreement, but the measured insertion loss is higher than predicted. The cause of the excess insertion loss has not been determined but it must be eliminated before thin-film circulators can be useful.

Finally, the results obtained using NiZn ferrite (Section 7.3.3) on a GaAs substrate containing FET test devices are shown in Fig. 7.31. In common with the YIG circulators, the isolation behaves as expected and the insertion loss is one or two dB higher than expected.

7.4.5. Conclusions

Although the feasibility of monolithic integration of ferrite devices with semiconductors has been established in principle, several problems remain to be solved before it becomes a practical, economical method of fabricating integrated mm-wave systems

(a) An excessive insertion loss of typically more than 1 dB was calculated; is this a design, material, or fabrication problem?
(b) There was cracking of the ferrite film due to shrinkage during RTA. This is due to densification of the ferrite film as it crystallizes. It may be possible to eliminate this effect through changes in the ferrite composition that minimize the crystallization temperature. The firing temperature of ceramic ferrite materials was recently reduced to below 900 °C, allowing the fabrication of low-temperature co-fired ceramic circulators and phase shifters [147].
(c) Thermal expansion coefficient mismatch effects may be minimized by using spinel or hexaferrite films on GaAs (Table 7.5), and by adjusting the ferrite composition to minimize the crystallization temperature.

Use of YIG on silicon was useful for initial demonstrations because of their high strength, but the processing advances described here enabled the fabrication of ferrite film circulators onto GaAs substrates without cracking the GaAs. It is recommended that future work should focus on the monolithic integration of self-biasing hexaferrite on GaAs for mm-wave applications.

Acknowledgment

This work was supported in part by the ARPA Ferrite Development Consortium, contract number MDA 972-93-H-0002, managed by Dr. Frank Patten and Dr. Stu Wolf. The authors thank Huey Daw-Wu for critical reading of the manuscript and the following scientists and engineers for their contributions to the developments described in this chapter: M. R. Daniel, N. J. Doyle, M. C. Driver, G. W. Eldridge, M. H. Hanes, R. L. Messham, and M. M. Sopira.

References

1. Neel, L. (1948). *Ann. Phys. (Paris)* **3**: 137.
2. Mee, J.E., Pullian, G.R., Archer, J.L., and Besser, P.J. (1969). *IEEE Trans. Magn.* **MAG-5**: 717.
3. Glass, H.L. (1987). *Microw. J.* **30**: 52.
4. Prinz, G.A. (1990). *Science* **250**: 1092.
5. Snoek, J.L. (1948). *Physica* **14**: 207.
6. Webb, D.C. (1988). *IEEE Trans. Magn.* **24**: 2799.
7. Itoh, T., Hori, S., Abe, M., and Tamaura, Y. (1991). *Jpn. J. Appl. Phys.* **29**: L1458.
8. Itoh, T., Hori, S., Abe, M., and Tamaura, Y. (1991). *J. Appl. Phys.* **69**: 5911.
9. Itoh, T., Zhang, Q., Abe, M., and Tamaura, Y. (1991). *J. Appl. Phys.* **70**: 6443.
10. Itoh, T., Hori, S., Abe, M., and Tamaura, Y. (1991). *IEEE Transl. J. Jpn.* **6**: 214.
11. Yoo, K.C. and Talisa, S.H. (1990). *J. Appl. Phys.* **67**: 5533.
12. Halpern, B.L. and Schmitt, J.J. (1994). *J. Vac. Sci. Tech.* **A 12**: 1623.

13. Dionne, G.F., Cui, G.C., McAvoy, T., Halpern, B.L., and Schmitt, J.J. (1995). *IEEE Trans. Magn.* **31**: 3853.
14. Halpern, B.L., Golz, J.W., Zhang, J.-Z., McAvoy, D.T., Srivatsa, A.R., and Schmitt, J.J. (1995). *Advances in Coating Technologies for Corrosion and Wear Resistant Coatings*, A.R. Srivatsa, C.R. Clayton, and J.K. Hirvonen, eds., The Minerals, Metals, & Materials Society, 99.
15. Golz, J., Zhang, J.Z., Han, H., Motherway, B., Srivatsa, A., Halpern, B.L., and Schmitt, J.J. (1995). *Advances in Coatings Technologies for Corrosion and Wear Resistant Coatings*, A.R. Srivatsa, C.R. Clayton, and J.K. Hirvonen, eds., The Minerals, Metals, & Materials Society, 37.
16. Herman, H. and Sampath, S. (1998). *Ind. Ceram.* **18**: 29.
17. Sampath, S., Gansert, R., and Herman, H. (1995). *JOM* **47**: 30.
18. Nathaway, R., Vispute, R.D., Chaudhari, S.M., Kanetkar, S.M., Ogale, S.B., Mitra, A., and Date, S.K. (1989). *J. Appl. Phys.* **65**: 3197.
19. Ogale, S.B. and Nawathey, R. (1989). *J. Appl. Phys.* **65**: 1367.
20. Ogale, S.B., Kanetkar, S.M., Chaudhari, S.M., Goodbole, V.P., Koinkar, V.N., Joshi, S., Nawathey, R., Vispute, R.D., Date, S.K., and Moghe, A.R. (1990). *Ferroelectrics* **102**: 85.
21. Koinkar, V.N., Nawathey, R., Chaudhari, S.M., Kanetkar, S.M., Date, S.K., and Ogale, S.B. (1989). *Proc. Fifth International Conference on Ferrites*, ICF-5, C.M. Srivastava and M.J. Patni, eds., **1**: 525.
22. Williams, C.M., Chrisey, D.B., Lubitz, P., Grabowski, K.S., and Cotell, C.M. (1994). *J. Appl. Phys.* **75**: 1676.
23. Carosella, C.A., Chrisey, D.B., Lubitz, P., Horwitz, J.S., Dorsey, P., Seed, R., and Vittoria, C. (1992). *J. Appl. Phys.* **71**: 5107.
24. Dorsey, P.C., Bushnell, S.E., Seed, R.G., and Vittoria, C. (1993). *J. Appl. Phys.* **74**: 1242.
25. Kidoh, H., Morimoto, A., and Shimizu, T. (1991). *Appl. Phys. Lett.* **59**: 237.
26. Simion, B.M., Thomas, G., Ramesh, R., Keramidas, V.G., and Pfeffer, R.L. (1995). *Appl. Phys. Lett.* **66**: 830.
27. Nawathey-Dikshit, R., Shinde, S.R., Ogale, S.B., Kulkarni, S.D., Sainkar, S.R., and Date, S.K. (1996). *Appl. Phys. Lett.* **68**: 3491.
28. Papakonstaninou, P., O'Neill, M., Atkinson, R., Salter, I.W., and Gerber, R. (1996). *J. Magnetism and Magnetic Materials* **152**: 401.
29. Welch, R.G., Neamtu, J., Rogalski, M.S., and Palmer, S.B. (1996). *Solid State Comm.* **97**: 355.
30. Welch, R.G., Neamtu, J., Rogalski, M.S., and Palmer, S.B. (1996). *Materials Lett.* **29**: 199.
31. Suzuki, Y., van Dover, R.B., Gyorgy, E.M., Phillips, J.M., Korenivski, V., Werder, D.J., Chen, C.H., Cava, R.J., Krajewski, J.J., Peck, Jr., W.F., and Do, K.B. (1996). *Appl. Phys. Lett.* **68**: 714.
32. Dorsey, P.C., Qadri, S.B., Grabowski, K.S., Knies, D.L., Lubitz, P., Chrisey, D.B., and Horwitz, J.S. (1997). *Appl. Phys. Lett.* **70**: 1173.
33. Karim, R. and Vittoria, C. (1997). *J. Magnetism and Magnetic Materials* **167**: 27.
34. Dorsey, P.C., Rappoli, B.J., Grabowski, K.S., Lubitz, P., Chrisey, D.B., and Horwitz, J.S. (1997). *J. Appl. Phys.* **81**: 6884.
35. Hoshi, Y., Shinozaki, H., Shimizu, H., Kato, K., and Kaneko, F. (1998). *J. Magnetism and Magnetic Materials* **177–181**: 241.
36. Westwood, W.D., Eastwood, H.K., and Sadler, A.G. (1971). *J. Vac. Sci. and Techn.* **8**: 176.
37. Zaquine, I., Benazizi, H., and Mage, J.C. (1988). *J. Appl. Phys.* **64**: 5822.
38. Morisako, A., Nakanishi, H., and Matsumoto, M. (1994). *J. Appl. Phys.* **75**: 5969.
39. Gu, B.X., Zhang, H.Y., Zhai, H.R., Lu, M., Zhang, S.Y., and Maio, Y.Z. (1994). *Appl. Phys. Lett.* **65**: 3404.
40. Acharya, B.R., Krishnan, R., Prasad, S., Venkataramani, N., Ajan, A., and Shringi, S.N. (1994). *Appl. Phys. Lett.* **64**: 1579.
41. Chen, Y., Sui, X., and Kryder, M.H. (1996). *IEEE Trans. Magnetics* **32**: 3313.

42. Acharya, B.R., Prasad, S., Venkataramani, N., Shringi, S.N., and Krishnan, R. (1996). *J. Appl. Phys.* **79**: 478.
43. Brandle, C.D. and Valentino, A.J. (1972). *J. Crystal. Growth* **123**: 478.
44. Adam, J.D., Krishnaswamy, S.V., Talisa, S.H., and Yoo, K.C. (1990). *J. Magnetism and Magnetic Materials* **83**: 419.
45. Grechishkin, R.M., Goosev, M.Yu., Ilyashenko, S.E., Neustroev, N.S. *J. Magnetism and Magnetic Materials* **157/158**: 305.
46. Leccabue, F., Panizzieri, R., Salviati, G., Albanese, G., and Sanchez Llamazares, J.L. (1986). *J. Appl. Phys.* **59**: 2114.
47. Naoe, M. and Matsushita, N. (1996). *J. Magnetism and Magnetic Materials* **155**: 216.
48. Lee, J., Lee, H.M., Kim, C.S., and Oh, Y. (1998). *J. Magnetism and Magnetic Materials* **177–181**: 900.
49. Matsumoto, M., Morisako, A., Haeiwa, T., Naruse, K., and Karasawa, T. (1991). *IEEE Transl. J. Magn. Jpn.* **6**: 648.
50. Fujii, E., Torii, H., and Aoki, M. (1989). *IEEE Transl. J. Magn. Jpn.* **4**: 512.
51. Deschandvres, J.L., Langlet, M., and Joubert, J.C. (1990). *J. Magnetism and Magnetic Materials* **83**: 437.
52. Tayama, M., Nikawa, K., and Okada, F. (1989). *Crystal Prop. Prep.* **27–30**: 1093.
53. Tanbakuchi, H., Nicholson, D., Kunz, B., and Ishak, W. (1989). *IEEE Trans. Magn.* **25**: 3248.
54. Mizunama, Y., Murakami, Y., Nakano, H., Ohgihara, T., and Okamoto, T. (1988). *IEEE Trans. Microwave Theory and Techniques* **36**: 1885.
55. Rainville, P.J. and Harackiewicz, F.J. (1992). *IEEE Microwave and Guided Wave Letters* **2**: 483.
56. Ishak, W.S. (1988). *Proc. IEEE* **76**: 171.
57. Bongianni, W.L. (1974). X-band signal processing using magnetic waves, *Microwave Journal* **17**: 49.
58. Castera, J.-P. (1984). *J. Appl. Phys.* **55**: 2506.
59. Webb, D.C. (1993). *IEEE MTT-S Digest,* 203.
60. Murakami, Y. (1993). *IEEE MTT-S Digest,* 207.
61. Okuda, T., Koshizuka, N., Hayashi, K., Takahashi, T., Kotani, H., and Yamamoto, H. (1987). *IEEE Trans. Magn.* **MAG-23**: 3491.
62. Shono, K., Kano, H., Koshino, N., and Ogawa, S. (1987). *IEEE Trans. Magn.* **MAG-23**: 2970.
63. Gomi, M., Okazaki, T., and Abe, M. (1987). *IEEE Trans. Magn.* **MAG-23**: 2967.
64. Deschanvres, J.L., Langlet, M., Labeau, M., and Joubert, J.C. (1990). *IEEE Trans. Magn.* **26**: 187.
65. Krumme, J.-P., Doorman, V., and Eckart, R. (1984). *IEEE Trans. Magn.* **MAG-20**: 983.
66. Deschanvres, J.L., Langlet, M., Bochu, B., and Joubert, J.C. (1991). *J. Magnetism and Magnetic Materials* **101**: 224.
67. Matsumoto, K., Sasaki, S., Yamanobe, Y., Yamaguchi, K., and Fujii, T. (1991). *J. Appl. Phys.* **70**: 1624.
68. Okuda, T., Katayama, T., Kobayashi, H., Kobayashi, N., Satoh, K., and Yamamoto, H. (1990). *J. Appl. Phys.* **67**: 4944.
69. Gomi, M., Tanida, T., and Abe, M. (1985). *J. Appl. Phys.* **57**: 3888.
70. Krumme, J.-P., Doorman, V., and Willich, P. (1985). *J. Appl. Phys.* **57**: 3885.
71. Nelson, H. (1974). *J. Crystal Growth* **27**: 1.
72. Glass, H.L. (1988). *Proc. IEEE* **76**: 151.
73. Cermak, J., Abraham, A., Fabian, T., Kabos, P., and Hyben, P. (1990). *J. Magnetism and Magnetic Materials* **83**: 427.
74. Glass, H.L. and Elliott, M.T. (1974). *J. Crystal Growth* **27**: 253.
75. Linares, R.C. (1968). *J. Crystal Growth* **3, 4**: 443.
76. Glass, H.L. (1976). *J. Crystal Growth* **33**: 183.

77. Oliver, S.A., Zavracky, P.M., McGruer, N.E., and Schmidt, R. (1997). *IEEE Microwave and Guided Wave Letters* **7**: 239.
78. Rachford, F.J., Levy, M., Osgood, R.M., Jr., Kumar, A., and Bakhru, H. (1999). *J. Appl. Phys.* **85**: 5217.
79. Buhay, H., Adam, J.D., Daniel, M.R., Doyle, N.J., Driver, M.C., Eldridge, G.W., Hanes, M.H., Messham, R.L., and Sopira, M.M. (1995). *IEEE Trans. Magn.* **31**: 3832.
80. Karim, R., Oliver, S.A., and Vittoria, C. (1995). *IEEE Trans. Magn.* **31**: 3485.
81. Adam, J.D., Buhay, H., Daniel, M.R., Eldridge, G.W., Hanes, M.H., Messham, R.L., and Smith, T.J. (1996). *IEEE MTT-S Digest,* 113.
82. Adam, J.D., Buhay, H., Daniel, M.R., Driver, M.C., Eldridge, G.W., Hanes, M.H., and Messham, R.L. (1995). *IEEE MTT-S*, 97.
83. von Aulock, W.H. (1965). *Handbook of Microwave Ferrite Materials,* New York: Academic Press.
84. Glass, H.L. and Liaw, J.H.W. (1978). *Mat. Res. Bul.* **13**: 353.
85. Balestrino, G., Martelluci, S., Paoletti, A., Paroli, P., Petrocelli, G., Tebano, A., Oliver, S.A., and Vittoria, C. (1995). *Solid State Communications* **96**: 997.
86. Dorsey, P.C., Lubitz, P., Harris, V.G., Chrisey, D.B., and Horwitz, J.S. (1995). *IEEE Trans. Magn.* **31**: 3455.
87. Tanaka, K., Omata, Y., Nishikawa, Y., Yoshida, Y., and Nakamura, K. (1991). *IEEE Translation J. Magnetics in Japan* **6**: 1001.
88. Oliver, S.A., Vittoria, C., Balestrino, G., Martellucci, S., Petrocelli, G., Tebano, A., and Paroli, P. (1994). *IEEE Trans. Magn.* **30**: 4933.
89. Srivastava, A.K., Hurben, M.J., Wittenauer, M.A., Kabos, P., Patton, C.E., Ramesh, R., Dorsey, P.C., and Chrisey, D.B. (1998). *J. Appl. Phys.* **85**: 7838.
90. Abe, M. and Tamaura, Y. (1984). *J. Appl. Phys.* **55**: 2614.
91. Abe, M., Tamaura, Y., Goto, Y., Kitamura, N., and Gomi, M. (1987). *J. Appl. Phys.* **61**: 3211.
92. Abe, M., Itoh, T., Tamaura, Y., Gotoh, Y., and Gomi, M. (1987). *IEEE Trans. Magn.* **MAG-23**: 3736.
93. Abe, M., Itoh, T., Tamaura, Y., and Gomi, M. (1988). *J. Appl. Phys.* **63**: 3774.
94. Talisa, S.H., Yoo, K.C., Abe, M., and Itoh, T. (1988). *J. Appl. Phys.* **64**: 5819.
95. Lubitz, P., Lawrence, S.H., Rachford, F.J., and Rappoli, B.J. (1994). *IEEE Trans. Magn.* **30**: 4539.
96. Gillies, M.F., Coehoom, R., van Zon, J.B.A., and Alders, D. (1998). *J. Appl. Phys.* **83**: 6855.
97. Toyama, M., Nikawa, Y., and Okada, F. (1989). *Proc. International Conference on Ferrites-5*, 1093.
98. Ovsiannikov, V.P., Lashkarev, G.V., Mazurenko, Y.A., and Bugaeva, M.E. (1997). *Electrochemical Society Proceedings* **25**: 1020.
99. Okuno, S.N., Hashimoto, S., and Inomata, K. (1992). *J. Appl. Phys.* **71**: 5926.
100. Matsushita, N., Nakagawa, S., and Naoe, M. (1992). *IEEE Trans. Magn.* **28**: 3108.
101. Naoe, M., Yata, M., and Matsumoto, K. (1987). *IEEE Trans. Magn.* **MAG-23**: 3429.
102. Tagami, M., Aoyama, M., Nishimoto, K., and Goto, F. (1985). *IEEE Trans. Magn.* **MAG-21**: 1164.
103. Naoe, M., Hoshi, Y., and Yamanaka, S. (1978). *J. Crystal Growth* **45**: 361.
104. Stearns, F.S. and Glass, H.L. (1975). *Mat. Res. Bull.* **10**: 1255.
105. Stearns, F.S. and Glass, H.L. (1976). *Mat. Res. Bull.* **11**: 1319.
106. Glass, H.L. and Liaw, J.H.W. (1978). *J. Appl. Phys.* **49**: 1578.
107. Glass, H.L. and Stearns, F.S. (1977). *IEEE Trans. Magn.* **MAG-13**: 1241.
108. Dötsch, H., Mateika, D., Röschmann, P., and Tolksdorf, W. (1983). *Mat. Res. Bull.* **18**: 1209.
109. Hylton, T.L., Parker, M.A., Coffey, K.R., and Howard, J.K. (1993). *J. Appl. Phys.* **73**: 6257.
110. Hylton, T.L., Parker, M.A., Coffey, K.R., and Howard, J.K. (1992). *Appl. Phys. Lett.* **61**: 867.

111. Yuan, M.S., Glass, H.L., and Adkins, L.R. (1988). *Appl. Phys. Lett.* **53**: 340.

112. Dorsey, P.C., Chrisey, D.B., Horwitz, J.S., Lubitz, P., and Auyeung, R.C.Y. (1994). *IEEE Trans. Magn.* **30**: 4512.

113. Dorsey, P.C., Qadri, S.B., Feldman, J.L., Horwitz, J.S., Lubitz, P., Chrisey, D.B., and Ings, J.B. (1996). *J. Appl. Phys.* **79**: 3517.

114. Krishnan, R.S., Srinivasan, R., and Devanarayanan, S. (1979). *Thermal Expansion of Crystals*, New York: Pergamon Press.

115. Oliver, S.A., Chen, M.L., Kozulin, I., McGruer, N.E., Zavracky, P.M., and Vittoria, C. (1998). in *Abstracts*, November, Miami, Florida, Abstract BF-03, 43rd Magnetism and Magnetic Materials Conference.

116. Tenzer, R.K. (1963). *J Appl. Phys.* **34**: 1267.

117. Morisako, A., Matsumoto, M., and Naoe, M. (1986). *IEEE Trans. Magn.* **MAG-22**: 1146.

118. Morisako, A., Matsumoto, M., and Naoe, M. (1987). *IEEE Trans. Magn.* **MAG-23**: 2359.

119. Matsuoka, M., Hoshi, Y., Naoe, M., and Yamanaka, S. (1982). *IEEE Trans. Magn.* **MAG-18**

120. Matsuoka, M., Naoe, M., and Hoshi, Y. (1985). *IEEE Trans. Magn.* **MAG-21**: 1474.

121. Morisako, A., Matsumoto, M., and Naoe, M. (1987). *IEEE Trans. Magn.* **MAG-23**: 56.

122. Matsuoka, M., Hoshi, Y., Naoe, M., and Yamanaka, S. (1984). *IEEE Trans. Magn.* **MAG-20**: 800.

123. Morisako, A., Matsumoto, M., and Naoe, M. (1988). *IEEE Trans. Magn.* **24**: 3024.

124. Sui, X., Kryder, M.H., Wong, B.Y., and Laughlin, D.E. (1993). *IEEE Trans. Magn.* **29**: 3751.

125. Hoshi, Y., Kubota, Y., and Ikawa, H. (1997). *J. Appl. Phys.* **81**: 4677.

126. Noma, K., Matsushita, N., Nakagawa, S., and Naoe, M. (1997). *J. Appl. Phys.* **81**: 4377.

127. Hoshi, Y., Kubota, Y., Onodera, H., Shinozaki, H., Shimizu, H., and Ikawa, H. (1997). *J. Appl. Phys.* **81**: 4690.

128. Lacroix, E., Gerard, P., Marest, G., and Dupay, M. (1991). *J. Appl. Phys.* **69**: 4770.

129. Gerard, P., Lacroix, E., Marest, G., Dupuy, M., Rolland, G., and Blanchard, B. (1990). *J. Magnetism and Magnetic Materials* **83**: 13.

130. Dorsey, P.C. and Vittoria, C. (1994). *J. Magnetism and Magnetic Materials* **137**: 89.

131. Morisako, A., Matsumoto, M., and Naoe, M. (1996). *J. Phys.* **79**: 4881.

132. Carey, R., Gago-Sandoval, P.A., Newman, D.M., and Thomas, B.W.J. (1993). *IEEE Trans. Magn.* **29**: 3799.

133. Donahue, E.J. and Schleich, D.M. (1992). *J. Appl. Phys.* **71**: 6013.

134. Suzuki, E., Hoshi, T., and Naoe, M. (1998). *J. Appl. Phys.* **83**: 6250.

135. Matsushita, N. and Naoe, M. (1993). *IEEE Trans. Magn.* **29**: 4089.

136. Noma, K., Matsushita, N., Nakagawa, S., and Naoe, M. (1996). *J. Appl. Phys.* **79**: 5970.

137. Morisako, A., Liu, X., Matsumoto, M., and Naoe, M. (1997). *J. Appl. Phys.* **81**: 4374.

138. Gerard, P., Lacroix, E., Marest, G., Blanchard, B., Rolland, G., Rolland, B., and Bechevet, B. (1989). *Solid State Communications* **71**: 57.

139. Schloemann, E.F. (1988). *Proc. IEEE* **76**: 188.

140. Weiss, J.A. and Dionne, G.F. (1989). *1989 IEEE MTT-S Intl. Microwave Symposium*, p. 145.

141. Baba, P.D., Argentina, G.M., Courtney, W.E., Dionne, G.F., and Temme, D.H. (1972). *IEEE Trans. Magnetics* **8**: 83.

142. Yang, C.J., Kim, S.W., and Kim, S.Y. (1994). *IEEE Trans. Magnetics* **30**: 4527.

143. Sriram, S., Messham, R.L., Smith, T.J., and Eldridge, G.W. (1996). *IEEE Trans. Electron Devices* **43**: 834.

144. Neidert, R.E. and Phillips, P.M. (1993). *IEEE Trans. Microwave Theory and Techniques* **41**: 1081.

145. Fathy, A., Delinger, E., Kalokitis, D., Pendrick, V., Johnson, H., Pique, A., Harshavardhan, K.S., and Belohoubek, E. (1995). *1995 IEEE MTT-S Digest*, p. 195.

146. Agarwal, A.K., Smith, T.J., Hanes, M.H., Messhams, R.L., Driver, M.C., and Nathanson, H.C. (1995). MICROX—An affordable silicon MMIC technology, *Workshop on Si RF Technology, 1995 IEEE MTT-S.*

147. Piloto, A.J., Partlow, D.P., Painter, C.J., Tolle, S.C., and Zaki, K. (1997). Advanced LTCC ferrite technolgy for low cost phase shifters in emerging RF systems, p. 332, *Proc. Intl. Symp on Microelectronics, Philadelphia, PA, Oct. 14–16, 1997.*

Ferroelectric Thin Films: Preparation and Characterization

S.B. KRUPANIDHI

Materials Research Center, Indian Institute of Science, Bangalore, India

8.1. Introduction

More emphasis in integrated circuit technology is being focused on microminiaturization, with the result that significant advantages in the application of small dimension ferroelectric thin films offer great potential for ready integrability [1]. Earlier attempts at deposition of ferroelectric thin films were limited mostly to development of thin-film capacitors that use large permittivity (k) materials such as barium titanate; as well, the process compleixities involved in depositing multicomponent material systems created another limitation. Current activity in ferroelectric thin-film research, motivated by the lastest advances in thin-film growth processes offers the opportunity to exploit several phenomena in ferroelectric materials including as polarization hysteresis [2, 3], pyroelectricity [4, 5], piezoelectricity [6–8], and electro-optic activity [9].

The primary impetus of recent research on ferroelectric thin films has been the substantial demand for the development of nonvolatile memory devices (also called FeRAMs, ferroelectric random access memories). They promise fast read-and-write cycles, low switching voltages (3–5 V and lower) non-volatility in the unpowered mode, long endurance (10^{12} cycles), and radiation hardness compatible with semiconductors that include GaAs [10]. Development of ferroelectric

ISSN 1079-4050

Vol. 28
ISBN 0-12-533028-6/$35.00

thin films with controlled properties at relatively lower growth temperature remains a major research task and several growth techniques are currently being explored in the hope of achieving these goals.

8.2. Growth Processes of Ferroelectric Thin Films

Several deposition techniques are being exploited for the growth of ferroelectric thin films; in general, numerous compositions are under exploration all over the world. Table 8.1 summarizes most of the current research activity, and under each technique the materials are listed that are now receiving attention vis-a-vis their development. Classification of these techniques is done mainly in terms of growth processes—physical vapor growth involving low energy bombardment and chemical routes involving no such bombardment.

The growth process with low energy ion bombardment includes magnetron sputtering [11], ion-beam sputtering [12] excimer laser ablation [13], electron cyclotron resonance (ECR) plasma-assisted growth [14] and plasma-enhanced chemical vapor deposition (PECVD) [15]. The techniques that do not involve bombardment include sol-gel [16], metalorganic decomposition (MOD) [17], solution growth [18], thermal and *e*-beam evaporations [19], flash evaporation [20], chemical vapor deposition (CVD) [21], metalorganic chemical vapor deposition (MOCVD) [22, 23] and molecular beam epitaxy (MBE) [24].

8.2.1. Physical Growth of Techniques involving Low Energy Ion Bombardment

8.2.1.1. Magnetron Sputtering

Sputter deposition of ferroelectric materials can be done either with a single target source consisting of multiple components or multiple elemental targets. Sputter deposition has always been the first growth attempted for the growth of ferroelectric thin films, as it has a respected industrial track record. Several modifications are necessary to accomplish stoichiometric ferroelectric oxide thin films as described in what follows.

a. Sputtering from a Single Target Source: Sputter deposition, with or without magnetron backing, had been the most popular dry growth technique for depositing ferroelectric thin films. The majority of these efforts described in the literature use compound ceramic targets of the desired composition as the starting material. Consistent success in the growth of stoichiometric ferroelectric thin films of Pb-based compounds (with this approach is somewhat limited due

TABLE 1.1
CLASSIFICATION OF GROWTH TECHNIQUES INVOLVED IN DEPOSITION OF DIFFERENT FERROELECTRIC THIN FILMS

Ferroelectric thin-film growth activity										
With bombardment					No bombardment					
Physical vapor				Chemical vapor						
Magnetron sputtering (dc and RF)	Ion-beam sputtering	Excimer laser ablation	ECR-aided	PECVD	Sol-Gel	MOD	Evaporation	CVD	MOCVD	MBE
BT	PZT	BiT	PT	PT	PT	PT	PG	PT	PT	BaMgF
PT	$KNbO_3$	PZT	BiT	BT	PZT	PZT	SbSI	PZT	PZT	
PZT	PLT	PLT	PZT	PZ	PLZT	PLZT	PZT	BST	BST	
PLZT	PZ	PG			PMN : PT	$LiNbO_3$	BT			
PMN : PT		KTM			BST					
BST		SBT			SBT					
PLT		SBN			SBN					
BiT		BaBT								
SBN										

to: a) limited control over the composition of the films caused by the large differences in sputter yields and as a result the preferential deficiency of volatile elements in the films; b) the presence of negative ions and reflected neutrals during sputtering of oxide targets, which can bombard the growing film uncontrollably and damage the film surface; and c) low deposition rates with oxide targets. However, efforts by some researchers continue to achieve ferroelectric thin films of usable quality from a single oxide target, and the most recent activity was related to the deposition of thin films of $BaTiO_3$-$SrTiO_3$ solid solution. Unlike the Pb-based perovskites, the Ba-related compounds do not seem to be affected by stoichiometric deviations. Good films of $(Ba,Sr)TiO_3$ are reported by single target sputtering. For example, Hwang *et al.* [25, 26] and Zafar *et al.* [27], reported on sputter-deposited stoichiometric BST thin films that possess useful device-related electrical properties.

Most recently, Ding *et al.* [28] attempted to fabricate crystalline thin films of PLT (28% La modified lead titanate) using a stoichiometric powder target of PLT on glass substrates. They adopted two sputtering processes: an *in-situ* sputtering process that maintains the substrate temperature at > 550 °C during the process in which the substrates were simultaneously maintained at relatively lower temperatures of 200 °C and then these films were subsequently annealed to realize the perovskite phase formation. Relationships between structure and processing temperature were also investigated. Studies revealed that the *in-situ* crystallized films exhibited relatively superior electro-optic properties.

b. Sputter deposition from Multiple Target Sources: Using reactive magnetron co-sputtering with multiple targets, and pure elemental targets, Wasa *et al.* [29] have grown excellent PLT (La-modified $PbTiO_3$) thin films. By using pure elemental targets rather than ceramic ones, they were able to deposit good-quality ferroelectric thin films. This was due to higher rates of deposition due to the reactive sputtering of metal targets and better control over the composition of the final film via the independent control of sputter rates of individual targets, thus allowing *in situ* alteration of composition during growth. The schematic of the technique is shown in the Fig. 8.1.

In general, the magnetron cathodes are vertically mounted at a convenient angle of inclination (20–30°) with respect to the central axis, such that all targets are focused to a common focal point while ensuring a large overlapping deposition area from each target. A substrate holder positioned in front of the target is capable of moving to and fro and also features simultaneous rotation, dc biasing, and substrate heating capabilities. Films were deposited at 500 °C at a pressure of 9 mtorr ($Ar: O_2 = 90: 10$), while the ratio of Pb/Ti was changed by charging the power density on the Pb target while keeping the power density on Ti unchanged. The effect of Pb content on crystallinity is shown in Fig. 8.1. It may be seen that the films consisting of large excess amounts of Pb (about 15%)

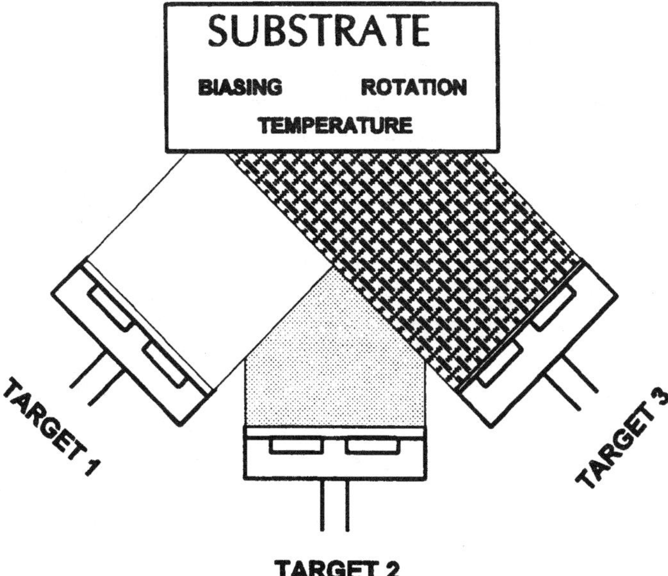

FIG. 8.1. Schematic diagram of multitarget magnetron sputtering.

initiated a perovskite phase along the (101) orientation, while the near-stoichiometric Pb content exhibited a much clearer tetragonal phase with a preferred orientation along the (001) and (100) directions. The group from Matsushita (Japan) [29] was the first to grow excellent quality epitaxial PLT thin films by this multimagnetron sputter deposition approach. In the preceding reference they detail the effect of stoichiometric condition on epitaxy, and subsequently on the electro-optic behavior of the PLT thin films.

8.2.1.2. Multi-Ion Beam Reactive Sputter (MIBERS) Deposition

The ion beam sputter deposition technique with concurrent low energy ion bombardment offers the following unique features: a) independent control of flux density and energy of the sputtered species; b) lower operating pressures during thin-films growth, which ensures better quality of films; and c) the possibility of independently controlled low energy ion bombardment (usually with reactive oxygen species) of the growing thin films. Besides incorporating reactive oxygen species in the films, this sort of bombardment also offers additional benefits such as increasing adatom mobility and proves extra energy to supplement thermal energy to the species during nucleation, resulting in crystallization of the thin films.

Figure 8.2 shows the schematic of the multi-ion beam reactive sputter (MIBERS) deposition system used to prepare PZT films. For these films [30] three independent metal targets of Pb, Zr, and Ti were individually sputtered by high-energy focused ion beam sources. Three individual metal targets about 7.5-cm diameter were coordinately arranged so that a flat profile of sputtered species was obtained. A fourth ion source was used to bombard the growing film, which was operated in defocused mode for obtaining a broad beam and was arranged to achieve a bombarding angle of 35° with respect to the normal of the substrate surface. An ion flux density measurement probe was placed adjacent to the substrates to measure the flux density of the bombarding ions at the substrate. This measurement, in conjunction with atomic flux density determined from the thickness monitor, establishes the ion/atom ratio, which is critical to achieve reproducible bombarding effects. The deposition rates of individual targets were measured as a function of ion beam voltage, while the ion beam current and oxygen partial pressures were kept constant. The beam voltage determines the energy of the sputtered ion, which is to be kept above a threshold level necessary to initiate the sputtering of each metal. Small quantities of molecular oxygen were

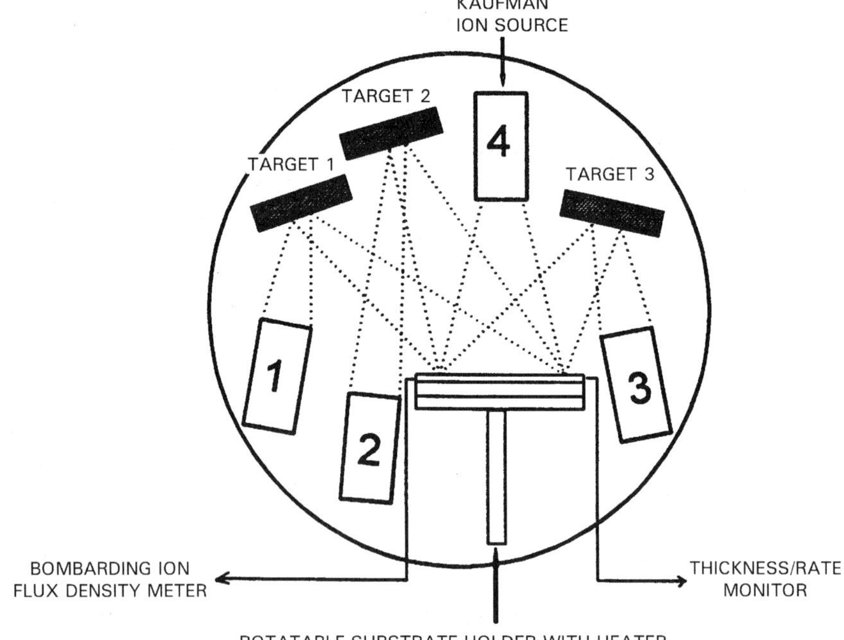

FIG. 8.2. Schematic diagram of multi-ion beam reactive sputtering (MIBERS) technique.

bled into the chamber during the deposition to create a reactive environment for the oxidation of the sputtered metal species. By adjusting the powers on individual targets, stoichiometric PZT films could be grown over large areas (10-cm diameter).

PZT (50/50) films were deposited on Pt-coated Si and bare Si substrates by using the MIBERS technique at room temperature and at a deposition rate of about 18 Å/min [31]. During the deposition, the growing films were directly bombarded with a low-energy oxygen ion (O_2^+/O^+) beam generated by a 3-cm Kauffman ion source in a single grid configuration. The beam was directed at the substrates with an angle of incidence of about 25° from the substrate surface normal. To compensate for the Pb re-sputtering during the direct bombardment and maintain a constant Pb content in the deposited films, the Pb fluence was increased by 12–19% with respect to the fluence used without direct bombardment. The exact increase in Pb fluence was determined by the bombarding conditions. The Pb fluence was adjusted by changing the voltage and current of the ion beam used for sputtering the Pb target. Nonbombarded films were also deposited under the same conditions in order to allow comparison with the bombarded films. It needs to be mentioned that the word "nonbombarded" means without direct ion bombardment. The intrinsic bombardment effect of ion beam sputter deposition due to the backscattered ions and sputtered neutrals, which in this case have energies of about 10 eV as measured, [15] is not specified, as it is common in both cases (with and without direct bombardment). As-grown films were annealed at temperatures from 550–700 °C in an oxidizing atmosphere to induce crystallization. The annealed films were characterized in terms of structure, morphology, and electrical properties to determine the effects introduced by the direct bombardment.

8.2.1.3. Pulsed Laser Ablation

Laser-induced vaporization (also called laser ablation) is another film deposition technique in which a plume of ionized and ejected material is produced by high-intensity laser irradiation of a solid target. Most commonly, UV excimer lasers are employed for this purpose and the wavelength of the radiation is tuned by the lasing gas composition, such as F_2 (157 nm), ArF (193 nm), KrF (248 nm), KrCl (308 nm), XeF (351 nm), and XeCl (308 nm). The KrF (248 nm) composition has been most dominantly employed due to its high-energy laser pulse output. Pulse-to-pulse duration can be 10-25 ns with repetition rates of up to several hundred hertz with energies approaching 500 mJ/pulse [32].

Laser ablation consists mainly of three processes: a) interaction of the laser beam with the target; b) adiabatic plasma expansion; and c) deposition of thin films. The ablation of the material is always normal to target surface irrespective of the angle of laser beam incidence. The generated plume is composed of neutrals, ionized atomic and mostly molecular species.

In spite of a few limitations of the technique, such as the occurrence of particulates on the film surface and unevenness of thickness, laser ablation offers several advantages, including: a) the film composition can be nearly identical to the target stoichiometry; b) deposition in a wide range of oxygen partial pressures; c) low crystallization temperatures due to high excitation energy of the photofragments in the laser produced plasma; d) high deposition rates; and e) deposition of materials with high melting temperatures. It is essential to mention that below the onset of the nonthermal ablation for certain fluence the material is removed from the target by a thermally assisted process, which leads to preferential evaporation of heavy and volatile species from the target. Typical energy densities of $> 2J/cm^2$ have been found to be characteristic at the onset of nonthermal ablation process.

A typical schematic of a laser ablation arrangement is shown in the Fig. 8.3. The output of a KrF excimer laser beam is focused by a uv-grade plano-convex lens and it is brought into the vacuum chamber through a quartz port. The incoming beam is incident on the rotating target at an angle of 45°. Films are deposited on the substrates, located in front of the target at a distance of 2–3 cm. This technique has been popularly employed for the successful deposition of high-Tc superconductors and is currently being exploited for the growth of device quality *in situ* stoichiometric ferroelectric thin films, such as $Bi_4Ti_3O_{12}$ [33, 34], $BaTiO_3$ [35], $SrTiO_3$ [36], $(Ba,Sr)TiO_3$ [37], PZT [38], PLT [39, 40], $SrBi_2Ta_2O_9$ [41] and $SrBi_2Nb_2O_9$ [42].

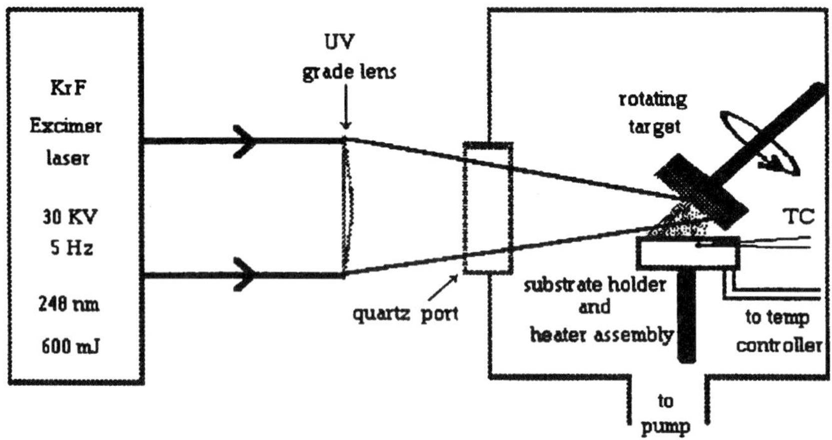

FIG. 8.3. Schematic diagram of pulsed laser ablation technique.

8.2.1.4. SOL-GEL Based Growth

Sol-gel technology is used as a method for depositing high-quality multicomponent oxide thin film, where the processing technology needs to be rapid and inexpensive. Materials that are used in modern device technology require high purity and close control over composition and microstructure. As the chemical reactants for sol-gel processing can be purified conveniently by distillation and crystallization, films of high purity can be fabricated by sol-gel processing. The basic principle involves first the polymerization of organometallic compounds such as alkoxides to produce a gel with a continuous complex network, and then this gel is dried and fired to displace the organic components, which results in the formation of the final inorganic oxide thin film. From a commercial viewpoint, alkoxides and acetates of most metals can be synthesized and are convenient starting materials with respect to both availability and cost. These precursors are taken in their stoichiometric ratios to form the basic spin-on solution. The viscosity of the solution is controlled through the addition of suitable organic solvents such as 2-methoxyethanol. The solution is then spin-coated on the substrate surface using a spin-coater rotating at 2000–6000 rpm. The organic film is then pyrolyzed to remove the organic components at relatively low temperatures ($\sim 300\ ^{\circ}C$). The process is repeated several times to obtain films of the required thickness. The grown oxide film is then annealed at the desired temperature to induce crystallinity. Sol-gel processing, used for the growth of thin films of several oxide materials, has led to very good quality thin films [43, 44]. Device-quality Pb-based perovskite thin films have been successfully grown by the sol-gel technique [45–48].

8.2.1.5 CVD and MOCVD Approach

Chemical vapor deposition (CVD) and/or metalorganic chemical vapor deposition (MOCVD) techniques have emerged as alternate approaches for the growth of ferroelectric oxide thin films [49]. These techniques have many potential advantages, including the ability to deposit high-quality, ultrathin layers on three-dimensional (3D) complex geometries, excellent composition control, and amenability to large-scale processing. The success of an MOCVD process depends critically on the availability of volatile and stable precursor materials. Although suitable metalorganic precursors of titanium are readily available, it may not be the case for the alkaline earth metals such as Sr and Ba. However, significant progress is evident in the literature for the growth of mostly Ba-related compounds such as ($BaTiO_3$ [50], $Bi_4Ti_3O_{12}$ [51], and $(Ba,Sr)TiO_3$ [52], while some activity has been reported for $(Pb,La)TiO_3$ [53] and $SrBiTaO_9$ [43, 44].

A good discussion of precursor preparation and subsequent delivery in a reaction zone of an MOCVD reactor has been detailed in a recent review [54], which describes the growth of $BaTiO_3$, $SrTiO_3$ and $(Ba,Sr)TiO_3$ thin films. Initial studies of MOCVD centered on the synthesis of $SrTiO_3$ using the reactants titanium isopropoxide (TPT), Sr $(dpm)_2$, and oxygen in the presence of water steam. The TPT is a liquid while the $Sr(dpm)_2$ is a solid. Because of the low volatility of the Sr-precursor, a source temperature in excess of 200 °C is required to deposit a film at reasonable growth rates. Care must be taken so as not to condense the precursor prior to the reaction zone. The reactor has two zones (a source zone and a reaction zone). The titanium precursor along with oxygen and water enter the reaction zone separately, while the solid Sr source is heated at a regulated temperature to accomplish a constant growth rate. Using this technique, Epitaxial $SrTiO_3$ thin films were deposited at 800 °C.

8.3. Processing of Ferroelectric Thin Films

In this section the processing of some of the ferroelectric thin films with the laser ablation technique has been described with particular emphasis on the structure processing relation in the ferroelectric thin films. The ongoing activities at the author's laboratory include development of barium strontium titanate ((Ba, Sr)TiO_3, (BST)) thin films for DRAM applications, lead zirconium titanate ($Pb(Zr, Ti)O_3$, (PZT)), layered structured strontium bismuth tantalate ($SrBi_2Ta_2O_9$, (SBT)), strontium bismuth niobate ($SrBi_2Nb_2O_9$,(SBN)) for non-volatile memory applications, and antiferroelectric materials like lead zirconate (PZ) for high-charge storage devices and microelectromechanical systems. Some of the highlights of materials processing that are involved in the development of the forementioned materials along with their required dielectric and electrical properties will be detailed in terms of composition. Generally perfect perovskite crystallization in ferroelectric oxide thin films can be induced by means of *in situ* crystallization and/or *ex situ* crystallization. In the former case, the substrates can be maintained at relatively higher temperatures (>550 °C). The nucleation and growth kinetics could be monitored via processing parameters, such as pressure, substrate temperature, and target composition, during deposition. In the latter case, the crystallization of perovskite thin films can be formed either by the conventional annealing method or a rapid thermal annealing (RTA) method. In the conventional annealing method, low-temperature processed thin films are subjected to high temperatures for longer times for the completion of crystallization.

Most recently, the rapid thermal annealing (RTA) technique that uses heat lamps has been popularly applied [55, 56] to IC processing for development of low resistance ohmic contacts. Some of the potential advantages over conventional furnace annealing as established in semiconductor processing are that rise time for heating to the desired temperature as well as of the overall annealing period are both very brief. An advantage of the short rise time may be reduction in surface damage and minimization/elimination of the film-substrate interaction even at annealing temperatures in excess of 700 °C.

8.3.1. FERROELECTRIC PZT THIN FILMS

The PZT ($PbZr_{0.52}Ti_{0.48}O_3$) has a perovskite structure and is a multiaxial polarizable system. The crystallinity of annealed PZT films deposited on platinum-coated silicon substrates at 2–5 J/cm^2 fluences and 1.0 mtorr partial pressure of oxygen is shown in Fig. 8.4. All the patterns of the *ex-situ* crystallized films consistently showed polycrystalline with mainly (100) and (110) orientation. The relative peak intensity ratio between the pyrochlore and perovskite peaks increased as the fluences were increased in these *ex situ* deposited films. Another noticeable thing is that the average crystallite sizes, calculated with Scherrer's formula, were increased with the increase in the fluences, ranging from 150–300 Å for fluences of 2–5 J/cm^2. The proportionality between the crystallite size of the films and fluence could be attributed to the fact that the ejected

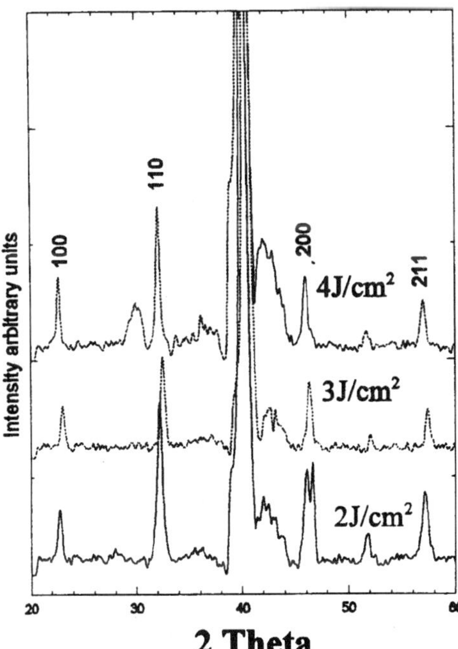

FIG. 8.4. X-ray diffraction patterns of PZT grown at different fluences.

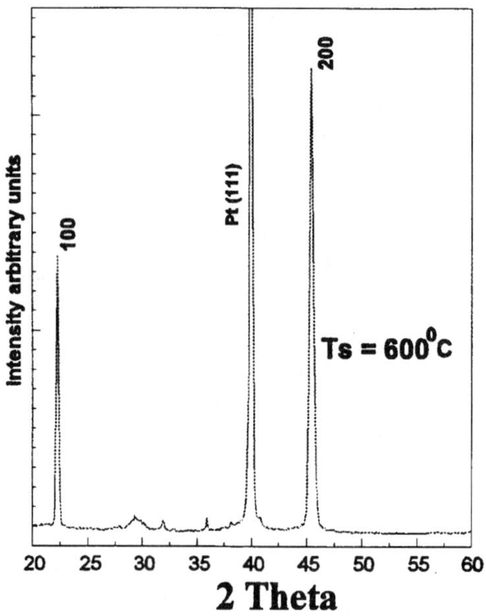

FIG. 8.5. Effect of substrate temperature on the phase formation of PZT thin films.

material in the plume would have higher energy (of the order of several eV) as the fluence went up and this tends to enhance the growth of crystallites. The crystallinity of *in situ*-grown PZT films is shown in Fig. 8.5. From the figure we see that the films deposited at 600 °C show perfect perovskite structure with preferred orientation along the (100). These observations also suggest that higher oxygen pressure in the range 100 mTorr is necessary to maintain stoichiometry in the PZT films deposited at higher temperatures and for films deposited at room temperatures, there is no need to use higher pressures. These results suggest the presence of several competing phenomena including, scattering of volatile species in the ablated plume, at low energies and low pressures and possible bombardments by the energetic species during ablation at higher energy densities, causing Pb deficiency due to preferential re-sputtering.

8.3.2. Paraelectric BST Thin Films

Barium strontium titanate [$(Ba_{0.5}Sr_{0.5})TiO_3$] has a perovskite structure and at room temperature it is paraelectric. The Curie temperature is −50 °C. Several deposition methods for BST films have been investigated including rf sputtering,

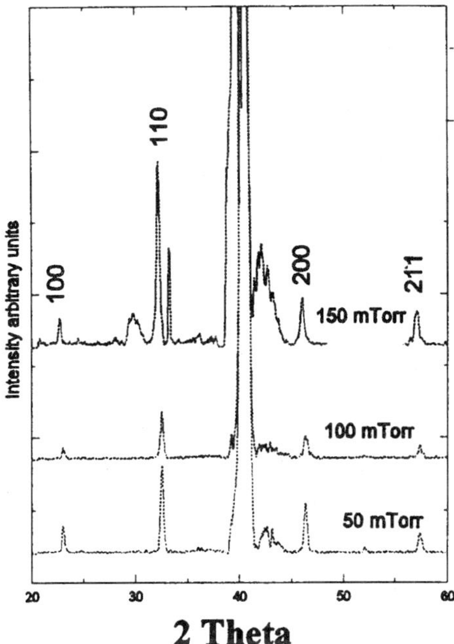

FIG. 8.6. X-ray pattern of BST thin films deposited at various oxygen partial pressures.

ion beam sputtering, as well as chemical routes that include both the MOD and the CVD technique.

Figure 8.6 shows the x-ray diffraction pattern for the BST films deposited by PLD [57] grown under different pressures, at 300 °C and were further annealed at 600 °C. At low pressures, the films tend to show only perovskite peaks. With the increase in pressure to approximately 100 mtorr, the perovskite phase reduces significantly while at even higher pressures, the films tend to show the presence of the pyrochlore phase [47]. The crystallite size analysis using Sherrer's equation reveals that there is a decrease in crystallite size from 600 Å at 50 mtorr to 350 Å at 150 mtorr. With this increase in pressure, the energies of the deposited species are reduced subsequently, hence affecting crystallinity, which accounts for the decrease in peak heights at higher oxygen pressures. The presence of a pyrochlore phase at even higher pressures originates from the absence in a stoichiometric percentage of (Ba + Sr) with respect to Ti as observed from the compositional analysis. These observations were found to be consistent with the earlier published literature. In addition, for films deposited at different substrate temperatures, it has been found that even at a deposition temperature of 450 °C no induced crystallinity was present and thus the films grown were amorphous in phase. For films grown above 500 °C, *in-situ* induced crystallinity was obtained, leading to (100) oriented films at 650 °C. Films grown below 500 °C were *ex-situ* annealed at 600 °C to induce crystallinity. For these films it was observed from the crystallite size analysis that there was a significant increase in grain size with an increase in substrate temperature. For room-temperature grown films, the crystallite size was around 269 Å while the grain size increased to 600 Å for the films deposited both at 300 °C and 450 °C. As observed from earlier reports, the variation can be attributed to the increase in the mobility of the deposited species at higher substrate temperatures.

8.3.3. ANTIFERROELECTRIC LEAD ZIRCONATE (PZ) THIN FILMS

Lead zirconate ($PbZrO_3$) is a typical antiferroelectric material at room temperature. An antiferroelectric material consists of dipoles that can be spontaneously polarized, but with neighboring dipoles polarized in antiparallel directions, so that the spontaneous macroscopic polarization of the material as a whole is zero. The AFE phase of PZ has an orthorhombic perovskite structure, with an antipolar arrangement along the pseudocubic [110] direction, whereas the field induced FE phase has a rhombohedral structure with the polar directions in the pseudocubic [111] for PZ crystal. The fabrication of PZ thin films has been done by many research groups using different thin-film techniques including sol-gel [58], reactive co-sputtering [59], multi ion beam sputtering [60] and, recently, pulsed laser ablation on Si substrates [61] and on Pt-coated Si substrates [62]. In all these techniques, except laser ablation, some sort of buffer layers containing titanium were employed to improve the adhesives of PZ thin films with the substrate. The structural analysis of PZ thin films done by XRD showed the same trend as that of PZT thin films. The effect of oxygen partial pressure during ablation showed a tremendous effect in the orientation of the films. Figure 8.7 shows the effect of oxygen partial pressure during ablation of the PZ thin films. The films deposited at 10 mtorr oxygen partial pressure showed the initiation of a perovskite phase along with an unwanted pyrochlore phase. As the pressure increased to 50 mTorr, perfect perovskite structure with a high orientation along (110) was formed. In the case of the PZT thin films, still lower pressures favored the perovskite phase formation [63]. This may be due to the presence of titanium, which favors the perovskite phase formation. This could be the reason for the usage of a Ti-containing buffer layer for the growth of PZ thin films by various groups. In the case of laser ablation, high energetic molecular species might be responsible for the perovskite phase formation without any buffer layer. It was shown earlier that oxidation of lead provides an excess energy (220 kJ/mol) for the crystallization reaction, [64] which could be a reason for the growth of the preferential orientation in *in situ* films.

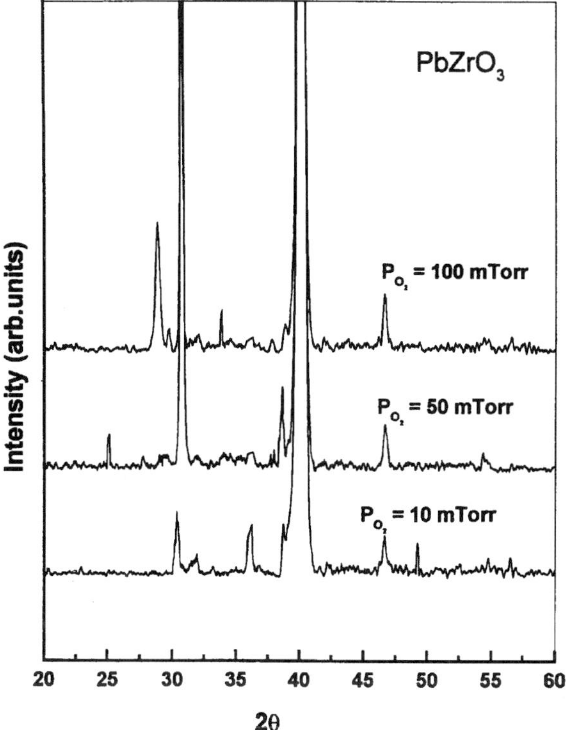

FIG. 8.7. X-ray pattern of *in situ* grown $PbZrO_3$ thin films.

Figure 8.8 shows EDAX data in terms of the ratio of the cationic species in the films with the variation of substrate temperature. It may be seen from the figure that there exists a decrease in Pb content in the films deposited at higher substrate temperature and lower pressure. However, as the ablation pressure was raised, the Pb content in the films became almost independent of growth temperature. This behavior may be associated with the modified sticking coefficient of Pb due to the oxidation at higher pressures, which resulted in a balanced stoichiometry at elevated substrate temperatures. These observations also suggest that higher oxygen pressures > 50 mtorr are necessary to maintain stoichiometry in the Pb-based films deposited at higher temperatures and also indicates no need for higher pressures for the films deposited at room temperature. In the case of films that contain nonvolatile elements like BST, the cationic ratio is almost constant throughout the deposition temperature range (30–600 °C). At higher pressures, there is an increase in the Ti content of the films, resulting in a pyrochlore phase, which was also revealed in the XRD pattern (Fig. 8.6).

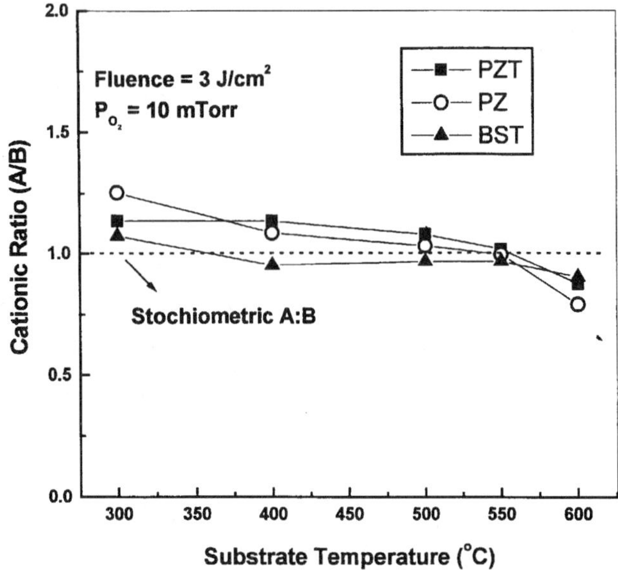

FIG. 8.8. Semiquantitative analysis of perovskite oxide thin films with growth temperature.

8.3.4. Ferroelectric Bi-L#ayered Structured Films

Another class of ferroelectrics that has recently attracted renewed interest due to their fatigue-free nature are the "Bi-layered Aurivillius compounds" for NVRAM applications. Examples of such ferroelectric materials are $SrBi_2Ta_2O_9$ (SBT), $SrBi_2Nb_2O_9$ (SBN), and $Bi_3Ti_4O_{12}$. The material class consists of Bi_2O_2 layers and double pervoskite-type TaO_6 octahedral units. Recently, several research groups have successfully prepared SBT films using MOD [65], PLD [41], and metal organic chemical vapor deposition (MOCVD) [66, 67]. Unfortunately, however, most of these preparation methods require high substrate temperatures, which is a disadvantage for device fabrication. However, with PLD the processing temperature can be minimized. The key advantage of the PLD in depositing SBT/SBN is the compositional fidelity between the target and the deposited film. This plays an important role in obtaining high-quality thin films for device application. Both SBT and SBN have two polarizable directions—one along the **a**-axis and the other along the **b**-axis.

Figure 8.9 shows the XRD patterns of Bi-layered structured thin films of $SrBi_2(Ta,Nb)_2O_9$ (50/50), grown at 400 °C and annealed at 750 °C/1 h. It was found that the phase starts evolving at temperatures >~400 °C, but the crystallinity was very poor. These films, after annealing at 750 °C for 1 hr show better

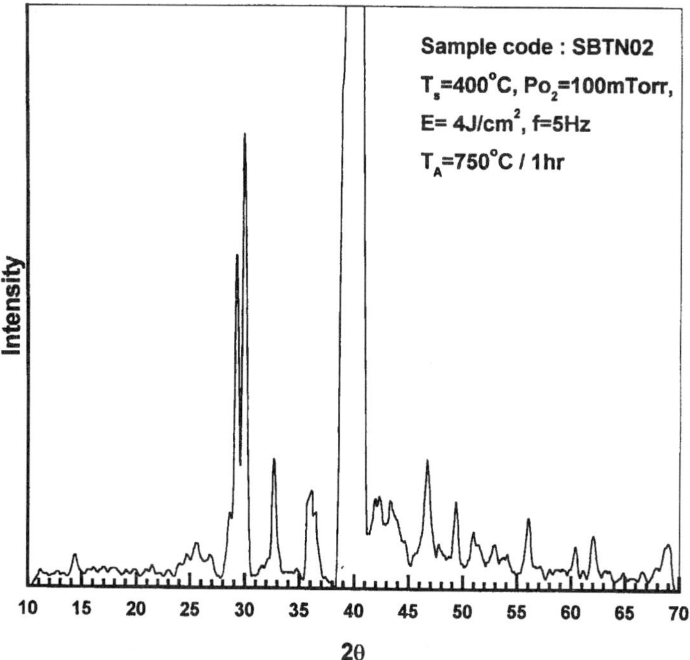

FIG. 8.9. X-ray pattern of SBTN thin films deposited at 400 °C and annealed at 750 °C.

crystallinity with random orientation [68]. On the other hand, films that were grown at a substrate temperature of 500 °C and above showed *in-situ* crystallization with a preferred orientation along the c-axis. However, it has also been found that the orientation of the *in situ* crystallized film changes from '**c**'-direction to '**a**'-direction as the substrate temperature is increased to 640 °C. It is worth mentioning here that, in SBT, the magnitude of the spontaneous polarization along the 'a' (or 'b') direction is more than that along the 'c'-axis of the pseudotetragonal unit cell. This result implies that the growth along the (008) direction is more favorable at lower temperatures (~500 °C); while the desired orientation is along the (200). The effect of the laser fluence on the crystallinity of the films that were grown at a temperature of 640 °C and a pressure of 100 mtorr shows that the appearance of the a-axis peaks (200) requires larger energy, either with the fluence, or substrate temperature. In conclusion, it can be said that, for bi-axial materials like SBT, the direction of the film growth is very important. For example, with Pt (111) the preferred growth direction is generally along the c-axis, unless very high energy was supplied. At the same time, the growth of the material with proper composition required a critical range of oxygen pressure.

8.3.5. Rapid Thermal Annealing Processed Ferroelectric Films

In the case of PZT thin films, conventional annealing for longer times at higher temperatures such as 650 °C, leads to other side effects, which include: a) unavoidable lead losses in PZT thin films as the Pb- re-evaporation and crystallization are two competing processes that occur simultaneously); and b) the presence of undesired film-substrate interface reactions. Such losses are more significant in thinner films (< 300 Å) and make it relatively inhibitive to crystallize without a controlled rate of Pb loss.

Figure 8.10a,b shows the x-ray diffraction patterns of the RTA-induced crystallization in PZT films annealed at different temperatures and for at different times. The RTA process was carried out in a commercial Heat Pulse System, model 210, manufactured by AG Associates, composed of several 1.5 kW tungsten/halogen lamps as the energy source. Figure 1.10a, b depicts a comprehensive summation of crystallization data obtained from a series of time-temperature combinations. It may be seen that at temperatures <600 °C, and with longer annealing times, films showed either an insufficient reaction or the presence of dominant pyrochlore phases. However, temperatures in excess of 600 °C are effective in imparting a pure perovskite phase even at annealing times as short as 10 sec. For conventional furnace annealing, it was noticed that with the annealing of PZT thin films, slightly larger amounts of excess Pb were needed to compensate for the loss accompying the crystallization. These observations were found consistent with our results obtained in PZT films deposited by the excimer laser ablation technique [38] and also prompted us to notice the differences in the temperature-time combinations based upon the Pb content present in the films prior to annealing. A significant point one needs to notice in the conventional annealing process is the amorphous phase transformation from a perovskite phase through a pyrochlore phase. However, in the RTA process, there is a sudden jump from the amorphous to the perovskite phase, which one sees reflected in the electrical properties [69].

8.4. Compound Phase Formation

The electrical properties of ferroelectric thin films are strongly tied to the microstructure and the crystal structure of the thin-film material. The thin-film composition and the method by which the thin film is processed control both microstructure and crystal structure. The relationships among composition, crystal structure, microstructure, and properties of the MIBERS-deposited ferroelectric thin films were explored in the La-doped $PbTiO_3$ (PLT) system, by observing the evolutionary development of these physical features during post-

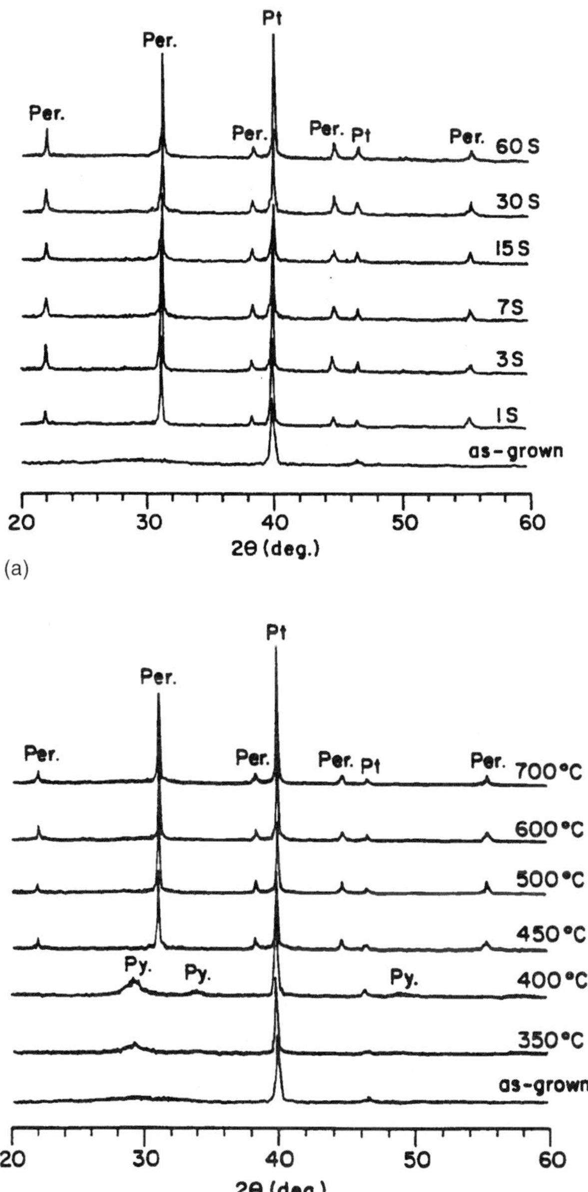

FIG. 8.10. (a) Phase evolution of PZT thin films with different annealing times in RTA process. (b) Effect of annealing temperatures on phase formation of PZT thin films during RTA process.

deposition annealing [70–72]. Particular emphasis was placed on determining how small changes in the composition affect the physical state of films annealed under constant conditions.

8.4.1. Effects of Anion Composition on Perovskite/Pyrochlore Phase Formation

Oxidation of the Pb in as-deposited PLT films controls the formation of perovskite and pyrochlore phases in the annealed films. Understanding this behavior requires knowledge of the steady-state reactive sputtering process and the reactions that occur during annealing. Once the dependence of the perovskite/pyrochlore formation upon Pb oxidation is clarified, perovskite films with a range of cation compositions can be prepared without the interference of pyrochlore formation.

During PLT deposition, the multi-ion beam reactive sputtering (MIBERS) technique [70–73] relies on oxygen background gases to provide oxidation of the depositing film. An oxide layer is formed at the surface of the Pb, La and Ti metal targets by adsorption, and an oxidation reaction subsequently follows. If the sputtering rate is slower than the rate of oxide formation at the target surface, metal oxide species are sputtered from the target. For sputtering rates that are faster than the oxidation rate, both metal and metal oxide species are sputtered and the relative amounts of each species are dependent on both the sputtering rate and the concentration of oxygen at the target surface.

Unlike La and Ti, the speciation of material sputtered from the Pb target is sensitive to the oxygen concentration (at the target surface) used for PLT deposition. The fraction of positive sputtered ions decreases as the oxygen concentration at the target surface is decreased. As the oxygen concentration at the target surface increases, the fraction of sputtered metal oxide species (e.g., PbO and PbO+) increases while the fraction of reduced metal species (e.g., Pb and Pb^{+2}) decreases; this results in a decrease in the total number of positive sputtered ions ejected from the target. The sensitivity of the Pb speciation to the oxygen concentration is attributed to the low free energy of PbO formation (as compared with La and Ti oxides), which produces an oxidation rate that is similar to the sputtering rate of the target surface.

These results reveal that the dependence of Pb speciation on oxygen concentration was used to deposit PLT films having two different oxygen contents. Films deposited at low P_{O2} (i.e., 8×10^{-3} Pa) were oxygen deficient and contained $PbO_{1-\delta}$; yet films deposited at high P_{O2} (i.e., 7.6×10^{-2} Pa contained PbO with no significant oxygen deficiency. Annealing the flims with the two different oxygen contents (but with equal cation compositions) produces two different crystalline phases. Oxygen-deficient films produce the perovskite

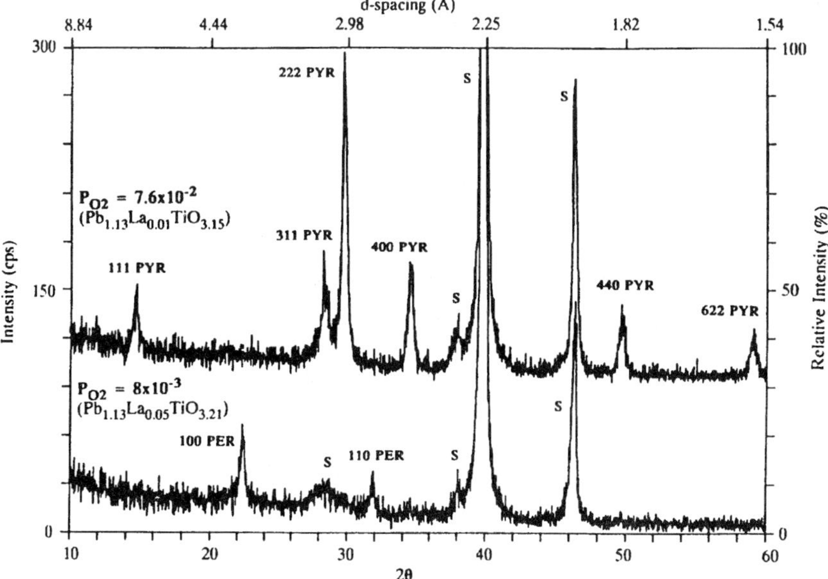

FIG. 8.11. X-ray pattern of PLT thin films deposited by MIBERS technique with various oxygen partial pressures.

phase, and fully oxidized films produce the pyrochlore phase, as indicated by the x-ray diffraction patterns shown in Fig. 8.11. The phase formation was found to be independent of cation composition for constant oxygen content.

8.4.2. EFFECTS OF CATION COMPOSITION ON MICROSTRUCTURE DEVELOPMENT

Microstructure evolution in perovskite PLT films is greatly dependent upon the as-deposited cation composition and the postdeposition annealing process. Annealing initiates crystallization, densification, and PbO evaporation; all these evolutionary mechanisms are composition dependent and interdependent. When comparing films annealed under different conditions it is necessary to characterize both the annealed composition and microstructure of the films for a proper comparison. If the annealing conditions are held constant, the relationships between the as-deposited composition and annealing induced microstructure development can be uncovered.

Crystallization of PLT films begins at annealing temperatures as low as 400 °C but the crystallization process is not completed unless the flims are annealed at temperatures of 600 °C or higher. Lead oxide evaporation occurs simultaneously with the perovskite crystallization; evaporation starts as low as 490 ± 50 °C and the rate of evarporation increases with increasing annealing temperature [73].

Both crystallization and PbO evaporation during annealing causes the evolution of the as-deposited microstructure. The annealed microstructure consists of fine grains, which assemble to form clusters that are separated by porous cluster boundaries. As crystallization and PbO evaporation proceeds during annealing, the cluster boundaries and fine-grain structure become more pronounced due to a thermal etching effect, which results from densification enhanced PbO loss at the grain and cluster boundaries. The annealing temperature does not change the cluster size, which suggests that the cluster boundaries observed after annealing develop from the low-density boundaries present in the as-deposited films. Because the film composition changes with increasing annealing temperature (due to PbO evaporation), it is difficult to differentiate between the effects of annealing temperature and composition on microstructure evolution.

Evaporation of PbO is not only temperature-dependent but is also dependent on the starting composition of the film. As shown in Fig. 8.12, PbO loss increases with an increasing $PbO_{1-\delta}$ excess in the as-deposited film. The PBO loss with a small $PbO_{1-\delta}$ excess is controlled by the vapor pressure of PbO in PLT while the vapor pressure of pure PbO limits PbO loss in films with a high $PbO_{1-\delta}$ excess. An increase in La concentration also results in an increase in PbO evaporation due to the preference for an increased number of cation vacancies at the higher La content.

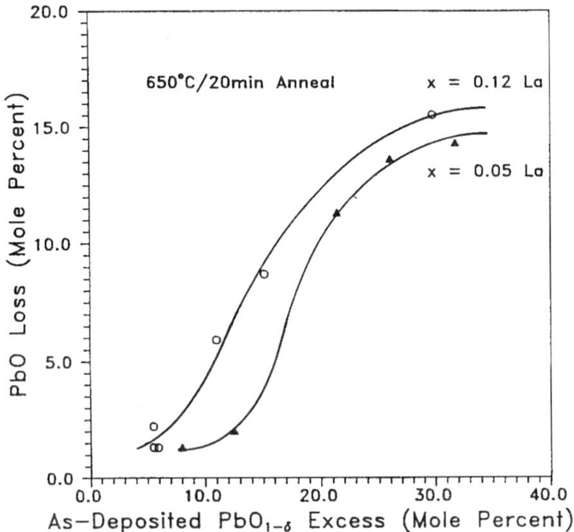

FIG. 8.12. PbO loss versus PbO excess in the film composition.

8.4.3. RELATIONSHIP BETWEEN COMPOSITION, MICROSTRUCTURE, AND ELECTRICAL PROPERTIES

An understanding of the relationship between composition and microstructure is highly advantageous when trying to engineer the properties of a thin film. By knowing the types of phases and the connectivity of those phases, mixing rules can be used to predict the property dependence on composition. For example, for the PLT films studied, the dependence of the electrical properties on composition can be understood by applying mixing rules to the simplified microstructure models for textured and nontextured films [71–73].

The dc resistivity measured through the thickness of the film is given as an example of an electrical property that illustrates the relationship between composition, microstructure, and properties [74]. As shown in Fig. 8.13, textured films exhibit a high dc resistivity (on the order of 10^{13} $\Omega \approx$ cm), which decreases slowly with increasing PbO excess. At the transition between the $\langle 100 \rangle$ texture and nontexture, the resistivity drops discontinuously to a low resistivity (on the order of 10^{9} $\Omega \approx$ cm). Because a mercury probe was used for the top electrode and the Pt substrate layer was used for the bottom electrode there are two curves (labeled Pt cathode and Hg cathode), which refer to measurements for opposite electric field polarities. The resistivity difference for the two polarities increases with increasing PbO content due to an increasing thin-film surface roughness (resulting from PbO evaporation) that alters which alters the contact between the Hg electrode and the PLT film.

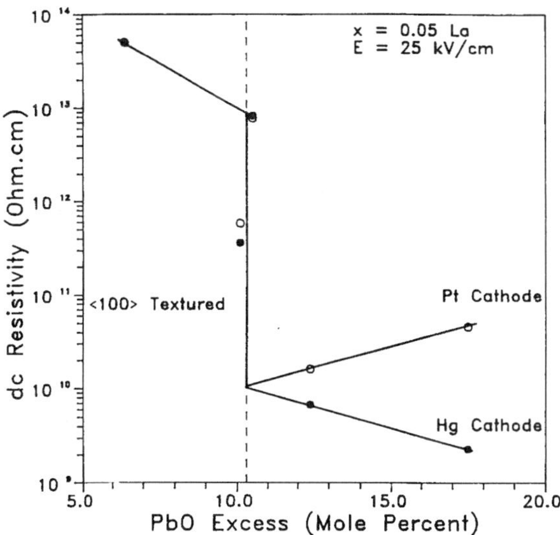

FIG. 8.13. Effect of excess PbO in PLT thin films on dc resistivity σ_{dc}.

8.5. Electrical Properties

The electrical characterization of ferroelectric thin films includes determination of property parameters through different electrical measurements. The investigation of electrical properties is based completely on industry requirements; for example, the advent of the nonvolatile random access memory (NVRAM) utilizes the principle of spontaneous polarization to store data. Micro-electro-mechanical systems (MEM) are based on the electrostrictive and piezoelectric property of ferroelectrics. Similarly, for integrated capacitors in dynamic random access memories (DRAM) almost linear dielectrics with high dielectric permittivity are used. With the advancement of technology a scale down of the device features is observed. For example, in the case of DRAM the progress in this field has brought orders of increase in cell density (a factor of four every generation) with little increase in the total device size (40% for every generation).

It is clear that technological demands are quite high and to meet these demands expensive research is required in both materials and processing. Before going into the details of the device as such, the basic interest is in the characterization of these materials. We follow a general trend for characterizing these materials through a series of tests. Ferroelectric thin films are generally characterized for both ac and dc responses. Dielectric studies involve both the frequency-domain and the time-domain response, while dc measurements are restricted to the leakage current flowing through the ferroelectric thin films under varying electric fields. Proper interpretation of these results provides clear and basic ideas about the material system. We start our discussion with the analysis of ferroelectric thin films based on their properties.

8.5.1. Dielectric Behavior in Ferroelectric Thin Films

The dielectric response of a material system involves the study under transient or alternating fields. This response can be characterized in one of two basically equivalent ways: 1) as the time dependence of the polarization or of the polarizing current under step-function excitation; or 2) as the frequency dependence of the polarization under alternating field excitation. Basically, both are Fourier transforms of each other for a linear system, and the choice between them is determined by the convenience of the measuring process and study of the departure from linearity of the materials. Thus, if one looks at the response in both domains repeated for different material systems under varying experimental conditions, one notices the apparent "universality" of data. The responses seem to follow a general universal trend, which might be expressed through certain power relationships. One of the forms of a universal relation in the time-domain response is the Curie-von Schweidler law. This law states that the discharge or depolarization currents of a wide range of dielectric materials follow the power law of time dependence [75]:

$$i(t) \propto t^{-n} \tag{8.1}$$

instead of the exponential relation, that corresponds to the simplest first-order differential equation decay, which characterises the Debye mechanism as

$$i(t) \propto \exp(t/\tau) \tag{8.2}$$

where τ refers to the relaxation time of the process. In what follows (section 8.5.1.1 and 8.5.1.2) we investigate the dielectric response observed in ferroelectric thin films in both the time- and frequency domain.

8.5.1.1. Time-Dependent Dielectric Response

The response of a dielectric system to a static field represents only one facet of the problem and for practical purposes a relatively insignificant one. Much more important experimentally, technologically and theoretically, is the time-dependent dielectric response. The technical significance of the time-dependent response is evident if we envisage the fact that most electrical applications of dielectrics involve the use of a step-function, delta-function or sinusoidally variable electric fields. The application of an electric field $E(t)$ to a dielectric system induces net polarization $P(t)$, which, however, does not have the same functional form as the driving field. The study of the time-dependent analysis rests on the inquisitiveness of the experimentalist to determine the functional form.

To enunciate the problem we consider a thin film in the form of a parallel plate capacitor with the dielectric material under study sandwiched between the two electrodes. We observe the response of this dielectric system to a step-function electric field, where the field is switched on at $t = 0$ and the response in the form of the current flowing through the system measured over a time scale. In observing the time-domain response of the dielectric system under study we encounter a typical characteristic curve on the time scale. The curve represented by Fig. 8.14 shows the response of a dielectric thin film over a time range giving rise to three distinguishable regions. This section deals with the study of the first region labeled region (I). The range of the first region is determined by the response time of the dielectric system and may extend from less than a microsecond to several days. This articular region is characterized by a decrease in the current flowing through the system during the charging process and is identified as the polarization current. The polarization current $dP(t)/dt$ characterizes the adjustment of the polarizing species to a step-function field and it must go to zero at infinitely long times. On the other hand, the steady current, or direct current (dc) arises from continuous movement of free charges across the dielectric

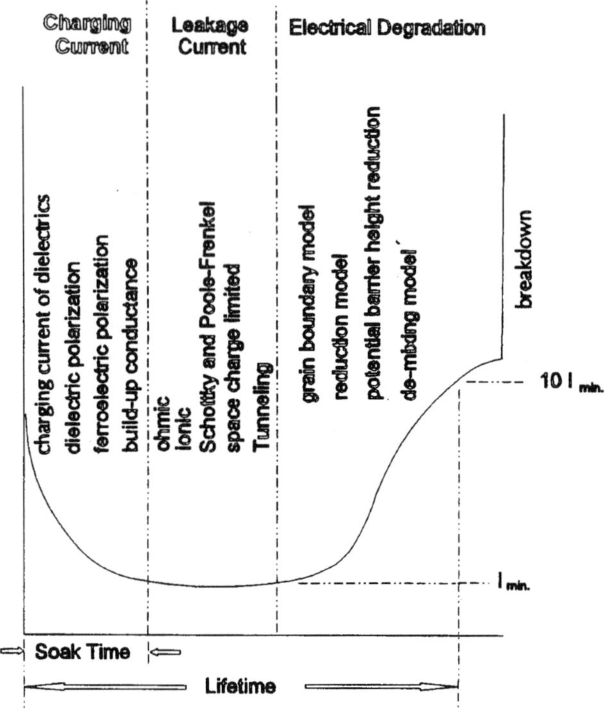

FIG. 8.14. Dielectric response in a ferroelectric thin films over time decades. (Bath-Tub Model).

material from one electrode to the other and this current does not change in any way the "center of gravity" of the charge distribution in the system. The universal power law as given in Eq. (8.1) is encountered for most dielectrics with values of n in the range $0 < n < 1$. Figure 8.15 shows the response of a dielectric material such as $(Ba,Sr)TiO_3$ under both charging and discharging conditions [57]. The exponent n shows a value in the range of 0.44 to 0.73 for different electric fields.

Several of the ideas from the past have been related to the origin of the "universal" response as represented by the Curie-von Schweidler law. Even though the law quite conclusively represents the behavior of a wide spectrum of disordered structures, both crystalline and glassy, and extending from good insulators to fast ionic conducting electrolytes, the universal character may originate from very different mechanisms, such as a space-charge model, many-body interaction, or distribution relaxation times (DRT). Jonscher [67] gave a comprehensive description of dielectric relaxation in solids, while, Waser *et al.* elucidated the relaxation mechanism in perovskite titanates [76].

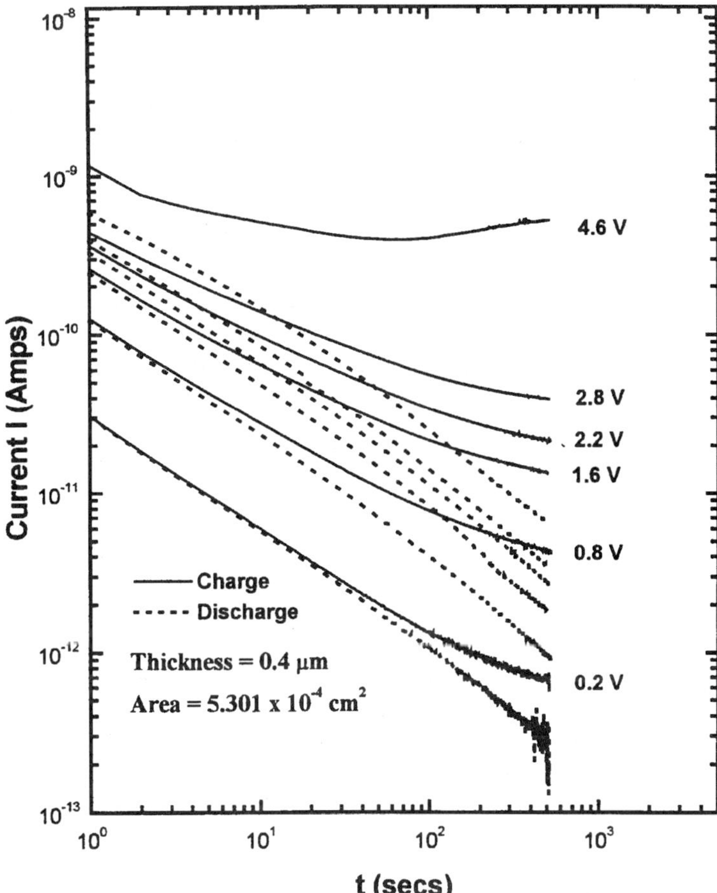

FIG. 8.15. Charging and discharging response in BST thin films.

8.5.1.2. Frequency-Domain Response

An alternative and powerful approach to the measurement of the dielectric response is in the frequency domain. In studying the frequency-domain response the practical range of frequencies used extend from very low values corresponding to 10^{-5} Hz to near gigahertz. The basic assumption, which goes into such a measurement, is that we consider the system under study to behave linearly and to respond to a harmonic excitation, that is, a sinusoidal wave. With the current advances in measurement technologies, the frequency-domain measurements are simpler and more precise. By contrast, time-domain measuring equipment has to be specially built, as there is not sufficient demand for commercial developments for the equipment to be made at reasonable prices, and the noise limitation inherent in the wideband measurement is severe.

From both an application and an engineering point of view the role of the frequency-domain response has been of tremendous significance. Relating to Jonscher's "universal" model [77, 78] for the dielectric constant as a function of radian frequency

$$\varepsilon^* = \varepsilon' - i\varepsilon'' = \varepsilon_\infty + \frac{\sigma}{i\varepsilon_o\omega} + \frac{a(T)}{\varepsilon_o}(i\omega^{n(T)-1}) \tag{8.3}$$

where ε_α is the "hig-frequency" value of the dielectric constant, $n(T)$ is the temperature-dependent exponent, which determines the "strength" of the ion-ion coupling and $a(T)$ determines the "strength" of the polarizability arising from the "universal" mechanism

$$\varepsilon_r' = \varepsilon_\infty + \sin(n(T)\pi/2)\omega^{n(T)-1}a(T)/\varepsilon_o \tag{8.4}$$

$$\varepsilon_r'' = \frac{\sigma}{\varepsilon_o\omega} + \cos(n(T)\pi/2)\omega^{n(T)-1}a(T)/\varepsilon_o \tag{8.5}$$

The first term in Eq. (8.4) characterizes the lattice response and that in Eq. (8.5) reflects the dc conduction part, while the second term in both equations refers to the charge-carrier contribution to the observed dielectric constant.

Apart from obtaining the dependence of the dielectric constant or frequency and temperature a further analysis includes a detailed study of the each of the terms in Eqs. (8.3) and (8.4). Because the time- and frequency-domain responses are Fourier transforms of each other, it is observed that the same "n" value characterizes both domain responses. Figure 8.16 illustrates the dielectric response characteristics of a BST and an SBT film taken at 100 °C. Both films show excellent match with the theoretically fitted curves for the real part of the dielectric constant (Eq. (8.4)). An important part of the dielectric response is in the complex impedance spectroscopy and the presentation of data. The complex parameters consist of the complex impedance (Z^*), complex admittance (Y^*), complex dielectric constant (E^*) and complex modulus (M^*). Proper choice of any two of these complex parameters to represent the dielectric data gives a glimpse of the capacitive components of the system and allows an inherent verification of the Debye process. (Note that frequency response data can always be correlated to that from an analogous electrical circuit consisting of an inductor (L), capacitor (C), resistor (R) or a combination of them [79]. For a pure Debye-type response the circuit consists of an R-C series network while the universal response corresponds to an R-C network in parallel. Thus, while analyzing the dielectric data of a sample one can eventually relate the response to the electrical analogous circuit for simplicity of understanding. The net outcome of this representation is that the response of each component of the film shows different

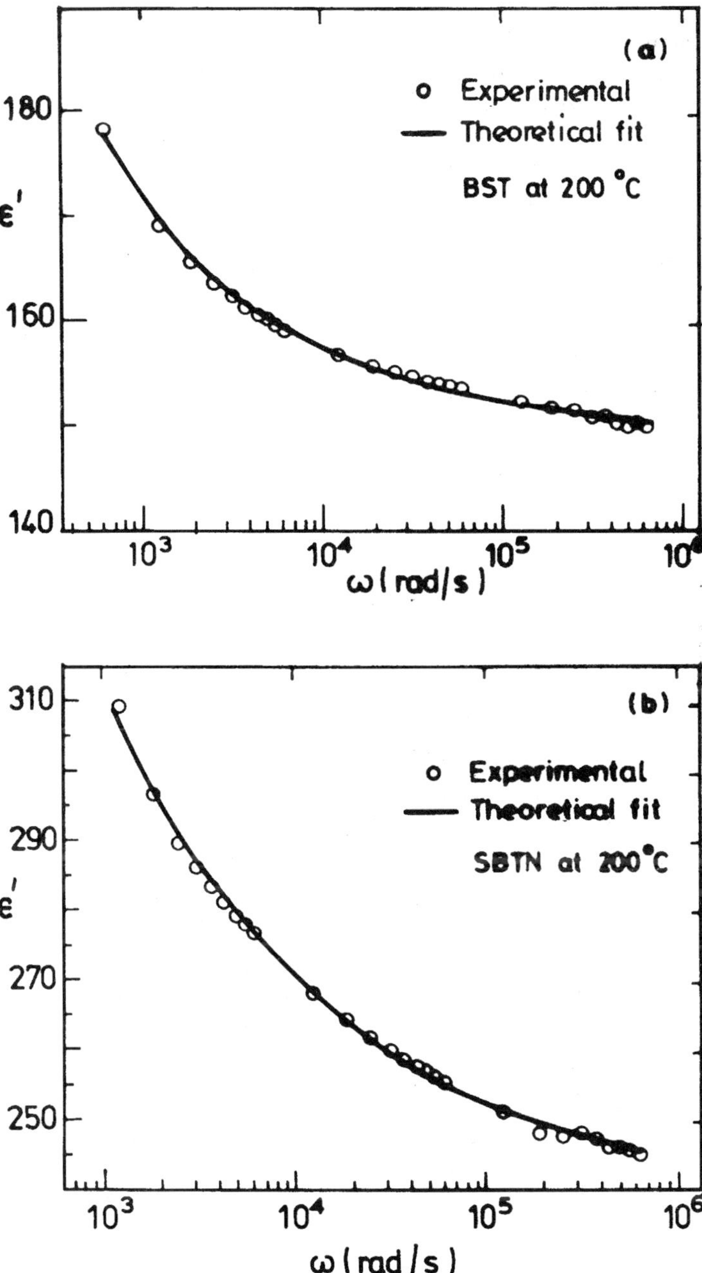

FIG. 8.16. Dielectric response of (a) BST and (b) SBTN thin films according to Joncher's law.

frequency dependence, and hence it is possible to delineate each component and its characteristics. Figure 8.17 shows the complex impedance plane (Cole-Cole) plot of a thin film composed of grains and grain boundary interfaces. Figure 8.18 shows the analysis result obtained from a BST thin-film sample at 280 °C, which shows a single semicircle exhibiting a response corresponding to the bulk grains.

The dc conductivity determined from the dielectric loss expression (Eq. (8.5)) gives a clear idea of the activation energy involved in the process and could possibly lead to further knowledge of the type of conduction mechanism that might be prevalent. Fig. 8.19 shows the Arrhenius plot of the dc conductivity extracted from the ac measurement conducted on a BST sample. The value of the activation energy computed (~0.97) may be related to the oxygen vacancy

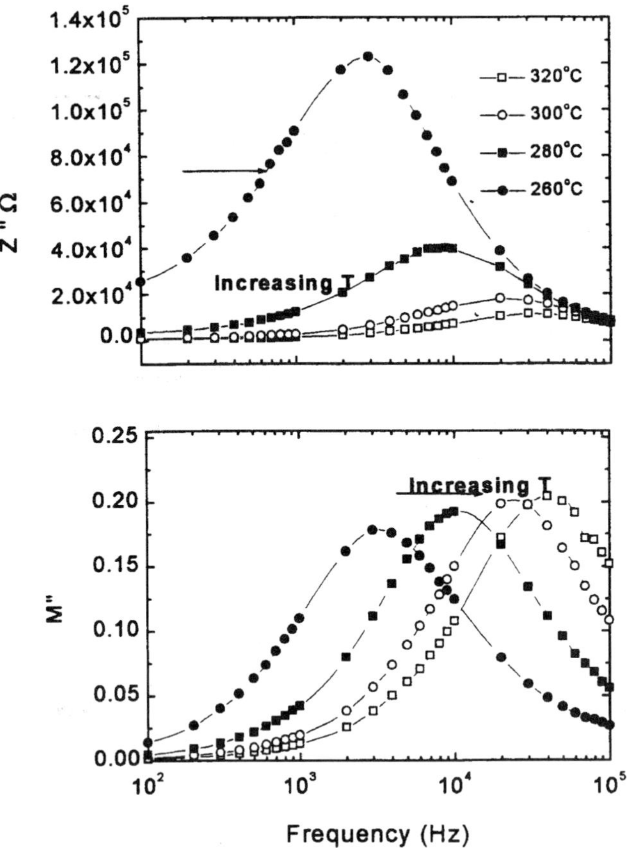

FIG. 8.17. Variation of Z'' and M'' with frequency in BST thin films.

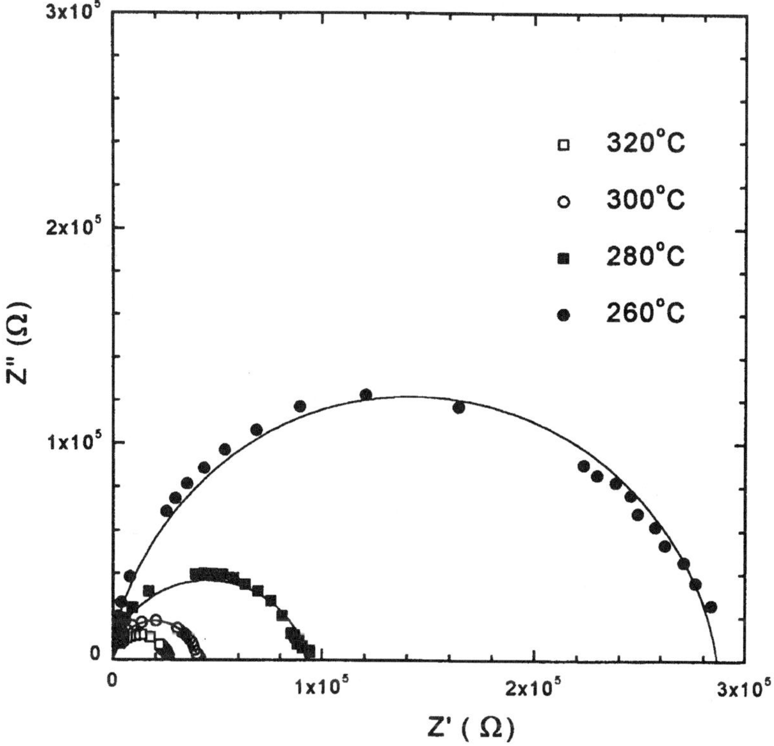

FIG. 8.18. Cole-Cole plot of Z′ vs Z″ in BST thin films.

motion through the thin film, which is in excellent agreement with earlier published data on bulk and thin films [80].

8.5.2. POLARIZATION HYSTERESIS AND CAPACITANCE VOLTAGE CHARACTERISTICS

The ferroelectric properties of different materials were evaluated by examining polarization versus applied electric field (*P-E*) hysteresis loops and are shown in Fig. 8.20 for samples (A): PZT (Zr/Ti = 65/35), (B): SBN, (C): PZ, respectively [38, 42, 81]. It should be noted that the saturated polarization (P_s) remanent polarization (P_r) and coercive fields (E_c) are materials properties. The double hysteresis behavior is representative of the antiferroelectric nature and is observed in the case of PZ thin films. The forward and reverse switching fields for the PZ thin films are 70 and 140 kV/cm, respectively. Recently, applications of high-

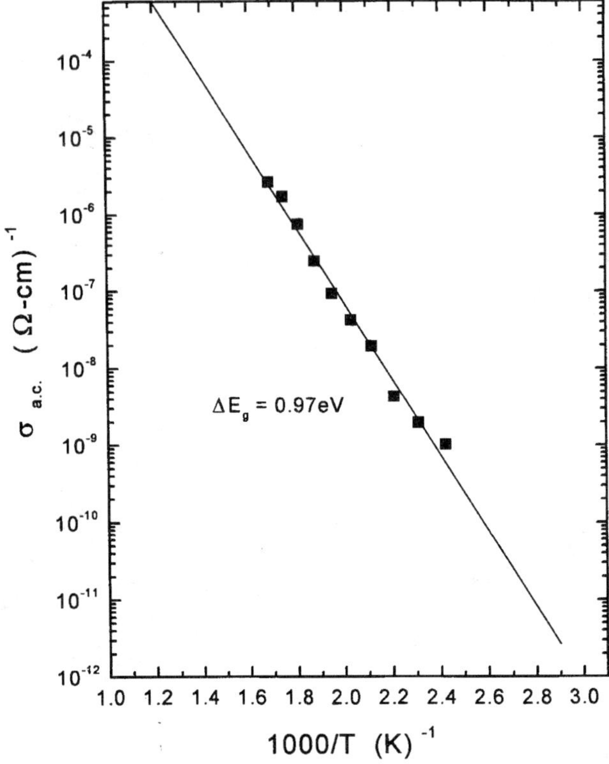

FIG. 8.19. Arrhenius plot of σ_{dc} vs (1000/T) for BST thin films.

resolution techniques such as scanning force microscopy (SFM), using nanoscale imaging methods, provided an opportunity to achieve unique insight into the polarization processes that occur in ferroelectric thin films at the nanoscale level [82].

For a ferroelectric thin film the hysteresis behavior is also reflected in the C-V characteristics, which show a butterfly loop with two peaks corresponding to the polarization switching in the films, while in the case of an antiferroelectric double butterfly loop corresponds to a double hysteresis loop. Figure 8.21 shows the C-V characteristics observed for: (a) ferroelectric PZT; (b) SBN; and (c) antiferroelectric PZ thin films. Apart from studying the capacitance behavior under a varying applied bias, the C-V curve can be used to measure the trapped charge present in a film [83]. For a paraelectric thin film, which shows a single peak at zero bias for both, sweeps (−ve to + ve and vice versa) the shift in the peaks can be used to determine the trapped charge at the interface during the sweep. Note

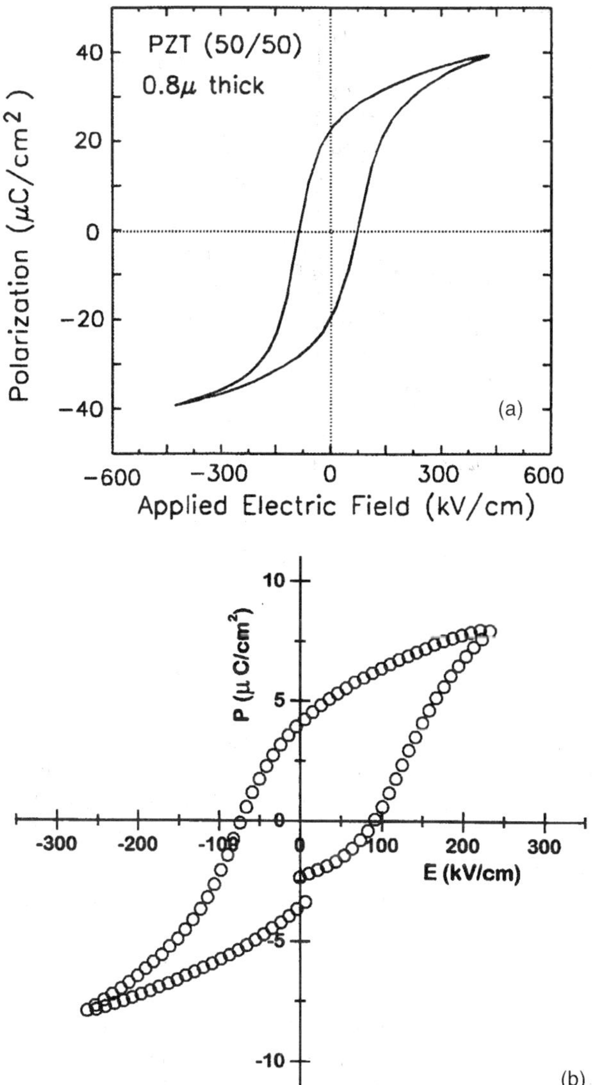

FIG. 8.20. Polarization vs applied electric field hysteresis response in (a) PZT, (b) SBTN.

that Kwak gave an account of the measurement of trapped interfacial charge density in BST thin films and a similar work was reported [84] on PZT, where C-V characteristics were used to determine the trap concentration after the application of repeated electrical stress.

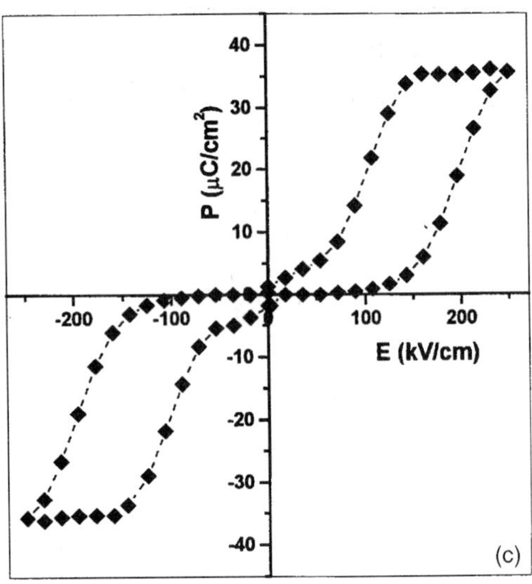

FIG. 8.20. (c) PZ, respectively.

8.5.3. SWITCHING PHENOMENON

The polarization-reversal effect, that is, the change of direction of the spontaneous polarization by an external electric field, is the most important property of ferroelectric material. Considerable research has been devoted to this effect. The most basic research is that of Merz [85], who reported on the switching transition of ferroelectric monocrystals. The switching time t_s for the polarization reversal of the ferroelectric thin-film capacitor is directly relevant to devices in microelectronics and should be as small as possible. Electrical experiments that consist of the reversal of the remanent polarization by alternate, sequentially applied voltage pulses, and subsequent observation of the resultant displacement currents, suggest that the polarization-reversal process in most ferroelectrics consists of the nucleation and subsequent growth of antiparallel domain.

Figure 8.22 shows the switching characteristics of PZT thin films grown by multi-ion beam reactive sputtering [86]. The figure shows that the switching time t_s, defined as the time from the onset to a point 90% down from the maximum value of the SP curve, is about 230 ns with the switched charge density of 18.5 $\mu C/cm^2$. Lohse *et al.* [87] showed that the switching time for SBT is on the order of 50 ns. However, in a comparison with the results of Scott *et al.* [88], this switching time appears rather large and may be attributed to the relatively large electrode area used. To quantify the dynamics of domain reversal further [89], the

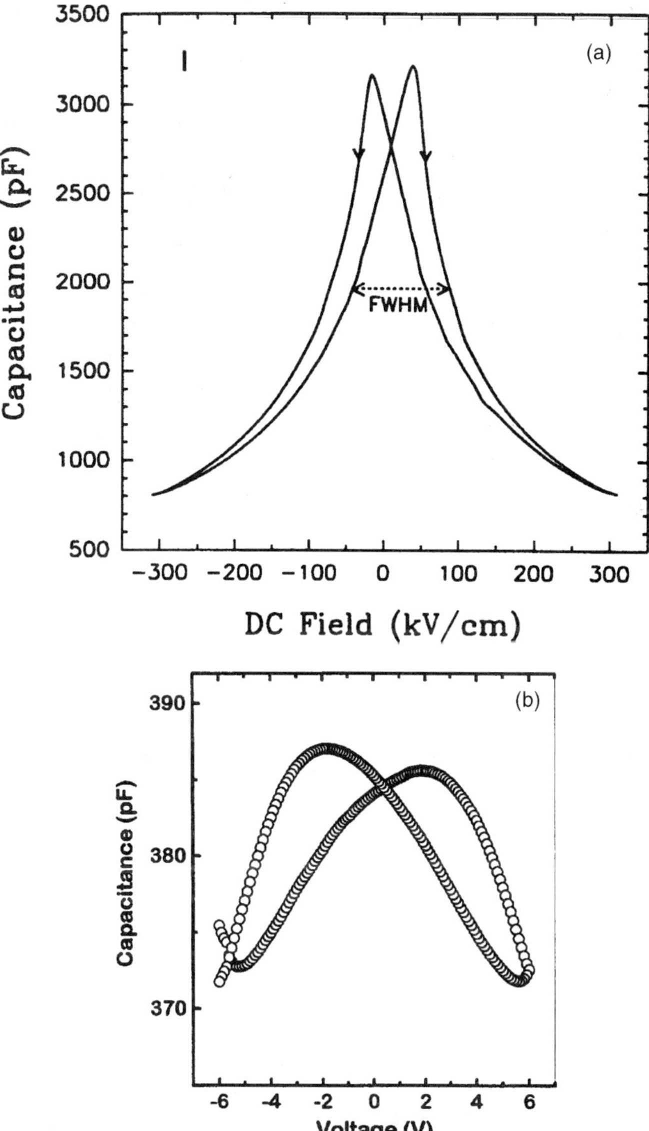

FIG. 8.21. Variation of capacitance versus voltage in (a) PZT, (b) SBTN.

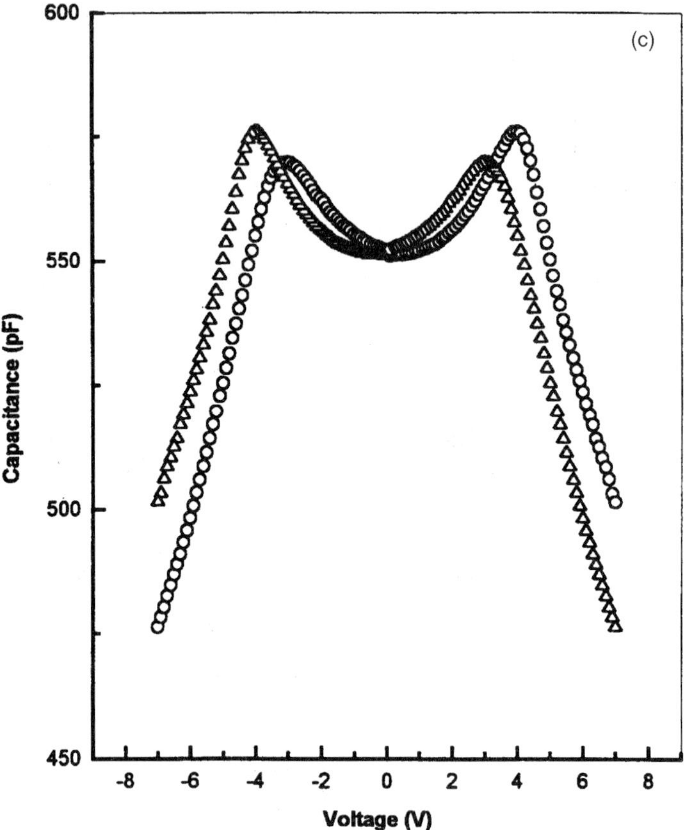

FIG. 8.21. (c) PZ thin films, respectively.

relation $t_s = t_o \exp(\alpha/E)$ was used to determine the activation field α for PZT thin films, which is about 284 kV/cm.

In the case of antiferroelectrics, the backward switching, that is, from the field-induced ferroelectric (FE) state to the antiferroelectric (AFE) state, gives an insight into the domain-reversal phenomenon. Figure 8.23 shows the backward switching characteristics in PZ thin films [62]. In this figure, the second current maximum shows the switching behavior of PZ thin films under different voltage pulses. The switching time between the field-induced FE to AFE phases was found to be about 50 ns, which was much less than that of a bulk La-modified antiferroelectric ceramics ($\sim$2 μs) [90].

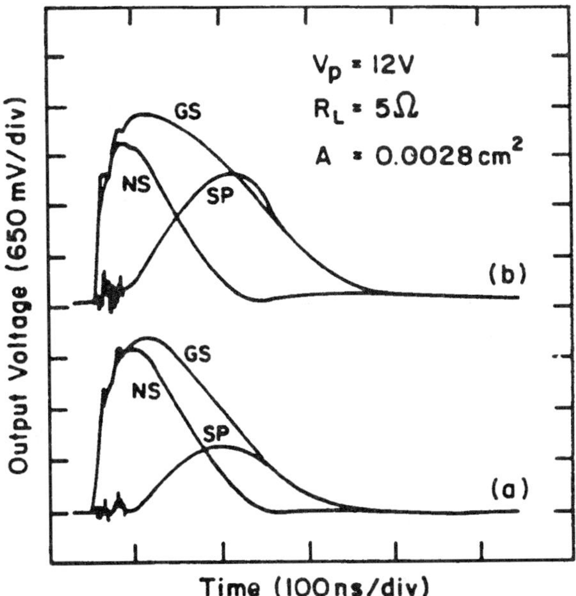

FIG. 8.22. Switching phenomenon in PZT thin films.

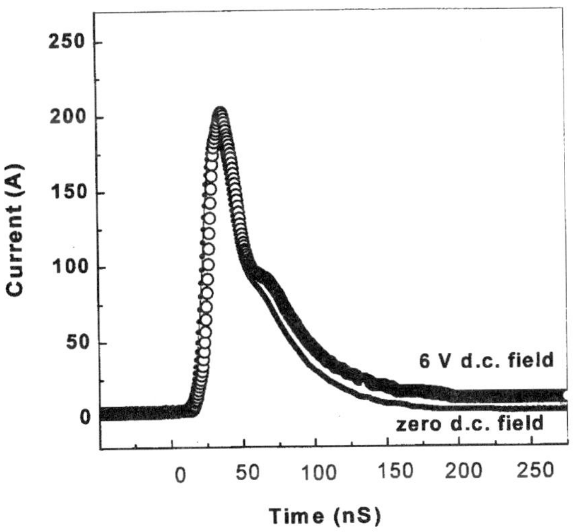

FIG. 8.23. Backward switching (FE → AFE) characteristics in PZ thin films.

8.5.4. Fatigue and Retention in Ferroelectric thin Films

Ferroelectric fatigue, which can be defined as the loss of switchable polarization with repeated hysteresis cycling, is a key issue for applications in nonvolatile memory devices. Several models have been proposed to explain the phenomenon of fatigue [91–94]. In general, in ferroelectric materials, fatigue is thought to arise from three different microscopic causes: 1) stress relaxation of the 90° domains to a 180° configuration as mechanical stresses are released externally and in doing so reorient and reduce the net polarization; the pinning of 180° domains by stress or charged defect is a related fatigue mechanism; 2) poling of a charged defect pair, such as a lead vacancy, an oxygen vacancy neighbor; for this case the dipoles become aligned with repetitive application of large electric fields, thus reducing or canceling some of the switchable polarization at the microscopic level; and 3) space-charge accumulation at the electrode/ferroelectric interface, which compensates for the applied voltage and acts as a detrimental screening of the external fields. Space-charge injection into the ferroelectric also leads to oxidation of the electrodes and valence conversion of the metal ions such as Ti^{+4} to Ti^{+3}. Figure 8.24a, b shows fatigue and retention behavior as a function of polarization for PZT (50/50), processed by the MIBERS technique for an applied field of ±170 kV/cm bipolar pulse cycling. It can be seen that the decay in polarization is less than 20% after 10^{10} switching cycles. Retention of the polarization of such films was measured by applying initially a −5 V write pulse and afterwards +5 V and −5 V read pulses. The loss in the stored charge is less than 10% after 10^5 s.

Different protocols have been used to overcome the problem related to fatigue. These include changing the electrode material from Pt to oxide-based conducting electrodes such as IrO_2, RuO_2, etc. [95–97]. Ramesh *et al.* [98] proved that laser-ablated Pb-thin films show better fatigue properties with (La,Sr)CoO_3 electrodes. No significant fatigue was found with such oxide electrodes even after 10^{12} cycles. The reasons why Pb- films with oxide electrodes have improved fatigue is still a subject of controversy. Alternatively, the use of ferroelectric-layered structured compounds [99] such as $SrBi_2Ta_2O_9$, $SrBi_2Nb_2)O_9$, and $SrBi_4Ti_3O_{12}$ have proven to be better substitutes for the PZT- and $BaTiO_3$-based ferroelectric capacitors. The enhanced properties of the layered compounds against their predecessors have been attributed to a lower oxygen vacancy concentration present in the film and the weakly pinned domain walls, which may be recovered through electric cycling of the ferroelectric film [94]. Significant research efforts are still needed to understand the basic sources of fatigue in Pb-based perovskite films.

8.5.5. Direct Current Leakage Characteristics

Region (II) as illustrated in Fig. 8.14 shows the true leakage characteristics of the thin film after the initial transient and corresponds to the dc conduction process

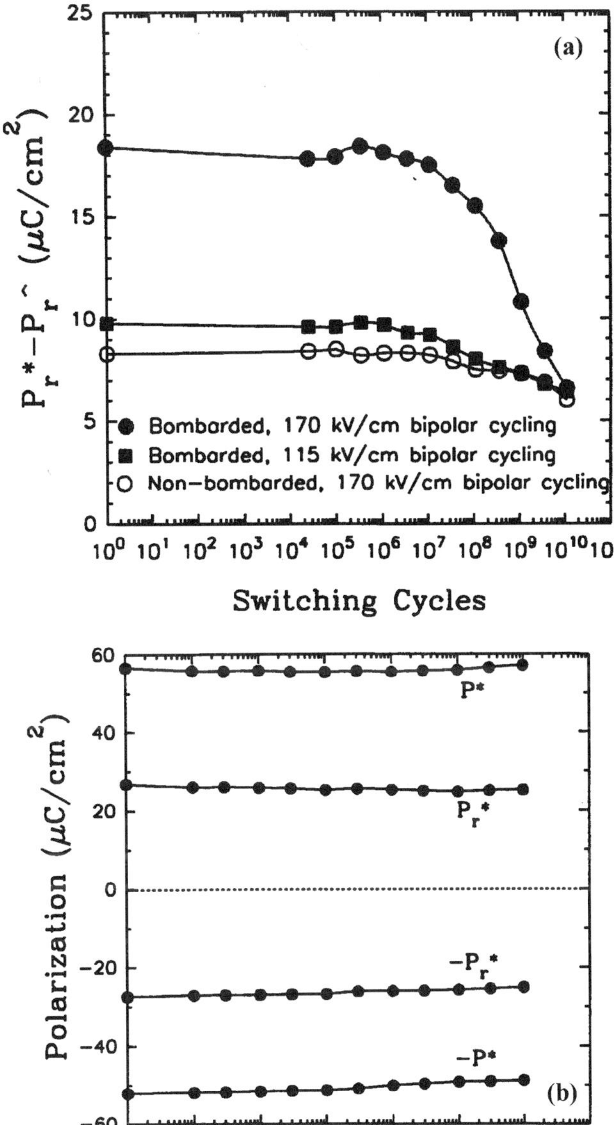

FIG. 8.24. (a) Fatigue and (b) retention behaviors in PZT thin films.

prevalent in the film. A study of the dc leakage property can result in determination of the conduction process involved and study of how to further improve the leakage properties required for device application [100]. For metal-insulator-metal configuration several works are available that show detailed study on ferroelectric thin films [62, 79, 101]. The vastness in the number of published literature in this topic specifically signifies the unavailability of a universal mechanism, which can be used to describe the conduction process in ferroelectric thin films.

Ferroelectric thin films pose an extremely complicated problem with respect to their defect chemistry. Apart from the thickness of the film, which has considerable influence on the electronic interface states, growth morphology, microstructure, crystallization conditions, nature of substrate, etc., the films also have several other parameters that add to the complexity. For example, the presence of ferroelectricity further complicates the situation through interactions with the intrinsic defects in the charge transfer process and space-charge formation. Comprehensive analysis of the defect chemistry of the ferroelectric perovskite structure titanates such as $BaTiO_3$, $SrTiO_3$, PZT by Waser [108], Raymond and Smyth and their co-workers [102] have revealed the following results.

- Undoped titanates $MTiO_3$ (M = Ba, Sr, Pb) are determined by a Schottky disorder, that is, formation of cation and anion vacancies. Oxygen vacancies $V_{\ddot{o}}$ are positively charged and act as donor-type native defects while cation vacancies act as acceptor-type native defects.
- For applications that require insulating properties, the perovskite titanate must be in an oxygen-excess state of non-stoichiometry.
- Nominally undoped titanate crystals and ceramics are usually governed by an unknown concentration of acceptor-type impurities.
- Heterovalent cations may be accommodated on alkaline earth sites or Ti sites and can act as foreign dopants. Excess electrons (when annealed under reducing atmospheres) either compensate donor impurities, or they are compensated by cation vacancies (when annealed under oxidizing atmospheres). Substitution accommodated acceptor impurities are usually compensated by oxygen vacancies.
- Donors (including oxygen vacancies) show shallow energy levels, while acceptor states are located deep in the bandgap.
- Oxygen vacancies are found to be relatively mobile ionic defects in comparison to cationic vacancies and involve a migration activation energy of 1 eV.

These arguments lead to the fact that the conduction process observed in ferroelectric thin films [79, 102, 103] is composed of electronic as well as ionic mobility.

8.5.5.1. Measurement of I-V Characteristics

For maintaining accuracy in the measurement of leakage characteristics both in the low-field as well as in the high-field region, it is necessary to reach saturation as depicted by region II in Fig. 8.14. Hu and Krupanidhi [104] have demonstrated that if an inadequately fast voltage ramp is employed for the I-V measurement there is an order of magnitude error in I_L. Figure 8.25 illustrates the result obtained for PZT films, using voltage ramps with different delay times. It is seen that for a higher delay time, after a critical voltage, the characteristics are shifted towards the degradation region (region III in Fig. 8.14), which is not intended. Hence this imposes the restriction on the delay times to be used for applied fields in excess of the critical value.

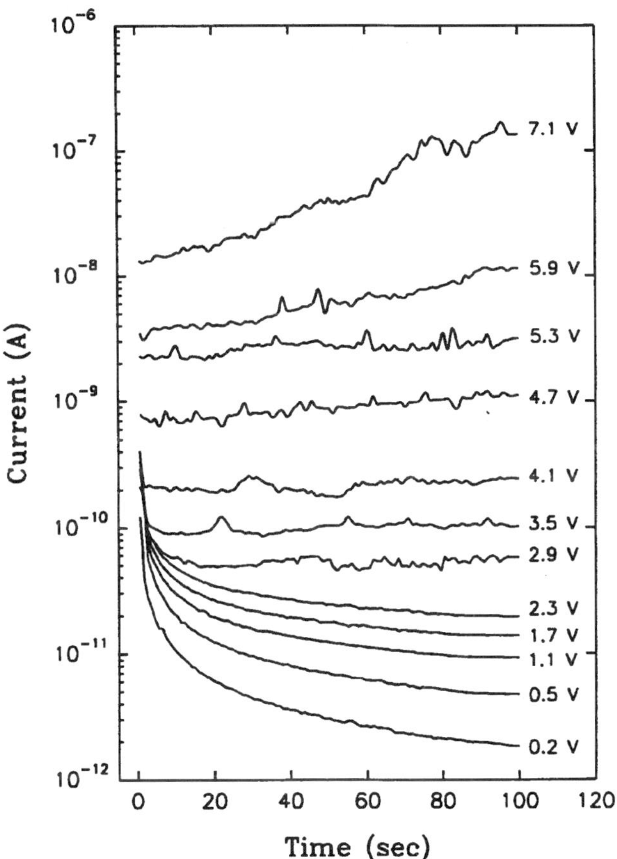

FIG. 8.25. Leakage current versus time behavior in PZT thin films with different voltage ramps.

In general, the leakage current through an MIM system is determined by the bulk of the film and by the electrode interfaces. The presence of ferroelectricity has also been shown to influence the leakage current. Practical study of the ferroelectric polarization of PZT on its leakage properties has been cited in Reference [104]. For this reference, apart from the influence of ferroelectric polarization on leakage current the simplified model appears to define the conduction mechanism. However, it is often found that more than one process is involved in the conduction process, while the features of only one mechanism are highlighted in different field regions [79].

8.5.5.2. Thermionic Emission and the Schottky Model

In the presence of an electronic energy barrier at the electrode/film interface the conduction mechanism is found to depend on the electrode metal work function and is governed by the Richardson-Schottky equation:

$$J = A^{**}T^2 \exp\left[\frac{\alpha E^{1/2} - W_b}{k_b T}\right] \qquad \alpha\sqrt{\frac{q^3}{\alpha\pi\varepsilon_r}\varepsilon_o} \tag{8.6}$$

Here, A^{**} is the effective Richardson constant, T is the absolute temperature, E is the applied electric field, W_b is the barrier height at zero field, k_b is the Boltzmann constant, q is the electronic charge, ε_r is the relative dielectric constant, and ε_0 is the permittivity of free space.

The Schottky effect leads to a field-dependent barrier height reduction and may be verified from Eq. (8.6) using the $\ln(\mathrm{J}/\mathrm{T}^2)$ vs. 1000/T and the $\ln(J/T^2)$ vs $\mathrm{E}^{1/2}$ plots. In either case, the results are straight lines giving slopes from which the value of the barrier height can be determined. The effect of electrodes on the leakage current of $SrTiO_3$ thin films has been demonstrated by Waser where the calculated barrier at the electrode/film interface is found to vary with the electrode metal work function. This form of the Schottky equation is, however, found not to give accurate values of the dielectric constant that can be determined from the slope of the $\ln(\mathrm{J}/\mathrm{T}^2)$ vs the $\mathrm{E}^{1/2}$ plot. In most cases the obtained values are orders of magnitude smaller than those obtained from optical measurements. Zafar *et al.* [105] used a modified Schottky equation based on the argument given by Simmons [106]. The argument states that for a film with thickness of orders of magnitude greater than the mean free path of the electrons traversing from one electrode to another the form of Schottky equation given by Eq (8.6) is not valid and a modification is necessary. In the experimental work, Zafar *et al.* show results that match the optical dielectric constant using the modified form.

8.5.5.3. Tunneling

Electrons may tunnel through a barrier that is sufficiently thin (usually < 10 nm) and energetically high. The tunneling process can originate form the following locations: 1) from the metal into the conduction band of the film; 2) from the trapping levels within the bulk; 3) directly between the valence band and the conduction band of the dielectric; 4) from the valence band of the dielectric into the metal electrodes; or 5) directly between the metal electrodes. For most practical cases where the thickness of the dielectric is $\gg 10$ nm, the most likely phenomenon is the tunneling at the electrode/film interface (Fowler-Nordheim emission). Scott [107] has reported this type of conduction mechanism to be dominant for applied electric fields 300–400 kV/cm in BST thin films, which has also been confirmed by Waser. [108].

8.5.5.4. Poole-Frenkel Effect

The conductivity of films that shows a hopping conduction of electronic carriers may be determined by equilibrium between absorption and emission of free carriers by traps. The high electric fields in the film disturb this equilibrium and shift it towards larger free carrier densities. Hwang illustrated evidence of the Poole-Frenkel effect on BST using IrO_2 electrodes [109].

8.5.5.5. Space-Charge Conduction

This mechanism assumes an injecting contact at the electrode/film interface, and based on the deduction of Child's law in an insulator we arrive at a functional form that was further modified by Lampart [110] to account for the presence of trap states within the bulk. The position of the trap levels in the bandgap can significantly influence the relationship. For the presence of shallow traps the SCLC expression assumes the following form:

$$J = \frac{9\mu k \varepsilon_o \theta E^2}{8L} \tag{8.7}$$

where k is the dielectric constant; ε_o is the permittivity of free space; μ is the mobility of the charge carriers; L is the thickness; and θ is the ratio of the total density of induced free carriers to the trapped carriers. For a distributed space-charge limited conduction the following relation holds:

$$I \propto \frac{V^{(l+1)}}{d^{2l+1}} \tag{8.8}$$

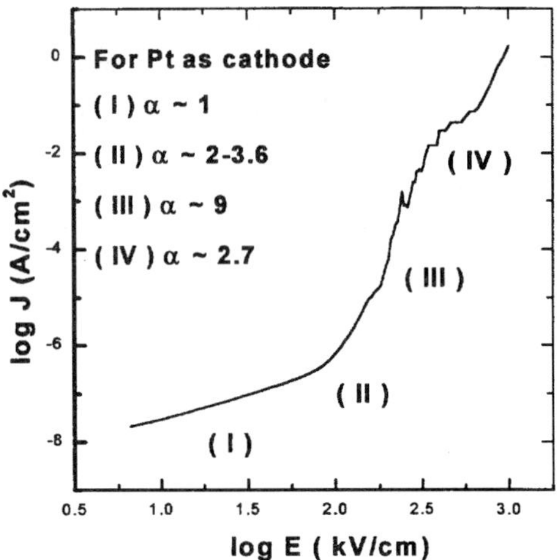

FIG. 8.26. The I-V characteristics in BST thin films.

where $l = T_t/T$, T_t is trap distribution temperature, T is absolute temperature V is the voltage applied across the thin film and d the thickness of the film. Evidence of the space-charge mechanism has been verified in PZT [104, 111], BST [112], and SBT [113] thin films where the modified space-charge law had been used to calculate the concentration as well as the energy level of the traps. Studies on PZT thin films by Scott revealed the presence of a space charge and that the conduction process follows the SCLC mechanism. Figure 8.26 shows the I-V characteristics obtained for a BST thin film [111] along with the specific slopes corresponding to different voltage regions of the curve.

8.5.6. Time-Dependent Dielectric Breakdown

This section deals with region (III) as depicted in Fig. 8.14 and corresponds to the time-dependent electrical breakdown in thin films under a continuous electrical stress. In summary, this process deals with a sudden or gradual increase of the leakage current, which may saturate at a certain higher value or may lead to total failure. In most cases the process is irreversible and leads to permanent damage of the film. From the application point of view, these data provide valuable information regarding the reliability and long-term behavior of the device integrated using a ferroelectric thin film. Due to the different complexities introduced in the thin-film form compared to bulk, it becomes difficult to form a basic foundation on the origin of the breakdown phenomena. Time-dependent dielectric breakdown (TDDB) is found to be characteristic of the intrinsic material, the processing and electrode materials. Several models describing the phenomena of TDDB are available in the literature, such as the grain boundary model [114, 115], and the reduction model [116–118], the grain-boundary potential barrier height model [119], and the de-mixing model [120]. Except for the grain boundary model, all other models are based on the mobile oxygen vacancies. Experiments conducted on bulk as well as thin films show oxygen-vacancy transport as one of the important issues in the breakdown process. Both Scott [103] and Waser [121] have related the oxygen-vacancy movement as an important parameter for resistance degradation in thin films. Krupanidhi and Peng [112] showed the effect of donor doping on the TDDB characteristics, which suppresses the concentration of oxygen vacancies and thereby improves properties (Fig. 8.27).

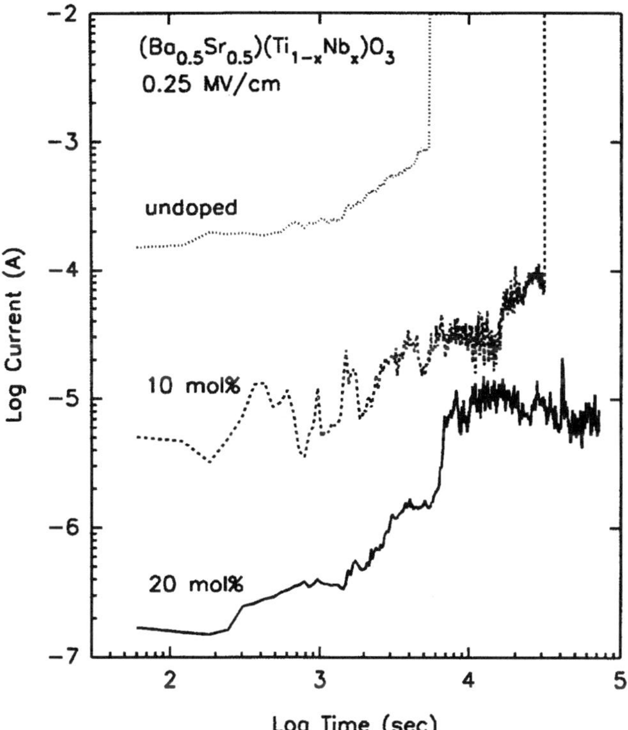

FIG. 8.27. Time-dependent dielectric breakdown phenomenon in BST thin films with doping concentration.

8.6. Process-Property Correlation: Low-Energy Oxygen Ion Beam Bombardment Effect

Concurrent low-energy ion bombardment of thin films during growth has long been recognized as an important tool in modifying the growth process, microstructure and properties of vapor-deposited films, while such a process is often referred to as ion-assisted deposition (IAD) [122]. Although the benefits of IAD are substantial, its application has been limited to the deposition of single-cation metal, semiconductor and oxide films. The lack of IAD used for multi-cation systems may be attributed to complexities introduced by preferential re-sputtering. Considering the PZT system, for instance, Pb- is volatile in comparison with Zr- and Ti; and direct bombardment of the growing PZT film results in preferential re-sputtering of the Pb. Figure 8.28 shows the percentage of re-sputtered Pb- (with respect to the total arrival rate of the Pb species) as a function of the ion energy of the direct bombardment in growing PZT films. The measurements were made with a constant ion fluence and deposition rate. From this figure it can be concluded that the application of direct bombardment requires an increase in Pb fluence to compensate for concurrent re-sputtering. With a single ceramic target, there is no convenient way to make such a compensation. Therefore, it is very difficult, if not impossible, to introduce direct ion-bombardment assistance into the single ion beam/single target configuration when attempting to grow stoichiometric ferroelectric PZT films. With the

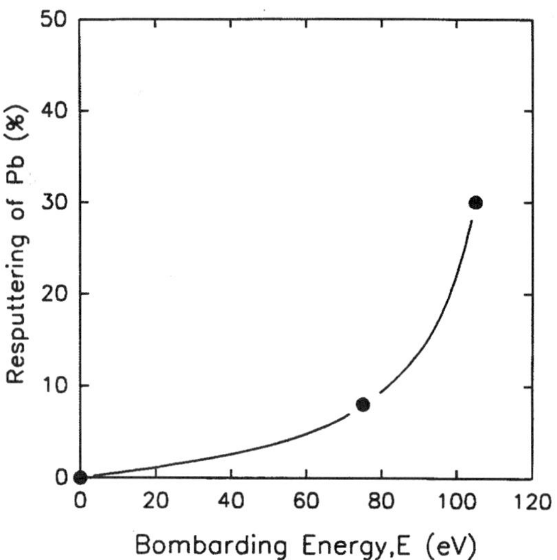

FIG. 8.28. The I-t response in PZT thin films with different voltage ramps.

MIBER technique compensation of preferential re-sputtering is not a problem, as each element is sputtered independently. Low-energy oxygen ion bombardment was successfully used to modify and enhance the physical properties of MIBER-deposited PZT thin films. In the following sections, the bombardment-induced effects are systematically described.

8.6.1. CRYSTALLIZATION ENHANCEMENT WITH PREFERRED ORIENTATION

Low-energy oxygen ion bombardment during film growth was found to have a pronounced effect on the growth process, structure and electrical properties of the PZT thin films, evidenced by the reduced perovskite formation temperature, induced preferential orientation, improved film morphology, and enhanced electrical properties. This type of bombardment-induced enhancement in crystallization was expected, as similar results have been widely reported for other materials. Excess quantities of Pb, as high as 20–25%, are usually needed to obtain the perovskite phase for postdeposition annealed PZT(50/50) films deposited on bare Si substrates. Nonbombarded films with nearly stoichiometric Pb content (approximately 3% excess Pb) deposited on bare Si substrates do not form the perovskite phase. However bombarded films of the same kind and with the same annealing exhibit a dominant perovskite phase. Figure 8.29 illustrates the crystallization enhancement with increasing bombarding ion flux density, which may be attributed to enhanced adatom mobility. Muller [123] visualized

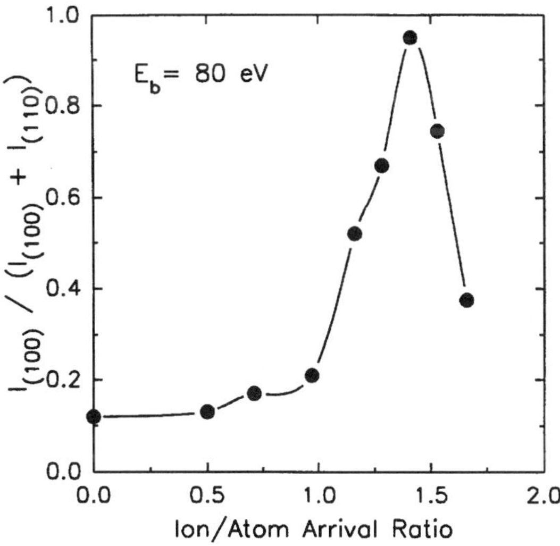

FIG. 8.29. Effect of Ion beam flux density on the orientational growth of PZT thin films.

the possible mechanism of this effect in molecular dynamics simulations of crystal growth. Low-energy ion bombardment induces local atomic rearrangement, allowing atoms to relax into lower energy sites. In this case, wherein the as-grown films are primarily amorphous, these local atomic rearrangements and relaxation may initiate nucleation of microcrystallites that are undetectable by x-ray diffraction. The presence of microcrystallites in the amorphous matrix of the as-grown films would necessarily affect crystallization during annealing. In addition, enhanced incorporation of oxygen in the films, introduced by reactive oxygen ion (O_2^+/O^+) bombardment, may be another reason for enhanced crystallization of the perovskite phase. It has been noted that sufficient oxygen concentration is crucial for PZT films in forming and maintaining the perovskite structure. [124].

8.6.2. Modification of Electrical Properties

8.6.2.1. Remanent Polarization and Coercive Field

As a natural consequence of structural modification, the impact of the low-energy oxygen ion bombardment on the electrical properties of the films is profound. Figure 8.30 shows the remanent polarization (P_r) and coercive field (E_c) as a function of annealing temperature (T_a) for both bombarded films and nonbom-

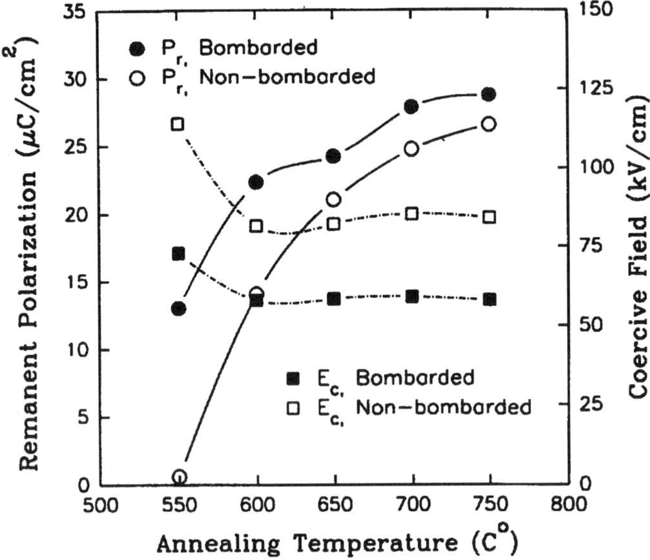

FIG. 8.30. Variation in P_r, E_c values in PZT thin films with and without bombardment.

barded films. A common trend is that, within the temperature range studied, P_r consistently increases with increasing T_a while E_c initially drops and afterwards flattens out. This is attributable to the development in the crystallinity of the perovskite phase. A more significant feature shown in the figure is that for a corresponding T_a, the bombarded films consistently show higher P_r and lower E_c than the nonbombarded films. The changes in P_r and E_c induced by the bombardment may be due to improved crystallization of the perovskite phase, an increased degree of (100) orientation, and improved electrode-film interfaces, which are attributable to denser and smoother film surfaces.

8.6.2.2. Dielectric Response

Figure 8.31 shows the changes of room-temperature low-field dielectric constant and dielectric loss with annealing temperature T_a for bombarded films and nonbombarded films. Films annealed above 600 °C are completely crystallized, and exhibit dielectric constants (between 700 and 1200) and dissipation factors (between 0.02 and 0.03), which are comparable to those observed in bulk ceramic PZT with the same composition. It can also be seen from this figure that for $T_a < 650$ °C the bombarded films have higher dielectric constants than is the case for the nonbombarded films. This corresponds to an enhanced development of the perovskite phase in the bombarded films. While the dielectric constant of

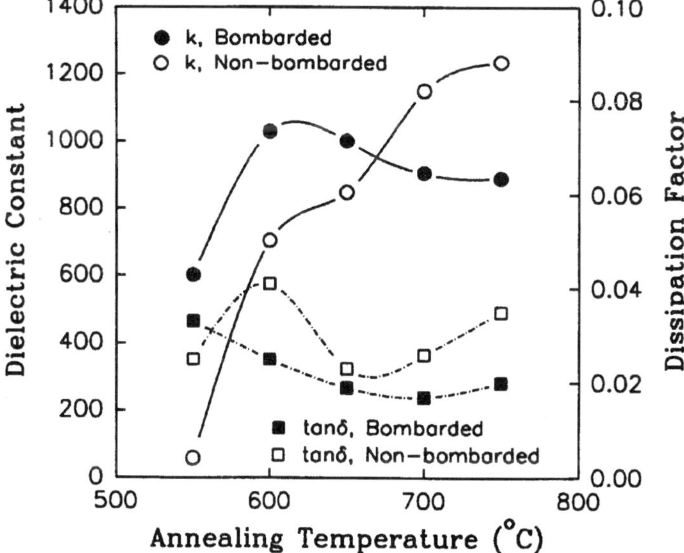

FIG. 8.31. Variation in dielectric constant and dissipation factors with and without bombarding.

nonbombarded films continues to increase with annealing temperature, the dielectric constant of the bombarded films decreases with an annealing temperature > 600 °C and becomes smaller than that of the nonbombarded films at higher temperatures. This behavior may be attributed to the development of a higher degree of (100) orientation in the bombarded films, which tends to lower the observed dielectric constant. In most cases, the bombarded films have a lower dielectric loss than the nonbombarded films.

8.6.2.3. Leakage Current and Time-Dependent Dielectric Breakdown

Figure 8.32 shows the results of current versus voltage (I-V) measurements of both the bombarded and nonbombarded films, annealed at 600 °C for 2 h. The ohmic resistivity of the bombarded films ($\sim 3 \times 10^{11}\ \Omega \cdot \text{cm}$) is about one order of magnitude higher than that of the nonbombarded films ($\sim \times 10^{10}\ \Omega \cdot \text{cm}$). The onseet voltage of the space-charge-limited conduction is much higher for the bombarded films (~12 V) than for the nonbombarded films (~3 V). The bombarded films also have much higher dielectric breakdown strengths than the nonbombarded films (~770 kV/cm and ~350kV/cm respectively). For films annealed at temperatures higher than 600 °C, the difference in I-V behavior between the bombarded films and the nonbombarded films is reduced. The

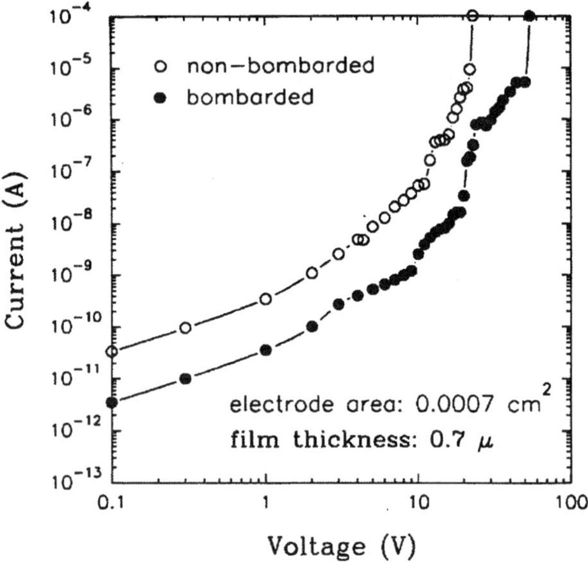

FIG. 8.32. Comparative I-V responses in PZT thin films with and without bombardment.

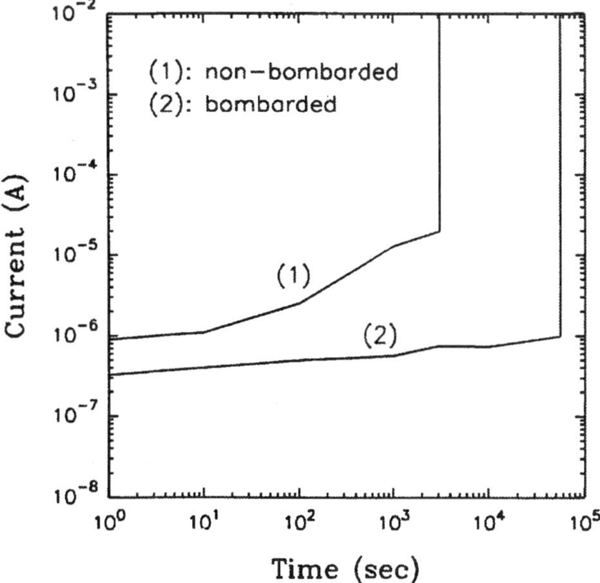

FIG. 8.33. Time-dependent dielectric breakdown phenomenon in PZT thin films with and without bombardment.

observed differences in the I-V behavior between these two kinds of films may be due to their differences in the microstructure and the degree of nonstoichiometry.

The time-dependent dielectric breakdown (TDDB) of ferroelectric films is believed to be closely related to the I-V behavior because both are dependent on the motion of charge carriers. Figure 8.33 shows the TDDB characteristics of both bombarded and nonbombarded PZT films annealed at 680 °C for 2 h. The bombardment-induced effect on TDDB is amazingly large. For a dc field of 450 kV/cm, the nonbombarded film breakdown is about 50 min, while the bombarded films do not break down for up to 925 min.

8.6.3. ELECTRON CYCLOTRON RESONANCE (ECR) PLASMA-SOURCE-ASSISTANCE-INDUCED EFFECTS

To illustrate the effect of ECR plasma during the development of electroceramic thin films, $SrTiO_3$ was chosen as the candidate for the present discussion [125]. Conventional processing of DRAM gate dielectrics employs vacuum techniques in which planar or three-dimensional (3D) capacitor structures are fabricated on the silicon substrate. The crystalline structure and composition are extremely sensitive properties that influence the electrical behavior of the complex oxide thin films. In the case of $SrTiO_3$ films deposited at a temperature of 400 °C, ECR plasma assistance enhanced crystallinity. Such enhancement in crystallinity is also reflected in the improved composition of $SrTiO_3$ films. The Sr/Ti ratio increased from 0.75 to approximately 1.0 with the presence of the ECR plasma. These factors result in increased ad-atom mobility due to low energy bombardment by a high density of species and improved reaction kinetics on the substrate surface due to the presence of activated species in the plasma.

8.6.3.1. Leakage Current Behavior

Figure 8.34 illustrates the effect of the ECR plasma presence on the leakage current behavior of $SrTiO_3$ with applied electric field, which is an essential requirement for high-permittivity dielectric thin films meant for developing DRAM-type devices. It may be seen that the I-V curve shows similar behavior that is near ohmic behavior in the low-field region, which is followed by an increase in current with higher electric fields, which may be attributed to the onset of a bulk-limited space-charge conduction process. This behavior is similar to that observed with other physical vapor growth processes described in this paper. However, two inherent features can also be seen from this figure. First, the leakage current in the near-ohmic region is reduced with the ECR plasma and second, the voltage for the space-charge onset is shifted to higher electric fields.

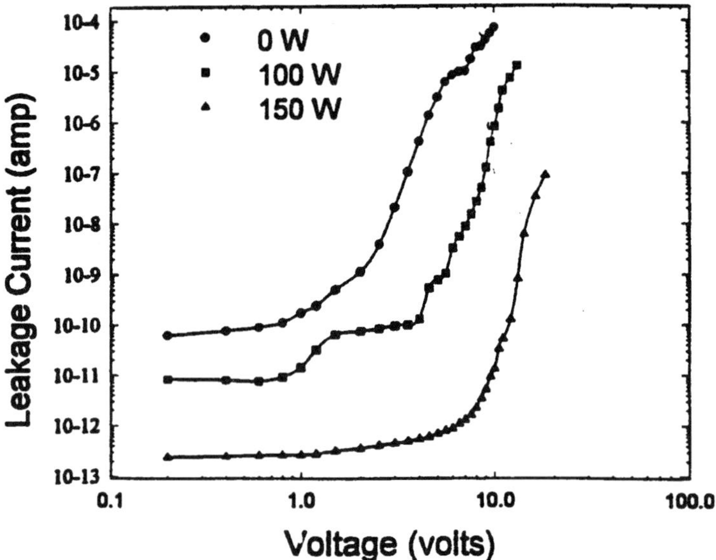

FIG. 8.34. Leakage current characteristics in $SrTiO_3$ thin films grown with ECR plasma presence.

These qualities establish the effect of the ECR plasma presence by improving the oxidation kinetics of the compounds via effective incorporation of the oxygen, which makes the films more insulating.

8.6.3.2. *Time-Dependent Dielectric Breakdown*

Figure 8.35 shows the time-dependent dielectric breakdown response for $SrTiO_3$ films under an applied constant electric field of 200 kV/cm. This level of the field was chosen intentionally to drive the films to near space-charge conditions and to observe the breakdown behavior. It may be seen from the figure that with the inclusion of the ECR plasma during the growth process, the TDDB is extended to a much longer duration, up to 5×10^4 s, whereas the breakdown occurred in the films without ECR plasma at much earlier periods of about 10^3 s. These behaviors once again are ascribed to the ECR plasma induced near perfect stoichiometry, denser grain-grain boundary structure and better oxygen incorporation. Several physical models have been described in the literature to explain the TDDB behavior, but our results appear to lean towards an oxygen-vacancy-dependent behavior.

8.6.3.3. *Charge Storage Density*

Dielectric layers meant for DRAM-type devices are expected to offer better charge storage densities in comparison with conventional gate dielectrics such as

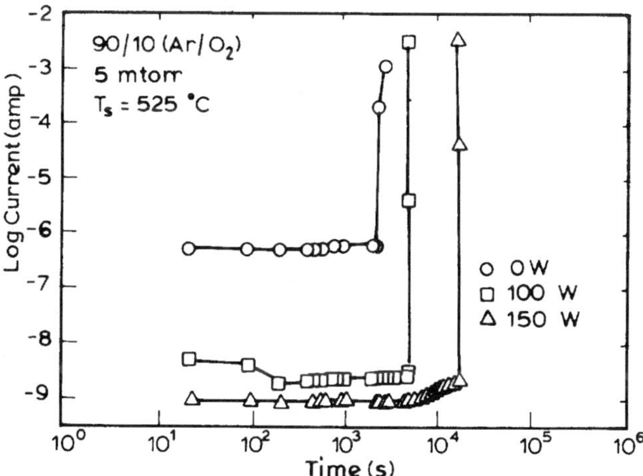

FIG. 8.35. Time-dependent dielectric breakdown effect in $SrTiO_3$ thin films grown with ECR plasma presence.

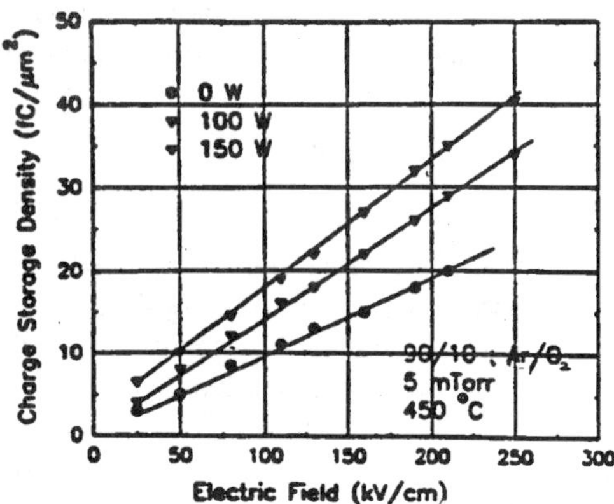

FIG. 8.36. Variation of charge storage density in $SrTiO_3$ thin films with applied electric fields for films deposited under different ECR plasma conditions.

SiO_2. Figure 8.36 shows the variation of charge storage density with the applied electric fields for films deposited under different ECR plasma conditions. The charge storage varies linearly with the applied electric field and the magnitude improves with increasing ECR power. It is also worth noticing that the films exhibit no breakdown even up to fields of 250 kV/cm and offer charge storage of as high as 40 fC/cm^2. These characteristics further establish the property enhancement via energetic particle bombardment. It was shown earlier that the presence of ECR plasma improved the structure and composition, both of which contribute to improvement in the dielectric behavior of the dielectric thin films. The improved charge storage density with the ECR plasma is also reflected in an improved dielectric constant.

8.7. Microstructure-Dependent Electrical Properties

Comparison of ferroelectric multicomponent oxide films derived by chemical routes and physical vapor growth process combined with low-energy ion bombardment reveals a close relationship among its microstructure and electrical properties, which is schematically illustrated for $(Ba,Sr)TiO_3$ (BST) films in Fig. 8.37. It may be realized that the microstructures of, for example BST, thin films could be classified into three types: (a) Multigrain structure (type I) throughout the film thickness; (b) columnar structure (Type II), which remains even after high-temperature annealing; and (c) highly dense columnar structure (Type III), which is encouraged via low-energy ion bombardment. Depending on the

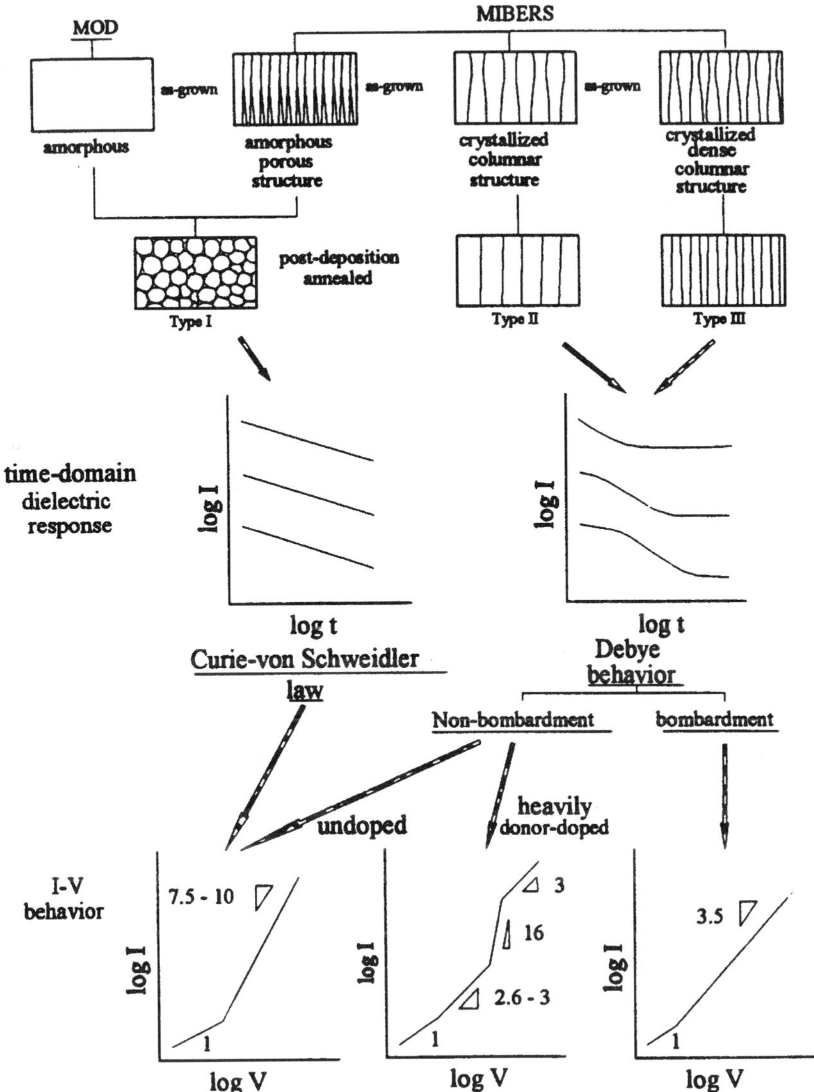

FIG. 8.37. Microstructural dependence on electrical properties.

microstructure, one may realize different time-domain responses, followed by different natures of conduction mechanisms.

Two types of I-V characteristics are observed in nonbombardment films depending on the chemistry of the films (doped or undoped) and substrate temperature during deposition. Only those films doped with higher donor concentration and deposited at high substrate temperatures show bulk-limited space-charge conduction with discrete shallow traps embedded in trap-distributed background at high electric fields. Bombarding the films at higher substrate temperatures undoubtedly improves the density, reduces the trap density and improves the time-dependent dielectric breakdown behavior. Such low-energy reactive ion-bombardment films exhibited either comparable or in some cases improved leakage behavior and transport mechanisms, making them comparable to high donor doping.

8.8. Summary

A comprehensive review is presented of the growth processes of ferroelectric thin films (both physical and chemical nature), which have received major attention. Almost every growth technique appears to offer ferroelectric thin-film properties that qualify for extensive research. Most recently, attention among the community has been focused on gaining a clearer understanding of these properties in close correlation with the processing and subsequent microstructure of these films. Also evident has been extensive work dealing with the effect of chemistry and exposing films to low-energy bombardment during growth in order to gain better control the gain structure and, subsequently, the electrical properties. However, irrespective of the nature of the application, the following issues require immediate attention: a) control of gain size and density; b) controlled process-compositioin-microstructure-property correlation; c) clearer understanding of film/substrate interfaces (both electronically and physically); d) better understanding of process dependent defects and their distribution; and e) time-dependent performance of the films and devices. In addition, there is also a need to continue to develop newer compositions that can cater to the needs of the latest device applications, such as MEMs, NVRAMs, DRAMs, and optoelectronic devices.

References

1. Francombe, M.H. (1972). *Thin Solid Films* **13**: 413.
2. Scott, J.F. and Paz de Arauji, C.A. (1989). *Science* **246**: 1400.
3. Miyasaka, Y. and Matsubara, S. (1991). *Proc. 7th Int. Symp. Applications of Ferroelectrics*, Piscataway, NJ: (IEEE Service Center), p. 121.
4. Whatmore, R.W. (1991). *Ferroelectrics* **118**: 241.
5. Watton, R. and Todd, M.A. (1991). *Ferroelectrics* **118**: 279.
6. Polla, D.L. (1995). *Microelect. Eng.* **29**: 51.

7. Nemirovsky, Y., Nemirovsky, A., Muralt, P., and Setter, N. (1996). *Sensors and Acutators* **A**, **56**: 239.
8. Uchino, K. (1990). *MRS Bulletin* **18**: 42.
9. Haertling, G.H. (1999). *J. Am. Ceram. Soc.* **82**: 797.
10. Dey, S.K. and Zuleeg, R. (1990). *Ferroelectrics*, **112**: 309.
11. Krupanidhi, S.B., Maffei, N., Sayer, M., and El-Assal, K. (1983). *J. Appl. Phys.* **54**: 6601.
12. Auciello, O., Kingon, A.I., and Krupanidhi, S. B. (1996). *MRS Bulletin* **21**: 25.
13. Aucielli, O. and Ramesh R. (1996). *MRS Bulletin* **21**: 31.
14. Mejia, S.R., McLeod, R.D., Kao, K.C., and Card, H.C. (1986). *Rev. Sci. Instrum.* **57**: 493.
15. Dey, S.K. and Alluri, P.V. (1996). *MRS Bulletin* **21**: 44.
16. Tuttle, B.A. and Schwartz, R.W. (1996). *MRS Bulletin* **21**: 49.
17. Haertling, G.H. (1990). *The 5th U.S.-Japan Seminar on Dielectric and Piezoelectric Ceramics*, Kyoto, Japan, December 1990.
18. Brinker, C.J., Hurd, A.J., Frye, G.C., Schunk, P.R., and Ashley, C.S. (1992). In *Chemical Processing of Advanced Materials*, L.L. Hench and J.K. West, eds., New York: John Wiley & Sons, Inc., p. 395.
19. Oikawa, M. and Toda, K. (1976). *Appl. Phys. Lett.* **29**: 491.
20. Philips, L.S. (1971). *Electron. Comp.* **12**: 523.
21. de Kejiser, M. and Dormans, G.J.M. (1996). *MRS Bulletin* **21**: 37.
22. Kojima, M., Okuyama, M., Nakagawa, T., and Hamkawa, Y. (1983). *Jpn. J. Appl. Phys.* **22**: **Suppl. 2**, 14.
23. Okada, M., Takai, S., Amemiya, M., and Tominaga, K. (1984). *Jpn. J. Appl. Phy.* **28**: 1030.
24. Sinharoy, S., Buhay, H., Francombe, M.H., Takei, W.J., Doyle, N.J., Riger, J.R., Lampde, D.R., and Sterke, E. (1991). *J. Vac. Sci. & Tech.*
25. Hwang, C.S., Lee, B.T., Kang, C.S., Kim, J.W., Lee, K.H., Cho, H., Horii, H., Kim, W.D., Lee, S.I., Roh, Y.B., and Lee, M.Y. (1998). *J. Appl. Phys.* **83**: 3708.
26. Hwang, C.S., Lee, B.T., Cho, H., Lee, K.H., Kang, C.S., Hideki, H., Lee, S.I., and Lee, M.Y. (1997). *Appl. Phys. Lett.* **71**: 371.
27. Zafar, S., Jones, R.E., Jiang, B., White, B., Chu, P., Taylor, D., and Gillespie, S. (1998). *Appl. Phys. Lett.* **73**: 175.
28. Ai-Li-Ding, Luo, W.G., Qiu, P.S., Feng, J.W., and Zhang, R.T. (1998). *J. Mater. Res.* **13**: 1266.
29. Wasa, K., Adachi, H., Hirochi, K., Matsushima, T., and Setsune, K. (1991). *J. Mater. Res.* **6**: 1595.
30. Hu, H., Kumar, V., and Krupanidhi, S.B. (1993). *J. Appl. Phys.* **74**: 3373.
31. Krupanidhi, S.B., Hu, H., and Kumar, V. (1992). *Ceramic Transactions* **25**.
32. Chrisey, D.B., and Hubler, G.K. (eds.) (1994). *Pulsed Laser Deposition of Thin Films*, New York: John Wiley & Sons.
33. Maffei, N. and Krupanidhi, S.B. (1992). *Appl. Phys. Lett.* **60**: 781.
34. Ramesh, R., Luther, K., Willkens, B., Hart, D.L., Wang, E., Taracon, J.M., Inam, A., Xu, X.D., and Venkatesan, T. (1990). *Appl. Phys. Lett.* **57**: 1505.
35. Davis, G.M. and Gower, M.C. (1989). *Appl. Phys. Lett.* **55**: 112.
36. Mohan Rao, G. and Krupanidhi, S. B. (1994). *J. Appl. Phys.* **75**: 2604.
37. Saha, S. and Krupanidhi, S.B. *J. Appl. Phys.* (in press).
38. Roy, D. and Krupanidhi, S.B. (1995). *J. Appl. Phys.* **76**: 980.
39. Tseng, Y.K., Liu, K.S., Jiang, J.D., and Lin, I.N. *Appl. Phys. Lett.* **72**: 3285.
40. Kang, Y.M., and Baik, S. (1997). *J. Appl. Phys.* **82**: 2532.
41. Hung-Ming Yang, Jian-Shing Luo, and Wen-Tai Lin, (1997). *J. Mater. Res.* **12**: 1145.
42. Bhattacharyya, S., Bharadwaja, S.S.N., and Krupanidhi, S.B. (1999). *Appl. Phys. Lett.*, **75**: 2656.
43. Zafar, S., Kaushik, V., Laberge, P., Chu, P., Jones, R.E., Hance, R.L., Zurcher, P. White, B.E., Taylor, D., Melnick, B., and Gillespie, S. (1997). *J. Appl. Phys.* **82**: 4469.

44. Dey, S.K., and Lee, J.J. (1992). *IEEE Trans. Electron Devices* **39**: 1607.
45. Lee, S.J., Kang, K.Y., Han, S.K., Jang, M.S., Chae, B.G., Yang, Y.S., and Kim, S.H. (1998). *Appl. Phys. Lett.* **72**: 299.
46. Kushida, K., Udaya Kumar, K.R., Krupanidhi, S.B., and Cross, L.E. (1993). *J. Am. Ceram. Soc.* **76**: 1345.
47. Kissurska, R.D., Brooks, K.G., Reaney, I.M., Pawlaczyk, C., Kosee, M., Setter, N. (1995). *J. Am. Ceram. Soc.* **78**: 1513.
48. Guanghuo Yi, Zheng Wu, and Sayer, M. (1988). *J. Appl. Phys.* **64**: 2717.
49. Yamaguchi, H., Lesaicherre, P., Sakua, T., Miyasaka, Y., Shintani, A., Yoshida, M. (1993). *Jpn. J. Appl. Phys.* **9B**: 4069.
50. Gorbenko, O.Y., Kaul, A.R., and Wahl, G. (1997). *Chem. Vapor Deposition* **3**: 193.
51. Fu, L., Liu, K., Zhang, B., and Chu, J. (1998). *Appl. Phys. Lett.* **72**: 1784.
52. Cho, H.J. and Kim, H.J. (1998). *Appl. Phys. Lett.* **72**: 786.
53. Chen, H.Y., Lin, J., Tan, K.L., and Feng, Z.C. (1996). *Thin Solid Films* **289**: 59.
54. Wessels, B.W. (1995). *Annu. Rev. Mater. Sci.* **25**: 525.
55. Singh, R. (1988). *J. Appl. Phys.* **63**: R58.
56. Hu, H., Shi, L., Kumar, V., and Krupanidh, S.B. (1992). In *Ceramic Transitions*, A.S. Bhalla and K.M. Nair, eds. Westerville, OH: American Society, p. 113.
57. Saha, S. and Krupanidhi, S.B. (1999). *Mat. Sci. & Eng.* B **57**: 135.
58. Tani, T., Li, J.F., Viehland, D., and Payne, D.A. (1994). *J. Appl. Phys.* **75**: 3017.
59. Yamakawa, K., Trolier-McKinstry, S., Dougherty, J.P., and Krupanidhi, S.B. (1995). *Appl. Phys. Lett.* **67**: 2014.
60. Kanna, I., Hayashi, S., Kitagawa, M., Takayama, R. (1995). *Appl. Phys. Lett.* **66**: 145.
61. Chattopadhyay, S., Ayyub, P., Palkar, V.R., Multani, M.S., Pai, S.P., Purandare, S.C., and Pinto, R. (1998). *J. Appl. Phys.* **83**: 7808.
62. Bharadwaja, S.S.N. and Krupanidhi, S.B. (1999). *J. Appl. Phys.*.
63. Muralt, P. *et al.* (1998). *J. Appl. Phys.* **83**: 3835.
64. Barin, I. (1989). *Thermochemical Data of Pure Substances*, Germany: VCH Verlags-gesellschaft.mbH.
65. Amanuma, K., Hase, T., and Miyasaka, Y. (1995). *Appl. Phys. Lett.* **66**: 221.
66. Li, T., Zhu, Y., Desu, S.B., Peng, C.H., and Nagata, M. 1996). *Appl. Phys. Lett.* **68**: 616.
67. Seong, N.J., Yoon, S.G., and Lee, S.S. *Appl. Phys. Lett.* **71**: 81.
68. Bhattacharyya, S., Bharadwaja, S.S.N., and Krupanidhi, S.B. *Appl. Phys. Lett.* (in press).
69. Hu, H., Peng, C.J., and Krupanidhi, S.B. (1993). *Thin Solid Films* **223**: 327.
70. Fox, G.R. and Krupanidhi, S.B. (1992). *J. Mater. Res.* **7**: 3039.
71. Fox, G.R. and Krupanidhi, S.B. (1993). *J. Mater. Res.* **8**: 2191.
72. Fox, G.R. and Krupanidhi, S.B. (1993). *J. Mater. Res.* **8**: 2203.
73. Fox, G.R. and Krupanidhi, S.B. (1994). *J. Mater. Res.* **9**, 699.
74. Fox, G.R. and Krupanidhi, S.B. (1993). *J. Appl. Phys.* **74**: 1949.
75. Joncher, A.K. (1983). *Dielectric Relaxation in Solids*, London: Chelsea Dielectrics Press.
76. Waser, R. (1997). *Integrated Ferroelectrics* **15**: 39.
77. Jonscher, A.K. (1977). *Nature* **253**: 231.
78. Dissado, L.A., Hill, R.M. (1979). *Nature* **279**: 685.
79. Sayer, M., McIntyre, D.S., Sedlar, M., Mansingh, A., Tondan, R., and Chivukula, V. (1995). *Integr. Ferroelectrics* **11**: 277.
80. Waser, R. (1995). *Science and Technology of Electroceramic Thin Films*, O. Aucielli and R. Waser, eds., Netherlands: Kluwer Academic Publishers, p. 223.
81. Bharadwaja, S.S.N., and Krupanidhi, S.B. (1999). *Mater. Sci. & Eng.* B **64**: 54.
82. Aucielli, O., Gruverman, A., Tokumoti, H., Prakash, S.A., Aggarwal, S., and Ramesh, R. (1998). *MRS Bulletin* **33**.

83. Dong-Hwa Kwak, Jang. B.T., Cha, S.Y., Lee, S.H., Lee, H.C., and Yu, B.G. (1996). *Integrated Ferroelectrics* **133**: 121.
84. Hong-Ming Chen and Lee, J. Ya-min (1998). *Appl. Phys. Lett.* **73**: 309.
85. Merz, W.J. (1954). *Phys. Rev.* **95**: 690.
86. Hu, H. and Krupanidhi, S.B. (1993). *Appl. Phys. Lett.* **62**: 651.
87. Lohse, O., Tiedke, S., Grossmann, M., and Waser, R. (1998). *Integrated Ferroelectrics* **22**: 123.
88. Scott, J.F., Kammerdiner, L., Parris, M., Traynor, S., Ottenbacher, V., Shawabkeh, A., and Oliver, W.F. (1998). *J. Appl. Phys.* **64**: 787.
89. Burfoot, J.C. and Taylor, G.W. (1979). *Polar Dielectrics and Their Applications*, Berkeley/Los Angeles: University of California Press.
90. Pan, W.Y., Gu, W.Y., and Cross, L.E. (1989). *Ferroelectrics* **99**: 185.
91. Duiker, H.M., Beala, P.D., and Scott, J.F., Paz de Araujo, C.A., Melnick, B.M., Cuchiara, J.D., and McMillan, L.D. (1990). *J. Appl. Phys.* **68**: 5783.
92. Dimos, D., Al-Shareef, H.N., Warren, W.L., and Tuttle, B.A. (1996). *J. Appl. Phys.* **80**: 1682.
93. Colla, E.L., Taylor, D.V., Tagantsev, A.K., and Setter, N. (1998). *Appl. Phys. Lett.* **72**: 2478.
94. Chen, T.C., Thio, C.L., and Desu, S.B. (1997). *J. Mater. Res.* **12**: 2628.
95. Lee, J.J., Thio, C.L., and Desu, S.B. (1995). *J. Appl. Phys.* **78**: 5073.
96. Al-Shareef, H.N., Aucielli, O., and Kingon, A.I. (1995). In *Science and Technology of Electroceramic Thin Films*, O. Auciello, and R. Waser, eds., *NATO ASI Series*, Series E: *Applied Sciences*, vol. 284, p. 133.
97. Fujisawa, H,. Hyoda, S., Jistsui, K., Shimizu, M., Niu, H., Okino, H., and Shiosaki, T. (1998). *Integrated Ferroelectrics* **21**: 107.
98. Ramesh, R., Aucielli, O., Keramidas, V.G., and Dat, R. (1995). In *Science and Technology of Electroceramic Thin Films*, O. Aucielli, and R. Waser, eds., Netherlands: Kluwer Academic Publisher, p. 1.
99. Paz de Arauji, C.A., Cuchiora, J.D., McMillan, C.D., Scott, M.C., and Scott, J.F. (1995). *Nature* **374**: 627.
100. Carrano, J., Sudhama, C., Chikarmane, V., Lee, J., Tasch, A., Shepherd, W., and Abt, N. (1991). *IEEE Ultrasonics, Ferroelectrics and Frequency Control* **38**: 690.
101. Grossmann, M., Hoffmann, S., Gusowski, S., Waser, R., Streiffer, S.K., Basceri, C., Parkar, C.B., Lash, S.E., and Kingon, A.I. (1998). *Integrated Ferroelectrics* **22**: 83.
102. Raymond, M.V. and Smyth, D.M. (1995). In *Science and Technology of Electroceramic Thin Films*, O. Auciello and R. Waser, eds., Netherlands: Kluwer Academic Publisher, p. 315–325.
103. Scott, J.F. (1994). *Integrated Ferroelectrics* **4**: 61.
104. Hu, H. and Krupanidhi, S.B. (1994). *J. Mater. Res.* **9**: 1484.
105. Zafar, S., Jones, B., Jiang, B., White, B., Kaushik, V., and Gillespie, S. (1998). *Appl. Phys. Lett.* **73**: 3533.
106. Simmons, J.G. (1965). *Phys. Rev. Lett.* **15**: 967.
107. Scott, J.F., Azuma, M. *et al.* (1992). *Proc. International Symposium on Applied Ferroelectrics*, p. 356.
108. Waser, R. *et al.* (1996). *Proc. Electroceram V*, Aveiro, Portugal, p. 293.
109. Hwang, C.S. *et al.* (1998). *J. Appl. Phys.* **83**: 3703.
110. Lampert, M. and Mark, P. (1970). *Current Injection in Solids*, New York: Academic Press.
111. Scott, J.F., Araujo, C.A., Melnick, B.M., McMillan, L.D., and Zuleeg, R. (1991). *J. Appl. Phys.* **70**: 382.
112. Krupanidhi, S.B., and Peng, C. (1997). *Thin Solid Films* **305**: 144.
113. Watanabe, K., Hartmann, A.J., Lamb, R.N., and Scott, J.F. (1998). *Integrated Ferroelectrics* **21**: 241.
114. Loh, E. (1982). *J. Appl. Phys.* **53**: 6229.
115. Neumann, H. and Arlt, G. (1986). *Ferroelectrics* **69**: 179.

116. Lehovec, K., and Shirn, G. A. (1962). *J. Appl. Phys.* **33**: 2036.
117. Lee, H.Y., and Burton, L.C. (1986). *IEEE Trans. Components, Hybrids, Manuf. Technol.* **CHMT-9**: 469.
118. Payne, D.A. (1968). *Proc. 6th Annual Reliability Physics Symposium*, IEEE, CA, p. 257.
119. Yoo, I.K., Stephenson, F.W., and Burton, L.C. (1987). *IEEE Trans. Components, Hybrids, Manuf. Technol.* **CHMT-10**, **2**: 274.
120. Waser, R. (1989). *Mater. Sci. Eng.* A. **109**: 171.
121. Waser, R. (1989). *J. Am. Ceram. Soc.* **72**: 2234.
122. Krupanidhi, S.B. (1995). *Science and Technology of Electroceramic Thin films*, O. Aucielli and R. Waser, eds., Netherlands: Kluwer Academic Publishers, pp. 23–51.
123. Muller, K.H., (1989). *Hand Book of Ion Beam Processing Technology*, J.J. Cuomo, S.M., Rossnagel, and H.R. Kaufman, eds., New Jersey: Noyes, p. 241.
124. Hu, H., Kumar, V., and Krupanidhi, S.B. (1992). *J. Appl. Phys.* **17**: 376.
125. Belsick, J.R., and Krupanidhi, S.B. (1993). *J. Appl. Phys.* **74**: 3438.

Integration Aspects of Advanced Ferroelectric Thin-Film Memories

DEBORAH J. TAYLOR

Motorola, Austin, Texas, USA

9.1. Introduction

This chapter discusses the highlights and solutions to some of the technical challenges that both research and development (R&D) and industrial groups have encountered when ferroelectric thin film capacitors have been integrated with semiconductor devices. In this chapter we take a practical approach to integration and process-related issues that pertain to ferroelectric thin film memories (i.e., Ferroelectric Random Access Memories or FeRAMs and Dynamic Random Access Memories or DRAMs). However, some of these issues can be more generally applied to other ferroelectric thin film applications (e.g., pyroelectric and infrared sensors, piezoelectric-acoustic components, and electro-optic devices), as ferroelectric thin films have also been integrated on semiconductor circuits for these applications.

This chapter complements Chapter 2, which is devoted to FeRAMs. It has been included because solutions to process-related integration issues are critical for making ferroelectric thin-film memories commercially successful and to thereby capture a larger share of the total memory market (which, incidentally, for the upcoming year is estimated to be worth over $60 billion [1]). This chapter is not intended to provide an exhaustive bibliography of the numerous publica-

Vol. 28
ISBN 0-12-533028-6/$35.00
ISSN 1079-4050

tions on ferroelectric thin films, nor is it a complete list of all the technical challenges that have arisen in integrating these films with semiconductor devices. Rather, it is a summary, or sampling, of recent literature that documents the integration problems that have challenged many researchers and the solutions they have suggested.

The following seven sections summarize recent literature published on the topic of integrating ferroelectric thin films with semiconductor devices. Section 9.2 introduces issues related to the design of the memory cells that are implemented in designing high-density FeRAMs and ultra-dense DRAMs. Sections 9.3 and 9.4 detail integration challenges and solutions for forming and patterning the capacitor stack. Section 9.5 discusses the damaging effects that hydrogen-containing ambients, typically present during the back-end processing of the silicon devices, have on ferroelectric capacitors. Section 9.6 focuses on the impact of ferroelectric processing on the silicon devices. Section 9.7 outlines the equipment issues for the commercial manufacturing of ferroelectric thin-film memories. Finally, Section 9.8 presents a summary and an outlook of the future. Due to intense industrial competition, some solutions to technical problems remain proprietary, causing an information gap in the literature over the last decade. The fact that only some published information is available on integration issues for low-density FeRAMs currently in production, points precisely to this gap.

9.2. Design Considerations

There are essentially two different integration designs involving memory cells with one transistor and one ferroelectric capacitor for FeRAMs [2, 3] and DRAMs [4–7] (Fig. 9.1). In both designs the ferroelectric capacitors are formed after the silicon transistors and before the interconnect metallization, as the ferroelectric crystallization temperature (e.g., usually about 550–800 °C) lies between the higher temperature treatment needed to form the silicon (Si) transistors and the lower temperature treatment used for the interconnection process. For example, typical thermal budgets used to activate the transistor's source/drain and phosphorus silicate glass (PSG) or boron phosphorus silicate glass (BPSG) flows are usually greater than 850 or 900 °C for 30 min in standard furnace anneals. Typical thermal budgets for final forming gas anneals in nitrogen/hydrogen (N_2/H_2), used to avoid Si spiking and aluminum (Al) hillocking of Al interconnects, are usually about 450 °C for 30 min [8]. The literature refers to the processing of the silicon devices (e.g., transistors and diodes) on the complementary metal oxide semiconductor (CMOS) technology as the "CMOS front-end" and the processing of the ferroelectric and interconnects as the "back-end." For both designs shown in Fig. 9.1, note that an interlevel dielectric or ILD (e.g., silicon-dioxide or SiO_2) is used between the CMOS front-end and the back-end. The intent is to isolate the Si devices from the processing used to form and pattern the ferroelectric capacitors. However, as discussed in Section 9.6, some researchers have found it necessary to add a barrier layer (e.g., titanium-dioxide or TiO_2) between the ILD and bottom electrode to further protect the Si devices from the ferroelectric capacitor diffusing towards them.

In these types of designs, the ferroelectric capacitor is either placed over the field oxide with a metal strap to connect the top electrode and the source/drain of

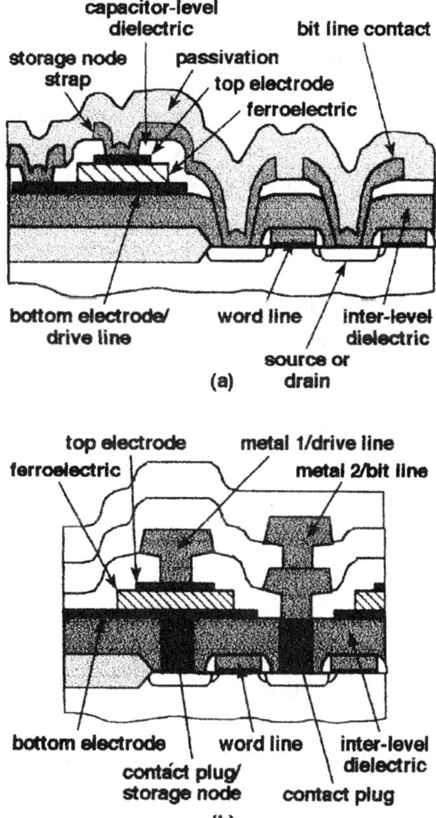

FIG. 9.1. Cross section of ferroelectric/CMOS memory cells where (a) the capacitor is placed over the field oxide and (b) the capacitor is placed directly on top of the transistor source/drain. Note that in (a) the triple-tiered capacitor stack is structured with an etch process that uses three separate photoresist masks, and in (b) the single-tiered capacitor stack is structured with an etch process that uses only one photoresist mask (adapted from Zurcher *et al.* [9], © IEEE, with permission).

the transistor (as shown in Fig. 9.1a) or it is placed directly on top of the source/drain with a conductive plug that connects the ferroelectric to the source/drain of the transistor (Fig. 9.1b). These basic cell designs can also have several different processing options with regard to the layout and structure of the ferroelectric capacitor. The ferroelectric capacitor can be structured with an etch process that uses up to three separate photoresist masks and, as such, could exhibit variations in overlap among the top electrode, the ferroelectric, and/or the bottom electrode. In Fig. 9.1a, the triple-tiered capacitor is structured with an etch process that uses three photoresist masks, while in Fig. 9.1b the single-tiered capacitor is structured with an etch process that uses only one photoresist mask.

Both memory cell designs have the potential to be viable in the commercial market for stand-alone and embedded memories (where memory is included on-chip rather than off-chip, resulting in improved performance but also an increase in silicon area). For stand-alone memories, high packing density (obtainable with a small cell size) is essential, and thus, the memory cell with the capacitor directly on top of the transistor (Fig. 9.1b) is, by necessity of real state, the required architecture for high-density FeRAMs and ultradense DRAMs. However, embedded memories can tolerate lower packing densities (i.e., a larger cell size), if, for example, the ferroelectric memory can be easily adapted to the other on-chip CMOS processes (i.e., logic circuits, analog components, etc.). Thus, for embedded memories and low-density FeRAMs, the memory cell with the capacitor over the field oxide (Fig. 9.1a) is a viable choice even with the triple-tiered capacitor. Since it offers a less stringent integration approach, it is therefore easier to maintain acceptable electrical properties of the capacitor (i.e., low leakage, high breakdown fields, higher charge, and higher reliability) [9].

Because several processing options exist for the layout and structure of the two identified cell designs, numerous technical challenges arise from the corresponding material interactions. In the next sections, several of these challenges and their solutions are discussed.

9.3. Capacitor Formation

As the electrodes of ferroelectric thin-film capacitors significantly influence the electrical properties of the capacitor, many groups have studied them in detail recently. For ferroelectric thin-film memories, the choice of the electrode material is very dependent on the device architecture. To be chemically compatible with silicon and the ferroelectric processing, and to be used in an industrial process, electrodes should meet most of the following criteria. (The list is a compilation of desirable characteristics mentioned in the following referenced chapters and articles [10–18].)

1. Electrodes should have high electrical conductivity.
2. They should have good adhesion to both the ferroelectric and underlying structure (e.g., Si, SiO_2, and the plug material).
3. They should be morphologically stable under the process conditions.
4. They should not react with the ferroelectric.
5. They should have uniform thickness and structure.
6. They should have low contact resistance (e.g., to underlying plug).
7. They should be good diffusion barriers for oxygen or other species (e.g., Si and the ferroelectric).
8. They should have suitable interfacial electronic properties [10]. For instance, for DRAM capacitors with high-permittivity ferroelectric thin films, the electrodes should not form a low-permittivity film in series with the high-permittivity film, as this effectively reduces the overall permittivity of the DRAM capacitor.
9. They should be able to be deposited using production tools.
10. They should be able to be etched down to small feature sizes (e.g., $<1\ \mu m$, for 1–4 Gbit DRAMs that use 0.18-μm CMOS technology) [11].
11. They may need to provide a Schottky barrier to minimize leakage across the dielectric [12–13].
12. The bottom electrode may possibly serve as a "template" to control the ferroelectric film microstructure or orientation [14–17]. These references show how different bottom electrodes affect the nucleation and microstructure of the ferroelectric film with respect to grain size, orientation, or phases present [17].
13. The top electrode should have good step coverage, especially for DRAMs [11].
14. The top electrode may be used to prevent hydrogen diffusion. Using Pt and Pd might not be suitable. In a hydrogen ambient (i.e., forming gas anneal) these noble metals can act as catalyzers to convert hydrogen molecules into atomic hydrogen, which can attack or deteriorate the ferroelectric thin film. (See Section 9.5 for more details.)

A variety of metal and metal oxide electrodes have been investigated for FeRAM and DRAM applications, where the ferroelectric thin film is an oxide deposited (by one of the deposition methods described in Chapter 1) on the bottom electrode. Examples of commonly used ferroelectric oxides are $PbZr_xTi_{1-x}O_3$ or PZT, $SrBi_2Ta_2O_9$ or SBT, and related ferroelectrics for FeRAMs; and $Ba_xSr_{1-x}TiO_3$ or BST and related ferroelectrics for DRAMs. Because the ferroelectric oxides are deposited in an oxygen-containing atmosphere and at relatively high temperatures (usually about 550–800 °C), the electrodes must be resistant to oxidation (they may not form an insulating layer), and they must be able to withstand the processing conditions of the oxide. For this reason, researchers have focused their studies on oxidation-resistant noble metals, such as Pt and Pd; metals such as Ru and Ir and their oxides which have a high conductivity or form a conductive oxide such as RuO_2 [15, 19–29] or IrO_2 [30–33]; multicomponent conductive oxides such as $YBa_2Cu_3O_7$ [34–35], $SrRuO_3$ [36–38], and $La_{1-x}Sr_xCoO_3$ [39–42]; Pt/oxidic hybrid electrodes [17, 27–30, 43] such as Pt/RuO_2 [27–29], Pt/$La_{1-x}Sr_xCoO_3$ [43], and Pt/IrO_2 [30]; and alloys such as Ru-Pt [44–45]. The choice of electrodes for FeRAM and DRAM applications and their deposition technique depend on device performance, processing and processing-related control requirements, environmental issues, and cost.

Throughout the last decade, a number of studies have been conducted on electrode materials, and Pt has been, by far, the most commonly studied material. (For other electrode materials, the interested reader is encouraged to consult some of the previous mentioned references.) Techniques used for the deposition of Pt have mostly included sputtering and *e*-beam evaporation. Table 9.1 lists some of the deposition techniques and conditions for Pt found in the literature, as well as two types of adhesion layers used and their thicknesses.

In the FeRAM and DRAM cells shown in Fig. 9.1, poor adhesion prevents the deposition of Pt directly onto the interlevel dielectric or ILD (e.g., SiO_2). To improve adhesion, an adhesion layer of Ti, for example, is often deposited onto the ILD prior to Pt deposition. However, minimizing the interaction between the adhesion layer and the ILD interface creates additional challenges in processing. In the most studied electrode and adhesion layer combination (i.e., Pt/Ti), hillocks protrude under thermal cycling and have proven to cause the capacitor to short electrically [20, 21, 48]. The height of the hillocks varies but in the worst cases have been observed to be equal to or even greater than the ferroelectric film thickness [47]. Additionally, data from Spierings *et al.* [48] show that hillock

TABLE 9.1
TECHNIQUES AND CONDITIONS USED FOR Pt DEPOSITION[a]

Deposition method	Deposition temperature (°C)	Thickness of Pt (Å)	Adhesion layer thickness (Å)	Ref.
Ion beam sputtering	25–600	500–4000	Ti (200–600	[20, 21]
DC magnetron sputtering	25	1500	Ti, Ta (500)	[46]
E-beam evaporation	25–250	250–2000	Ti (50–2000)	[47]
Sputtering	25	170–1000	Ti (0–200)	[48]
RF sputtering	25	1400–2400	Ti (70–1000)	[49]
DC sputtering	25	2000	Ti (1000)	[50]
CVD	350	1500–2200	None	[51, 52]

[a] Adapted from H.N. Al-Shareef and A.I. Kingon [17] with permission from Gordon and Breach Publishers.

formation is due to Pt stress relief. According to researchers [17, 48], this can be decreased by changing the Pt deposition technique, deposition conditions, annealing atmosphere, and film thickness.

If not controlled, Pt and Ti can interdiffuse in inert atmospheres during ferroelectric deposition and/or crystallization temperatures [7, 47, 49, 50, 53, 54]. If the Ti diffuses to the Pt/ferroelectric interface, and the electrode is in an oxygen environment, then TiO_x may form [54]. The amount of TiO_x that forms depends on the relative thickness of Pt/Ti and the thermal deposition or annealing details [50]. This formation of TiO_2 can be advantageous for the formation of some ferroelectric thin films [48, 49, 55], and detrimental to others [56, 57].

In the case of PZT, the formation of TiO_x on the Pt surface can sometimes be advantageous. Hase *et al.* [55] reported that the formation of TiO_x on the surface and in the grain boundaries of the Pt was a nucleation site for the perovskite PZT phase. They demonstrated this effect with an experiment where PZT was deposited on Pt, with and without a Ti adhesion layer. With the Ti layer, the PZT had a high nucleation site density and a homogeneous (single-phase) microstructure. Without the Ti layer, the PZT had a low nucleation site and an inhomogeneous microstructure, which consisted of perovskite islands or rosettes surrounded by a pyrochlore matrix. Similar findings to those of Hase *et al.* have also been reported [48, 49].

In the case of SBT the formation of TiO_x on the top surface of a Pt electrode has been observed to be detrimental. In the capacitor stack Pt/SBT/Pt/Ti/SiO_2, Melnick *et al.* [56] have shown that the charge-voltage hysteresis loop of SBT is affected by the Ti adhesion thickness. As the Ti adhesion layer increased from 0 to 200 Å the value of the remanent polarization for SBT decreased from 13 to 4 $\mu C/cm^2$ while the coercive field changed slightly. This decrease appears to be due to Ti diffusing to the SBT/bottom electrode interface and may be best explained by Auciello *et al.* [57], who noticed that Bi does not get fully incorporated into the films when Ti remains at the SBT/bottom electrode interface. Using the mass spectroscopy of recoiled ions (MSRI) technique, they examined the initial stages of growth of SBT on different bottom electrode layers, and confirmed that heating of a Pt/Ti/SiO_2/Si substrate to 700 °C (prior to the SBT deposition) resulted in Ti and Si diffusing to the SBT/Pt interface. At different temperatures, Im *et al.* [58] found that when SBT was deposited on a Pt/Ti bottom electrode, Sr and Ta species were always readily incorporated in the growing film, but at certain temperatures the incorporation of Bi was negligible. More specifically, they saw at 700 °C (as shown in Fig. 9.2) that Bi was not fully incorporated into the film. This is possibly due to preferential binding of oxygen to Ti and Si species on the Pt surface, which scavenges the oxygen needed to fix the Bi to the Pt surface and results in Bi evaporation due to its high vapor pressure. These results are very significant because some researchers are depositing SBT in oxygen environments at 700–800 °C. Nevertheless, there are a few ways to avoid the formation of extraneous TiO_x.

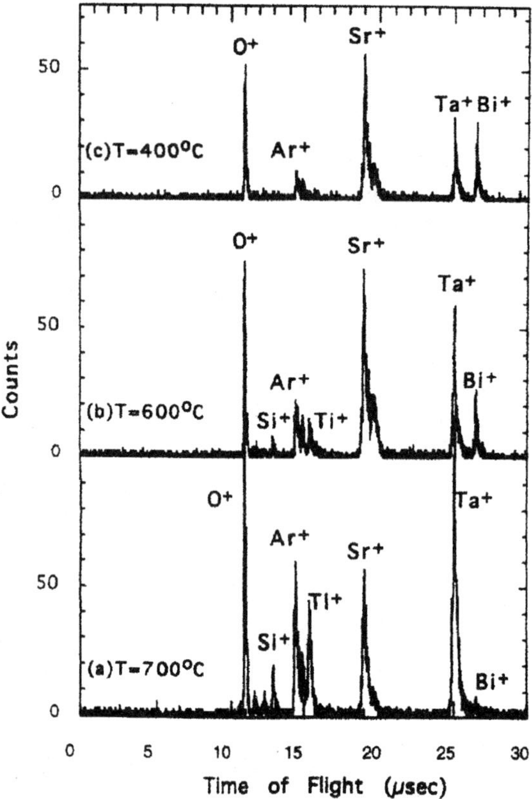

FIG. 9.2. Mass spectroscopy of recoiled ions (MSRI) spectra of SBT film on $Pt/Ti/SiO_2/Si$ substrate under $PO_2 = 5 \times 10^{-4}$ torr at (a) 700 °C, (b) 600 °C, and (C) 400 °C (from Im *et al.* [58], © *Applied Physics Letters*, with permission).

1. Deposit SBT at lower temperatures (<700 °C) [58].
2. Incorporate excess Bi into films during growth. Researchers sometimes add up to 10% extra Bi when processing SBT films because the diffusion of Bi through the bottom electrode during the firing of the capacitor stack at 800 °C [59, 60] has been found to affect some of the electrical properties of SBT) [61].
3. Use an alternative diffusion barrier or adhesion layer, for example, stoichiometric TiN, TaN, TiAlN, TaSiN, or *in situ* formed M-O-Si (where M is Ir or Pt) barrier layers by annealing M/Si or M/WSi_x samples in an oxygen-containing ambient [62–64]. Although several researchers mention using TiN as a diffusion barrier, recent work by Auciello *et al.* [65] using the MSRI technique shows TiN not to be an effective barrier, as it starts to

oxidize when exposed to oxygen at 200–300 °C. As an alternative barrier, Auciello *et al.* suggest TiAlN as it does not show any strong evidence of oxidation below 600 °C.

In addition to the integration issues already discussed here, the diffusion of Ti to the surface can cause adhesion problems. Bruchaus *et al.* [49] observed that the loss of Ti could degrade the adhesion of Pt to SiO_2. However, by increasing the thickness of Ti to about 1000 Å (when the thickness of Pt is about 1400 Å), Bruchaus *et al.* showed that the adhesion quality improved. It is interesting to note that studies by Tisone and Drobeck [53] show that Ti can diffuse to the top of the Pt surface along its grain boundaries.

In addition to the effect the constituents of a ferroelectric capacitor (the electrodes and the ferroelectric thin film) have on material interactions, the constituents also significantly affect the electrical behavior of the capacitor. These electrical properties include the value of the remanent polarization, the saturation polarization, the coercive field, the fatigue behavior, and the leakage current. In PZT-based thin films, for example, several groups [15, 22, 24, 26] have shown that various electrical properties are affected when Pt or RuO_x $(0 \leq x \leq 2)$ electrodes are used. As shown in Fig. 9.3 [26], when a Ru top and RuO_2 bottom electrode (hereafter abbreviated to Ru/RuO_2) are used instead of a Pt top and Pt bottom electrode (hereafter abbreviated to Pt/Pt) the value of the remanent polarization is significantly reduced and its average coercive voltage is somewhat larger. Taylor *et al.* [26] suggest that this could be due to a better crystallization or a more favorable orientation of the PZT grown on Pt as compared to RuO_2. Furthermore, the asymmetrical charge-voltage hysteresis loop of the PZT with Ru/RuO_2 electrodes is indicative of a different top and bottom electrode.

Researchers have also studied the effect that electrodes have on the endurance of PZT. They have seen that PZT with RuO_2 [15, 21, 22, 24, 26, 66] and other oxidic [34, 41, 66] electrodes (as compared to Pt electrodes) exhibit little or no fatigue. Some models have been suggested for the cause of fatigue of PZT thin-film capacitors with Pt electrodes. They are based on concepts of space charge-motions and domain pinning [67–69] and describe fatigue as being related to the formation of dendrite oxygen deficient filaments, that is, to be a bulk phenomenon. However, these studies have little physical evidence to support them [70]. Another model explains fatigue as being caused by defect entrapment at the ferroelectric-electrode interface [71]. This model gives a reasonable explanation of why the electrical endurance of PZT is longer with oxidic electrodes as compared to Pt electrodes. Assuming that the defects are oxygen vacancies, these can be eliminated at the interface between the PZT and the oxidic electrodes but not at the PZT-Pt interface.

Research and industrial groups are now supplementing electrical studies with analytical studies to better understand the physical effects of the capacitors.

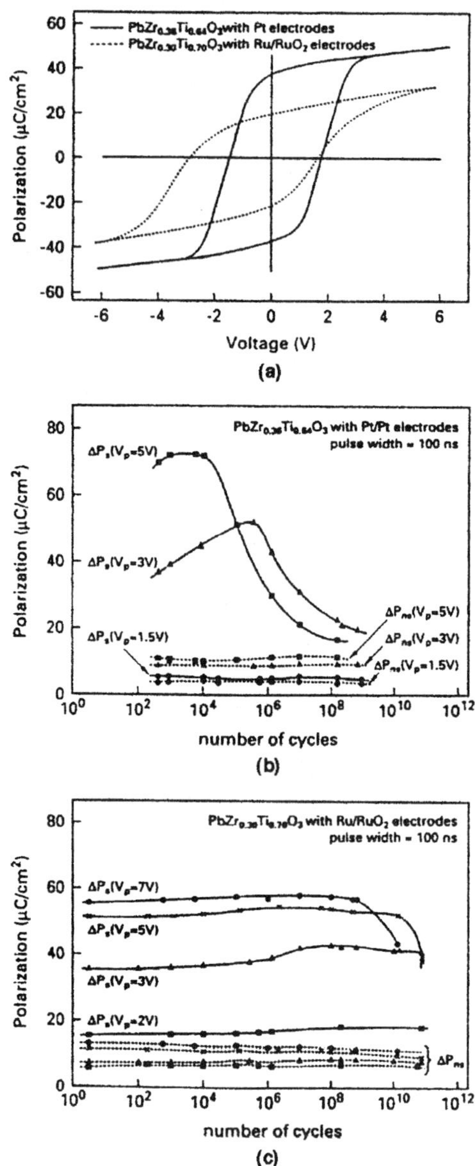

FIG. 9.3. (a) Charge-voltage hysteresis curves of PZT with Pt/Pt and Ru/RuO_2 electrodes measured using a sinusoidal voltage (1 kHz) on a capacitor of area 50,000 μm^2 with a 220-nF reference capacitor; (b) fatigue behavior of PZT with Pt/Pt electrodes at $V_p = 1.5$, 3.0 and 5.0 V; (c) fatigue behavior of PZT with Ru/RuO_2 electrodes at $V_p = 1.5$, 3.0, 5.0, 7.0 V (from Taylor *et al.* [26], © *Thin Solid Films*, with permission).

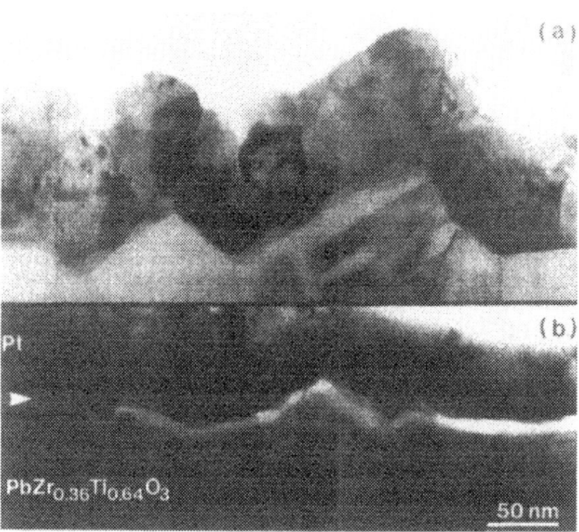

FIG. 9.4. XTEM images of PZT with Pt top electrode interface of (a) fresh and (b) fatigued Pt/PZT/Pt/Ti capacitors (from Taylor *et al.* [72], © *Integrated Ferroelectrics*, with permission).

Taylor *et al.* [26, 72], supplemented the understanding of any physical effects that may occur during fatigue in Pt/PZT/Pt [72] or Ru/PZT/RuO_2 [26] capacitors with analytical analysis (i.e., transmission electron microscopy or TEM, scanning electron microscopy or SEM and scanning auger microprobe or SAM) on fresh and fatigued capacitors. For fresh and fatigued Pt/PZT/Pt capacitors, representative cross-sectional TEMs (XTEMs) images were made (Fig. 9.4). From these images they observed a thin interface (<1 nm) between the PZT and Pt of the fresh capacitor and a noticeable interface layer for the fatigued capacitor. Using high magnification on the fatigued capacitor (Fig. 9.5), they found a 5–10-nm-thick amorphous layer at the interface of the PZT with the top and bottom electrode. The interface layers, they noted, demonstrated that the changes in the PZT/Pt interfaces are related to fatigue, even though amorphization during TEM specimen preparation (i.e., ion-milling) could not be completely excluded.

For fresh and fatigued Ru/PZT/RuO_2 and Pt/PZT/Pt capacitors, Taylor *et al.* [26, 72] performed SEM and SAM studies. These studies found Ru to migrate from the top electrode into the PZT and form channel-like structures locally in the PZT, causing fatigue. Yet for Pt, no electromigration of Pt into PZT was observed. They ascribed the two different fatigue behaviors of PZT with Pt- and Ru-based electrodes to different causes. More specifically, they attributed fatigue in Pt/PZT/Pt [72] capacitors to interface effects and the fatigue observed in Ru/PZT/RuO_2 capacitors to an interface-generated bulk effect [26].

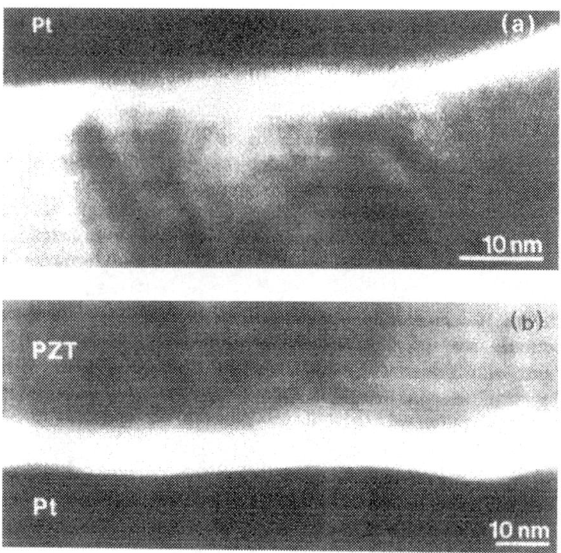

FIG. 9.5. Higher magnification images of the interfacial layers between the fatigued PZT and the (a) Pt top electrode (b) Pt bottom electrode of a Pt/PZT/Pt/Ti capacitor (from Taylor *et al.* [72], © *Integrated Ferroelectrics*, with permission).

It should be noted that while RuO_2 and other oxidic electrodes (e.g., IrO_2) improve the endurance of PZT, capacitor stacks with RuO_2 electrodes have higher capacitor leakage [17, 26, 71]. Therefore it is not entirely clear which electrode material is optimal for the capacitors. As a possible solution, researchers are now studying hybrid electrodes made up of Pt and a conductive oxide electrode (e.g., RuO_2 or IrO_2) [27–30]. Such hybrids take advantage of the favorable characteristics of Pt (i.e., lower capacitor leakage) and the conductive oxide electrode (i.e., improved endurance).

Researchers have also investigated SBT-based (e.g., SBT doped with Nb) films in addition to PZT-based films for nonvolatile memory applications. In contrast to the PZT-based films that require oxidic-based electrodes for improved endurance, SBT shows little fatigue with Pt electrodes [73–74]. The SBT-based thin films (like PZT-based thin films) also show other suitable electrical properties for ferroelectric memories (for instance, acceptable polarization retention and leakage levels for memories, little imprint, etc.). It is, therefore, not surprising that both PZT and SBT-based thin films are currently being manufactured for low-density FeRAMs [75–81]. To illustrate, low-density devices such as the Matsushita Electronic Corporation or MEC 64-kbit-embedded memory microcontroller [75–77] and the Panasonic-Symetrix Corporation 64-kbit-embedded memory microcontroller [80] are now commercially available using SBT-based thin films. And Rohm Co., meanwhile, has produced a 1-kbit-embedded memory microcontroller using PZT- based thin films [79].

For FeRAMs and DRAMs that use the stacked contact architecture shown in Fig. 9.1b, the memory cell area can be significantly reduced. However, additional integration or process-related challenges accompany this improvement. These challenges occur because the bottom electrode is deposited on a conductive plug structure where the electrode, diffusion barrier/adhesive layer, and plug must remain conductive even after they are exposed to the high processing temperatures of the ferroelectric oxide. Consequently, the electrode/barrier/plug cannot form an insulating oxide as in the case of Pt/Ti. As a result, other electrode/barrier combinations (e.g., Pt/TiN [82], Pt/TaN [82], Pt/TiAlN [65], Pt/IrO_2 [30], and RuO_2/TiN [6, 83]) have been investigated, and Pt/TiAlN and the metal oxide electrode combinations show the most promise. According to Sameshima *et al.* [82] and others, Pt/TiN and Pt/TaN oxidize, while according to T. Nakamura *et al.* [30–31], P.-Y. Lesaicherre *et al.* [6] and O. Auciello *et al.* [65], respectively, IrO_2, RuO_2, and TiAlN provide effective oxygen barriers.

It is important to note that since TiN has been an industrial standard as a diffusion barrier between metals and Si, researchers have investigated alternative bottom electrode/barrier/plug structures using it [4, 5, 11, 84]. One such structure includes Pt/TiN on a poly-Si plug with SiO_2 on the sidewalls of the Pt/TiN bottom electrode to prevent direct oxidation of the TiN [4, 5]. According to researchers, the TiN in this structure did not oxidize. As Summerfelt [11] mentions, however, this structure may not be suitable for 0.18-μm- (or 0.13-μm-) DRAM generations, as the lack of capacitance on the sidewalls makes it difficult to achieve the desired cell capacitance. Another structure investigated is a recessed plug where the TiN is part of the poly-Si plug and the Pt electrode is deposited on top of the recessed plug [84]. At first this appears to be a viable solution because the Pt protects the TiN from oxidizing and the Pt sidewalls could contribute to the cell capacitance. However, as Summerfelt suggests, the diffusion of oxygen through the bottom electrode (i.e., Pt) might still cause a problem because it could create an increase in resistivity of the TiN layer and may also form TiO_2 [85, 86].

Many combinations of patterning the capacitor stack are possible. Some groups, for example, pattern the bottom electrode before depositing the ferroelectric and the top electrode, and others deposit the entire stack and pattern the stack in several different ways. For instance, they may pattern the whole stack, or the top electrode and ferroelectric together and the bottom electrode later, or each layer separately.

Among the advantages of first depositing blanket films of the capacitor stack (i.e., the bottom electrode, ferroelectric, and top electrode) and then patterning the top electrode with the ferroelectric and the bottom electrode are the following:

1. good conformality between the ferroelectric and bottom electrode because the bottom electrode is unpatterned when the ferroelectric is deposited;
2. no adverse effects of crystallization of the ferroelectric thin film due to etch damage (e.g., surface roughness) as described by Menk *et al.* [87], as the bottom electrode is unpatterned when the ferroelectric is deposited; and
3. no chemical residue is left behind on the bottom electrode because the bottom electrode is unpatterned when the ferroelectric is deposited; residues on the bottom electrode have been shown to effect the electrical properties of a patterned capacitor negatively [87].

However, two possible disadvantages of this type of processing are the following:

1. ferroelectric etch damage, caused by exposing the ferroelectric thin film to the bottom electrode etch process; and
2. electrical shorting of the capacitor, caused by the redeposition of etched materials along the capacitor sidewalls during plasma etch. (The next section discusses this issue further.)

Finally, from the preceding discussion it may seem logical to first pattern the bottom electrode. However, following this approach may have the following disadvantages:

1. poor conformality between the ferroelectric and the bottom electrode because, for example, a patterned bottom electrode could introduce surface roughness from surface damage; and
2. an additional compound buffer layer (e.g., TiO_2, TiN, ZrO_2 or ZrN) may be necessary between PZT and SiO_2 or Si_3N_4 in order to prevent these materials from interacting [88].

9.4. Electrode and Capacitor Patterning

The patterning of electrodes and ferroelectric layers is a significant integration issue that continually needs to be addressed as the dimensions of feature sizes shrink [11, 13, 87, 89]. In most of the literature, the issue of patterning electrodes is discussed extensively. Equally important, however, is the patterning of both the ferroelectric layer and electrodes. For ultralarge scale integration (ULSI) applications, the etch must have sufficient etch profile and critical dimension control, selectivity with respect to the underlying layers and the photoresist, a residue-free etch process (i.e., residue can be eliminated after the photoresist strip process), and a high etch rate. Additionally, it must not damage the underlying circuit components [87, 89].

In the last decade, significant progress with lithographically defined photoresist and etching has been reported for both Pt- and Ru-based electrodes. The etching of these electrodes is to some extent similar to the etching of other noble metal and metal-oxide electrodes. In comparing the two electrodes, Ru-based electrodes are easier to pattern because several volatile compounds such as RuO_{4-x} and Ru + O + F compounds (i.e., RuF_5 and $Ru(CO)_5$) can be generated [11, 13]. The relatively easy patterning of Ru led some research groups in 1994 and 1995 to successfully achieve 0.18-μm features with tall storage nodes, of approximately 0.2 μm [6, 90–92]. Several disadvantages of etching Ru are:

1. the by-product RuO_{4-x} is particularly dangerous because it is a highly toxic and flammable gas (i.e., RuO_{4-x} has a melting temperature of 25.4 °C and decomposes at 108 °C) [11, 13];
2. Ru-based electrodes, as compared to Pt or Ir, potentially have a greater likelihood of wafer cross-contamination, according to Summerfelt [11], as RuO_{4-x} is volatile and can readily decompose on contact;
3. Ru-based electrodes are susceptible to pinholes and notching, according to DeOrnellas *et al.* [89], because of their ease of oxidation; and
4. Ru-based bottom electrodes oxidize to RuO_2, which has a higher resistivity than pure Ru or Pt and leads to a larger time constant, thereby affecting device speed.

For the etching of Pt-based electrodes, research and industrial groups generally consider physical-only sputtering (e.g., ion milling) and chemically assisted physical or plasma etching (e.g., reactive ion etching or RIE [8, 13, 93–97]) as suitable etch technologies for studying small geometries (i.e., <0.5 μm). Some groups use wet (or chemical-only) etches [98–101] as a means to study isolated structures so as to reduce development time and cost. However, this type of etch is unsuitable for features much smaller than 0.5 μm as a result of undercutting due to its isotropic nature. By contrast, RIE is the most promising etch technique for manufacturers because it combines the advantages of sputtering by energetic ions (i.e., anisotropy) and etching through the chemical reactivity of species produced in the plasma (i.e., compared to other etches, RIE has relatively high etch rate selectivities) [8, 102]. For recent summaries on state-of-the-art dry etching techniques for ferroelectric capacitors, the author recommends the following articles: Achard *et al.* [8] (for PZT-based capacitors); Menk *et al.* [87] (for SBT and PZT-based capacitors); Chung *et al.* [103] (for PZT-based capacitors); Farrell *et al.* [104] (for BST-based capacitors); and DeOrnellas *et al.* [89] (for SBT-, PZT-, and BST-based capacitors).

A major disadvantage of etching ferroelectric capacitor stacks with Pt electrodes common to both ion milling and RIE (which both have a physical-etch component) is the redeposition of etched material, which forms "fences" or "veils" that remain on the top electrode and capacitor sidewalls. Several groups

[13, 60, 104–105] have observed these fences, which are shown in Fig. 9.6 for Pt/SBT/Pt capacitors. In some instances, the fences could cause the following:

1. cracking or poor coverage of subsequent processing (e.g., with the overlying dielectric and metal layers); and
2. electrical shorting if, for example, the redeposited capacitor materials are conducting on the sidewalls of the capacitor.

Several etch techniques have been developed to reduce or eliminate fences. Two of these are discussed in what follows for ion-milling and RIE. Note that with both of these etch techniques a photoresist mask layer is used. Recently, though, several groups have reported success in producing "fence-free" capacitors using a very thin oxide hardmask layer, which is patterned by a photoresist mask. The SEM micrographs of fence-free capacitors etched with hardmask layers are impressive; however, many details of this etch are proprietary.

To reduce fences in the case of ion milling, Taylor *et al.* [60] utilized a high incident angle (Fig. 9.6c) wherein the wafer was physically tilted normal at an angle with respect to the ion mill to keep the sidewall higher than the rate of redeposition. Due to the etch angle and the wafer's rotation, however, a tapered sidewall developed. This sidewall, they noted, did not electrically affect the capacitor properties but it could reduce the packing density of high-density FeRAMs or ultradense DRAMs (e.g., for capacitors with one dimension $<0.3\ \mu m$).

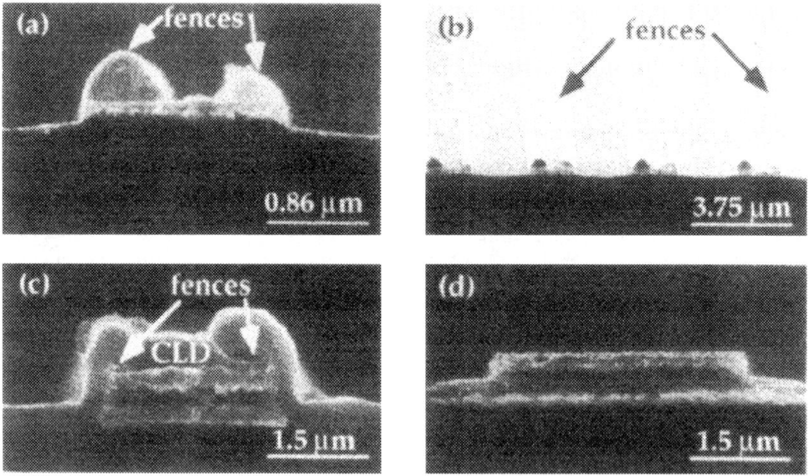

FIG. 9.6. SEM micrograph of Pt/SBT/Pt/TiO_2 capacitor (a) with "fences" (cross section), (b) with "fences" (tilt view), (c) with "fences" and capacitor level dielectric (cross section), and (d) without "fences" (cross section) (from Taylor *et al.* [61], © Materials Research Society, with permission).

In the case of RIE, Kotecki [13] eliminated fences by etching the fence layer and eroding the side of the photoresist during the etch process. While the fences were etched "clean," the capacitors developed sloped sidewall angles [104], which resulted in loss of control of the critical feature sizes. According to Kotecki, only limited tapering of the sidewall angle can be tolerated as the dimensions of the features shrink to 0.18 μm (i.e., about the expected feature size for DRAMs with 1–4 Gbit densities). In the case of low-density FeRAMs, though, a sloped sidewall angle (i.e., profiles of 60° or greater) can be tolerated because, as DeOrnellas *et al.* [89] describes, feature sizes are typically on the order of about 2.0–3.0 μm for 64–256 kbit devices. Important to note, several groups have reported that RIE processing can cause physical damage to the films. Some researchers have seen damage at the near surface region that has caused changes in the electrical properties of the capacitors (e.g., Pt/PZT/Pt capacitors) [87]. When the capacitors were annealed in oxygen-containing atmospheres, however, the electrical properties were recovered [97, 106]. Other researchers have noted that they were able to recover this damage on large capacitors but not on submicron capacitors [107].

In addition to fences, another critical issue vis-a-vis etching capacitor stacks that has plagued some groups is the delamination of the electrodes from the barrier/adhesive layer and/or the delamination of the ferroelectric from the electrodes. The exact cause of the delamination can be difficult to isolate as it can be related to a host of material interactions (e.g., stress or diffusion) that arise from the processing of the capacitor stack (e.g., temperature budget). When the cause has been considered to be thermally related, groups have worked toward reducing their thermal budget. The European ESPRIT project called [Ferroelectric Layers for Memory Applications and Sensors (FELMAS)] [8] was able to limit delamination by using low substrate temperatures that minimized thermal expansion effects. A review of recent literature shows that other groups [99, 108] have used rapid thermal processing anneals instead of furnace anneals to reduce the thermal budget.

In a manufacturing environment, plasma etching is the technique that most groups agree shows the most promise. Recently, Tegal Corporation [89] pursued research into plasma etching and reported significant progress because they demonstrated 0.2-μm-Pt "post" electrode patterns after the photoresist strip process with minimal residue and large profile angles (Fig. 9.7). According to DeOrnellas *et al.*, the main challenges of plasma etching for ferroelectric devices are profile control, elimination of residue after resist strip, and critical dimension control (especially for the smaller features sizes required for larger memories). DeOrnellas *et al.* also mention that one of the upcoming strategic goals for Tegal Corporation is to produce a clean anisotropic dense area profile and an open area profile with critical dimension of <0.05 μm so as to make plasma etching a suitable manufacturing process for sub-0.5-μm FeRAMs and DRAMs. Other tool

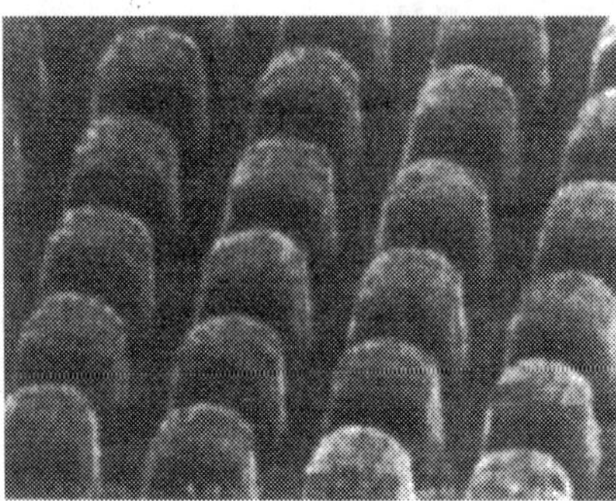

FIG. 9.7. Plasma etch of a 200-nm Pt "post" electrode pattern after photoresist strip. Note, the post shows minimal residue even with large profiles (from DeOrnellas *et al.* [89], © Integrated Ferroelectrics, with permission).

companies are also involved in developing plasma etch tools for ferroelectric thin films.

As a final note on the subject of electrode and capacitor patterning, we mention that several groups have recently started to investigate alternatives to conventional photolithography and etching methods for high-density ferroelectric memories in ULSI applications. One of these methods, pioneered by Uchida *et al.* [109], includes preparing metalorganic decomposition (MOD) photosensitive solutions (i.e., adding a photoresist to the ferroelectric precursors) and etching the capacitor stacks optically. Another alternative, developed by Alexe *et al.* [110], is the self-assembly of bismuth-containing nanoscale electrodes (i.e., Bi_2O_3) with layered perovskite ferroelectic thin films. More specifically these electrodes self-form or assemble during the pulsed-laser deposition (PLD) growth of Bi-rich $Bi_4Ti_3O_{12}$ on epitaxial conductive $La_{0.5}Sr_{0.5}CoO_3$(LSCO) when the LSCO is held at a high temperature. That is, self-organized arrays of epitaxial mesas (i.e., conducting δ-Bi_2O_3 electrodes) form when excess Bi segregates from the ferroelectric thin film and migrates to the surface. As a result, when the capacitor stack δ-$Bi_2O_3/Bi_4Ti_3O_{12}$/LSCO was electrically tested (i.e., Hg dot contacts were deposited on top of the δ-Bi_2O_3 electrode) it showed ferroelectric behavior. Although self assembly is still in the early development stage and many questions arise concerning whether it could be a controllable or manufacturable process, several advantages of this intriguing technology are emerging. As described by Scott *et al.* in a recent review article [111], this process could make possible the scaling of ferroelectric memories to 1-Gbit densities through the reduction of cell size (Alexe *et al.* [110] report cells that are 200 nm in size), and eliminate submicron photolithography for ferroelectric thin films containing bismuth. For more information on alternatives to conventional lithography, the author recommends consulting the preceding references.

9.5. Hydrogen-Containing Ambient

Several groups have observed a noticeable deterioration of electrical properties and even compositional or chemical changes when ferroelectric capacitors were exposed to hydrogen-containing ambients [60, 97, 112–115]. Typically, hydrogen may be present during the following plasma deposition and etch back-end processes:

1. deposition of the capacitor level dielectric that is used to encapsulate the completed capacitors prior to metallization;
2. etching of contact holes toward both the electrodes of the capacitors and the source/drain of the transistors;
3. deposition of intermetal dielectrics;
4. deposition and etch in the final dielectric passivation layer; and
5. post metallization or "forming gas" anneal (e.g., typically an anneal at 450 °C for 30 min in N_2/H_2 ambient), which is used to recover process-induced gate oxide damage by saturating Si dangling bonds at the gate oxide interface and reducing contact interfaces.

For ferroelectric thin-film capacitors where sensitivity to hydrogen is exhibited, the following solutions have been suggested.

1. Annealing in oxygen—several groups [60, 97, 113] have seen that the recovery of hydrogen damage can be achieved by an oxygen anneal. For example, Fig. 9.8 shows the degradation of a Pt/SBT/Pt capacitor following the deposition of a capacitor level dielectric (i.e., chemical vapor deposited or CVD SiO_2) and the RIE etch of contact holes, which is then recovered with a subsequent oxygen anneal [60].
2. Utilization of an IrO_2 top electrode—Fujisaki *et al.* [116] found that hydrogen degradation is significantly reduced if IrO_2 (instead of Pt or Pd) is used as a top electrode for PZT. More specifically, they found IrO_2/PZT/Pt capacitors to be stable against hydrogen annealing up to 400 °C [116]. Furthermore, they theorized that the degradation of Pt/PZT/Pt capacitors annealed in hydrogen is a result of the Pt top electrode acting as a catalyzer, which in the hydrogen-containing atmosphere causes hydrogen molecules to be adsorbed to the catalyzer's surface (i.e., Pt). The molecular hydrogen then decomposes into atomic hydrogen, which as an active radical can attack or deteriorate the PZT even at very low temperatures (i.e., 300 °C or less). However, Fujisaki *et al.* found that the bottom Pt electrode does not act like a catalyzer when PtO covers the grains. For more details on their model, see the forementioned reference. Note, however, that according to Auciello *et al.* [65], the catalysis of hydrogen by Pt does not explain the degradation mechanism. For instance,

for SBT using the MSRI technique, Im *et al.* [117] found that degradation after hydrogen annealing is due mainly to the degradation of the near surface region of the SBT layer, caused by the depletion of Bi in this region as a result of hydrogen annealing. Furthermore, from preliminary data Auciello *et al.* [65] have observed with cross-sectional TEM that the depletion of Bi in SBT extends in from the surface, $\sim$ 50–100 Å. Note that Auciello *et al.* plan to confirm these results with other cross-sectional TEM and profile methods. Yet for PZT (unlike SBT), Aggarwal *et al.* [118] (using Raman spectroscopy) and Auciello *et al.* [65] (using the MSRI technique) show that there is no depletion of Pb in the near surface region as a result of hydrogen annealing. Rather, Raman spectroscopy indicates that the remanent polarization is reduced during hydrogen annealing because the Pb ion gets into the PZT lattice in between the Ti^{4+} or Zr^{4+} and O^{2-}, and inhibits ion displacement, and thus polarization.

3. Development of etch processes that do not contain hydrogen have been suggested [9].
4. Utilization of certain deposition techniques such as sputtering spin-on-glass (SOG) [87] or special plasma-enhanced chemical vapor deposition (PECVD) [114, 119], believed to reduce the amount of process damage by changing process parameters (e.g., hydrogen content and chemistry, the level of stress and preparation temperature of the deposited layer, the purity of the materials, etc. [8]) have been suggested.
5. Changing the conditions of the forming gas anneal—some have suggested this may be acceptable for stand-alone memories. However, for embedded memories this change could be very challenging because in these applications, transistor properties usually are more tailored toward analog devices [9].

Finally, it should be noted that the industry has also developed proprietary solutions, evidenced by the fact that low-density FeRAMs are now commercially available but the complete details of their processing steps have not been disclosed.

9.6. Impact of the Ferroelectric Processing on Silicon Devices

In some semiconductor fabrication facilities, groups are depositing and structuring ferroelectric capacitors on top of silicon devices (e.g., transistors, diodes, etc). In so doing they have introduced new or “novel” elements into the semiconductor process that have the potential to cause degradation of the silicon devices. For example, if BST and its electrodes are introduced, then its ions (Ba, Sr) and the

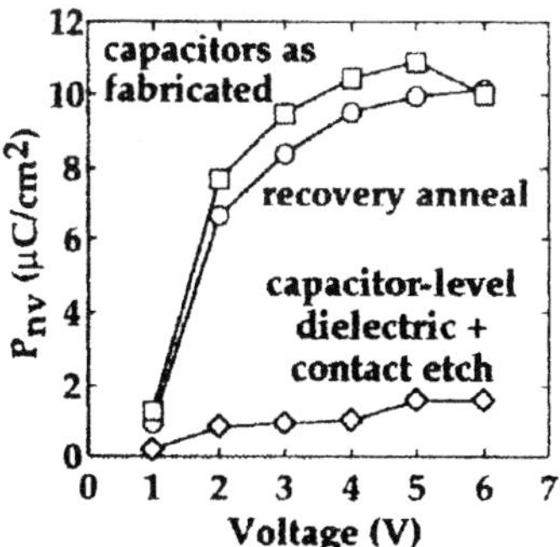

FIG. 9.8. The nonvolatile polarization (P_{nv}) as a function of voltage for Pt/SBT/Pt/TiO_2 capacitor after fabrication of the capacitor, after deposition of a capacitor level dielectric and contact etch, and after a subsequent recovery anneal in oxygen (from Taylor *et al.* [61], © Materials Research Society, with permission).

electrode material (e.g., Pt, Ru or Ir) have the potential to degrade some of the critical device properties of the silicon devices. Therefore, with the introduction of any novel elements to a fabrication facility, it is essential to identify and access the impact of short and long-term damage to silicon while taking cautionary measures to limit or prevent cross-contamination of the wafers, which could induce pollution of subsequent process equipment and wafers. Achard *et al.* [8], for example, with secondary ion mass spectroscopy (SIMS) detected a significant amount of lead contamination (i.e., 7 atomic % on the Si surface at 20 ppm at 60-nm depth) on the backside of a wafer after PZT sol-gel deposition and annealing at 700 °C. Similarly, other groups have detected other elements on the backside of wafers after BST or SBT deposition.

To reduce damage to Si transistors and prevent cross contamination, Summerfelt [11] suggests the following steps when introducing novel elements to a fabrication facility.

1. Identify and in most cases quantify the degradation behavior vs the amount of added impurity.
2. Develop wet or dry cleaning processes ("cleans") before critical processes, in order to remove impurities.
3. Determine which processes can transfer impurities, and make a quantitative analysis.

It is well known that Pt can harm transistors and for this reason, some groups have used barrier layers (e.g., TiO_2 [120]) and proprietary solutions to protect their transistors and ensure their reliability. One technique that different groups have used to study some of the short-term effects of the novel elements on the Si device is to prepare Si devices both with and without ferroelectric capacitors and then compare their electrical properties. Notably, recent studies in the literature show little or no difference in many of the CMOS transistor or diode characteristics when these devices have been processed alone or with ferroelectric capacitors [9, 97, 113]. Furthermore, it should be noted that these studies have focused on the short-term effects, and the literature needs more fully to address some of the long-term effects (e.g., the reliability of the ferroelectric/CMOS devices).

Finally, as low levels of contaminants can easily degrade transistor gate properties, the need to isolate the CMOS front-end from the ferroelectric back-end is critical. For this reason, fabrication facilities that process ferroelectric memories have adopted strict guidelines and dedicated specific tools to the processing of novel materials. Regrettably, published literature related to this topic is scant. It is hoped that more will be accessible in the near future.

9.7. Equipment Issues

Until recently, one of the largest hurdles in moving R&D efforts into manufacturing has been development of suitable processing and associated tools for film deposition, etch, etc. Recently, a number of equipment suppliers have actively started to develop processes and tools in collaboration with industrial groups. Together these groups are often under contract to keep information proprietary as this information can give their industry an edge on their competitors. However, some articles [11, 13, 89, 121–122] have recently been published that, to some extent, address equipment issues for large-scale commercialization of ferroelectric memories, especially ultra-dense (or high-permittivity) DRAMs.

In an article [13] on DRAMs in *Semiconductor International*, Kotecki summarizes the additional equipment needed to introduce BST (for the 1-Gbit-DRAM generation and beyond) into a fabrication facility that is already equipped with standard DRAM tools, running wafers with 300-mm diameter. More specifically, he suggests the following equipment be included:

1. Electrode sputtering deposition system;
2. Metalorganic chemical vapor deposition (MOCVD) reactor for deposition of the high-dielectric layer; and
3. RIE system to pattern the electrode.

With regard to manufacturers of these tools, Kotecki points out that there are now numerous companies that manufacture electrode sputtering deposition systems. Several companies are developing MOCVD reactor capability (e.g., Varian Associates [121] and Aixtron AG [122] have developed MOCVD reactors capable of depositing BST [123] on 200- and 300-mm wafers, respectively) and several companies (e.g., Tegal Corporation [89]) have demonstrated etch processes for Pt.

In order to make BST a commercial success, several groups mention that additional work may still be needed with high-dielectric deposition equipment, electrode etch equipment, and even shared critical semiconductor processing equipment (e.g., photolithographic steppers). If the procedures and knowledge base can be established on how ferroelectric and traditional semiconductor materials can share equipment without cross contamination, significant cost savings could occur, as semiconductor fabrication facility might need only a single stepper and this could thereby save on both equipment costs and floor space.

9.8. Summary and Outlook

This review presents highlights of the latest research on the integration of ferroelectric thin films with silicon devices for nonvolatile and volatile memory applications, based upon representative work chosen from the literature in this field.

From a historical perspective we see that significant progress has been made over the last 10 years by R&D and industrial groups in addressing important integration and process-related issues. The integration of low-density FeRAMs has been demonstrated, and small-scale production has been possible since 1991, when Ramtron Corporation made commercially available a nonvolatile 4096-bit FeRAM (FRAM®). In the same year, National Semiconductor Corporation demonstrated that ferroelectric thin films could be integrated on CMOS to make a 1-transistor-1-capacitor DRAM memory cell. Significant accomplishments have been ongoing for stand-alone and embedded ferroelectric memory applications (used, e.g., in smartcards and RF ID tags).

Concrete illustrations of some low-density and embedded FeRAMs that are commercially available are: the Ramtron Corporation 4-kbit, 16-kbit-, and 64-kbit-PZT-based RAM (FRAM®) [124]; the Matsushita Electronic Corporation (or MEC) 64-kbit-embedded SBT-based memory microcontroller [75–77, 125]; the Panasonic-Symetrix Corporation 64-kbit-embedded SBT-based memory on an 8-bit RISC processor [80]; and the Panasonic-Symetrix Corporation 1-kbit-(13.56 MHz) SBT-based RAM [126] which is the major component of the "loyalty card" [127]. This card is marketed in Asia and consumers use it to store points that they collect with major Japanese airlines and major gasoline and oil companies, for example, the Shell Company. For these cards, over one million memories per month have been produced since July 1998 [127]. Some of the other groups that have produced prototypes of low-density FeRAM are: NEC

Corporation, which has developed a 256-byte-embedded memory microcontroller [78]; Rohm Corporation, which has produced a 1-kbit-embedded PZT-based memory microcontroller [79]; and Motorola, Inc. developer of an 8-kbit-embedded memory microcontroller using their HC11E9 microcontroller [81]. (Note that for some of the examples of ferroelectric memories mentioned here and in what follows, some companies consider the type of ferroelectric material that they are using confidential, hence for these products or prototypes this information is not included.)

As integration challenges are addressed for low-density FeRAMs, groups are racing to achieve higher density FeRAMs and ultradense DRAMs (with high-permittivity materials). For example, with regard to FeRAMs: Ramtron Corporation.-Fujitsu, Ltd. has developed a 1-Mbit-PZT-based RAM (FRAM®) [128]; MEC has a 256-Kbit-RAM prototype [129]; and Hyundai Electronics Industry Corporation. Symetrix Corporation has a 256-kbit-SBT RAM prototype [130]. Then with regard to high-permittivity DRAM, several groups have developed 64-kbit or greater BST DRAM prototypes.

A much better understanding of the physics of ferroelectric thin films has accompanied this notable progress. However, many questions remain for research and industrial groups to address as they continue to study ferroelectric memories now and into the twenty-first century.

Acknowledgment

It is a pleasure to acknowledge O. Auciello, C.A. Paz de Araujo, D. Williams, B. Luderman, and N. Bartush for their many valuable discussions and contributions.

References

1. Auciello, O., Scott, J.F., and Ramesh, R. (1998). *Physics Today* **51**(7): 22.
2. Cuppens, R., Larsen, P.K., and Spierings, G.A.C.M. (1992). *Microelectron. Eng.* **19**(1–4): 245.
3. Sumi, T., Moriwaki, N., Nakane, G., Nakakuma, T., Judai, Y., Uemoto, Y., Nagano, Y., Hayashi, S., Azuma, M., Fujii, E., Katasu, S., Otsuki, T., McMillan, L., Paz de Araujo, C., and Kano, G. (1994). *ISSCC Technol. Dig.*, 268.
4. Eimori, T., Ohno, Y., Kimura, H., Matsufusa, J., Kishimura, S., Yoshida, A., Sumitani, H., Maruyama, T., Hayashide, Y., Moriizumi, K., Katayama, T., Asakura, M., Horiikawa, T., Shibano, T., Itoh, H., Sato, K., Namba, K., Nishimura, T., Satoh, S., and Miyoshi, H. (1993). *IEDM Technol. Dig.*, 631.
5. Ohno, Y., Horikawa, T., Shinkawata, H., Kashihara, K., Kuroiwa, T., Okudaira, T., Hashizume, Y., Fukumoto, K., Eimori, T., Shibano, T., Arimoto, K., Itho, H., Nishimura, T., and Miyoshi, H. (1994). *VLSI Technol. Symp. Dig.*, 149.

6. Lesaicherre, P.Y., Yamamichi, S., Yamaguchi, H., Takemura, K., Watanabe, H., Tokashiki, K., Satoh, K., Sakuma, T., Yoshida, M., Ohnishi, S., Nakajima, K., Shibahara, K., Miyasaka, Y., and Ono, H. (1994). *IEDM Technol. Dig.*, 831.
7. Fazan, P. (1994). *Integrated Ferroelectrics* **4**: 247.
8. Achard, H., Mace, H., and Peccoud, L. (1995). *Microelectron. Eng.* **29**(1–4): 19.
9. Zurcher, P., Jones, R.E., Chu, P.Y., Taylor, D.J., White, B.E., Jr., Zafar, S., Jiang, B., Lii, Y.-J.T., and Gillespie, S.J. (1997). *IEEE Trans. on Components, Packaging, and Manufacturing Technol. Part A* **20**(2): 175.
10. Bell, J.M., Knight, P.C., and Johnston, G.R. (1996). in *Ferroelectric Thin Films: Synthesis and Basic Properties*, C.A. Paz de Araujo, J.F. Scott , and G.W. Taylor, eds., vol. 10, part 1, Amsterdam: Gordon and Breach Publishers, p. 93.
11. Summerfelt, S.R. (1997). in *Thin Film Ferroelectric Materials and Devices*, R. Ramesh, ed., Boston: Kluwer Academic Publishers, p. 1.
12. Dietz, G.W., Antpohler, W., Klee, M., and Waser, R. (1995). *J. Appl. Phys.* **78**: 6113.
13. Kotecki, D.E. (1996). *Semiconductor International* **19**(12): 109.
14. Vijay, D., Kwok, C.K., Pan, W., Yoo, I.K., and Desu, S.B. (1992). *Proc. Eighth IEEE ISAF*, 408.
15. Bernstein, S., Wong, T.Y., Kisler, Y., and Tustison, R. (1993). *J. Mat. Res.* **8**: 12.
16. Auciello, O., Gifford, K.D., and Kingon, A.I. (1994). *Appl. Phys. Lett.* **64**: 2873.
17. Al-Shareef, H.N. and Kingon, A.I. (1996). in *Ferroelectric Thin Films: Synthesis and Basic Properties*, C.A. Paz de Araujo, J.F. Scott, and G.W. Taylor, eds., vol. 10, part 1, Amsterdam: Gordon and Breach Publishers, p. 193.
18. Varniere, F., Kim, B.E., Agius, B., Bisaro, H., Olivier, J., Chevrier, G., Achard, H., Mace, H., and Peccoud, L. (1995). in *Ferroelectric Thin Films IV*, B.A. Tuttle, S.B. Desu, R. Ramesh, and T. Shiosaki, eds., vol. 361, Mat. Res. Soc. Symp. Proc, Pittsburgh, p. 235.
19. Green, M.L., Gross, M.E., Papa, L.E., Schnoes, K.J., and Brasen, D.B. (1985). *J. Electrochem Soc.* **132**: 2677.
20. Hren, P.D., Rou, S.H., Al-Shareef, H.N., Gifford, K.D., Auciello, O., and Kingon, A.I. (1992). *Integrated Ferroelectrics* **2**: 311.
21. Al-Shareef, H.N., Gifford, K.D., Rou, S.H., Hren, P.D., Auciello, O., and Kingon, A.I. (1993). *Integrated Ferroelectrics* **3**: 321.
22. Kwok, C.K., Vijay, D.P., Desu, S.B., Parikh, N.R., and Hill, E.A. (1993). *Integrated Ferroelectrics* **3**: 121.
23. Lesaicherre, P.Y., Yamamichi, S., Yamaguchi, H., Takemura, K., Watanabe, H., Tokashiki, K., Satoh, K., Sakuma, T., Yoshida, M., Ohnishi, S., Nakajima, K., Shibahara, K., Miyasaka, Y., and Ono, H. (1994). *IEDM Technol. Dig.*, 831.
24. Obhi, J.S., Patel, A., and Tossell, D.A. (1994). *British Cer. Proc.* **52**: 57.
25. de Keijser, M., Dormans, G.J.M., and van Veldhoven, P.J. (1994). *Integrated Ferroelectrics* **5**: 221.
26. Taylor, D.J., Geerse, J., and Larsen, P.K. (1995). *Thin Solid Films* **263**: 221.
27. Chung, I., Lee, J.K., Lee, W.I., Chung, C.W., and Desu, S.B. (1995). in *Ferroelectric Thin Films IV*, B.A. Tuttle, S.B. Desu, R. Ramesh, and T. Shiosaki, eds., vol. 361, Mat. Res. Soc. Symp. Proc., Pittsburgh, p. 249.
28. Al-Shareef, H.N., Chen, Y.L., Auciello, O., and Kingon, A.I. (1995). in *Ferroelectric Thin Films IV*, B.A. Tuttle, S.B. Desu, R. Ramesh, and T. Shiosaki, eds., vol. 361, Mat. Res. Soc. Symp. Proc, Pittsburgh, p. 229.
29. Al-Shareef, H.N., Bellur, K.R., Auciello, O., and Kingon, A.I. (1995). *Appl. Phys. Lett.* **66**: 239.
30. Nakamura, T., Nakao, Y., Kamisawa, A., and Takasu, H. (1994). *Appl. Phys. Lett.* **65**: 1522.
31. Nakamura, T., Nakao, Y., Kamisawa, A., and Takasu, H. (1994). *Jpn J. Appl. Phys.* **33**(9B), 5207.
32. Aoki, K., Fukuda, Y., Numata, K., and Nishimura, A. (1995). *Jpn. J. Appl. Phys.* **34**: 5250.

33. Jung, S.W., Bung, I., Lee, J., and Lee, J. (1998). in *Ferroelectric Thin Films VI*, R.E. Treece, R.E. Jones, Jr., C.M. Foster, S.B. Desu, and I.K. Yoo, eds., vol. 493, Mat. Res. Soc. Symp. Proc, Pittsburgh, p. 201.
34. Al-Shareef, H.N., Auciello, O., and Kingon, A.I. (1995). in *Science and Technology of Electroceramic Thin Films*, O. Auciello and R. Waser, eds., vol. 284, NATO/ASI Series E, Dordrecht, Netherlands: Kluwer Academic Publishers, p. 133.
35. Lichtenwalner, D.J., Dat, R., Auciello, O., and Kingon, A.I. (1994). *Ferroelectrics* **152**: 97.
36. Bensch, W., Schmale, H.W., and Reller, A. (1990). *Solid State Ionics* **43**: 171.
37. Eom, C.B., Van Dover, R.B., Phillips, J.M., Fleming, R.M., Cava, R.J., Marshall, J.H., Werder, D.J., Chen, C.H., and Fork, D.K. (1993). in *Ferroelectric Thin Films III*, E.R. Myers, B.A. Tuttle, S.B. Desu, and P.K. Larsen, eds., vol. 310, Mat. Res. Soc. Symp. Proc., Pittsburgh, p. 145.
38. Isobe, C., Ami, T., Hironaka, K., Watanabe, K., Sugiyama, M., Nagel, N., Katori, K., Ikeda, Y., Gutleban, C.D., Tanaka, M., Yamoto, H., and Yagi, H. (1997). *Integrated Ferroelectrics* **14**(1–4): 95.
39. Cheung, J.T., Morgan, P.E.D., and Neugaonkar, R. (1993). *Integrated Ferroelectrics* **3**(2): 147.
40. Zhang, J., Cui, G., Gordan, D., Van Buskirk, P., and Steinbeck, J. (1993). in *Ferroelectric Thin Films III*, E.R. Myers, B.A. Tuttle, S.B. Desu, and P.K. Larsen, eds., vol. 310, Mat. Res. Soc. Symp. Proc., Pittsburgh, p. 249.
41. Dat, R., Lichtenwalner, D.J., Auciello, O., and Kingon, A.I. (1994). *Appl Phys. Lett.* **64**: 2873.
42. Lee, J., Ramesh, R., and Keramidas, V.G. (1995). in *Ferroelectric Thin Films IV*, B.A. Tuttle, S.B. Desu, R. Ramesh, and T. Shiosaki, eds., vol. 361, Mat. Res. Soc. Symp. Proc., Pittsburgh, p. 67.
43. Ramesh, R., Lee, J., Sands, T., Keramidas, V.G., and Auciello, O. (1994). *Appl. Phys. Lett.* **64**: 2511.
44. Katori, K., Nagel, N., Watanabe, K., Tanaka, M., Yamoto, H., and Yagi, H. (1997). *Integrated Ferroelectrics* **17**(1–4): 443.
45. Watanabe, K., Tanaka, M., Nagel, N., Katori, K., Sugiyama, M., Yamoto, H., and Yagi, H. (1997). *Integrated Ferroelectrics* **17**(1–4): 451.
46. Takemura, K., Sakuma, T., Matsubara, S., Yamamichi, S., Yamaguchi, H., and Miyasaka, Y. (1994). *Integrated Ferroelectrics* **4**(4): 305.
47. Eichorst, D.J., Blanton, T.N., Barnes, C.L., and Bosworth, L.A. (1994). *Integrated Ferroelectrics* **4**: 239.
48. Spierings, G.A.C.M., Von Zon, J.B.A., Klee, M., and Larsen, P.K. (1992). *Integrated Ferroelectrics* **3**(3): 283.
49. Bruchhaus, R., Pitzer, D., Eibl, O., Scheithauer, U., and Hoesler, W. (1992). in *Ferroelectric Thin Films II*, A.I. Kingon, E.R. Myers, and B. Tuttle, eds., vol. 243, Mat. Res. Soc. Symp. Proc., Pittsburgh, p. 123.
50. Olowolafe, J.O., Jones, R.E., Campbell, A.C., Maniar, P.D., Hedge, R.I., and Mogab, C.J. (1992). in *Ferroelectric Thin Films II*, A.I. Kingon, E.R. Myers, and B. Tuttle, eds., vol. 243, Mat. Res. Soc. Symp. Proc., Pittsburgh, p. 355.
51. Kwon, J.H. and Yoon, S.-G. (1997). *Thin Solid Films* **303**: 136.
52. Ahn, J.H., Choi, G.P., Choi, W.Y., Lee, W.J., Yoon, S.G., and Kim, H.G. (1998). *Integrated Ferroelectrics* **21**: 185.
53. Tisone, T.C. and Drobeck, J. (1972). *J. Vac. Sci. Tech.* **9**(1): 271.
54. Sreenivas, K., Reaney, I., Maeder, T., and Setter, N. (1994). *J. Appl. Phys.* **75**: 232.
55. Hase, T., Sakuma, T., Amanuma, K., Mori, T., Ochi, A., and Miyasaka, Y. (1994). *Integrated Ferroelectrics* **8**(1–2): 89.
56. Melnick, B., Chu, P., Zurcher, P., White, B.E., Jr., Zafar, S., Taylor, D.J., Roberts, D., Raymond, M., Alluri, P., Remmel, T., Kottke, M., Apen, E., Chen, W., Liu, R., Fejes, P., Tracy, C., Jones, B., and Gillespie, S. (1998). "Integration of high-density ferroelectric memories with submicron silicon CMOS technology, presented at ISIF 1998.

57. Auciello, O., Krauss, A.R., Im, J., and Schultz, J.A. (1998). in *Annual Review of Material Science*, vol. 28, Palo Alto: Annual Reviews, p. 375.
58. Im, J., Krauss, A.R., Dhote, A.M., Gruen, D.M., Auciello, O., Ramesh, R., and Chang, R.P.H. (1998). *Appl. Phys. Lett.* **72**: 2529.
59. Amanuma, K., Hase, T., and Miyasaka, Y. (1995). in *Ferroelectric Thin Films IV*, B.A. Tuttle, S.B. Desu, R. Ramesh, and T. Shiosaki, eds., vol. 361, Mat. Res. Soc. Symp. Proc., Pittsburgh, p. 21.
60. Chu, P.Y., Jones, R.E., Jr., Zurcher, P., Taylor, D.J., Jiang, B., Gillespie, S.J., Lii, Y.T., Kottke, M., Fejes, P., and Chen, W. (1996). *J. Mat. Res.* **11**(5): 1065.
61. Taylor, D., Jones, R.E., Jr., Lii, Y.T., Zurcher, P., Chu, P.Y., and Gillespie, S.J. (1996). in *Ferroelectric Thin Films V*, S.B. Desu, R. Ramesh, B.A. Tuttle, R.E. Jones, and I.K. Yoo, eds., vol. 433, Mat. Res. Soc. Symp. Proc., Pittsburgh, p. 97.
62. Summerfelt, S.R. (1996). U.S. Patent No 5,504,041 (2 April 1996); 5,585,300 (17 Dec. 1996).
63. Hara, T., Tanaka, M., Sakiyama, K., Onishi, S., Ishihara, K., and Kudo, J. (1997). *Jpn. J. Appl. Phys. Lett.* **36-2**: 893.
64. Saenger, K.L., Grill, A., and Kotecki, D.E. (1998). in *Ferroelectric Thin Films VI*, R.E. Treece, R.E. Jones, Jr., C.M. Foster, S.B. Desu, and I.K. Yoo, eds., vol. 493, Mat. Res. Soc. Symp. Proc, Pittsburgh, p. 143.
65. Private communication with O. Auciello.
66. Shepard, W.H. (1991). in *Ferroelectric Thin Films*, E.R. Myers and A.I. Kingon, eds., vol. 200, Mat. Res. Soc. Symp. Proc., Pittsburgh, p. 277.
67. Smyth, D.M. (1990). *Ferroelectrics* **116**: 117.
68. Duiker, H.M., Beale, P.D., Scott, J.F., Paz de Araujo, C.A., Melnick, B.M., Cuchiaro, J.D., and McMillan, L.D. (1990). *J. Appl. Phys.* **68**. 5783.
69. Scott, J.F., Paz de Araujo, C.A., Melnick, B.M., McMillan, L.D., and Zuleeg, R. (1991). *J. Appl. Phys.* **70**: 382.
70. Spierings, G.A.C.M., Ulenaers, M.J.E., Kampschoer, G.L.M., van Hal, H.A.M., and Larsen, P.K. (1991). *J. Appl. Phys.* **70**: 365.
71. Desu, S.B. and Yoo, I.K. (1993). *Integrated Ferroelectrics* **3**: 365.
72. Taylor, D.J., Larsen, P.K., Dormans, G.J.M., and De Veirman, A.E.M. (1995). *Integrated Ferroelectrics* **7**: 123.
73. Paz de Araujo, C.A., Cuchairo, J.D., Scott, M.C., and McMillan, L.D. (1993). International Patent Publication, No. WO 93/12542 (24 June 1993).
74. Amanuma, K., Hase, T., and Miyasaka, Y. (1995). *Appl. Phys. Lett.* **66**: 221.
75. Sumi, T., Judai, Y., Hirano, K., Ito, T., Mikawa, T., Takeo, M., Azuma, M., Hayashi, S., Uemeto, Y., Arita, K., Nasu, T., Nagano, Y., Inoue, A., Matsuda, A., Fuji, E., Shimada, Y., and Otsuki, T. (1996). *Jap. J. Appl. Phys.* **35**(2b): 1516.
76. Fukushima, T., Kawahara, A., Nanba, T., Matsumoto, M., Nishimoto, T., Ikeda, N., Judai, Y., Sumi, T., Arita, K., and Otsuki, T. (1996). *VLSI Technol. Symp. Dig.*, 46.
77. Asari, K., Hirano, H., Honda, T., Sumi, T., Takeo, M., Moriwaki, N., Nakane, G., Nakakuma, T., Chaya, S., Mukunoki, T., Judai, Y., Azuma, M., Shimida, Y., and Otsuki, T. (1998). *IEICE Trans. Electronics* **E81- C**(4): 488.
78. Miwa, T., Yamada, J., Okamato, Y., Koike, H., Toyoshima, H., Hada, H., Hayashi, Y., Okizawa, H., Miyasaka, Y., Kunio, T., Miyamoto, H., Gomi, H., and Kitajima, H. (1998). *Proc. 1998 IEEE Custom Integrated Circuits Conference*, 435.
79. Takasu, H., Nakamura, T., and Kamisawa, A. (1998). *Integrated Ferroelectrics* **21**: 41.
80. Paz de Araujo, C., Otsuki, T., Cuchairo, J., and McMillan, L.D. (1998). *Proc. Seventh Biennial IEEE Nonvolatile Memory Tech. Conference*, 27.
81. Jones, R.E. Jr. (1998). *Proc 1998 IEEE Custom Integrated Circuits Conference*, 431.

82. Sameshima, K., Nakamura, T., Hoshiba, K., Kamisawa, A., Atsuki, T., Soyama, N., and Ogi, K. (1993). *Jpn J. Appl. Phys.* **32**: 4144.
83. Takemura, K., Yamamichi, S., Lesaicherre, P.-Y., Tokashiki, K., Miyamoto, H., Ono, H., Miyasaka, Y., and Yoshida, M. (1995). *Jpn. J. Appl. Phys.* **34**(9B): 5224.
84. Snadhu, G.S. and Fazan, P.C. (1995). U.S. Patent No. 5,381,302 (10 Jan. 1995).
85. Grill, A., Kane, W., Viggiano, J., Brady, M., and Laibowitz, R. (1992). *J. of Mat. Res.* **7**(12): 3260.
86. McIntyre, P.C. and Summerfelt, S.R. (1997). *J. Appl. Phys.* **82**: 4577.
87. Menk, G.E., Desu, S.B., Pan, W., and Vjay, D.P. (1996). in *Ferroelectric Thin Films V*, S.B. Desu, R. Ramesh, B.A. Tuttle, R.E. Jones, and I.K. Yoo, eds., vol. 433, Mat. Res. Soc. Symp. Proc., Pittsburgh, p. 189.
88. Parikh, N., Stephen, J.T., Swanson, M.L., and Myers, E.R. (1990). in *Ferroelectric Thin Films*, E.R. Myers and A.I. Kingon, eds., vol. 200, Mat. Res. Soc. Symp. Proc., Pittsburgh, p. 193.
89. DeOrnellas, S., Rajora, P., and Cofer, A. (1997). *Integrated Ferroelectrics* **17**(1–4): 395.
90. Yuuki, A., Yamamuka, M., Makita, T., Horikawa, T., Shibano, T., Hirano, N., Maeda, H., Mikami, N., Ono, K., Ogata, H., and Abe, H. (1995). *IEDM Technol. Dig.*, 115.
91. Nishioka, Y., Shiozawa, K., Oishi, T., Kanamoto, K., Tokuda, Y., Sumitanai, H., Aya, S., Yabe, H., Itoga, K., Hifumi, T., Muromoto, K., Kuroiwa, T., Kawahara, T., Nishikawa, K., Oomori, T., Fujino, T., Yamamoto, S., Uzawa, S., Kimata, M., Nunoshita, M., and Abe, H. (1995). *IEDM Technol. Dig.*, 903.
92. Yamaguchi, H., Iizuka, T., Koga, H., Takemura, K., Sone, S., Yabuta, H., Yamamichi, S., Lesaicherre, P.-Y., Suzuki, M., Kojima, Y., Nakajima, K., Kasai, N., Sakuma, T., Kato, Y., Miyasaka, Y., Yoshida, M., and Nishimoto, S. (1996). *IEDM Technol. Dig.*, 675.
93. Poor, M.R., Hurd, A.M., Fleddermann, C.B., and Wu, A.Y. (1990). in *Ferroelectric Thin Films*, E.R. Myers and A.I. Kingon, eds., vol. 200, Mat. Res. Soc. Symp. Proc., Pittsburgh, p. 211.
94. Vijay, D.P., Desu, S.B., and Pan, W. (1993). *J. Electrochem. Soc.* **140**(9): 2635.
95. Saito, K., Choi, J.H., Fukuda, T., and Ohue, M. (1992). *Jap. J. Appl. Phys.* **31–2**(9A): L1260.
96. Van Glabbeck, J.J., Spierings, G.A.C.M., Ulenaers, M.J.E., Dormans, G.J.M., and Larsen, P.K. (1993). in *Ferroelectric Thin Films III*, E.R. Myers, B.A. Tuttle, S.B. Desu, and P.K. Larsen, eds., vol. 310, Mat. Res. Soc. Symp. Proc., Pittsburgh, p. 127.
97. Dormans, G.J.M., Larsen, P.K., Spierings, G.A.C.M., Dikken, J., Ulenaers, M.J.E., Cuppens, R., Taylor, D.J., and Verhaar, R.D.J. (1995). *Integrated Ferroelectrics* **6**(1–4): 93.
98. Trolier, S., Geist, C., Safari, A., Newnham, R.E., and Xu, Q.C. (1986). *Proc. Eighth IEEE ISAF*, 707.
99. Baude, P.F., Ye, C., and Polla, D. (1993). in *Ferroelectric Thin Films III*, E.R. Myers, B.A. Tuttle, S.B. Desu, and P.K. Larsen, eds., vol. 310, Mat. Res. Soc. Symp. Proc., Pittsburgh, p. 139.
100. Mancha, S. (1992). *Ferroelectrics* **135**(1–4): 131.
101. Asselanis, D. and Mancha, D. (1998). U.S. Patent No. 4, 759, 823 (26 July 1998).
102. Steinbruchel, C. (1992). *Mat. Sci. Technol* **8**: 565.
103. Chung, C.W., Song, I., and Lee, J.S. (1998). in *Ferroelectric Thin Films VI*, R.E. Treece, R.E. Jones, Jr., C.M. Foster, S.B. Desu, and I.K. Yoo, eds., vol. 493, Mat. Res. Soc. Symp. Proc, Pittsburgh, p. 119.
104. Farrell, C.E., Milkove, K.R., Wang, C., and Kotecki, D.E. (1997). *Integrated Ferroelectrics* **16**(1–4): 1.
105. Yoo, I.K. and Desu, S.B. (1992). *Mat. Sci. and Eng.* **B13**: 319.
106. Pan, W., Desu, S.B., Yoo, I.K., and Vijay, D.P. (1994). *J. Mat. Res.* **9**: 2976.
107. Auciello, O. (1998). *Integrated Ferroelectrics* **21**: xi.
108. Vasant Kumar, C.V.R., Sayer, M., Pascual, R., Amm, D.T., Wu, Z., and Swanston, D.M. (1991). *Appl. Phys. Lett.* **58**: 1161.

109. Uchida, H., Soyama, N., Kageyama, K., Ogi, K., Scott, M.C., Cuchiaro, J.D., Derbenwick, G.F., McMillan, L.D., and Paz de Araujo, C.A. (1997). *Integrated Ferroelectrics* **16**(1–4): 41.

110. Alexe, M., Scott, J.F., Curran, C., Zakharov, N.D., Hesse, D., and Pignolet, A. (1998). *Appl. Phys. Lett.* **73**(11): 1592.

111. Scott, J.F., Alexe, M., Zakharov, N.D., Pignolet, A., Curran, C., and Hesse, D. (1998). *Integrated Ferroelectrics* **21**: 1.

112. Ishihara, K., Ishikawa, T., Hamada, K., Onishi, S., Kudo, J., and Sakiyama, K. (1995). *Integrated Ferroelectrics* **6**(1–4): 301.

113. Moazzami, R., Maniar, P.D., Jones, R.E., and Mogab, C.J. (1994). *VLSI Technol. Symp. Dig.*, 55.

114. Maniar, P.D., Moazzami, R., Jones, R.E., Campbell, A.C., and Mogab, C.J. (1993). in *Ferroelectric Thin Films III*, E.R. Myers, B.A. Tuttle, S.B. Desu, and P.K. Larsen, eds., vol. 310, Mat. Res. Soc. Symp. Proc., Pittsburgh, p. 151.

115. Zafar, S., Kaushik, V., Laberge, P., Chu, P., Jones, R.E., Hance, R.L., Zurcher, P., White, B.E., Taylor, D.J., Melnick, B., and Gillespie, S.J. (1997). *J. Appl. Phys.* **82**(9): 4469.

116. Fujisaki, Y., Kushida-Abdelghafar, K., Miki, H., and Shimamoto, Y. (1998). *Integrated Ferroelectrics* **21**: 83.

117. Im, J., Auciello, O., Krauss, A.R., Gruen, D.M., Chang, R.P.H., Kim, S.H., and Kingon, A.I. (1999). *Appl. Phys. Lett.* **74**(8): 1162.

118. Aggarwal, S., Perusse, S.R., Tipton, C.W., Ramesh, R., Drew, H.D., Venkatesen, T., Romero, D.B., Podobedov, V.B., and Weber, A. (1998). *Appl. Phys. Lett.* **73**: 1973.

119. Rod, B., Moore, R., McCullen, J., and Terrel, J. (1994). *Integrated Ferroelectrics* **4**: 155.

120. Nasby, R.D., Schwank, J.R., Rodgers, M.S., and Miller, S.L. (1992). *Integrated Ferroelectrics* **2**(1–4): 91.

121. Hwang, C.S., Park, S.O., Cho, K.-J., Kang, C.S., Kang, H.-K., Lee, S.I., and Lee, M.Y. (1995). *Appl. Phys. Lett.* **67**: 2819.

122. Deschler, M., Schumacher, M., Woelk, E., Schmitz, D., Strauch, G., Heuken, M., and Juergensen, H. (1998). *Integrated Ferroelectrics* **21**: 381.

123. Roeder, J.F., Bilodeau, S.M., Carl, R.J., Jr., Gardiner, R.A., and Van Buskirk, P.C. (1997). *Integrated Ferroelectrics* **18**: 109.

124. Eastep, B., MacWilliams-Brooks, J., and Mitra, S. (1999). A demonstration of low voltage performance from scaled PLZT films, on a fully integrated 64 K FRAM®, Presented at ISIF 1999.

125. Shimada, Y., Arita, K., Fujii, E., Nasu, T., Nagano, Y., Noma, A., Izutsu, Y., Nakao, K., Tanaka, K., Yamada, T., Uemeto, Y., Asari, K., Nakane, G., Inoue, A., Sumi, T., Chaya, S., Nakakuma, Y., Hirano, H., Judai, Y., Sasai, Y., and Otsuki, T. (1999). Advanced LSI embedded with FeRAM for contactless IC cards and its manufacturing technology, Presented at ISIF 1999.

126. Coombe, G.B., Meester, T.A., Cordoba, M.V., Kamp, D.A., Derbenwick, G.F., Nakane, G., Inoue, A., Sumi, T., and Otsuki, T. (1999). An RF ID tag with an embedded 1 K FeRAM using SBT, Presented at ISIF 1999.

127. Private communication with Carlos Paz de Araujo.

128. Kraus, W., Lehman, L., Wilson, D., Yamazaki, T., Ohno, C., Nagai, E., Yamazaki, H., and Suzuki, H. (1999). A 42.5 mm^2 1 Mb nonvolatile ferroelectric memory utilizing advanced architecture for enhance reliability, Presented at ISIF 1999.

129. Amanuma, K. and Kunio, T. (1997). *Jpn. J. Appl Phys.* **35**: 5229.

130. Kamp, D.A., Cordoba, M.V., Hodges, D.E., Derbenwick, G.F., Kye, H.W., Kang, W.S., and Kang, N.S. (1999). A 3 V 1T/1C_256 Kbit FeRAM using SBT, Presented at ISIF 1999.

Subject Index

Note: Page numbers followed by f or t refer to the figure or table on that page.

N

T

U

V

Recent Volumes In This Series

Maurice H. Francombe and John L. Vossen, *Physics of Thin Films,* Volume 16, 1992.

Maurice H. Francombe and John L. Vossen, *Physics of Thin Films,* Volume 17, 1993.

Maurice H. Francombe and John L. Vossen, *Physics of Thin Films, Advances in Research and Development, Plasma Sources for Thin Film Deposition and Etching,* Volume 18, 1994.

K. Vedam (guest editor), *Physics of Thin Films, Advances in Research and Development, Optical Characterization of Real Surfaces and Films,* Volume 19, 1994.

Abraham Ulman, *Thin Films, Organic Thin Films and Surfaces: Directions for the Nineties,* Volume 20, 1995.

Maurice H. Francombe and John L. Vossen, *Homojunction and Quantum-Well Infrared Detectors,* Volume 21, 1995.

Stephen Rossnagel and Abraham Ulman, *Modeling of Film Deposition for Microelectronic Applications,* Volume 22, 1996.

Maurice H. Francombe and John L. Vossen, *Advances in Research and Development,* Volume 23, 1998

Abraham Ulman, *Self-Assembled Monolayers of Thiols,* Volume 24, 1998

Subject and Author Cumulative Index, Volumes 1–24, 1998.

Ronald A. Powell and Stephen Rossnagel, *PVD for Microelectronics: Sputter Deposition Applied to Semiconductor Manufacturing,* Volume 26, 1998.

Jeffrey A. Hopwood, *Ionized Physical Vapor Deposition*, Volume 27, 2000.

ISBN 0-12-533028-6

90076 >

9 780125 330282